Conversion Factors for Units of Molecular Energy

	$J \cdot mol^{-1}$	$cal \cdot mol^{-1}$	$cm^3 \cdot atm \cdot mol^{-1}$
1 $J \cdot mol^{-1}$ =	1	2.390057×10^{-1}	9.86914
1 $cal \cdot mol^{-1}$ =	4.18400	1	41.2925
1 cm^3 $atm \cdot mol^{-1}$ =	0.1013259	2.42175×10^{-2}	1
1 $kWh \cdot mol^{-1}$ =	3,600,000	860,421	3.55289×10^{7}
1 $cm^{-1} \cdot molecule^{-1}$ =	11.96258	2.85912	118.0602
1 $eV \cdot molecule^{-1}$ =	96484.6	23060.5	952,225

	$kWh \cdot mol^{-1}$	$cm^{-1} \cdot molecule^{-1}$	$eV \cdot molecule^{-1}$
1 $J \cdot mol^{-1}$ =	2.77778×10^{-7}	8.35940×10^{-2}	1.036431×10^{-5}
1 $cal \cdot mol^{-1}$ =	1.162222×10^{-6}	3.49757×10^{-1}	4.33642×10^{-5}
1 cm^3 $atm \cdot mol^{-1}$ =	2.81461×10^{-8}	8.47024×10^{-3}	1.050173×10^{-6}
1 $kWh \cdot mol^{-1}$ =	1	300,938	37.3115
1 $cm^{-1} \cdot molecule^{-1}$ =	3.32294×10^{-6}	1	1.239838×10^{-4}
1 $eV \cdot molecule^{-1}$ =	2.68014×10^{-2}	8065.57	1

Chemical Thermodynamics

Chemical Thermodynamics

Peter A. Rock

Department of Chemistry
University of California, Davis

University Science Books
Mill Valley, California

University Science Books
20 Edgehill Road, Mill Valley, CA 94941

Library of Congress Catalog Card Number 82-051233
ISBN 0-935702-12-1

PRINTED IN THE UNITED STATES OF AMERICA
10 9 8 7 6 5 4 3 2 1

Production Credits

Design and production: Greg Hubit Bookworks
Copy editing: Betty Adam
Technical Illustrations: Renaissance Studios
Typesetting: Syntax International
Printing and binding: Maple-Vail Book Manufacturing Group

Chapter-opening Photo Credits

Chapter 1 M. C. Escher, *Waterfall*; collection Haags Gemeentemuseum—The Hague
Chapter 2 Courtesy of National Bureau of Standards
Chapter 3 Courtesy of Pacific Gas and Electric Company
Chapter 4 Courtesy of General Motor Corporation
Chapter 5 Courtesy of Pacific Power and Light
Chapter 6 Courtesy of D. D. Osheroff and W. O. Sprenger, Bell Laboratories
Chapter 7 Courtesy of NASA
Chapter 8 Courtesy of General Electric Research and Development Center
Chapter 9 Courtesy of Pacific Gas and Electric Company
Chapter 10 Courtesy of the Nitrogen Group of the Union Chemicals Division, Union Oil Company of California
Chapter 11 Courtesy of Envirogenics Systems Company, El Monte, California
Chapter 12 Courtesy of United Technologies
Chapter 13 Courtesy of Envirogenics Systems Company, El Monte, California
Chapter 14 Redrawn from *Introduction to Molecular Spectroscopy* by Gordon M. Barrow. Copyright © 1962 by McGraw-Hill Book Company. Used with permission of McGraw-Hill Book Company.

Preface

THIS BOOK was written as an introduction to the subject for undergraduates. It differs from other texts on chemical thermodynamics in that the fundamental concepts of thermodynamics (temperature, internal energy, and entropy) are introduced *in depth* in the first four chapters, accompanied by only the essential minimum of mathematics. The foundations of thermodynamics thus are learned at the outset without contending simultaneously with extensive manipulations of partial derivatives.

The basic mathematical methods of classical chemical thermodynamics are developed in Chapter 5. Chapter 6 is devoted to the Third Law of Thermodynamics, and Chapters 7 through 13 involve numerous applications of the principles of chemical thermodynamics to a wide variety of problems. Chapter 14 is a brief introduction to statistical thermodynamics, with an emphasis on the calculation of third-law entropies and equilibrium constants from molecular data.

A distinction is made in the text between those thermodynamic functions that are indispensable to chemical thermodynamics (temperature, internal energy, and entropy) and the thermodynamic functions that are introduced for convenience (e.g., enthalpy, Helmholtz energy, and Gibbs energy). Careful attention has been given to the description of standard states of substances and to the role that standard states play in the analysis of chemical equilibria. The *activity function* is extensively employed in the thermodynamic analysis of chemical equilibria. A thermodynamically rigorous characterization of equilibrium constants for chemical reactions is emphasized throughout the text.

There are many solved examples in the body of the text, and each chapter contains an extensive problem set, with problems arranged roughly in order of increasing difficulty. The key to the understanding of chemical thermodynamics is to work problems involving a wide variety of applications of the basic principles. For this reason, the student is urged to work at least half of the problems at the end of each chapter.

The general approach to thermodynamics that I have used in this book was influenced by many people. The development of the energy and entropy concepts was influenced most strongly by *The Feynman Lectures on Physics*

(Feynman 1963).* The development of the temperature concept and the discussion of magnetic systems is based primarily on the treatment by M. W. Zemansky in *Heat and Thermodynamics* (Zemansky 1964). The major influence on my approach to the thermodynamics of solutions and chemical reactions is that presented in the classic text *Thermodynamics* by G. N. Lewis, M. Randall, K. S. Pitzer, and L. Brewer (Lewis 1961). Others who have influenced my thinking about thermodynamics are my former instructors Professors L. V. Coulter, K. Eriks, W. F. Giauque, R. E. Powell, and D. A. Shirley. I have had many enjoyable (and some embarrassing) discussions on thermodynamics with my colleagues R. K. Brinton, G. A. Gerhold, H. Hope, J. E. Keizer, D. A. McQuarrie, C. P. Nash, and D. H. Volman, and my former students J. C. Hall, J. J. Kim, R. Murray, J. M. Postma, and L. F. Silvester.

I also thank D. A. McQuarrie, J. Ledbetter, D. A. Brant, R. C. Chang, D. H. Volman, and J. E. Keizer for reading all or parts of the manuscript and for providing many helpful comments and criticisms.

Special thanks go to my publisher, Bruce Armbruster, who was always encouraging, helpful, and patient. It was a pleasure to work with Greg Hubit, of Bookworks, who handles the complex task of book production with finesse.

Finally, I give very special thanks to Elaine Rock who assumed the arduous task of typing several drafts of the manuscript.

* See Appendix III for references

Contents

Chemical Thermodynamics

1

The Purpose and Scope of Thermodynamics

The fascination of a growing science lies in the work of the pioneers at the very borderland of the unknown, but to reach this frontier one must pass over well travelled roads; of these one of the safest and surest is the broad highway of thermodynamics.

G. N. Lewis and M. Randall,
Thermodynamics. New York:
McGraw-Hill, 1923.

1-1 *The Laws of Thermodynamics Constitute General Restrictions on Macroscopic Energy Transfers*

Thermodynamics is the science of the utilization and conversion of energy. The laws of thermodynamics are formulated in terms of completely general restrictions on macroscopic (i.e., matter in bulk) energy transfers in nature. No one has ever produced a device that operates in violation of any of the laws of thermodynamics. If such a device were produced, then modern civilization would be profoundly changed. The evolution of modern civilization was made possible by increases in the quality and quantity of energy utilization. The Industrial Revolution was powered by Watt's steam engines.

The laws of thermodynamics, like all scientific laws, consist of concise summaries of numerous experimental observations. However, the laws of thermodynamics differ from most other scientific laws in that the thermodynamic laws usually are stated in terms of the *impossibility* of achieving certain types of results; that is, the thermodynamic laws tell us what cannot be accomplished in the realm of energy transfers, no matter how ingenious we are or how hard we try. A similar situation is found in the Special Theory of Relativity, wherein it is asserted that the velocity of light cannot exceed a certain limiting value (about 3.00×10^8 meters per second).

In effect, thermodynamics tells us that it is a complete waste of time to try to beat the laws of thermodynamics. Innumerable attempts have been made to bypass the laws of thermodynamics, but not a single such attempt has succeeded. The laws of thermodynamics permit no exceptions. The discovery of even a single exception to a thermodynamic law would lead to a collapse of the foundations of thermodynamics and would, in turn, necessitate a far-reaching scientific reassessment of nature. An assessment of the foundations of thermodynamics led Albert Einstein to the following conclusion:

> Thermodynamics is the only science about which I am firmly convinced that, within the framework of the applicability of its basic principles, it will never be overthrown.

1-2 *There Are Three Thermodynamic Laws*

An understanding of the laws of thermodynamics requires an understanding of temperature, work, heat, energy, and entropy. The approach used in this chapter will be simply to state the three laws of thermodynamics in a variety of ways; then in Chapters 2 through 6 we shall discuss the major concepts embodied in the laws, before proceeding in Chapters 7 through 14 to apply the laws to a wide variety of problems. In most cases, the emphasis of the applications will be on the chemical aspects of the problem.

Stated in terms of the operation of machines, the laws of thermodynamics are as follows:

First Law

It is impossible to construct a device that operates in a cyclical manner and performs work without the input of energy to the device.

Second Law

It is impossible to construct a device that operates in a cyclical manner and takes in energy as heat from a high-temperature reservoir and performs work without, in some part of the cycle, discharging energy as heat to a low-temperature reservoir. In other words, all heat engines require a temperature difference in order to operate.

An alternative statement of the second law in terms of the operation of machines is: *It is impossible to construct a device that can transfer energy as heat from a low-temperature reservoir to a reservoir at a higher temperature without the input of energy as work.* (You have to pay the power company to run your refrigerator.)

Third Law

It is impossible to construct a device capable of reducing the temperature of any macroscopic system to the absolute zero of temperature.

The first two laws of thermodynamics sometimes are expressed in terms of the operation of *perpetual motion machines.*

First Law

It is impossible to construct a perpetual motion machine of the first kind.

A perpetual motion machine of the first kind is a machine that can output energy as work continuously without the need for any input of energy to the machine.

Second Law

It is impossible to construct a perpetual motion machine of the second kind.

A perpetual motion machine of the second kind is a machine that can continuously convert all the energy taken in as heat into work.

Stated in terms of the thermodynamic functions *energy, entropy, and temperature*, the laws of thermodynamics for any actual (i.e., nonhypothetical) process are as follows:

First Law

The total energy is conserved.

Second Law

The total entropy increases.

Third Law

The absolute temperature remains above zero degrees.

1-3 *Classical Thermodynamics Uses a Macroscopic, Phenomenological Approach*

If this book is your first serious encounter with the laws of thermodynamics, then the statements of the laws of thermodynamics in Section 1-2 probably appear cryptic. There is no cause for alarm. The concepts embodied in the laws are discussed thoroughly in the next three chapters. At this juncture it is sufficient to reread the statements of the laws several times, think about their significance, and then note the points where you need clarification and amplification. Thermodynamics is a science that on first encounter seems unusual. The "unusual" nature of thermodynamics arises from the fact that the foundations of classical thermodynamics are independent of the molecular theory of matter. The classical thermodynamic viewpoint is macroscopic and phenomenological, as opposed to microscopic (molecular) and mechanistic. The model-independent or abstract nature of classical thermodynamics is a direct consequence of the statement of the laws in very general terms.

The prime objective of thermodynamics is to discover the universal laws governing the utilization and conversion of energy on a macroscopic level. The general nature of thermodynamic laws enables the utilization of the laws to test any proposed molecular level model for matter. The thermodynamic behavior predicted by any molecular theory *must* be consistent with the laws of thermodynamics. If a molecular level theory is inconsistent with the laws of thermodynamics, then the theory is untenable. However, consistency of a theory with the laws of thermodynamics does not guarantee the validity of the theory. The model-independent nature of classical thermodynamics was obtained at a high price; namely, classical thermodynamics cannot be used to provide any information regarding the molecular level origins of the differences in the properties of one substance versus the properties of another substance. The derivation of the thermodynamic properties of a macroscopic system from the molecular properties of its constituent molecules and/or ions is the objective of *statistical thermodynamics.* Time-dependent macroscopic phenomena are treated in *nonequilibrium thermodynamics.* The major emphasis in this book is on classical *chemical thermodynamics.* Chemical thermodynamics involves the extension of thermodynamics to systems that undergo chemical changes.

1-4 *The Laws of Thermodynamics Govern the Response of Chemical Equilibria to Changes in Temperature, Pressure, and Composition*

The utility of thermodynamics in engineering is suggested by the statements of the laws of thermodynamics in terms of the operation of machines. Indeed, we shall see that thermodynamics provides equations that enable us to calculate the maximum possible efficiency of any engine. The utility of thermodynamics in chemistry is not obvious from the statements of the laws of thermodynamics given in Section 1-2. However, we shall see that thermodynamics is of major importance in chemistry. In particular, *chemical thermodynamics* offers the following:

1. It provides us with the necessary and sufficient conditions for the establishment of a chemical equilibrium in a system, including systems involving chemical reaction(s).

2. It enables us to rule out on thermodynamic grounds some chemical processes that, under the given conditions, might otherwise be regarded as possible.
3. It provides us with general thermodynamic equations that enable us to predict quantitatively the effect of a change in the temperature, the pressure, or the composition on a reaction equilibrium.
4. It provides us with general thermodynamic equations that enable us to convert thermodynamic data in one form into another form that is more useful for the problem at hand (e.g., the conversion of data for the energy change and the entropy change of a chemical reaction into an equilibrium constant for the reaction or the conversion of an equilibrium constant and concentration data for a reaction into an electrochemical cell voltage.)

Many other examples of the utility of chemical thermodynamics will be described in Chapters 6 through 14, but the above examples are sufficient to convey a sense of the importance of chemical thermodynamics in the study of chemical reaction equilibria.

It will prove helpful to your study of chemical thermodynamics to recognize that you already possess a more than passing acquaintance with the laws and principles of chemical thermodynamics, which was obtained in your introductory chemistry and physics courses. Further, the laws of thermodynamics have such widespread applicability to our everyday lives that all of us have at least some intuitive notions, based on our personal experiences, of the restrictions that the laws place on human endeavors. It was noted by P. W. Bridgman,* a great experimental thermodynamicist, that the laws of thermodynamics convey a stronger sense of their human origins than other laws of science.

> It must be admitted, I think, that the laws of thermodynamics have a different feel from most of the other laws of physics. There is something more palpably verbal about them—they smell more of their human origin. The guiding motif is strange to most of physics: namely, a capitalizing of the universal failure of human beings to construct perpetual motion machines of either the first or the second kind. Why should we expect nature to be interested either positively or negatively in the purposes of human beings, particularly purposes of such unblushingly economic tinge?

1-5 *The United States Utilizes about 84 Quadrillion Kilojoules of Energy Annually*

In 1980 the United States utilized 84 quadrillion kilojoules of energy (1 quadrillion kJ $= 10^{15}$ kJ), which corresponds to a per person rate of energy

* P. W. Bridgman, *The Nature of Thermodynamics.* New York: Dover Publications, Inc. (n.d.).

TABLE 1-1

U.S. Energy Utilization by Type in 1980

Energy source	*Amount of energy utilized/10^{15} kJ*	*Percent of total*
Petroleum liquids	40	48
Natural gas	21	25
Coal	17	20
Nuclear power	2.9	3.5
Hydroelectric power	2.9	3.5
	84	100

use (power use) of

$$\frac{84 \times 10^{15}\ \text{kJ}\cdot\text{y}^{-1}}{(220 \times 10^6\ \text{people})(3.15 \times 10^7\ \text{s}\cdot\text{y}^{-1})} = 12\ \text{kJ}\cdot\text{s}^{-1}\cdot\text{person}^{-1}$$

A kilojoule per second is equal to a kilowatt, kW. A power usage of 12 kW per person is equivalent to one hundred twenty 100-W light bulbs burning *continuously* for each person in the United States. An annual energy use of 12 kW per person is also equivalent to 62 barrels of crude oil per person per year.

At the present time, the United States produces about 80% of its annual energy requirements; the 20% shortfall is made up for the most part by imported oil. Oil furnishes about 48% of the total United States energy supply (Table 1-1), and about 43% of the oil used in the United States is imported. At $32 per barrel, the annual cost of the oil imported by the United States is $86 billion. The major ways in which energy is utilized in the United States are shown in Table 1-2. Note that the major areas of energy use are industry (35%), transportation (22%), and space heating (18%).

TABLE 1-2

Distribution of Energy Use in the United States

Energy use	*Percent of total*
Industrial	35
Transportation	22
Space heating	18
Feedstock (for chemical syntheses)	5
Military	4
Water heating	4
Air conditioning	3
Refrigeration	2
Cooking	2
Miscellaneous	5
	100

A truly major challenge facing the United States in the next decade is to achieve energy independence and thereby eliminate the present massive annual outflow of capital for the purchase of imported oil. Although the United States reserves of oil and gas are only sufficient to last about 25 years at present rates of consumption, the United States has massive coal resources equivalent to over 300 years of supply at the present rate of *total* United States annual energy use. The present total United States energy requirements could be supplied by solar energy collection systems, which would occupy less than 0.3% of the total United States land surface. The achievement of United States energy independence will not be easy, but there is no doubt that thermodynamics will play a central role in the struggle to achieve this goal.

Problems

1. Explain why it is that all other factors being equal, the heavier the car, the lower the gas mileage.

2. Make a list of all the different energy sources that you have at least heard or read about.

3. Where does your body derive the energy necessary to maintain your body temperature?

4. Explain why the air temperature in your car tires increases when the car is driven.

5. Explain the observation that when a piece of metal and a piece of wood both at about 65°F or less are picked up simultaneously, one in each hand, the piece of metal feels cooler than the piece of wood.

6. Explain why the temperature of a drill bit or a saw blade increases while in use.

7. Explain why it usually takes much less energy per mile to ride a bike than to walk. Are there any conditions under which the reverse is true?

8. Over 90% of the energy utilized in the United States each year is derived from the combustion of fossil fuels. List the major fuels that are used as energy sources in the United States.

9. Make a complete list of all the motors in your home and compare your result with the results of your classmates. Note that searching out all the motors requires some care. For example, a "frost-free" ice- and cold-water-dispensing refrigerator has a compressor motor, three or more fan motors, a motor to drive the ice maker, a motor to push the cubes out of the storage bin, and a motor associated with the cold-water reservoir.

10. Make a study of the amount of energy utilized per hour of operation by the major energy-consuming devices in your home. This study can be

done by observing the increase in energy use on the power company meters when a device is turned on.

11. Use your most recent utility bill to determine the cost per kilojoule of electricity and natural gas in your area. Compare your results with the same quantities for 1 year ago. (*Note:* Natural gas use in the United States is billed in *therms*; 1 therm $= 10^5$ Btu $= 1.05 \times 10^5$ kJ.)

12. Explain why it takes longer to cook food by boiling it in water at a high elevation (say, in the mountains) than by boiling it in water at sea level.

13. Make a complete list of all the motors in a passenger car or other road vehicle to which you have access.

14. Explain the function of a car radiator. Why is it necessary to have a radiator in a car?

15. Explain why it is that when a movie film is run in reverse, the resulting action almost always evokes laughter.

16. What is your interpretation of the following quote from the *Rubaiyat of Omar Khayyam* (translated by Edward Fitzgerald)?

> The Moving Finger writes; and, having writ,
> Moves on: nor all the Piety nor Wit
> Shall lure it back to cancel half a Line,
> Nor all thy Tears wash out a Word of it.

17. Do you agree with the statement, "Certain occurrences are inevitable—there is absolutely no way to prevent them from happening."? Cite two or more examples to support your position. Do you agree with the statement, "Everything is relative—there are no absolutes."? Cite two or more examples to support your position.

18. Do you agree with the statements, "Anything is possible." "Nothing is impossible."? Do you think it is possible to freeze water at 10°C and 1 atm? Do you think it is possible to take the combustion products of, say, a burned log and reassemble them into the original log?

19. Why is an assessment of the relative energy utilization efficiencies of competing devices an important prepurchase consideration? Do you know the meaning of an EER (energy efficiency rating) of, say, 8 for a device?

20. Why do you think it is that in a region where the prevailing winds are west to east across a mountain range there is much more precipitation on the western slopes of the mountain range than on the eastern slopes of the mountain range?

21. Explain why it is not possible to lower the temperature in your kitchen by leaving the refrigerator door open.

22. Briefly describe your understanding of how an automobile air conditioner works.

23. The three laws of thermodynamics are sometimes paraphrased in the following cynical terms:

I. You never get something for nothing.
II. You never get more than you pay for, and you usually get less.
III. Perfection is unattainable.

Interpret these statements in terms of the statements of the thermodynamic laws given in this chapter.

24. It is observed that undisturbed hot boiled tap water freezes faster than undisturbed cold water (if you don't believe it, then try it). Can you explain this observation? (*Hint:* Boiled tap water produces much clearer ice than cold tap water).

25. Without reference to any texts or notes, write your understanding of the following terms:

pressure	energy	fever
work	hot	calorimeter
heat	cold	equilibrium
temperature	entropy	chemical equilibrium
thermometer	motor	friction
efficiency	heat engine	refrigerator

File your results until you have covered through Chapter 8 of this text and then review and revise (if necessary) your statements.

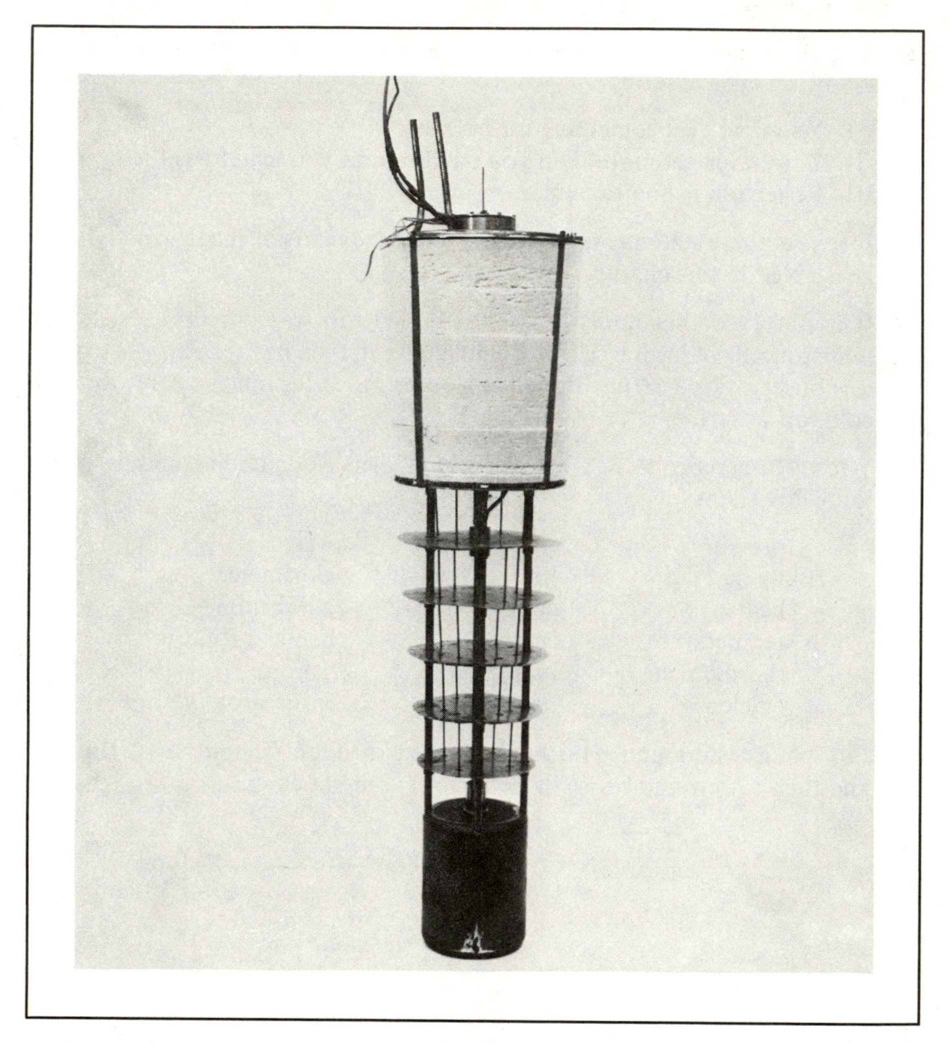

A photograph of a constant volume gas thermometer that is currently being used at the U.S. National Bureau of Standards. The gas thermometer bulb consists of a right circular cylinder, with a volume of 433 cm^3, made of 80% platinum, 20% rhodium and is suspended in a heavy inconel metal case shown at the bottom. A uniform annulus of 0.25 mm between the bulb and the case provides the counter-pressure system. The bulb is connected to a constant-volume valve and diaphragm pressure transducer (neither shown) by a tube of 90% platinum, 10% rhodium about 0.9 mm in internal diameter and 60 cm in length (top of tube can be seen at the top center above the insulation).

2

Temperature

> *. . . [temperature] turns out to be a much more fundamental quantity than energy; so much so, that the physicist is gratified to be given, in every particular case, energy as a function of it, rather than vice versa, which would be quite unnatural.*
>
> Erwin Schrödinger, *Statistical Thermodynamics.* New York: Cambridge University Press, 1946.

2-1 *A Thermometer Is a Device Used to Measure Temperature*

Temperature is a fundamental concept that is used in the thermodynamic analysis of the behavior of matter. It is not a simple matter to answer in concrete terms the question, What is temperature? Our approach to temperature will be to describe, by means of specific examples, how temperature is measured; that is, we shall introduce the temperature concept in *operational* (i.e., how we measure it) terms. Thus our development of the temperature concept is tied closely to the way in which we measure the property that we call *temperature*. In the discussion to follow it is helpful to recall the temperature measurements that you undoubtedly have made on numerous occasions. Although the fundamental basis of the temperature concept is subtle, the measurement of temperature is simple and straightforward.

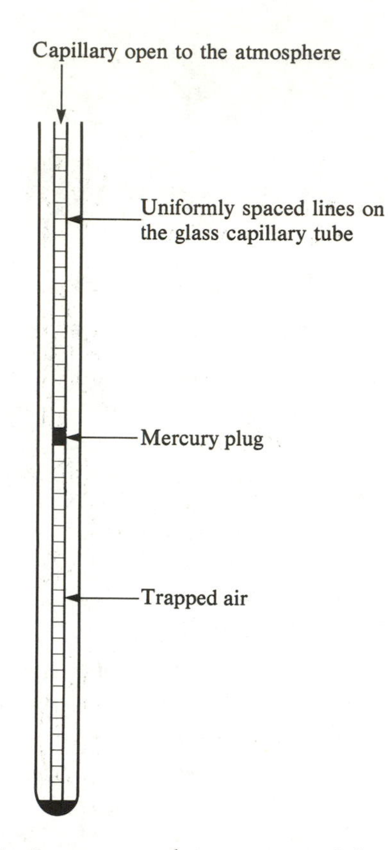

Figure 2-1. A simple constant pressure gas thermometer consisting of a small movable plug of liquid mercury in a glass capillary tube. As the pressure of the trapped air below the mercury plug increases, the plug moves upward until the pressure of the trapped air equals the atmospheric pressure acting on the top side of the movable mercury plug.

The thermometer that we shall use in our initial description of temperature measurements is shown in Figure 2-1. The thermometer is a piece of heavy-walled glass capillary tubing containing a small movable plug of mercury. The capillary tube is open to the atmosphere and thus the position of the mercury plug is determined by the fact that the pressure of the air trapped below the mercury plug equals the atmospheric pressure when the mercury plug is stationary (neglecting the small pressure contribution due to the mercury plug itself). If the pressure of the trapped air exceeds atmospheric pressure, then the mercury plug moves upward until the pressure of the trapped air equals the atmospheric pressure.

Suppose that we bring the thermometer shown in Figure 2-1 in contact with an ice-water mixture and record the position of the mercury plug relative to the marks on the capillary tube. If we now transfer the thermometer to a sample of boiling water, then we find that the mercury plug moves to a higher position in the capillary. Provided that the atmospheric pressure is constant, we find by experiment that the position of the mercury plug is the same for any ice-water mixture into which the thermometer is placed. A similar result is obtained for the higher position of the mercury plug when the thermometer is placed in any sample of boiling water. When the thermometer is squeezed in the armpit of a healthy human being, the position of the mercury plug is 37% of the *total* distance between the position of the mercury plug when the thermometer is placed in the ice-water mixture and the position of the mercury plug when the thermometer is placed in boiling water.

All that is necessary to establish a numerical temperature scale is to assign arbitrary numerical values to the temperatures of two systems that yield reproducible positions of the mercury plug on our thermometer scale. The Celsius scale, named after the eighteenth-century Swedish astronomer Anders Celsius, is established by assigning a temperature of exactly 0.01°C (where °C denotes degrees Celsius) to a physically enclosed ice-water–water vapor mixture, and a temperature of exactly 100°C to boiling water at an atmospheric pressure of *one standard atmosphere* (760 torr). On the Celsius scale, the normal freezing point of water is 0°C and the body temperature of a normal, healthy human being is 37°C. The measured temperatures of several systems on the Celsius scale are shown in Figure 2-2.

2-2 *The Thermodynamic (Kelvin) Temperature Scale Is an Absolute Temperature Scale*

The volume V of the air trapped below the mercury plug in the thermometer shown in Figure 2-1 is equal to

$$V = Ah \tag{2-1}$$

where A is the cross-sectional area of the capillary and h is the height of the bottom surface of the mercury plug above the bottom of the capillary. If we plot the volume of the trapped air in our thermometer versus the corresponding Celsius temperature of the seven highest temperature systems noted in Figure 2-2, then we obtain a result like that shown in Figure 2-3. The most interesting feature of the plot shown in Figure 2-3 is the intercept on the temperature axis. The value of this intercept is -273.15°C. All Celsius-scale thermometers of the type shown in Figure 2-1 yield the same value of the $V = 0$ intercept on the temperature axis. These results suggest that an *absolute temperature scale* can be established by adding 273.15 to the Celsius temperature scale:

$$T/\mathrm{K} = t/^\circ\mathrm{C} + 273.15 \tag{2-2}$$

where the units of the absolute (Kelvin) temperature scale are degrees kelvin, denoted K. Thus the temperature of an ice + water mixture at 1 atm on the Kelvin temperature scale is

$$T/\mathrm{K} = 0^\circ\mathrm{C}/^\circ\mathrm{C} + 273.15$$
$$T = 273.15\ \mathrm{K}$$

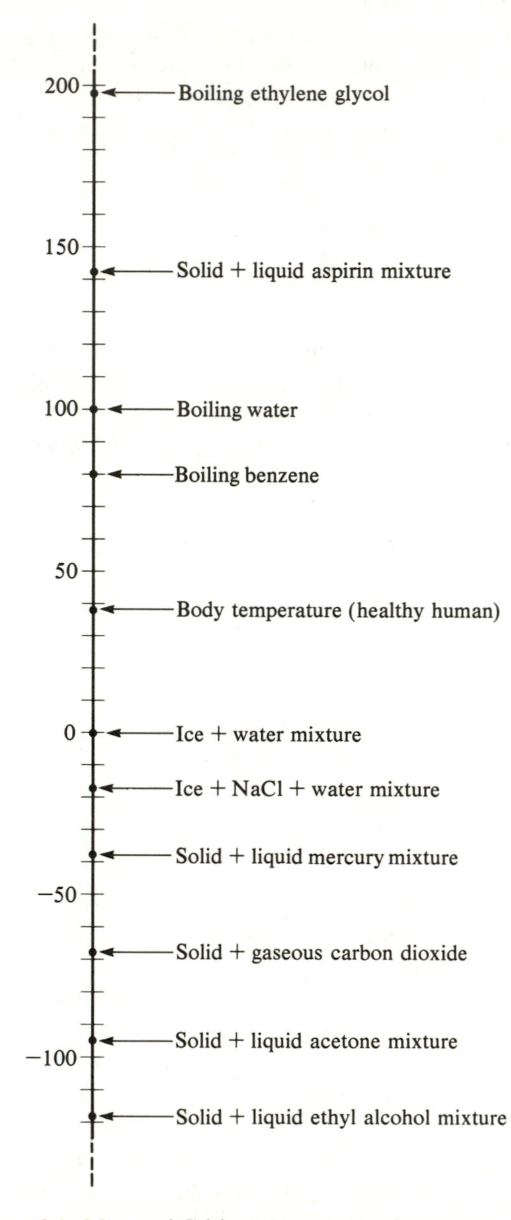

Figure 2-2. Measured Celsius temperatures of several systems.

Similarly the temperature of boiling water at one atmosphere pressure on the Kelvin temperature scale is $T = (100.00 + 273.15)\ \text{K} = 373.15\ \text{K}$.

The Kelvin temperature scale is called an *absolute temperature scale* because the lowest possible temperature on this scale is zero degrees kelvin, 0 K, and all temperatures on the Kelvin absolute temperature scale are *positive*. We distinguish the absolute Kelvin temperature from the Celsius temperature by

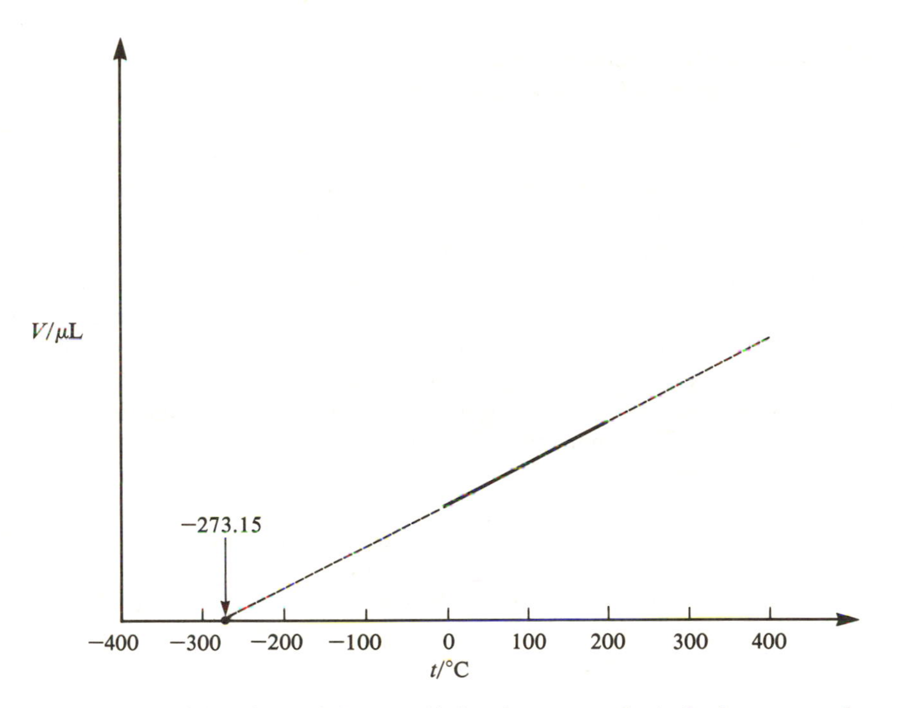

Figure 2-3. A plot of the volume of air trapped below the mercury plug in the thermometer shown in Figure 2-1 versus the Celsius temperature of various systems. Note the linear behavior of V versus t. The value of the intercept on the temperature axis is $-273.15°C$.

using the symbol T for the absolute temperature and the symbol t for the Celsius temperature.

EXAMPLE 2-1

Compare the size of a Kelvin degree with the size of a Celsius degree.

Solution: The Kelvin and the Celsius temperature scales differ only in the location of the zero points on the two scales. Thus the numerical *difference* in the temperatures of two systems is exactly the same on the Kelvin and Celsius temperature scales. Therefore the Kelvin and the Celsius degrees are of equal magnitude.

There is a very significant complication involved in the measurement of absolute temperatures by the method outlined above. The complication is that when we change the nature of the gas trapped below the mercury plug in the thermometer (say, from air to argon or methane), then we obtain slightly different values for the temperature of a particular system. This is an undesirable situation because we want our measurements to yield temperatures that are independent of the particular properties of the gas trapped in the thermometer.

or

$$P = \frac{F}{A}$$

The force due to a mass m in a gravitational field with an acceleration due to gravity of g is $F = mg$; thus

$$P = \frac{mg}{A}$$

Multiplying the numerator and denominator on the right-hand side of the above equation by h yields

$$P = \frac{mgh}{Ah}$$

The volume of a cylinder of cross-sectional area A and height h is $V = Ah$; thus

$$P = \frac{mgh}{V}$$

The density ρ is *defined* as the mass per unit volume; thus

$$\text{density} = \frac{\text{mass}}{\text{volume}}$$

or

$$\rho = \frac{m}{V}$$

Substitution of this expression for ρ into the preceding equation for P yields

$$P = \rho g h \tag{2-4}$$

Thus we see that the pressure exerted by a column of liquid is directly proportional to the height h of the column of liquid.

2-5 *The Ideal Gas Temperature Scale Is the Absolute Thermodynamic Temperature Scale*

We are now in a position to define the *ideal gas temperature scale.* Let P_T be the measured pressure of the gas trapped in a particular constant volume gas thermometer when the thermometer bulb is in contact with a substance at the absolute temperature T, and let $P_{273.16}$ be the measured pressure of the trapped gas when the same constant volume gas thermometer is in contact with water at its triple point. The ideal gas temperature scale is defined by the following equation:

$$T = (273.16\ \text{K}) \lim_{P_{273.16} \to 0} \left(\frac{P_T}{P_{273.16}}\right) \tag{2-5}$$

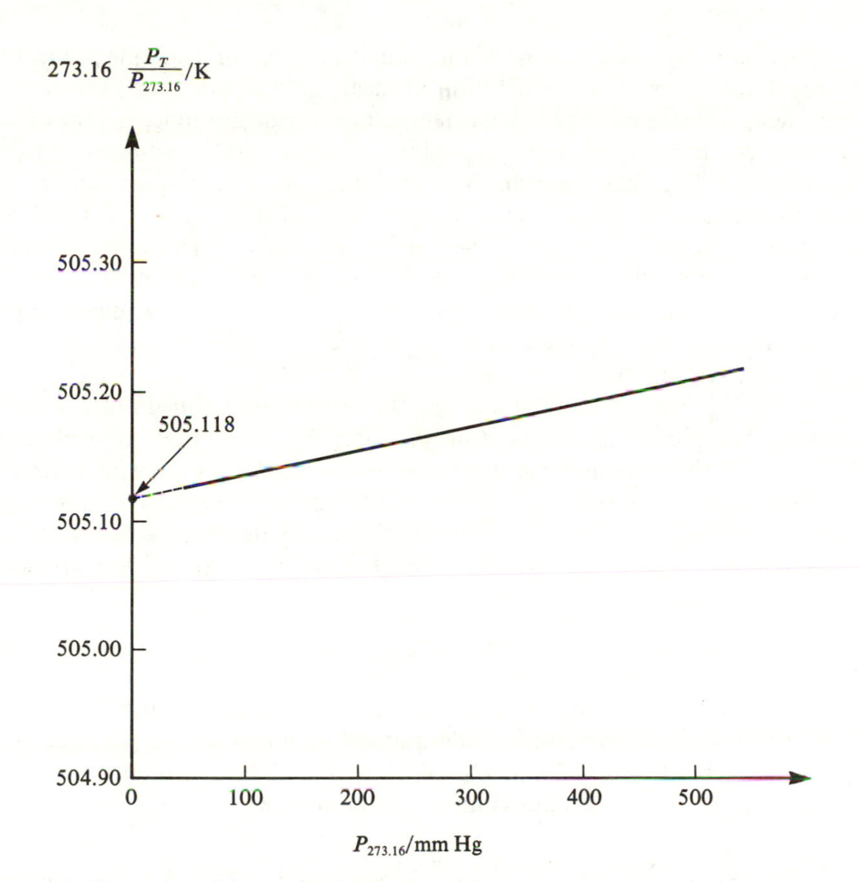

Figure 2-5. Illustration of the graphical method used to determine the ideal gas temperature of the melting point of tin. Measured values of $273.16P_T/P_{273.16}$, each obtained with a different number of moles of gas in the constant volume gas thermometer, are plotted against the corresponding values of $P_{273.16}$. The resulting plot is extrapolated to find the value of $273.16P_T/P_{273.16}$ when $P_{273.16} = 0$. The value of the intercept found by extrapolation (505.118) is equal to the ideal gas temperature of the melting point of tin.

The precise experimental determination of an ideal gas temperature of, say, the melting point of tin (i.e., the temperature of a mixture of solid and liquid tin) requires a *series of pairs of pressure measurements.* The first pair of pressure measurements yields a value of the ratio $P_T/P_{273.16}$. The number of moles of gas trapped in the gas thermometer is then lowered by removing some of the gas and then the value of the ratio $P_T/P_{273.16}$ is remeasured. The procedure is repeated several times with progressively smaller amounts of gas trapped in the gas thermometer. The measured values of the ratio $P_T/P_{273.16}$, each multiplied by 273.16, are then plotted against the corresponding values of $P_{273.16}$. The ideal gas temperature is equal to the value of intercept on the $273.16P_T/P_{273.16}$ axis (Figure 2-5).

There is a very good reason for adopting the rather cumbersome procedure described above for the determination of ideal gas temperatures. The reason is that temperatures on the ideal gas temperature scale are independent of the particular gas (say, argon, nitrogen, helium, etc.) in the thermometer bulb of the constant volume gas thermometer. This is clearly a desirable situation, because the absolute thermodynamic temperature must be independent of the particular physical properties of any substance. If our measured thermodynamic temperatures depended on the nature of the substance used in the thermometer, then such temperatures would lack fundamental significance, because different thermometers would yield different values of the temperature for the same system.

It is not difficult to understand why the procedure outlined above for the determination of ideal gas temperatures is effective. Recall that an ideal gas is gas for which there are no intermolecular interactions and for which the gas molecules themselves occupy a completely negligible fraction of the total volume occupied by the gas. It is an experimentally demonstrable result that fixed-composition gases exhibit ideal gas behavior *in the limit of zero pressure.* The P, V, T relation for an ideal gas is

$$PV = nRT \tag{2-6}$$

where P, V, and T are the gas pressure, the gas volume, and the thermodynamic gas temperature, respectively; n is the number of moles of the gas; and R is the gas constant, $R = 0.08206\ \text{L}\cdot\text{atm}\cdot\text{K}^{-1}\cdot\text{mol}^{-1}$. Thus, in the limit of zero pressure we have for a constant volume gas thermometer

$$T = (273.16\ \text{K}) \lim_{P_{273.16}\to 0} \left(\frac{P_T}{P_{273.16}}\right) = 273.16\ \text{K}\left[\frac{nRT/V}{nR(273.16)/V}\right] = T$$

In effect, the method that we have used to introduce the absolute thermodynamic (ideal gas) temperature scale involves the postulate that Eq. 2-6 holds in the limit of zero pressure for all fixed-composition gases.

All temperatures on the ideal gas temperature scale are necessarily positive and, in addition, *the temperature* $T = 0$ *is undefined* (see Eq. 2-5). We designate the unit of temperature on the ideal gas temperature scale as the kelvin; the SI unit designation for temperature is K (without any degree sign). Although the ideal gas temperature scale is independent of the properties of any particular gas, the scale does depend on the properties of gases in general. Thus, the lowest temperature at which precision measurements can be made with a (helium) gas thermometer is about 1 K.

The accurate determination of the temperatures of reproducible fixed points (e.g., melting points and boiling points) on the ideal gas temperature scale is of considerable importance, because known temperatures for reproducible fixed points can be used to calibrate other types of thermometers, such as a resistance thermometer, that are much easier to use for temperature measurements than an ideal gas thermometer.

2-6 *The International Practical Temperature Scale (IPTS) Forms the Experimental Basis for Practical Temperature Measurements*

A set of internationally agreed upon temperatures for primary and secondary fixed reference points, together with agreed upon interpolating instruments and interpolating functions, constitute the *International Practical Temperature Scale* (IPTS). The current version of the IPTS was accepted by the International Committee for Weights and Measures in 1968 and is referred to as IPTS-68. The IPTS was provisionally extended from 13.81 K to 0.519 K in 1976. The 1976 *e*xtended *p*rovisional *t*emperature scale is designated as EPT-76. A reevaluation of the IPTS is planned for the 1980s. IPTS-68 "has been chosen in such a way that the temperature measured on it closely approximates the thermodynamic temperature; the difference is within the limits of the present accuracy of measurement" [The International Practical Temperature Scale of 1968, *Metrologia* 5; 35–44 (1969)]. The thermodynamic temperature scale is the ideal gas temperature scale.

The defining fixed points of the IPTS-68 are given in Table 2-1. Temperatures in degrees Celsius (°C) are related to the corresponding ideal gas temperature by the equation

$$t/°\mathrm{C} = T/\mathrm{K} - 273.15 \qquad (2\text{-}7)$$

TABLE 2-1
Defining (Primary) Fixed Points of the IPTS-68

System	T_{68}/K
Triple point of H_2	13.81
Equilibrium vapor pressure of liquid H_2 at a pressure of $\frac{25}{76}$ standard atmosphere (33,330.6 $\mathrm{N \cdot m^{-2}}$)	17.042
Normal[a] boiling point of hydrogen	20.28
Normal boiling point of neon	27.102
Triple point of oxygen	54.361
Normal boiling point of oxygen	90.188
Triple point of water	273.16 (exact)
Normal boiling point of water	373.15
Normal freezing point of tin	505.118[b]
Normal freezing point of zinc	692.73
Normal freezing point of antimony	903.89[c]
Normal freezing point of silver	1235.08
Normal freezing point of gold	1337.58

[a] *normal* means the 1 standard atmosphere = $1.01325 \times 10^5\ \mathrm{N \cdot m^{-2}}$ value.
[b] May be used as an alternative to the normal boiling point of water.
[c] A secondary point.

The assigned temperatures on EPT-76 extend the IPTS-68 scale down to 0.519 K using as fixed points the onset of superconductivity for several metals [Pb (7.200 K), In (3.415 K), Al (1.180 K), Zn (0.851 K), and Cd (0.519 K)], together with the normal boiling point of helium (4.222 K) (see *Chemical and Engineering News*, August 7, 1978, pp. 21–22).

The IPTS-68 interpolating instruments, that is, the working thermometers that are calibrated using the defining fixed points are as follows:

Range	*Thermometer*
13.81 K–903.89 K	Platinum resistor
903.89 K–1337.58 K	Platinum vs. 90% platinum + 10% rhodium thermocouple
1337.58 K and above	Optical pyrometer (determination of temperature based on distribution of radiation emitted by object)

The exact mathematical forms of the interpolating functions that must be used are also defined by IPTS-68. For example, the interpolating function for the Pt versus Pt + Rh thermocouple thermometer is

$$\varepsilon = a + bt_{68} + ct_{68}^2 \tag{2-8}$$

where ε is the measured electromotive force at the Celsius temperature t_{68} (68 for IPTS-68) and the constants a, b, and c are obtained from measured values of ε at the normal freezing points of Sb, Ag, and Au, for which the values of t_{68} are known. The value of t_{68} for the Sb sample used must be checked with a platinum resistance thermometer.

Any system with a measurable property that varies monotonically with temperature and that gives reproducible values of the measured property when the temperature is cycled (i.e., the property does not exhibit hysteresis) can be used as a thermometer. In addition to the types of thermometers already mentioned in this chapter, there are several other types in use, such as liquid-in-glass capillary (length), thermistors (resistance), quartz crystal (frequency of mechanical oscillation), acoustic (sound velocity in a metal rod), magnetic (magnetic susceptibility), and gas-ionization thermometers (current). A more complete list of thermometers is given in Table 2-2. The minimum uncertainties associated with different types of thermometers are shown in Figure 2-6.

Our discussion of temperature is based on an *operational* (how we measure it) *approach*. In effect, temperature has been described in terms of a set of operations that describe how we go about measuring temperature. The operational approach is used extensively in thermodynamics and has the advantage of linking fundamental concepts directly to experimental measurements. Definitions, no matter how clever, useful, or convenient, do not create either facts or concepts. Fundamental concepts cannot be defined; they can only be described.

TABLE 2-2
Temperature Measuring Instruments

Name	*Phys. qty.*	*Form*	*Range*	*Resolution*
Magnetic Thermometer	Magnetic susceptance	Paramagnetic salt such as Cerous Magnesium Nitrate in form of single spherical crystal	0.01 K to 1.5 K	0.001 K
Acoustic Thermometer	Sound velocity (through wave-length & freq.)	Acoustical resonant cavity	0.1 K to 50 K	0.0001 K
Vapor Thermometer	Pressure	Saturated vapor filled metal bulb with capillary	0.2–2 K; He_3 scale 1–5.2 K; He_4 scale	0.001 K in standard thermometry
Germanium Thermometer	Electrical resistance	Encapsulated germanium chips with four terminals	1.5 K to 100 K	0.0001 K in laboratory standards
Carbon Resistance Thermometer	Electrical resistance	Carbon resistor (½ or ¼ W, as used in electronics)	1.5 K to 100 K	0.001 K, immediately after calibration
Gas Thermometer	Pressure (generally at constant volume)	Gas filled bulb with capillary	4 K to 1338 K	0.001 K at lower end. 0.2 K at upper end
Thermistor	Electrical resistance	Semiconducting metal oxide chip or bead with two leads	4 K to 200°C	0.01 K (with selected units. 0.001 K achievable in controlled baths)
Quartz Thermometer	Frequency of mechanical oscillations	Quartz crystal with Y cut	11 K to 250°C	0.001 K
Resistance Thermometer	Electrical resistance	Platinum wire (for highest accuracy the wire is coiled & strain-free mounted in a glass tube or a metal capsule and has four terminals. For industrial applications the wire coil is clamped or embedded in insulating material). Standard values: 0.25 Ω, 25 Ω, 100 Ω at 0°C	14 K to 1064°C	0.00001 K in some standards. 0.1 K typically in industrial instruments
Thermocouple	Thermal emf	Two different metal or alloy wires joined at both ends forming a circuit, which is opened at one point for measuring emf One end is at a reference temperature, the other at temperature under measurement.	20 K to 2400°C	0.1 K in lab standards. 1 K typically in industrial instruments
Acoustic Thermometer*	Sound velocity (through length & transit time)	Metallic rod (Al, W, Rn, Tn, Mo)	30 K to 3100°C	1%
Nuclear Quadrupole Resonance Thermometer	Frequency of electromagnetic field	Potassium chlorate specimen in the tank coil of an oscillating circuit	30 K to 300 K	Not yet determined
Liquid-in-glass Thermometer	Thermal expansion	Glass bulb filled with mercury or toluene (also ethyl alcohol and xylol)	125 K to 600°C	0.01 K (for narrow ranges)
Bimetallic Thermometer	Differential thermal expansion	Two adherent strips, coiled to form a spiral or helix. Also two rods	125 K to 400°C	1 K (for narrow ranges)
Noise Thermometer	Noise voltage	Resistor in the form of a thin Pt filament (dia. = 0.01 mm)	0°C to 1064°C	Not yet determined
Total Radiation Pyrometer	Total radiance	Total radiance detector (thermopile)	0°C to 5000°C and up	Several K
Spectrally Selective Pyrometer	Emf or resistance	Photodiode or photoconductive radiation sensor	0°C to 5000°C and up	2 K
Automatic Monochro-matic Optical Pyrometer	Spectral con-centration of radiance (ratios)	Photoelectric detector (photomultiplier or photodiode)	750°C to 5000°C and up	0.03 K in lab standards 0.25 K in industrial instruments
Manual Monochro-matic Optical Pyrometer	Spectral con-centration of luminance (ratios)	Human eye (visual observation)	750°C to 5000°C and up	1.5 K
Two (or three) Color Pyrometer	Ratio of spectral concentrations of radiance	Photoelectric detector (photomultiplier or photodiode)	750°C to 5000°C and up	Several K

**Note:* This is an acoustical thermometer with a sensor different from the preceding one.
Source: Temperature Measurement Guide, © Chilton Company, Philadelphia.

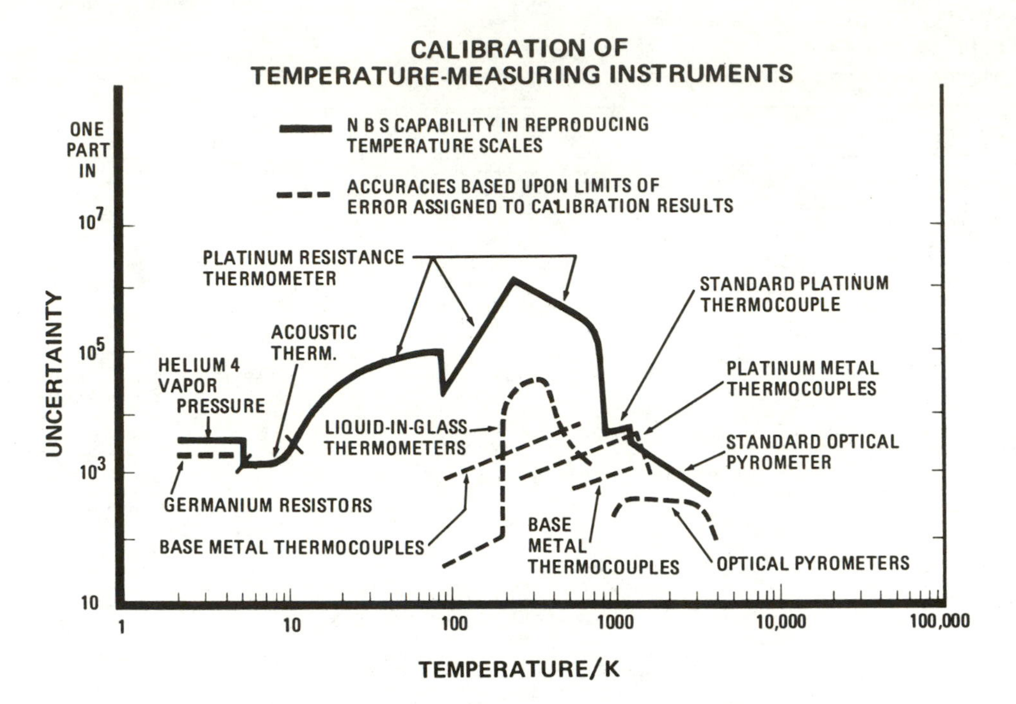

Figure 2-6. Minimum measurement uncertainties associated with temperature measurements made with different types of thermometers. (*Source:* NBS Technical Note No. 262.)

Problems

1. Given that the freezing and boiling points of water on the Celsius and Fahrenheit temperature scales are (0°C, 100°C) and (32°F, 212°F), respectively, derive the following relationships:

$$t/°\mathrm{C} = \tfrac{5}{9}[(t'/°\mathrm{F}) - 32]$$
$$T/\mathrm{K} = 255.37 + \tfrac{5}{9}(t'/°\mathrm{F})$$

where t is the temperature in degrees Celsius and t' is the temperature in degrees Fahrenheit. Determine the value of t'(°F) when $T = 0$ K.

2. Use the relationships given in Problem 1 to compute the Celsius and Kelvin temperatures that correspond to 98.6°F.

3. What temperature is numerically identical on the Celsius and Fahrenheit temperature scales?

4. The temperature 0°F was chosen originally as the temperature of a mixture of salt (NaCl), ice, and water. (This salt-ice-water system is used as the freezing medium in the preparation of homemade ice cream). Compute the Celsius and Kelvin temperatures of the salt-water-ice system.

5. The Rankine temperature scale is an *absolute* temperature scale that has the same type of relationship to the Fahrenheit scale as the Kelvin scale has to the Celsius scale. Determine the temperature of the freezing point of water on the Rankine scale.

6. Why is it that all the systems noted in Table 2-1 involve at least two phases of a substance?

7. Suggest a type of thermometer not mentioned in the text. What is the temperature range of applicability of your thermometer?

8. Can you think of any property of a living system that could be used as a thermometer?

9. How would you determine the temperature of the Sun? To what part of the Sun would your results apply?

10. What is the lowest Celsius temperature that can be measured with a mercury-in-glass capillary thermometer?

11. Explain why a mercury-in-glass thermometer with a uniform-bore capillary that is correct at 0°C and 100°C reads 49.9°C when the temperature is actually 50.0°C. Note the volume of a sample of mercury V_t at Celsius temperature t, relative to the volume V_0 at 0°C, is given by

$$V_t = V_0(1 + 0.18182 \times 10^{-3}t + 0.0078 \times 10^{-6}t^2)$$

12. The highest temperature reached on Earth is about 10^9 K (nuclear explosion); the lowest temperature reached on Earth is about 10^{-6} K. What types of thermometers would you suggest for measuring such extremes of temperature?

13. A quantity of interest in the assessment of the performance of refrigeration devices is the wet-bulb temperature. The wet-bulb temperature is measured by surrounding the bulb of a conventional mercury-in-glass thermometer with a piece of wet cotton and then swinging the thermometer back and forth in air. It is observed that the wet-bulb temperature is always less than or equal to the dry-bulb temperature. Explain.

14. Explain why water with a dissolved colored dye is not particularly suitable as a thermometer fluid for a liquid-in-glass thermometer used for atmospheric temperature measurements.

15. The lower fixed point on the Celsius temperature scale is the triple point of water. What is the difference in degrees Kelvin between the triple point and the freezing point of water? Why is the triple point of water a better choice as a fixed temperature point than is the freezing point of water?

16. The thermometric property is the quantity that is measured with a particular thermometer. Thus *pressure* is the thermometric property of a constant volume gas thermometer. List the thermometric properties of the following

types of thermometers:
(a) Liquid-in-glass capillary thermometer
(b) Platinum resistance thermometer
(c) Thermocouple thermometer
(d) Optical pyrometer
(e) Gas-ionization thermometer
(f) Magnetic thermometer
(g) Thermistor thermometer
(h) Bimetallic coil
(i) Constant-pressure gas thermometer

17. Solar furnaces are designed to focus sunlight (say, with a parabolic mirror) onto a relatively small area. Given that the surface temperature of the sun is about 6000 K, estimate the highest possible temperature that can be obtained in a solar furnace.

18. There are no substances capable of existing as solids above 4300 K. Describe how you would make temperature measurements at $T > 4300$ K.

19. The dependence of the length of a brass rod on the temperature is given by

$$l_t = l_0(1 + 0.1781 \times 10^{-4}t + 0.098 \times 10^{-7}t^2)$$

where l_t is the length at Celsius temperature t and l_0 is the length at 0°C. Suppose that a brass rod is used as a thermometer for which it is assumed that $l_t \propto t$. If the brass rod thermometer is calibrated at 0°C and 100°C, then what will the measured temperature be when the actual temperature is 50.00°C?

20. A bimetallic strip is a composite material made up of two strips of different metals fastened together. When such a bimetallic strip is wound in a spiral and subjected to a change in temperature, the strip changes its curvature because of the different rates of expansion with temperature of the two metals. Sketch designs for (a) a flat spiral thermometer and (b) a single-helix thermometer based on bimetallic strips. Note that such thermometers do not involve a liquid phase.

21. Liquid-in-glass thermometers are of two types: mercury in glass and alcohol plus red dye in glass. The dependence of volume on temperature of mercury and alcohol are given by

$$V_t = V_0(1 + 0.18182 \times 10^{-3}t + 0.0078 \times 10^{-6}t^2) \qquad \text{(mercury)}$$
$$V_t = V_0(1 + 1.012 \times 10^{-3}t + 2.20 \times 10^{-6}t^2) \qquad \text{(alcohol)}$$

where V_0 is the volume when $t = 0$°C. Compute the percent change in the volume of these two liquids over the temperature range 0°C to 100°C. Which of the two liquids is the more sensitive when used to detect a change in temperature in the range 0°C to 100°C?

22. Platinum resistance thermometers possess a resistance-temperature relationship that is highly reproducible. The relationship is described by the

Callendar–Van Dusen (CVD) equation

$$t = \frac{100(R_t - R_0)}{R_{100} - R_0} + \delta\left(\frac{t}{100} - 1\right)\frac{t}{100} + \beta\left(\frac{t}{100} - 1\right)\left(\frac{t}{100}\right)^3$$

where t is in degrees Celsius and R_t, R_{100}, and R_0 are the thermometer resistances at t, 100°C, and 0°C, respectively. The constants δ and β are determined by calibration at the sulfur (444.674°C) and oxygen (−182.962°C) points, respectively ($\beta = 0$ for $t > 0$°C).

(a) Some typical data for a platinum resistance thermometer are $R_0 = 25.523\ \Omega$, $\delta = 1.491$, $\beta = 0.1108$, and $(R_{100} - R_0)/100R_0 = 0.00392532$. Using the CVD equation, compute the temperature in degrees kelvin for systems that yield values of $R_t = 26.5045\ \Omega$ and $25.0372\ \Omega$, respectively.

(b) Derive an expression for dR_t/dt from the CVD equation and evaluate it, for the particular values of the constants given, at $t = -100$°C, 0°C, and 100°C.

23. Suppose we calibrate the platinum resistance thermometer described in Problem 22 at the normal freezing and boiling points of water, and we *assume* that the resistance-temperature relationship is of the form

$$R_t = R_0(1 + \alpha t)$$

rather than that given in the problem. What will this thermometer yield for the temperature of systems with ideal gas temperatures of 0.00°C, 25.00°C, 50.00°C, 75.00°C, and 100.00°C, respectively? Plot the temperature discrepancy Δt versus the temperature over the range 0°C–100°C. What would the curve look like if we use the CVD equation rather than the above expression to calculate t from R_t?

24. To what precision must the resistance of a platinum resistance thermometer be measured in order to yield t within ±0.0001°C at around 25°C? (See Problem 22.)

25. Make a complete list of the number of temperature-sensing devices in your home and explain how each different type of device works. If you do not know how a particular device works, then consult the references in Appendix III.

Pacific Gas and Electric Company's Rock Creek hydroelectric power plant in northern California's Plumas County has a generating capacity of 182,000 kW. The plant, which began operation in 1950, draws water from the north fork of the Feather River to generate electricity.

3

Energy and the First Law of Thermodynamics

There is a fact, or if you wish, a law, governing all natural phenomena that are known to date. There is no known exception to this law—it is exact as far as we know. The law is called the conservation of energy. It states that there is a certain quantity, which we call energy, that does not change in the manifold changes which nature undergoes. That is a most abstract idea, because it is a mathematical principle; it says that there is a numerical quantity which does not change when something happens.

R. P. Feynman, R. B. Leighton, and M. Sands, *The Feynman Lectures on Physics*, Vol. I. Reading, Mass.: Addison-Wesley Publishing Company, Inc., 1965.

THE TERM *energy* is used so extensively that every civilized adult has at least some idea of the meaning of the term. The frequency with which the term *energy* is used has increased dramatically since the 1973–1974 embargo on oil exports to the United States and to some other countries by the *O*rganization of *P*etroleum *E*xporting *C*ountries (OPEC), which in 1973 consisted of Saudi Arabia, Iran, Venezuela, Nigeria, Libya, Kuwait, Iraq, United Arab Emirates, Algeria, Indonesia, Qatar, Ecuador, and Gabon. Prior to the 1973–1974 OPEC oil embargo, the price of OPEC oil was about \$2 per barrel (42 gallons), but by 1980 the OPEC contract price per barrel of oil had risen to \$34. The OPEC oil embargo, and the subsequent rapid escalation of oil prices, made it painfully clear that countries that must import large quantities of oil are economically vulnerable to outside actions over which they have little or no control. The 1973–1974 OPEC oil embargo generated a worldwide interest in energy considerations that has continued unabated* into the 1980s. We hear endless talk of the "energy crisis," "energy conservation," "energy supplies," and even "soft energy," to name just a few examples. Yet with all this talk, if you ask the question "What is energy?", you will find a truly amazing range of responses, many of which if not outright unintelligible at least border on unintelligibility. Why is it that such a strange situation exists? Why is it so difficult to answer the question "What is energy?"?

3-1 *The Total Amount of Energy Never Changes*

Energy is a fundamental abstract concept. Our approach toward developing an understanding of the energy concept, which is in essence a mathematical concept, will involve a description of the mathematical properties of the *energy function U*, together with descriptions of the mathematical formulas that we use to calculate the amounts of the various kinds of energy. Different forms of energy are distinguished by the different formulas used in the calculations. It is important to recognize that energy is not a "thing." There are no energy meters; that is, there are no meters capable of the direct measurement of energy. There are electric current meters and gas flow meters and volume meters and mass meters and pressure meters and length meters and time meters and thermometers, but none of these are energy meters. A determination of the amount of energy transferred from one system to another system, or the amount of energy converted from one form to another form within the same system, always involves measurements of the appropriate physical parameters followed by calculations involving the appropriate energy formula(s).

The most significant mathematical property of the energy function is that of conservation. The *principle of energy conservation*, which has passed innumerable experimental tests and failed none, amounts to the statement that, for any real process whatsoever, the total amount of energy remains *exactly* the same at the end of the process as it was at the beginning of the process. Energy can be transferred and energy can be transformed, but energy cannot be created and energy cannot be destroyed.

* In March, 1983, the OPEC oil price was dropped to \$29 per barrel as a result of an oil glut arising from a combination of overproduction and decreased usage.

To clarify how the energy concept is utilized in science, we shall describe an analogy devised by the physicist R. P. Feynman (Feynman, Leighton, and Sands, 1965). The analogy consists of a story about a little boy, his set of blocks, and his mother's efforts to keep track of his blocks. The boy's blocks consist of a set of 28 identical pieces. At the end of each day the mother checks her son's room to make sure that all the blocks are in the room. One day she makes a count of the blocks and finds that only 26 of them are visible. A search of the room leads to the discovery of a locked box that the mother suspects may contain the missing blocks, but she cannot find the key, and her son does not remember where the key is located. Not wishing to press the issue, the mother waits for a day when all the blocks can be accounted for, and then she weighs both the empty box and one of the blocks. Her results lead her to formulate the following equation to determine whether or not all the blocks are in the room:

$$\begin{pmatrix}\text{number of blocks}\\ \text{visible in room}\end{pmatrix} + \frac{(\text{weight of box}) - (\text{weight of empty box})}{(\text{weight of one block})} = 28$$

For several days the equation above works perfectly, but one day it fails. An intensive search of her son's room leads the mother to conclude that possibly the missing blocks are in a cylindrical pail partly filled with a thick murky liquid in the corner of the room, but the mother does not want to dirty her hands in the liquid. Therefore, she decides to wait until her first equation works again; when it finally does, then she measures the height of liquid in the pail and the increase in height produced by submerging one of the blocks. These results lead her to propose the following revised equation:*

$$\begin{pmatrix}\text{number of blocks}\\ \text{visible in room}\end{pmatrix} + \frac{(\text{weight of box}) - (\text{weight of empty box})}{(\text{weight of one block})} + \frac{\begin{pmatrix}\text{height of liquid}\\ \text{in pail}\end{pmatrix} - \begin{pmatrix}\text{height of liquid}\\ \text{in pail with}\\ \text{no blocks}\end{pmatrix}}{\begin{pmatrix}\text{increase in height}\\ \text{of liquid produced}\\ \text{by one block}\end{pmatrix}} = 28$$

For some time the revised equation works well, but one day it too fails. A very thorough search of the room fails to yield the missing blocks. At this point the mother becomes quite perplexed, and while standing in front of the window she notices that the missing blocks are lying outside on the ground below the window. Counting the blocks on the ground outside she finds that when these

* It should be evident at this point that this woman, in addition to being a mother, is also a first-rate experimental and theoretical physicist.

are included, she can once again account for all the blocks. Her equation now reads

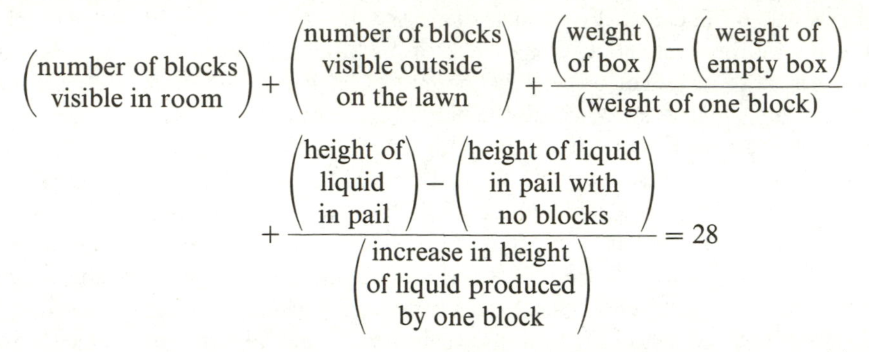

$$\begin{pmatrix}\text{number of blocks}\\ \text{visible in room}\end{pmatrix} + \begin{pmatrix}\text{number of blocks}\\ \text{visible outside}\\ \text{on the lawn}\end{pmatrix} + \frac{\begin{pmatrix}\text{weight}\\ \text{of box}\end{pmatrix} - \begin{pmatrix}\text{weight of}\\ \text{empty box}\end{pmatrix}}{(\text{weight of one block})}$$

$$+ \frac{\begin{pmatrix}\text{height of}\\ \text{liquid}\\ \text{in pail}\end{pmatrix} - \begin{pmatrix}\text{height of liquid}\\ \text{in pail with}\\ \text{no blocks}\end{pmatrix}}{\begin{pmatrix}\text{increase in height}\\ \text{of liquid produced}\\ \text{by one block}\end{pmatrix}} = 28$$

For quite some time this equation works nicely until one day a very surprising thing happens—the left-hand side of her equation adds to 30. Because she had never observed this result before, the mother is completely baffled. Discussing her observations with her neighbor she finds that her neighbor's little boy also has a set of blocks just like her son's, and a check by his mother leads to the discovery that two of her son's blocks are missing. Because her neighbor's boy had been playing with the mother's son in his room the previous day, the mother felt confident that she could account for the excess blocks.

The key point of the Feynman analogy is that the blocks are really incidental to the basic development of the equations—the crux of the development is the assumed principle of conservation. There are no blocks involved in the energy conservation principle, but that is the only essential difference between the Feynman analogy and the way in which the energy conservation principle is used in science.

3-2 *Energy Has Many Different Forms and There Is a Different Formula for Each Form*

To understand and to use the principle of energy conservation, we need to know the formulas for the various forms of energy. Although the formulas are known for many of the forms of energy, the formulas for all the forms of energy are not known. Fortunately, we can apply the principle of energy conservation to a problem provided that we know the formulas for all the forms of energy that may increase or decrease in the problem. That this is indeed so is clear from the Feynman analogy. We would not need to know the formula for calculating the number of blocks submersed in the murky liquid as long as there was no possibility of any of the blocks being in there. Further, there are undoubtedly other hiding places for the blocks, each with an associated formula for calculating the number of blocks that were not discovered because they were not in use when the conservation principle was tested. Thus the application of the energy conservation principle to a particular *system* only requires a knowledge of the formulas for the forms of energy that can increase or decrease within the system, together with formulas for calculating the amount of energy that enters the system from the *surroundings* or that exits the system to the surroundings. We define the boundaries of the system (the boy's room in the Feynman analogy); everything outside the system is the surroundings.

The known forms of energy are as follows:

gravitational energy	chemical energy
kinetic energy	surface energy
electrical energy	electromagnetic energy
nuclear energy	heat
elastic energy	mass energy

Not all these forms of energy are completely separable from one another. Thus chemical energy involves the kinetic energy of the electrons and the electrical energy of interaction between atoms in a molecule or ion and the electrical energy of interaction between molecules and ions. Further, elastic energy and surface energy are, as we shall see, special forms of chemical energy.

You already know the formulas for several of the given energy forms from previous courses in physics and chemistry. The general term for a form of energy that depends on the position of an object relative to something else is *potential* energy. The gravitational potential energy of a mass m in the earth's gravitational field is given by the formula

$$\text{gravitational potential energy} = mgh \tag{3-1}$$

where h is the distance of the object from the surface of the earth and g is the gravitational acceleration in the region. The formula (Eq. 3-1) involves the assumption that g is a constant over the range that h is varied. If g is not constant, then the formula is more complicated; but such complications are fine points that need not concern us at this stage. The kinetic energy of a mass m moving with a velocity v is given by

$$\text{kinetic energy} = \frac{mv^2}{2} \tag{3-2}$$

Suppose that we have an object of mass m at height h moving with a velocity v, and also suppose that the only two forms of energy that we need to consider are gravitational potential energy and kinetic energy. Application of the principle of energy conservation yields the equation

$$mgh + \frac{mv^2}{2} = \text{constant} \tag{3-3}$$

Equation 3-3 tells us that if the height h decreases, then the velocity v must increase because m and g are constants and the sum of the two energy terms is equal to a constant.

EXAMPLE 3-1

Suppose a stationary lead ball of 1.00-kg mass held 2.00 m above the floor is allowed to fall freely to the floor. Calculate the velocity of the mass at the instant of impact. Take $g = 9.81 \text{ m}\cdot\text{s}^{-2}$.

Solution: To solve this problem we use Eq. 3-3:

$$mgh + \frac{mv^2}{2} = \text{constant}$$

In the initial state we have $v = 0$ and $h = 2.00$ m; thus

$$mgh + 0 = \text{constant}$$

and

$$(1.00 \text{ kg})(9.81 \text{ m}\cdot\text{s}^{-2})(2.00 \text{ m}) = 19.62 \text{ kg}\cdot\text{m}^2\cdot\text{s}^{-2} = \text{constant}$$

(Note that the units of energy used here are kilogram meters squared per square second, $\text{kg}\cdot\text{m}^2\cdot\text{s}^{-2}$. The SI unit of energy is the joule, J; $1 \text{ J} = 1 \text{ kg}\cdot\text{m}^2\cdot\text{s}^{-2}$.) At the instant of impact we have $h = 0$ and $v > 0$; thus

$$mgh + \frac{mv^2}{2} = 0 + \frac{(1.00 \text{ kg})v^2}{2} = 19.62 \text{ kg}\cdot\text{m}^2\cdot\text{s}^{-2}$$

or

$$v = (2 \times 19.62 \text{ m}^2\cdot\text{s}^{-2})^{1/2} = 6.26 \text{ m}\cdot\text{s}^{-1}$$

When the lead ball comes to rest on the floor, $h = 0$ and $v = 0$. What happened to the energy?

Einstein used the energy conservation principle to explain the *photoelectric effect*. The photoelectric effect involves the ejection of electrons from a metal surface by incident light. The key experimental observations are as follows:

1. Only light with a frequency greater than or equal to a certain threshold value ν_0, which is different for different metals, can eject electrons from the surface of the metal.
2. The kinetic energy ($mv^2/2$) of the ejected electron depends only on the frequency of the incident light and, in particular, is independent of the *intensity* of the incident light.

A photoelectric effect plot (kinetic energy of the ejected electrons versus the frequency of the incident light) for a lithium surface is shown in Figure 3-1. Einstein knew that Max Planck had explained successfully the properties of blackbody radiation by invoking the postulate that the energy of a photon of frequency ν is given by

$$\text{photon energy} = E = h\nu \tag{3-4}$$

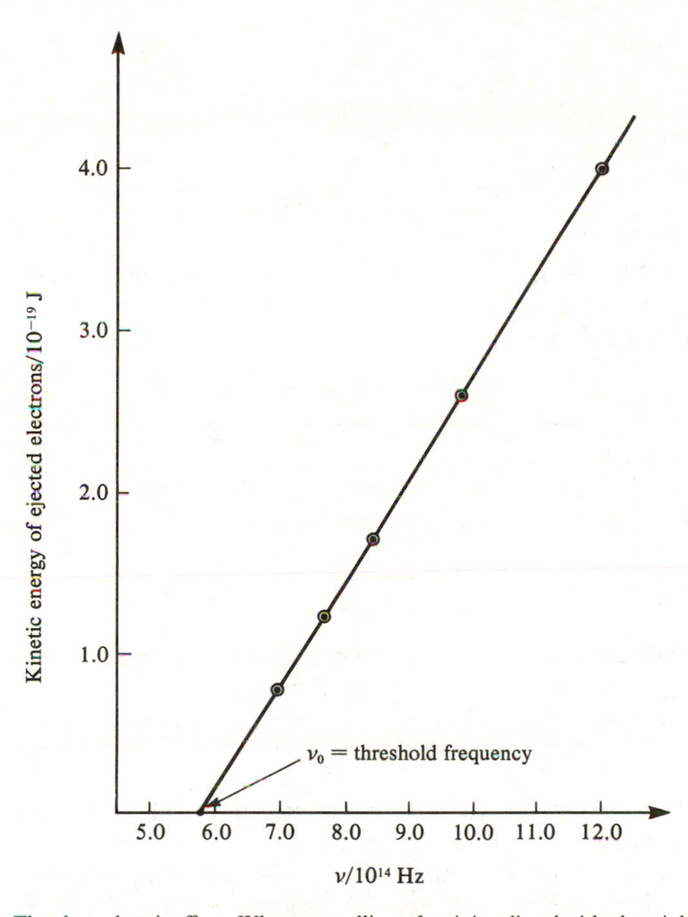

Figure 3-1. The photoelectric effect. When a metallic surface is irradiated with ultraviolet radiation, electrons are ejected from the surface. Here we plot the kinetic energy of the ejected electrons versus frequency of the ultraviolet radiation for lithium. In 1905, Einstein, using Planck's quantum hypothesis $E = h\nu$, was able to explain the linear dependence of the kinetic energy of the ejected electrons on the frequency of the incident ultraviolet radiation. The frequency below which no electrons are ejected is called the *threshold frequency* and is a natural consequence of Einstein's theory. Einstein's theory of the photoelectric effect led to our picture of electromagnetic radiation as a beam of photons with the energy of a particular photon of frequency ν being equal to $h\nu$.

Let the minimum amount of energy required to eject an electron from the metal surface be denoted by Φ; then, provided, $h\nu > \Phi$, we have, on application of the energy conservation principle to the photoelectric effect experiment,

$$h\nu = \Phi + \frac{mv^2}{2} \qquad (h\nu \geqslant \Phi) \tag{3-5}$$

Because $\Phi = h\nu_0$ (where ν_0 is the threshold frequency) is a constant for a particular metal surface, Einstein was able to explain the fact that a plot of the kinetic energy of the ejected electrons ($mv^2/2$) versus the frequency ν of the incident light is a straight line:

$$\frac{mv^2}{2} = h\nu - \Phi = h\nu - h\nu_0$$

or

$$\frac{mv^2}{2} = h(\nu - \nu_0) \qquad (\nu \geqslant \nu_0)$$

Furthermore, the slope of the $mv^2/2$ versus ν plot was predicted to be, as found experimentally, independent of the metal surface and equal to Planck's constant $h = 6.626 \times 10^{-34}$ J·s.

EXAMPLE 3-2

The allowed energies in attojoules (1 aJ = 10^{-18} J) of an electron in a hydrogen atom are given by the equation

$$E_n = -\frac{2.19 \text{ aJ}}{n^2} \qquad (n = 1, 2, 3, 4, \ldots)$$

Calculate the frequency and the wavelength of the photon emitted from a hydrogen atom when the electron goes from the $n = 2$ to the $n = 1$ energy level. The wavelength of a photon is related to the frequency of the photon by $\lambda = c/\nu$, where c is the velocity of light.

Solution: From the energy conservation principle, we know that the decrease in the energy of a hydrogen atom that results from the electron dropping from the $n = 2$ to the $n = 1$ energy level must be compensated exactly by a release of energy. The energy is released in the form of a photon with energy $h\nu$; thus, we can write

$$\text{(decrease in H atom energy)} + \text{(photon energy)} = 0$$

or

$$(E_1 - E_2) + h\nu_{21} = 0$$

The value of ν_{21} is given by

$$\nu_{21} = \frac{E_2 - E_1}{h} = \frac{[(-2.19/2^2) - (-2.19/1^2)] \times 10^{-18} \text{ J}}{6.63 \times 10^{-34} \text{ J·s.}}$$

$$= 2.48 \times 10^{15} \text{ s}^{-1}$$

The wavelength of the photon emitted is

$$\lambda = \frac{c}{\nu} = \frac{3.00 \times 10^8 \text{ m·s}^{-1}}{2.48 \times 10^{15} \text{ s}^{-1}} = 1.21 \times 10^{-7} \text{ m}$$

A photon with 1.21×10^{-7} m wavelength lies in the far ultraviolet region of the electromagnetic spectrum.

The most widely known energy formula is the Einstein mass-energy equation

$$E = mc^2 \tag{3-6}$$

where c is the velocity of light ($3.00 \times 10^8\ \text{m}\cdot\text{s}^{-1}$), m is the increment in mass, and E is the associated energy increment. The essence of the Einstein mass-energy equation is that if the total energy of a body changes, then the mass of the body changes by a proportional amount, where the proportionality constant is c^2. Thus when a chemical reaction occurs that leads to the evolution of an amount of energy,

$$\Delta U = U_{\text{products}} - U_{\text{reactants}}$$

Then the *total* mass of the reaction products is less than the total mass of reactants by an amount

$$\Delta m = m_{\text{products}} - m_{\text{reactants}}$$

that is equal to

$$\Delta m = \frac{\Delta U}{c^2}$$

The application of the energy conservation principle to nuclear decay processes led, in conjunction with Einstein's mass-energy relation, to the prediction and subsequent experimental verification of the existence of neutrinos. The missing (i.e., unaccounted for) energy in the decay processes was equated to (at that time unknown) neutral particles named *neutrinos* of mass $\Delta U/c^2$, where ΔU was the unaccounted for energy. The discovery of neutrinos is an example of use of the energy conservation principle to extend our understanding of matter.

EXAMPLE 3-3

When a positron and an electron come together, they annihilate each other; the result of the annihilation is two identical photons. Compute the frequency of the photons produced in the positron-electron annihilation reaction. The rest ($v = 0$) mass of an electron and the rest mass of a positron is 9.11×10^{-31} kg.

Solution: Application of the energy conservation principle to the positron-electron annihilation process yields

$$2m_e c^2 = 2h\nu$$

where $2m_e c^2$ is the energy loss when an electron and a positron are annihilated, and $2h\nu$ is the energy of the two photons produced in the annihilation. Thus

$$\nu = \frac{m_e c^2}{h} = \frac{(9.11 \times 10^{-31}\ \text{kg})(3.00 \times 10^8\ \text{m}\cdot\text{s}^{-1})^2}{(6.63 \times 10^{-34}\ \text{J}\cdot\text{s})} = 1.24 \times 10^{20}\ \text{s}^{-1}$$

It might be argued that the total energy of a system is given by Einstein's mass-energy equation $E = mc^2$ and that it is therefore possible to specify energy in absolute terms. However, Einstein's equation does not give the total energy of a system, but rather it gives the amount by which the energy of a particular system increases (or decreases) when the mass of the system increases (or decreases); or, alternatively, when a particular system loses an amount of energy equal to $\Delta U = U_2 - U_1$, then the mass of the system decreases by $\Delta m = m_2 - m_1$, where

$$\Delta U = c^2 \Delta m$$

Because we are unable to specify the absolute energy of a system, we are free to choose arbitrarily a zero of energy for any system. The particular choice made for the zero of energy is solely a matter of convenience. Only energy changes have operational (i.e., experimental) significance. Energy changes are not directly measurable, but energy changes can be calculated using appropriate energy formulas that involve measurable quantities.

3-3 *The Change in a Thermodynamic State Function Is Independent of the Manner in Which the Change in Thermodynamic State Is Accomplished*

In classical thermodynamics a macroscopic (i.e., matter in bulk) point of view is adopted and attention is focused on the properties of the system as a whole. It is necessary to carry out appropriate experiments to determine the thermodynamic state functions necessary and sufficient for the thermodynamic description of the system.

A state of a system is said to be defined when all the thermodynamic state functions have definite, time-independent values that are also independent of the history of the system. For a given change in the thermodynamic state of a system, the change in any thermodynamic state function of the system has a definite value independent of how the change in state was accomplished. We shall require, therefore, that any thermodynamic state function X of the system satisfy the condition that dX be an exact differential, that is, that the integral of dX be independent of path

$$\int_1^2 dX = X_2 - X_1 = \Delta X \tag{3-7}$$

In other words, the value of ΔX depends only on the initial and final thermodynamic states of the system; the value of ΔX is independent of the way in which the particular change in state is accomplished.

Some examples of thermodynamic state functions are volume, pressure, and temperature. If we know the values of the initial and final temperatures of a system, then we know the value of $\Delta T\,(=T_2 - T_1)$; it makes no difference how the value of T was changed from T_1 to T_2. The same is true for changes in the volume ΔV and for changes in the pressure ΔP. Not all thermodynamic functions have the mathematical property defined by Eq. 3-7. Two key examples of thermodynamic functions, the values of which depend on the way in which a change in state is accomplished, are work w and heat q. As we shall see, it is not possible, in general, to calculate the amount of work or heat involved when a system undergoes a change in thermodynamic state, *unless* the way in which the change in state is accomplished (i.e., the mathematical path connecting the initial and final states) is known. Thermodynamic functions whose differentials dX are not exact are not thermodynamic *state* functions.

Thermodynamic state functions are classified as either *intensive*, that is, independent of the mass of the system (e.g., pressure and temperature), or *extensive*, that is, directly proportional to the mass of the system (e.g., volume and internal energy). If we combine two identical thermodynamic systems, say system a and system b, then the resulting combined system has the same values of the intensive thermodynamic state functions as the separate systems (e.g., $T = T_a = T_b$), whereas the values of all the extensive state functions for the combined system are exactly twice those for either of the two separate systems (e.g., $V = V_a + V_b = 2V_a = 2V_b$).

3-4 *The Internal Energy U Is a Thermodynamic State Function*

The principle of energy conservation implies that a change in the thermodynamic state of a system involves a definite change in the energy of the system. Furthermore, experiments show that the magnitude of the energy change for a given change in state is proportional to the mass of the system. The foregoing considerations enable us to state the *First Law of Thermodynamics* as follows:

> *The internal energy U is a thermodynamic state function that is subject to a conservation principle.*

The internal energy of a system does not include the gravitational potential energy or the kinetic energy associated with the motion of the center of mass of the system.

For any change in state occurring in an *isolated system*, that is, a system closed off from its surroundings by walls through which matter or energy cannot pass, we have from the first law

$$dU_{\text{sys}} = 0 \qquad \text{(isolated system)} \tag{3-8}$$

For any change in state of a system that involves the transfer of matter or energy from the surroundings to the system, or vice versa, we have from the first law

$$dU_{\text{sys}} + dU_{\text{sur}} = 0 \tag{3-9}$$

For any *cyclic* process, that is, one in which the system is finally returned to its initial state, we have from the first law

$$\oint dU_{\text{sys}} = 0 \qquad \text{(cyclic process)} \tag{3-10}$$

where the circle on the integral sign denotes a cyclic (closed path) process.

The First Law of Thermodynamics actually goes beyond energy conservation in that the first law involves, in addition to energy conservation, the postulate that the internal energy is a thermodynamic state function. The internal energy is not the same as the total energy. The thermodynamic state of, say, a piece of iron remains unchanged when the iron is lifted in the gravitational field or when the iron is moving through space. Thus we do not include the kinetic energy ($mv^2/2$) or the gravitational potential energy (mgh) in the internal energy of a thermodynamic system.

3-5 *A Mathematical Relationship Between Thermodynamic State Functions Is Called an "Equation of State"*

A fundamental postulate of thermodynamics is the *state principle* (Kirkwood and Oppenheim):

> *For any real system there exists a positive integer n such that if n intensive thermodynamic state functions are fixed, then all other intensive thermodynamic state functions are fixed.*

The state principle apparently cannot be deduced from any more fundamental macroscopic principle; it must be regarded as a separate postulate based firmly on experimental observations. The state principle leads directly to the concept of an *equation of state;* that is, if X_i represents an intensive thermodynamic state function of a system, then there exists for defined states of a system an equation of the form

$$X_{n+1} = f(X_1, X_2, \ldots, X_n) \tag{3-11}$$

If the value of one extensive state function is given, then the mass of the system is determined.

The best known example of an equation of state is the ideal gas equation of state

$$PV = nRT \tag{3-12}$$

where P is the gas pressure, V is the gas volume, T is the absolute thermodynamic temperature, n is the number of moles of the gas, and R is the gas constant. If we divide both sides of Eq. 3-12 by V and then make the substitution $\rho = n/V$, where ρ is by definition the gas density, then we obtain the equation of state in the form

$$P = R\rho T \tag{3-13}$$

Equation 3-13 has the form of Eq. 3-11 in that all the thermodynamic state functions in Eq. 3-13 are intensive state functions. The equation of state (Eq. 3-13) tells us that of the three state functions P, ρ, and T only two are independent. We can specify arbitrarily the pressure and the temperature of an ideal gas, but once P and T are specified, then the value of the gas density ρ is automatically determined by Eq. 3-13:

$$\rho = \frac{P}{RT}$$

Thus only two of the set of three intensive state functions (P, ρ, T) are independent variables. Furthermore, if, in addition to the specification of any two of the intensive thermodynamic state functions, we also specify the value of any one extensive thermodynamic state function, say, the volume V, then the mass of the gas is also determined, because

$$n = \rho V$$

and

$$n = \frac{m}{M} = \frac{\text{mass of gas}}{\text{molar mass of gas}}$$

Thus, a fixed mass of an ideal gas has only two independent thermodynamic state functions, and the specification of any two state functions is sufficient to fix the values of *all* the other state functions, including the internal energy of the gas. Some additional examples of equations of state are given in Table 3-1.

TABLE 3-1
Some Examples of Equations of State

1. $PV = nRT$ (ideal gas)
   n = number of moles $R = 8.314\ \text{J}\cdot\text{K}^{-1}\cdot\text{mol}^{-1}$
2. $\left(P + \frac{n^2a}{V^2}\right)(V - nb) = nRT$ (van der Waals gas)
   a and b are constants characteristic of a particular gas.
3. $\left(P + \frac{n^2A}{TV^2}\right)(V - nB) = nRT$ (Berthelot gas)
   A and B are constants characteristic of a particular gas.
4. $P(V - nb')e^{na'/RTV} = nRT$ (Dieterici gas)
   a' and b' are constants characteristic of a particular gas.
5. $\mathcal{T} = KT\left(\frac{L}{L_0} - \frac{L_0^2}{L^2}\right)$ (ideal elastic substance)
   $\mathcal{T}$ = tension, L = length, L_0 = length at zero tension, and K = constant.
6. $\mathcal{T}' = (A - BT)(L - L_0)$ (Hooke's law spring)
   A and B are constants characteristic of a particular spring alloy.
7. $V = V_0(1 + \alpha t - \beta P)$ (simple solid or liquid)
   V_0, α, and β are constants characteristic of a particular solid or liquid and t is the temperature in degrees Celsius.
8. $\mathcal{E} = a + bT + cT^2 - \frac{\Delta V}{n\mathcal{F}}(P - 1)$ (electrochemical cell)
   a, b, c, and $\mathcal{F}$ are constants and $\mathcal{E}$ is the cell voltage.
9. $\chi = \frac{C}{T}$ (Curie's law paramagnetic solid at constant P)
   χ is the magnetic susceptibility and C is a constant.
10. $\sigma = (115.42 - 0.14530T)\rho^{3/2}$ (surface tension of water in $\text{dyne}\cdot\text{cm}^{-1}$)
    σ is the surface tension and ρ is the density of water.

Although thermodynamic theory requires that an equation of state exist, the particular form of the equation cannot be obtained from thermodynamic theory. This is because an equation of state expresses the behavior of a particular system in contrast to other systems and hence must be obtained from experimental observations or molecular theory or a combination of the two. An experimentally determined equation of state can be confidently regarded as valid only within the ranges of the values of the variables used to establish the equation.

In the thermodynamic analysis of chemical systems of fixed composition it is usually (but by no means always) sufficient to specify the values of two thermodynamic state functions to determine completely the state of a system, for example, P and T or P and V or T and U (relative to an arbitrary zero). Because thermodynamic state functions are macroscopic variables, it must always be remembered that thermodynamic equations are valid only for systems composed of sufficient numbers of molecular species to warrant the macroscopic approach. This condition is imposed on our equations, excepting the energy conservation principle, because, for very small numbers of molecules, fluctuation phenomena may make it impossible to specify definite values for thermodynamic state functions of the system. In this book we shall confine our attention to macroscopic systems.

3-6 *A Note on Units*

The SI (International System) unit of pressure is the pascal (Pa). A pascal is defined as follows (pressure = force/area):

$$1 \text{ pascal} = 1 \text{ Pa} = \frac{1 \text{ newton}}{(\text{meter})^2} = 1 \text{ N}\cdot\text{m}^{-2}$$

The newton (N) is the SI unit of force and the meter (m) is the SI unit of length. A newton is defined as ($F = ma$)

$$1 \text{ newton} = 1 \text{ N} = 1 \text{ kg}\cdot\text{m}\cdot\text{s}^{-2}$$

One standard atmosphere (1 atm) is equal to

$$101.325 \text{ kilopascal} = 101.325 \times 10^3 \text{ Pa} = 1 \text{ atm}$$

The various common pressure units are related as follows:

$$101.3 \text{ kPa} = 1 \text{ atm} = 760 \text{ torr} = 14.7 \text{ psi} = 1.013 \text{ bars}$$

where 1 psi is one pound per square inch and 1 bar = 10^5 N·m^{-2}.

The SI unit of energy is the joule (J). The joule is defined as (energy = force × distance)

$$1 \text{ J} = 1 \text{ newton}\cdot\text{meter} = 1 \text{ N}\cdot\text{m}$$

A newton is a kilogram-meter-per second squared (1 N = 1 kg·m·s^{-2}) thus

$$1 \text{ J} = 1 \text{ N}\cdot\text{m} = 1 \text{ kg}\cdot\text{m}^2\cdot\text{s}^{-2}$$

The various common energy units are related as follows:

$$1 \text{ calorie} = 4.184 \text{ joule} = 4.184 \times 10^7 \text{ erg}$$

or

$$1 \text{ cal} = 4.184 \text{ J} = 4.184 \times 10^7 \text{ erg}$$

and

$$1 \text{ British thermal unit} = 1 \text{ Btu} = 1.05 \text{ kJ}$$

The SI unit of volume is the cubic meter (m^3). Note that (1 dm = 0.1 m = 10 cm)

$$1 \text{ m}^3 = 10^3 \text{ dm}^3 = 10^6 \text{ cm}^3$$

We define the liter as

$$1 \text{ liter} = 1 \text{ L} = 1000 \text{ cm}^3 = 1 \text{ dm}^3 = 10^{-3} \text{ m}^3$$

A more complete set of units, unit prefixes, and unit conversion factors are given inside the back cover of the text. The formal definitions of SI units are given in Appendix II. The values of various physical constants are given inside the front cover of the text.

EXAMPLE 3-4

Compute the density and the volume of 1.000 mol of an ideal gas at a temperature of 273.15 K and a pressure of 101.3 kPa.

Solution: From Eq. 3-13 we have

$$\rho = \frac{P}{RT} = \frac{101.3 \times 10^3 \text{ Pa}}{(8.314 \text{ J}\cdot\text{K}^{-1}\cdot\text{mol}^{-1})(273.15 \text{ K})}$$

$$\rho = \frac{101.3 \times 10^3 \text{ N}\cdot\text{m}^{-2}}{(8.314 \text{ N}\cdot\text{m}\cdot\text{mol}^{-1})(273.15)} = 44.61 \text{ mol}\cdot\text{m}^{-3}$$

The density is the number of moles per unit volume

$$\rho = \frac{n}{V}$$

and thus

$$V = \frac{n}{\rho} = \frac{1.000 \text{ mol}}{44.61 \text{ mol}\cdot\text{m}^{-3}} = 2.242 \times 10^{-2} \text{ m}^3$$

but $1 \text{ L} = 10^{-3} \text{ m}^3$, and the volume of the gas in liters is

$$V = 2.242 \times 10^{-2} \text{ m}^3 \left(\frac{1 \text{ L}}{10^{-3} \text{ m}^3}\right) = 22.42 \text{ L}$$

3-7 *In a Reversible Process a System Is in Thermodynamic Equilibrium Throughout the Process*

A thermodynamic equilibrium state of a system is defined as one in which the thermodynamic state functions have defined time-independent values that are independent of the method by which the system was brought to that state.

The existence of equilibrium states can be deduced from experimental determinations of the independent thermodynamic state functions of the system. The existence of a thermodynamic equilibrium state implies the following in such a state:

1. There are no unbalanced forces either within the system or between the system and its surroundings (mechanical equilibrium).
2. The temperature of the system is uniform throughout, and the temperature of the system is also the same as that of the surroundings, if the system and surroundings are separated by diathermic (heat conducting) walls (thermal equilibrium).
3. There are no net chemical changes (including diffusion) taking place that lead to a change in the chemical composition of the system (chemical equilibrium).

Consider a system initially in a state of thermodynamic equilibrium that is to undergo a change to some other equilibrium state. If this process is not carried out in such a way that the thermodynamic state functions of the system have well-defined values throughout the process, then we cannot employ the equation of state for the system in such a process, because the equation of state holds only for equilibrium states. We now define a reversible process. *A reversible process is one in which the system throughout the process is never more than infinitesimally removed from a state of thermodynamic equilibrium* and to which, therefore, the equation of state is applicable. An equation of state relates states traversed by a system in a reversible process. Although all naturally occurring processes are more or less irreversible, and the conditions for complete reversibility can never be attained completely in the laboratory, it is assumed that by means of appropriate experimental devices and techniques, the conditions for reversibility can be approximated to any desired degree of accuracy.

3-8 *Work and Heat Are Modes of Energy Transfer*

When a force F acts on a system and displaces the system by an amount dX, the work done on the system δw is given by

$$\delta w = -F\,dX \tag{3-14}$$

In general, the work done depends on the path along which the process is carried out, and the path must be specified before w,

$$w = -\int_1^2 F\,dX \tag{3-15}$$

can be calculated. When the path must be specified before the integral in Eq. 3-15 can be evaluated, then the integral in Eq. 3-15 is called a *line integral.* If the work done is independent of the path, for example, the work done in lifting a weight in a gravitational field,

$$w = -\int_1^2 F\,dX = \int_1^2 mg\,dh = mg(h_2 - h_1)$$

then the force F is said to be *conservative.* A conservative force always can be expressed as the negative derivative of a potential energy Φ with respect to the displacement:

$$F = -\frac{d\Phi}{dX} \qquad \text{(conservative force)} \tag{3-16}$$

and thus for a conservative force

$$w = -\int_1^2 F\,dX = \int_1^2 \frac{d\Phi}{dX}\,dX = \int_1^2 d\Phi = \Phi_2 - \Phi_1 \tag{3-17}$$

The British thermodynamicist R. P. Joule was the first to show that when a definite amount of work is done on a system enclosed by *adiabatic walls* (e.g., wood or plastic, as opposed to metal), a definite change in thermodynamic state of the system occurs, as evidenced by measurements of the temperature and the volume of the initial and the final states of the system. Also, the amount of work done is independent of the way in which the work is performed on the system. Thus, utilizing the internal energy function U, we can write

$$\Delta U = U_2 - U_1 = w = -\int_1^2 F\,dX \qquad \text{(adiabatic process)} \tag{3-18}$$

Equation 3-18 is based on the energy conservation principle. Note that work w has the units of energy (work = force × distance = newton-meters = joules). *Work is a mode of energy transfer* that occurs by virtue of the existence of unbalanced forces between the system and the surroundings.

When an amount of work w is done on a system (e.g., by compression of the system), the internal energy U of the system increases by an amount ΔU that is equal to the amount of work w that is done on the system. Our sign convention for work is to *take work done on the system as positive.*

The concept of mechanical equilibrium requires that the transfer of energy as work from the surroundings to the system (or vice versa) involves the existence of unbalanced forces between the system and the surroundings. When the forces are only infinitesimally unbalanced throughout the process in which energy is transferred as work, then the process is said to be *reversible.*

It is found by experiment that the amount of work done on a system enclosed by *diathermic walls* (e.g., rigid metal sheets) depends on how that work is done. That is, the change in the thermodynamic state of a system in a diathermic enclosure caused by the performance of work on the system depends on how the work is performed. It is also found by experiment that a change in the thermodynamic state of a system in a diathermic enclosure can be produced without performing work on it, namely, by raising or lowering the temperature of the surroundings. We therefore conclude that there exists a mode of energy transfer between systems that does not require the existence of unbalanced forces between system and surroundings. Rather, such energy transfer occurs when the system is separated from its surroundings by a diathermic enclosure and a temperature difference exists between the system and the surroundings. Energy transfer that occurs by virtue of the existence of a temperature difference is called *heat* and is represented by the symbol q. *Heat is the mode of energy transfer that occurs by virtue of the existence of a temperature gradient. Heat input to a system is taken as positive.*

Heat and work are defined only for processes; heat and work are modes of energy transfer. A system cannot possess either heat or work. Heat and work are not thermodynamic state functions. There is no function of state that represents the heat or the work. When a system undergoes a particular change in state, the values of q and w for the process depend on the path. Different paths between the same initial and final states yield different values of q and w, whereas the value of ΔU (and the change in any other state function) is independent of the path. Because the values of q and w depend on the path, the differentials of q and w are not exact. To emphasize the inexactness of dq and dw, we shall write the differentials of q and w as δq and δw (and not as dq or dw), respectively. Furthermore, we have

$$\int_1^2 \delta q = q \qquad (\textit{not } q_2 - q_1)$$

$$\int_1^2 \delta w = w \qquad (\textit{not } w_2 - w_1)$$

The differentials δq and δw are inexact differentials.

3-9 *In Any Process the Total Energy Is Conserved*

Consider a process in which forces act on a system in an enclosure with heat-conducting (diathermic) walls. A combination of the internal energy concept with the conservation principle applied to work and heat transfers yields, for an infinitesimal process,

$$dU = \delta q + \delta w \tag{3-19}$$

Integration of Eq. 3-19 yields for a finite process

$$\Delta U = q + w \tag{3-20}$$

In Eq. 3-19, δq represents the *sum* of all the heat transfers in the process, δw represents the sum of all the work transfers (which may be of several types) in the process, and dU represents the change in internal energy of the system. Comparison of Eq. 3-18 with Eq. 3-19 shows that *an adiabatic process is one for which* $\delta q = 0$.

Equation 3-19 constitutes a mathematical statement of the First Law of Thermodynamics. The passage from state 1 to state 2 for any system involves a definite change in U,

$$\int_1^2 dU = U_2 - U_1 = \Delta U \tag{3-21}$$

independent of the way in which the change in state is accomplished. However, for different paths between the same initial and final states, q and w are different. Recall that q and w are not state functions of the system, but rather q and w are modes of energy transfer; q and w are defined only for processes.

Because q and w depend on the path, it is necessary to specify the path before q and w can be calculated. If the process is not reversible, then the thermodynamic state functions will not, in general, be defined during the process (e.g., it may not be possible to specify the macroscopic temperature or pressure of the system due to the existence of temperature and pressure gradients), and therefore it will not be possible, in general, to calculate q and w for such a process, unless one or the other of these quantities is zero. Nonetheless, whether the process itself is reversible or irreversible is immaterial to the value of $\Delta U = U_2 - U_1$ for the process, provided that the initial and final states are defined, because U is a thermodynamic state function. Thus, provided that we have available the equation of state for the system, we can devise a reversible path between the same initial and final states as in the irreversible process, compute the values of q and w for the hypothetical reversible path, and equate the sum $q + w$ to ΔU *for the actual irreversible process.* The calculated values of q and w for the reversible process have no necessary relationship to the actual irreversible process, but the sum $q + w$ for the hypothetical reversible process is exactly the same as the sum $q' + w'$ for the actual irreversible process.

3-10 *When a Force Acts to Cause a Displacement of a System, Energy Is Transferred as Work*

Consider the problem of calculating the work done on a gas when the gas is compressed by an external force F. From Figure 3-2 and Eq. 3-15 we have (pressure = force per unit area)

$$\delta w = -F\,dX = -\left(\frac{F}{A}\right)A\,dX = -P_{\text{ext}}\,dV \tag{3-22}$$

where $P_{\text{ext}} = F/A$ is the external pressure acting on the gas, because F is an externally applied force. The minus sign in Eq. 3-22 arises from our convention of taking work done *on* the system as positive (compression of the gas gives $dV < 0$). In a reversible process, the pressure of the gas P can be equated to the externally applied pressure P_{ext} because in such a case P and P_{ext} can differ at most infinitesimally ($P = P_{\text{ext}} + dP$). When $P_{\text{ext}} = P$, then we can use the equation of state of the system to eliminate P (or V) from the expression

$$\delta w = -P\,dV \tag{3-23}$$

Consider the problem of calculating the work done on an ideal gas in the reversible isothermal (constant temperature) compression of the gas from an initial volume of V_1 to a final volume of V_2. The process is reversible and thus Eq. 3-23 for a reversible process is applicable:

$$\delta w = -P\,dV$$

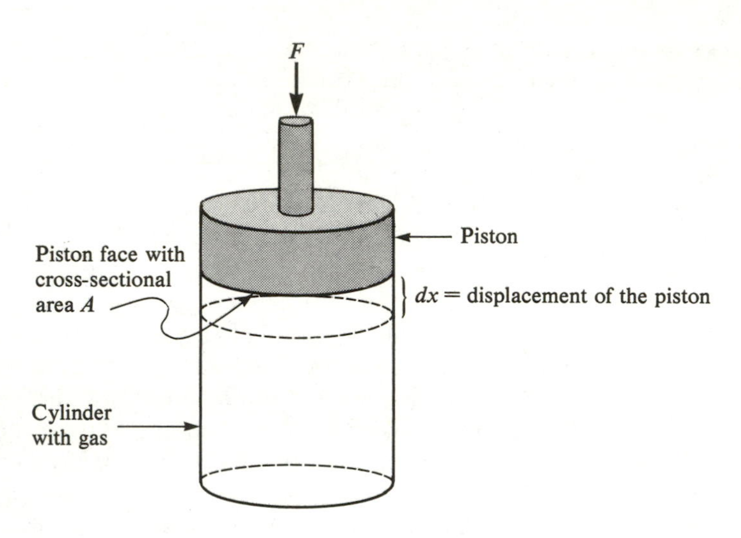

Figure 3-2. The displacement of the piston a distance dX by the action of the force F on the piston involves the transfer of an amount of energy as work from the surroundings to the gas equal to

$$\delta w = -F\,dX$$

Because the process is reversible, we can use the equation of state for the gas, $P = nRT/V$, and thus

$$\delta w = -\frac{nRT}{V}\,dV = -nRT\,d\ln V \tag{3-24}$$

Integration of Eq. 3-24 from V_1 to V_2 yields for the work done on the gas:

$$w = -\int_{V_1}^{V_2} nRT\,d\ln V = -nRT\int_{V_1}^{V_2} d\ln V = -nRT\ln\left(\frac{V_2}{V_1}\right) \tag{3-25}$$

where we were able to move T through the integral sign because T is constant in an isothermal process. Because the process is a compression, $V_2 < V_1$ and $\ln(V_2/V_1) < 0$; thus $w > 0$. Note that for a given value of the ratio (V_2/V_1) the work done on the gas is greater the higher the Kelvin temperature; also, the work done on the gas is directly proportional to the number of moles of gas n that is compressed.

EXAMPLE 3-5

Calculate the work in joules necessary to compress reversibly 0.200 mol of an ideal gas at 300 K from an initial volume of 4.00 L to a final volume of 2.00 L.

Solution: The work involved in the isothermal, reversible compression of an ideal gas is given by Eq. 3-25:

$$w = -nRT\ln\left(\frac{V_2}{V_1}\right)$$

Note that we could also have calculated w as follows:

$$-P_{ext}\Delta V = -P_2(V_2 - V_1) = -(1.23 \text{ atm})(4.00 - 2.00) \text{ L}$$

$$w = -2.46 \text{ L}\cdot\text{atm}$$

convert from the energy unit liter atmospheres to joules by using of two values of the gas constant R:

$$R = 8.314 \text{ J}\cdot\text{K}^{-1}\cdot\text{mol}^{-1} = 0.08206 \text{ L}\cdot\text{atm}\cdot\text{K}^{-1}\cdot\text{mol}^{-1}$$

$$w = -(2.46 \text{ L}\cdot\text{atm})\left(\frac{8.314 \text{ J}}{0.08206 \text{ L}\cdot\text{atm}}\right) = -249 \text{ J}$$

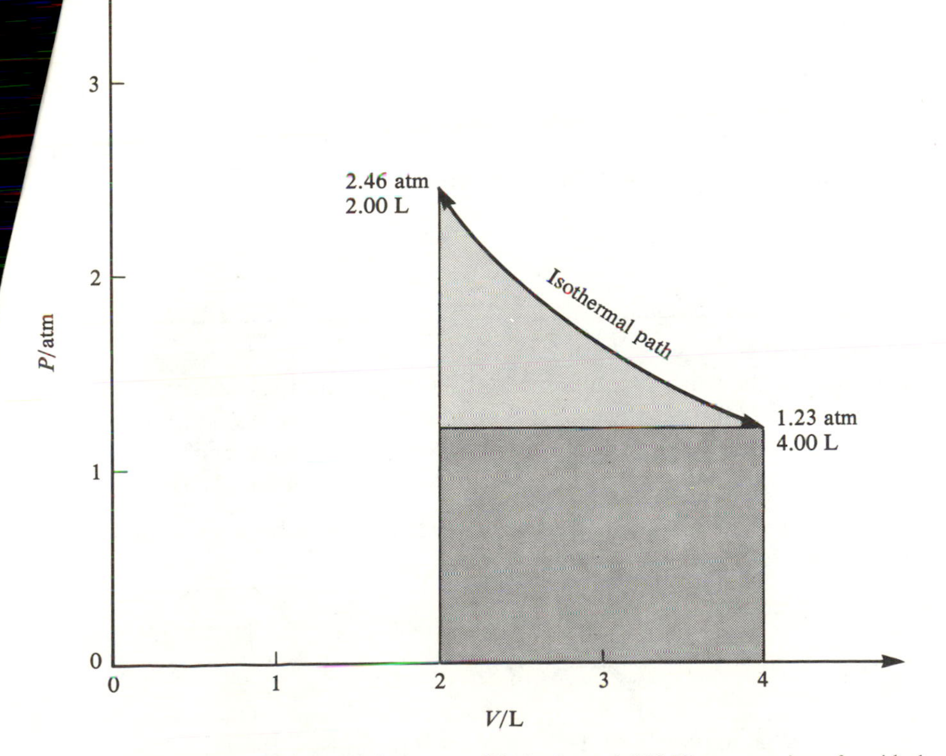

Figure 3-3. Comparison of the work for a reversible isothermal (300 K) compression of an ideal gas from 1.23 atm, 4.00 L to 2.46 atm, 2.00 L with the irreversible isothermal, isobaric expansion of an ideal gas initially at 2.46 atm, 2.00 L for which P_{ext} is discontinuously decreased to 1.23 atm, at which pressure the gas expands to a final volume of 4.00 L.

and

$$w = -(0.200 \text{ mol})(8.314 \text{ J}$$

Thus 346 J of energy is transfer
gas (the system) in the compressi

Consider an ideal gas at equilibrium
a piston as shown in Figure 3-2. Suppose
the piston is suddenly decreased to a new
equilibrium pressure P of the gas. The decrea
force across the piston and, because $P > P_{ext}$
until a new equilibrium state is attained when t
the new value of P_{ext}. We can compute the wo
reversible, *isobaric* (constant pressure) expansion o

$$\delta w = -P_{ext}\,dV$$

Integration of the above equation from V_1 to V_2 with
for the work

$$w = -\int_{V_1}^{V_2} P_{ext}\,dV = -P_{ext}\int_1^2 dV = -P_{ext}(V_2 - V_1)$$

where the value of P_{ext} is equal to P_2, the final equilibrium p
Thus for an ideal gas Eq. 3-26 can be written as

$$w = -P_2\Delta V = -nRT_2\left(\frac{V_2 - V_1}{V_2}\right)$$

EXAMPLE 3-6

Calculate the work in joules when 0.200 mol of an ideal gas at 30
volume of 2.00 L is subjected to a sudden decrease in the external p
from 2.46 atm to 1.23 atm, and the gas expands irreversibly aga
constant external pressure of 1.23 atm to a final volume of 4.00 L.

Solution: The work involved in the irreversible isobaric expansion of
ideal gas is given by Eq. 3-27:

$$w = -P_{ext}\Delta V = -nRT_2\left(\frac{V_2 - V_1}{V_2}\right)$$

and

$$w = -(0.200 \text{ mol})(8.314 \text{ J}\cdot\text{K}^{-1}\cdot\text{mol}^{-1})(300 \text{ K})\left(\frac{4.00 \text{ L} - 2.00 \text{ L}}{4.00 \text{ L}}\right) = -249 \text{ J}$$

The negative value of w tells us that 249 J of energy was transferred
as work from the gas to the surroundings in the irreversible, isobaric

Pressure-volume work ($-\int P\,dV$) is equal to the area under the curve for the process plotted on the P-V plane. The path on the P-V plane for the particular reversible, isothermal compression of an ideal gas described in Example 3-5 is shown in Figure 3-3. For comparison we have also marked off the area corresponding to the work done in the irreversible gas expansion described in Example 3-6. However, it is important to recognize that we cannot show the path on the P-V plane for the irreversible expansion of the gas, because the P-V path is not defined in the irreversible expansion. In other words, P and V for the gas do not have defined values *during* the irreversible expansion. Note that the work done by the gas ($-w$ = 249 J) in the *irreversible* expansion from 2.46 atm, 2.00 L to 1.23 atm, 4.00 L at P_{ext} = 1.23 atm is *less* than the work done by the gas in the isothermal *reversible* expansion of the gas between the same initial and final states ($-w$ = 346 J). The value of w for a reversible compression is equal to the value of $-w$ for a reversible expansion between the same two states, as seen by interchanging V_1 and V_2 in Eq. 3-25.

EXAMPLE 3-7

Given the equation of state $V = V_0(1 + \alpha t - \beta P)$ for a liquid where V_0, α and β are constants: (a) Derive an expression for the work done when the pressure on a liquid is increased reversibly and isothermally from P_1 to P_2. (b) Use your equation to evaluate the work done when 2.00 mol of water ($\beta = 4.67 \times 10^{-5}$ atm^{-1}) is subjected to an increase in pressure from 1.00 atm to 2.00 atm at 25°C.

Solution: (a) The work done is given by

$$w = -\int_1^2 P_{ext}\,dV$$

The process is reversible and thus $P_{ext} = P$. Further from the equation of state we have

$$dV = d[V_0(1 + \alpha t - \beta P]$$

The process is isothermal and thus

$$dV = -V_0\beta\,dP$$

because V_0, α, and β are constants. Hence

$$w = \int_{P_1}^{P_2} V_0\beta P\,dP = \frac{V_0\beta}{2}(P_2^2 - P_1^2)$$

(b) For 2.00 mol of water, V_0 = 0.0360 L and thus

$$w = \frac{(0.0360 \text{ L})(4.67 \times 10^{-5} \text{ atm}^{-1})}{2}[(2.00 \text{ atm})^2 - (1.00 \text{ atm})^2]$$

and

$$w = 2.52 \times 10^{-6} \text{ L·atm}$$

Converting liter atmospheres to joules we obtain

$$w = (2.52 \times 10^{-6}\ \mathrm{L\cdot atm})\left(\frac{8.314\ \mathrm{J}}{0.08206\ \mathrm{L\cdot atm}}\right) = 2.55 \times 10^{-4}\ \mathrm{J}$$

The small value of w for the compression of a condensed phase relative to that for a gas phase is a consequence of the much lower compressibility of a condensed phase and thus a much lower displacement by a given applied force. The compressibility of water vapor at 1 atm and 300 K is 10,000 times greater than the compressibility of liquid water at the same conditions.

3-11 *It Is Necessary to Include All the Relevant Work Terms in the First-Law Expression, $dU = \delta q + \delta w$*

Some of the more commonly encountered types of work and the associated expressions for δw are assembled in Table 3-2. These expressions can be combined with the appropriate equations of state (see Table 3-1) to calculate w for various types of work.

When the First Law of Thermodynamics is applied to a system, it is absolutely essential to include all the relevant work terms in δw. For example, a reversible electrochemical cell of fixed composition involves both *ex*pansion work ($-P\,dV$) and *elec*trochemical work ($-\mathscr{E}\,dZ$, where $\mathscr{E}$ is the cell voltage

TABLE 3-2
Work Terms (Reversible Processes)

Type of work	*Expression for δw_i*
Expansion or compression of a system (pressure-volume work)	$\delta w = -P\,dV$
Electrochemical work (passage of a charge dZ through a cell with cell voltage $\mathscr{E}$)	$\delta w = -\mathscr{E}\,dZ$
The work done in expanding the surface area (A) of a system with surface tension σ	$\delta w = \sigma\,dA$
The work done in lifting a system of mass m to a height h in a gravitational field with gravitational acceleration g	$\delta w = mg\,dh$
The work done in stretching or compressing a spring or elastic material with tension $\mathscr{T}$ and length L	$\delta w = \mathscr{T}\,dL$
The magnetic work done on a substance with magnetic polarization M in a magnetic field of strength $\mathscr{H}$;	$\delta w = \mathscr{H}\,dM$
The electrical work done on a substance with electric polarization P in an electric field of strength E; ε_0 is a constant	$\delta w = \varepsilon_0 E\,dP$

and Z is the electric charge that passes through the cell). Thus the expression for δw for the cell is

$$\delta w = \delta w_{\text{exp}} + \delta w_{\text{elec}} = -P\,dV - \mathscr{E}\,dZ$$

and the first-law expression for the cell is

$$dU = \delta q + \delta w = \delta q - P\,dV - \mathscr{E}\,dZ$$

If surface effects are significant for a liquid or a solid, then we have to consider the work term associated with a change in the surface area as well as a change in volume, and thus the first-law expression becomes

$$dU = \delta q - P\,dV + \sigma\,dA$$

It is a matter for experiment to determine the number of independent intensive state functions that must be considered in the thermodynamic analysis of a system. In general, because of the state principle, *the number of independent intensive state functions that are involved is equal to the number of different work terms plus one.*

The first law for a system in which only pressure-volume work ($\delta w = -P_{\text{ext}}\,dV$) is possible takes the form

$$dU = \delta q - P_{\text{ext}}\,dV \tag{3-28}$$

Such a situation prevails, for example, in a chemical system of fixed mass and composition in which surface and field (electric, magnetic, etc.) are unimportant. For an adiabatic ($\delta q = 0$) reversible change in state, Eq. 3-28 becomes

$$dU = -P\,dV$$

where P can be expressed as a function of V using the equation of state. Whereas for an adiabatic, irreversible process

$$dU = -P_{\text{ext}}\,dV$$

where P_{ext} cannot in general be expressed as a function of V using the equation of state.

An apparent failure of the first law should be regarded with skepticism: first, because such apparent failures will arise whenever the experimenter fails to consider all the relevant independent variables of the system; second, because a verifiable failure is unprecedented.

3-12 *Heat Capacity Is the Measure of the Capability of a System to Take in Energy as Heat*

An important thermodynamic property of systems is *heat capacity. We define the heat capacity at constant X* (where X is a thermodynamic state function) *of a system as*

$$C_X = \lim_{\Delta T \to 0} \left(\frac{q_X}{\Delta T}\right) = C_X(X, T) \tag{3-29}$$

where q_X is the heat absorbed by the system at constant X when the temperature of the system is increased by ΔT. The heat capacity is an extensive property, and C_X will in general be a function of X and T. The larger the heat capacity of a system, the smaller the increase in temperature of the system arising from the input of a given amount of energy as heat. The heat capacity is necessarily a positive quantity (this will be proved in Chapter 4). Although the heat capacity cannot be negative, systems with an infinite heat capacity are known.

For a system in which only $P\,dV$ work is possible, we have from the first law

$$\delta q = dU + P_{\text{ext}}\,dV$$

and at constant volume ($dV = 0$)

$$\delta q_V = dU$$

and thus

$$q_V = \Delta U \tag{3-30}$$

Substitution of Eq. 3-30 into Eq. 3-29 yields

$$C_V = \lim_{\Delta T \to 0}\left(\frac{q_V}{\Delta T}\right) = \lim_{\Delta T \to 0}\left(\frac{\Delta U}{\Delta T}\right)_V = \left(\frac{\partial U}{\partial T}\right)_V \tag{3-31}$$

The derivative $(\partial U/\partial T)_V$ is a *partial derivative;* it means that in the differentiation of the expression for U with respect to T, the variable V must be held constant. Partial derivatives are described in detail in Chapter 5.

The experimental determination of C_V for a system can be accomplished by introducing a known amount of energy as heat (e.g., by electric heating) into the system held at fixed volume and by determining the associated temperature increase, which is kept as small as practicable, consistent with the defining equation ($\lim \Delta T \to 0$). Note that the heat capacity has the units joules per kelvin (i.e., $\text{J}\cdot\text{K}^{-1}$).

EXAMPLE 3-8

A 1.00-mol sample of water vapor is enclosed in a rigid container equipped with a 9.97 Ω electric resistance heater. When a current of 0.200 A is passed through the resistor for 100.0 s, the temperature of the apparatus was observed to increase by 0.465 K. Separate measurements on the evacuated container show that the heat capacity of the container and resistance heater is 60.50 $\text{J}\cdot\text{K}^{-1}$. Compute the value of C_V for the water vapor.

Solution: The electrical energy converted to heat by passing an electric current through the resistor (Joule heating) is

$$\Delta U = I^2Rt$$

where I is the current, R is the resistance, and t is the time that the current I

flows through the resistor. From Eq. 3-31 we have for small ΔT

$$C_V = \left(\frac{\Delta U}{\Delta T}\right)_V$$

The total heat capacity C_V is equal to

$$C_V = C_{V,\mathrm{H_2O}(g)} + C_{V,\mathrm{apparatus}} = C_{V,\mathrm{H_2O}(g)} + 60.50\ \mathrm{J\cdot K^{-1}}$$

Thus

$$C_{V,\mathrm{H_2O}(g)} = \frac{I^2Rt}{\Delta T} - 60.50\ \mathrm{J\cdot K^{-1}}$$

With I in amperes, R in ohms, and t in seconds, I^2Rt has the units joule; thus

$$C_{V,\mathrm{H_2O}(g)} = \frac{(0.200\ \mathrm{A})^2(9.97\ \Omega)(100.0\ \mathrm{s})}{(0.465\ \mathrm{K})} - 60.50\ \mathrm{J\cdot K^{-1}}$$

$$C_{V,\mathrm{H_2O}(g)} = 25.3\ \mathrm{J\cdot K^{-1}}$$

The heat capacity at constant volume per mole of water vapor $\bar{C}_{V,\mathrm{H_2O}(g)}$ (where the bar denotes *per mole*) is

$$\bar{C}_{V,\mathrm{H_2O}(g)} = \frac{C_{V,\mathrm{H_2O}(g)}}{n}$$

where n is the number of moles of water. Thus

$$\bar{C}_{V,\mathrm{H_2O}(g)} = \frac{25.3\ \mathrm{J\cdot K^{-1}}}{1.00\ \mathrm{mol}} = 25.3\ \mathrm{J\cdot K^{-1}\cdot mol^{-1}}$$

A calorimeter is a device used to determine the amount of heat evolved or absorbed in a process. Constant volume processes can be carried out in the laboratory in *bomb calorimeters* (Figure 3-4). In such experiments a stainless steel bomb containing the reactants (e.g., solid benzoic acid and oxygen gas at about 30 atm) and equipped with appropriate electrical leads for resistance heating and electrical ignition of the reaction is immersed in water containing a sensitive thermometer. The bomb and fluid are enclosed within adiabatic walls (e.g., thick plastic). Because the combined system (calorimeter plus water bath) is isolated, we have

$$\Delta U_{\mathrm{tot}} = \Delta U_{\mathrm{sys}} + \Delta U_{\mathrm{sur}} = 0$$

where ΔU_{sur} represents the energy change in all parts of the system exclusive of the chemicals, which we regard here as the system of interest, placed in the bomb. Combination of the above equation with Eq. 3-31 applied to a small

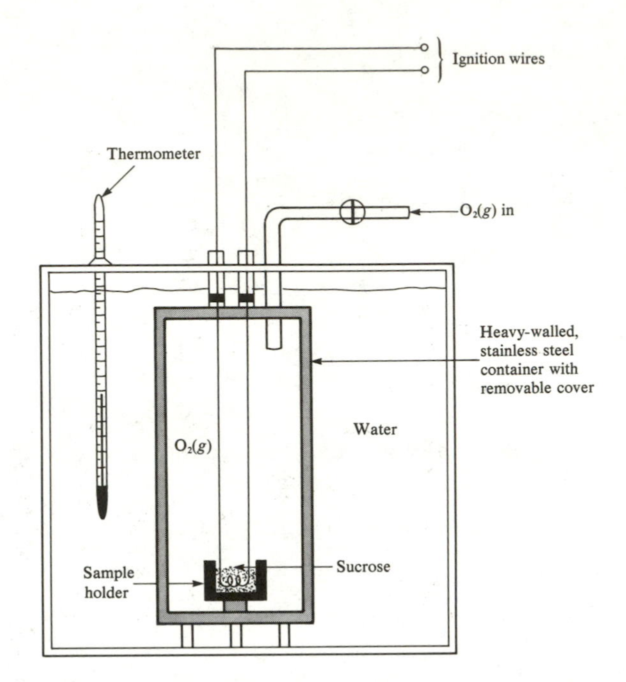

Figure 3-4. Combustion of sucrose in a sealed reaction vessel (bomb calorimeter). The sample (in this case, sucrose) is placed in the sample holder together with a coil of ignition wire. The container cover is then screwed into place and the vessel is pressurized with excess oxygen gas to about 30 atm. The calorimeter is then placed in an insulated water bath equipped with a thermometer. A short burst of current heats the ignition wire to red heat and thereby initiates the combustion reaction.

temperature change ($\Delta U = C_V \Delta T$) yields

$$\Delta U_{\text{sys}} = -\Delta U_{\text{sur}} = -C_V(T_2 - T_1) \tag{3-32}$$

where the last equality holds if $T_2 - T_1$ is small, that is, if C_V can be regarded as a constant. The value of C_V is determined electrically (Example 3-8) for the calorimeter plus contents in a separate experiment. The combustion reaction

$$C_6H_5COOH(s) + \tfrac{15}{2}\,O_2(g) \longrightarrow 7\,CO_2(g) + 3\,H_2O(\ell)$$

is initiated by a short burst of current through the ignition wire, which triggers the reaction. The energy evolved as heat by the combustion gives rise to a temperature increase. The measured value of $\Delta T = T_2 - T_1 > 0$ is used in Eq. 3-32 to compute ΔU. Various small corrections must be applied to convert ΔU_{sys} to ΔU_{rxn}, the energy change for the combustion reaction. For example, a correction must be applied for the energy released by the ignition process. This energy release is determined in a separate experiment in which the ignition circuit is triggered in the absence of the reactants.

3-13 *The Enthalpy Change for a Constant Pressure Process Is Equal to the Heat Transferred in the Process*

Although bomb calorimetry is an important source of thermodynamic data on chemical systems, chemists are much more interested in processes taking place at constant pressure (e.g., open to the atmosphere) than in processes occurring at constant volume. For this reason it is convenient to *define* a thermodynamic state function that has the following property for a constant pressure process,

$$\Delta X = q_P$$

analogous to $\Delta U = q_V$ for constant volume processes. From the first law we have

$$\delta q = dU + P\,dV$$

Adding and subtracting VdP to the above equation yields

$$\delta q = dU + P\,dV + VdP - VdP$$

or

$$\delta q = d(U + PV) - VdP \qquad (3\text{-}33)$$

From Eq. 3-33 it can be deduced that the function $U + PV$ has the desired property; that is,

$$d(U + PV) = \delta q_P$$

and

$$\Delta(U + PV) = q_P \qquad (3\text{-}34)$$

where q_P is the heat absorbed or evolved when the process occurs at constant pressure. The function $U + PV$ occurs frequently in chemical thermodynamics and it is convenient to give it a special symbol and name. We define the *enthalpy* H as

$$H = U + PV \qquad (3\text{-}35)$$

and from Eq. 3-34 we have

$$\Delta H = q_P \qquad (3\text{-}36)$$

Both U and V are extensive state functions (P is intensive), and thus the enthalpy is an extensive thermodynamic function of state. Equation 3-36 holds for both reversible and irreversible processes; however, Eq. 3-36 does not hold, in general, for processes involving work terms in addition to $-P\,dV$. For example, if

$$dU = \delta q - P\,dV - \mathscr{E}\,dZ \qquad \text{(electrochemical cell)}$$

then at constant pressure and cell voltage

$$\Delta H = q_P - \mathscr{E}Z \qquad (3\text{-}37)$$

Absolute enthalpies cannot be determined because absolute energies cannot be determined. We are, therefore, free to pick an arbitrary zero for the enthalpy of any system. Note, however, that we cannot arbitrarily pick *both* a zero of enthalpy and a zero of energy, because that would violate Eq. 3-35. The enthalpy is a derived quantity because it is defined in terms of other more fundamental quantities (U, P, and V). *The enthalpy is a convenience function*, and if H as defined by Eq. 3-35 turns out to lack certain desired quantities, as is the case in systems involving magnetic work, then we are free to change the definition of H. However, such changes must be treated with great care to avoid inconsistencies.

The most important type of heat capacity in chemical thermodynamics is the *heat capacity at constant pressure* C_P. From Eq. 3-29 we have

$$C_P = \lim_{\Delta T \to 0} \left(\frac{q_P}{\Delta T} \right) \tag{3-38}$$

Substitution of the relation $\Delta H = q_P$ into Eq. 3-38 yields

$$C_P = \lim_{\Delta T \to 0} \left(\frac{\Delta H}{\Delta T} \right)_P = \left(\frac{\partial H}{\partial T} \right)_P \tag{3-39}$$

Further, because

$$dH = \delta q_P = C_P dT \qquad \text{(constant } P\text{)}$$

we have

$$\Delta H = \int_{T_1}^{T_2} C_P dT \qquad \text{(constant } P\text{)} \tag{3-40}$$

and if C_P is independent of temperature, then integration of Eq. 3-40 yields

$$\Delta H = C_P \Delta T \qquad \text{(constant } P \text{ and } C_P\text{)} \tag{3-41}$$

It is very difficult to determine C_V for a liquid or a solid, because to measure C_V, energy must be added as heat while keeping the volume of the liquid or solid constant. However, it is easy to add heat to a condensed phase while keeping the pressure constant. The simplest way to measure C_P is by electrical resistance heating.

EXAMPLE 3-9

The molar heat capacity at constant pressure $\bar{C}_P$ of liquid water over the temperature range 273.15 K to 373.15 K is constant and equal to 75.40 $J \cdot K^{-1} \cdot mol^{-1}$. Compute the enthalpy change ΔH of 25.17 mol of water (1 lb) when the temperature is increased from 273.15 K to 373.15 K.

Solution: Because C_P is constant, we can use Eq. 3-41:

$$\Delta H = C_P \Delta T = n\bar{C}_P \Delta T$$

thus

$$\Delta H = (25.17 \text{ mol})(75.40 \text{ J} \cdot \text{K}^{-1} \cdot \text{mol}^{-1})(373.15 \text{ K} - 273.15 \text{ K}) \left(\frac{1 \text{ kJ}}{1000 \text{ J}} \right)$$

$$\Delta H = 189.8 \text{ kJ}$$

A British thermal unit (Btu) is the amount of heat required to raise the temperature of 1.00 lb of water by 1.00°F. A degree Kelvin is $\frac{5}{9}$ of a degree

Fahrenheit; therefore

$$1\ \text{Btu} = \left(\frac{189.8\ \text{kJ}}{100\ \text{K}}\right)\left(\frac{5\ \text{K}}{9^\circ\text{F}}\right) = 1.054\ \text{kJ}$$

The value of ΔH for a chemical reaction, for example, an acid-base reaction,

$$\text{NaOH}(aq) + \text{HCl}(aq) \longrightarrow \text{H}_2\text{O}(\ell) + \text{NaCl}(aq)$$

or a precipitation reaction,

$$\text{NaCl}(aq) + \text{AgNO}_3(aq) \longrightarrow \text{AgCl}(s) + \text{NaNO}_3(aq)$$

or a dissolution reaction,

$$\text{NaCl}(s) \xrightarrow{\text{H}_2\text{O}(\ell)} \text{NaCl}(aq)$$

is readily determined in an adiabatic calorimeter open to the atmosphere. A simple adiabatic calorimeter is a Dewar flask equipped with a thermometer and a stirrer to mix the reactants (Figure 3-5). Because the calorimeter is adiabatically isolated and the process takes place at constant pressure, we have

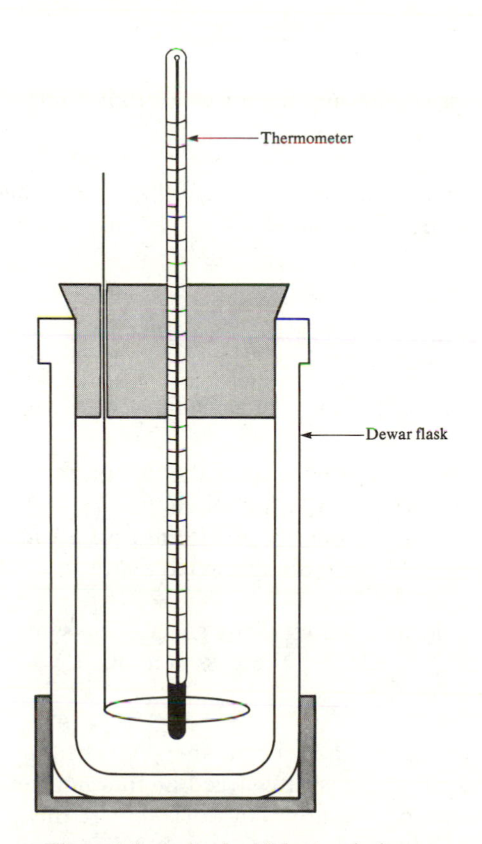

Figure 3-5. A simple adiabatic calorimeter.

$q_P = 0$, and thus

$$0 = \Delta H_{\text{tot}} = \Delta H_{\text{rxn}} + \Delta H_{\text{cal}} \tag{3-42}$$

where rxn denotes the enthalpy change of the reaction and ΔH_{cal} denotes the enthalpy change of the calorimeter and contents. If the temperature change in the calorimeter is small, then C_P can be taken as a constant and, using Eq. 3-41, Eq. 3-42 becomes

$$\Delta H_{\text{rxn}} = -C_{P,\text{cal}}\Delta T \tag{3-43}$$

The value of $C_{P,\text{cal}}$ can be determined either by electrical heating or by running a reaction with a known ΔH_{rxn} and measuring the resulting ΔT.

The description of calorimetric experiments outlined in this chapter has involved only an elementary discussion of the basic principles. Calorimetry is an important source of thermodynamic data on chemical systems and the associated experimental apparatus has been extensively developed to yield data that in some cases are reliable to better than 0.1% (see Rossini, 1956). We shall continue our analysis of thermochemistry in Chapter 8.

Problems

1. Calculate the maximum amount of work that could be obtained from a metric ton (1000 kg) of water that falls a distance of 100 m.

2. Suppose that it was found by experiment that 0.385 mol of a gas at 10.0 atm and 300 K occupied a volume of 1.00 L. Is the gas ideal?

3. Show that

$$w = nRT\ln\left(\frac{P_2}{P_1}\right)$$

for the reversible isothermal compression of an ideal gas from P_1 to P_2.

4. (a) Compute the work done when 2.00 mol of an ideal gas expands isothermally and reversibly to a final volume $V_2 = 3V_1$ at 300 K.
(b) Compute the work done when 2.00 mol of an ideal gas expands isothermally at 300 K into an evacuated space to a final volume of $V_2 = 3V_1$.
(c) The initial and final states of the gas are the same in parts (a) and (b), and thus $\Delta U_a = \Delta U_b$ for the gas. Reconcile your results in parts (a) and (b) with the first law.

5. A refrigerator operates in a cyclic manner. The heat q_1 removed at the low temperature in one cycle is discharged at the high temperature by virtue of the performance of an amount of work w. Use the first law to show that the heat ejected at the high temperature, $-q_2$, is equal to $q_1 + w$.

6. If energy is added as heat to an ice-water mixture at 0°C, the net result is the melting of some of the ice and the temperature remains at 0°C. What is the value of C_P for an ice-water mixture?

7. The molar heat capacity at constant pressure for $NH_3(g)$ at 1 atm pressure over the temperature range 298.15 K (25°C) to 873.15 K (600°C) is given by

$$\bar{C}_P = 29.75 + 2.51 \times 10^{-2}T - 1.55 \times 10^{5}T^{-2}$$

with T in kelvins; $\bar{C}_P$ has the units $J \cdot K^{-1} \cdot mol^{-1}$. Compute ΔH for 1.00 kg of $NH_3(g)$ when the temperature is increased from 25°C to 600°C.

8. Our sign convention for work is to take *work done on the system as positive;* this sign convention for work is not adopted universally. For example, in most engineering texts the convention is to take work done *by* the system as positive, and the first law is written as

$$dU = \delta q - \delta w$$

Show, by considering pressure-volume work, that the value of dU is the same for either sign convention, as, of course, it must be. The reason for the different sign convention for work in engineering is that in engineering the emphasis is on the work done *by the system* (e.g., engines), whereas in physical chemistry the emphasis is on the properties of substances.

9. The *specific heat* c_{sp} of a substance is defined as the heat capacity at constant pressure *per gram of substance;* thus

$$c_{sp} = \frac{C_P}{m}$$

where m is the mass of the substance in grams.

The specific heat of ice is $2.09\ J \cdot K^{-1} \cdot g^{-1}$ and the specific heat for liquid water is $4.18\ J \cdot K^{-1} \cdot g^{-1}$. The enthalpy of fusion of ice is $6.01\ kJ \cdot mol^{-1}$ and the enthalpy of vaporization of liquid water is $44.0\ kJ \cdot mol^{-1}$. Compute the value of ΔH for 100 g of water that is initially ice at −10°C and 1 atm and that is converted to water vapor at 100°C and 1 atm. Does your result for ΔH depend on the way in which the above process is carried out?

10. Consider the calorimeter circuit in Figure 3-6. A constant current I is passed through the circuit for 352.50 s; from potentiometric measurements it was found that while the current was flowing, $\mathscr{E}_s = 1.2373$ V and $\mathscr{E}_h = 0.9878$ V. If $R_s = 100.325\ \Omega$, compute the electrical energy added as heat to the calorimeter. Also compute the heater resistance R_h. If the observed temperature change during the above time was 0.0204 K, compute the heat capacity of the calorimeter and contents. Suppose now a sample container is opened in the calorimeter and 0.2527 g is allowed to dissolve in a solvent in the calorimeter. The observed temperature change was −0.3047 K. Compute ΔH of solution per gram for the solid.

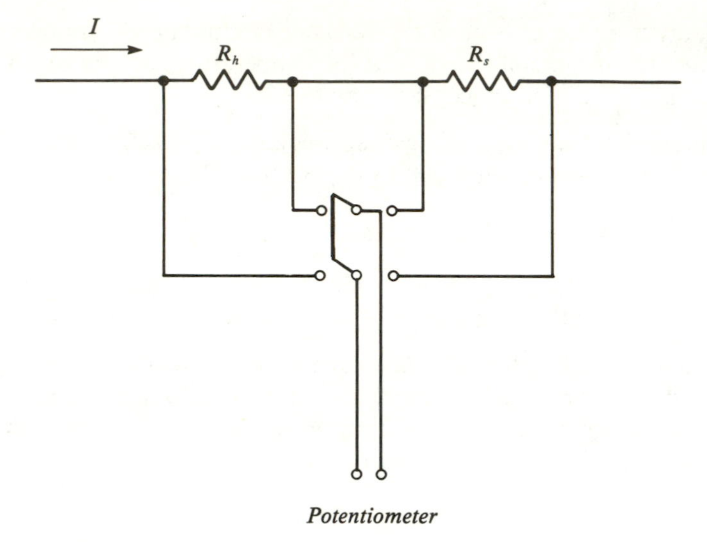

Figure 3-6

11. Show that $\Delta H = q_P$ holds for an irreversible constant pressure process involving only pressure-volume work.

12. Show that

$$\Delta H = q_P - \mathscr{E}Z$$

for an electrochemical cell at constant voltage $\mathscr{E}$ and constant pressure.

13. According to the kinetic theory of gases the translational kinetic energy of a monatomic gas is given by

$$U = \tfrac{3}{2}nRT$$

where n is the number of moles of gas.

(a) Compute ΔU and ΔH when the temperature of 2.00 mol of the gas is increased from 300 K to 600 K.

(b) Obtain expressions for C_V and C_P for the gas.

(c) Compute ΔU and ΔH when the pressure on 2.00 mol of the gas is increased from 2.00 atm to 4.00 atm at 300 K.

14. Explain why the temperature decreases when a gas undergoes an adiabatic expansion against an external pressure.

15. Suppose n mol of an ideal gas is allowed to expand isothermally at 300 K from an initial state of 10.0 atm, 1.00 L to a final state of 1.00 atm, 10.0 L by three different paths:

(a) Reversibly.

(b) The external pressure is decreased instantaneously from 10.0 atm to 1.00 atm and the gas expands against a constant external pressure of 1.00 atm.

(c) The external pressure is decreased instantaneously from 10.0 atm to 5.0 atm and the gas expands to 2.00 L against $P_{ext} = 5.00$ atm; when the gas comes to equilibrium at 5.00 atm, 2.00 L, the pressure is decreased instantaneously from 5.00 atm to 1.00 atm and the gas expands to 10.0 L against $P_{ext} = 1.00$ atm.

Calculate the work done in each of the three cases. What would be the limiting value of w if the external pressure was decreased instantaneously in an increasing number of steps with smaller ΔP increments for each step?

16. Given that the internal energy of a fixed mass of ideal gas depends only on the temperature, show that

$$dU = C_V dT$$

$$dH = C_P dT$$

and

$$C_P = C_V + nR$$

17. Calculate the values of q (the heat evolved or absorbed) for the three cases in Problem 15.

18. Obtain an expression for the work done on isothermally and reversibly stretching: (a) a Hooke's law spring and (b) an ideal elastic cylinder from L_0 to $2L_0$. Does it make a difference what the temperature of the substance is? Suppose the extended material is allowed to snap suddenly back to L_0 without restriction. How much work is done?

19. A 10-cm cube of iron ($\rho = 7.86\ \mathrm{g \cdot cm^{-3}}$) is floating on mercury ($\rho = 13.59\ \mathrm{g \cdot cm^{-3}}$) in air. Sufficient water ($\rho = 1.00\ \mathrm{g \cdot cm^{-3}}$) is poured into the vessel to cover the iron completely; will the iron cube rise or sink in the mercury? By how many centimeters?

20. Given that the equation of state for a gas is

$$P(V - nb) = nRT \qquad b = \text{constant}$$

obtain an expression for the work done by the gas in a reversible, isothermal expansion from V_1 to V_2. Given that the internal energy of a fixed mass of the gas depends only on T, obtain expressions for ΔU and ΔH for the gas in the reversible, isothermal expansion. Compare the results to those obtained for an ideal gas.

21. The van der Waals constants (see Table 3-1) for $O_2(g)$ are

$$a = 1.360\ \mathrm{L^2 \cdot atm \cdot mol^{-2}} \qquad b = 3.183 \times 10^{-2}\ \mathrm{L \cdot mol^{-1}}$$

Suppose 2.000 mol of $O_2(g)$ expands reversibly and isothermally at 300.0 K from an initial volume of 1.00 L to a final volume of 10.0 L. Calculate the work done by the gas and compare with the work done by an ideal gas that expands reversibly and isothermally between the same initial and final states. What is the molecular level origin of the difference in the w values for the two cases?

22. Suppose we have 8.0 mL of a solution with $C_P = 33.5\ \mathrm{J \cdot K^{-1}}$ at a temperature of 12.0°C. If the temperature of the solution is measured by inserting the bulb of a mercury-in-glass thermometer, initially at 22.5°C, with an effective bulb heat capacity of $40.0\ \mathrm{J \cdot K^{-1}}$, what will the thermometer read for the temperature of the above solution?

23. From the equation of state $PV = nRT$, we know that PV = constant for the isothermal expansion of an ideal gas. Given the results in Problem 16, show that

$$PV^{C_P/C_V} = \text{constant}$$

for the reversible adiabatic expansion of an ideal gas.

24. For the reversible adiabatic expansion of an ideal gas we have (Problem 23)

$$PV^{\gamma} = \text{constant} \qquad \gamma = \frac{C_P}{C_V}$$

Show that the work done on an ideal gas in a reversible adiabatic expansion from P_1, V_1 to P_2, V_2 is

$$w = \left(\frac{P_2V_2 - P_1V_1}{\gamma - 1}\right)$$

25. Compute the energy released when 1000-Å diameter spherical water droplets combine to form 1.00 mol of bulk liquid water. Take the surface tension of water as $0.073\ \mathrm{N \cdot m^{-1}}$.

26. Given that the magnetic work done on a particular sample is

$$w = \int_{M_1}^{M_2} \mathscr{H}\, dM$$

where $\mathscr{H}$ is the magnetic field strength and M is the magnetization, obtain an expression for w for a magnetic solid that obeys Curie's equation

$$M = c\left(\frac{\mathscr{H}}{T}\right) \qquad c = \text{constant for a particular material}$$

Assume that the field is increased reversibly and isothermally from $\mathscr{H} = 0$ to $\mathscr{H}$.

27. It is observed that a mass suspended by a rubber band *rises* when the temperature of the rubber band increases; whereas when the same mass is held up by a metallic spring, the mass moves to a lower position (the spring extends further) when the spring is heated. Explain these observations using the appropriate equations of state.

28. Prove that if X and Y are thermodynamic state functions, then $X + Y$, and X/Y are also thermodynamic state functions.

29. Show that the amount of heat required to raise the temperature *of the air* at constant pressure P in a house from T_1 to T_2 is

$$q = \frac{PV\bar{C}_P}{R} \ln \frac{T_2}{T_1}$$

where P is the air pressure, V is the house volume, and $\bar{C}_P$ is the molar heat capacity of air. Note that the moles of air in the house continuously decreases as T increases, because P is constant and thus some of the expanded air escapes through cracks. Explain why the value of q required to raise the *house* temperature from T_1 to T_2 is much greater than the value of q calculated from the above equation.

30. Derive the following for the amount of heat that must be removed from a house to decrease the air temperature from T_2 to T_1, taking into account the fact that as the house air is cooled, outside air at a temperature T_0 enters the house to keep P constant:

$$q = \frac{PV\bar{C}_P}{R} \ln \left(\frac{T_1}{T_2}\right) + n'\bar{C}_P(T_1 - T_0)$$

where n' is the number of moles of outside air that enters the house during the cooling process.

31. The dew point of air is the temperature at which air becomes saturated with moisture. Explain why the minimum evening temperature in the winter in mild climates is usually equal to the dew point.

32. It is observed that the air moving up a geographical slope undergoes a decrease in temperature, whereas air moving down a geographical slope undergoes an increase in temperature. Assuming that the heating effect of a downslope wind is due to the conversion of gravitational potential energy to internal energy, estimate the increase in temperature of air that moves downslope 2000 m. Take $\bar{C}_P = \frac{7}{2}R$ for air. Compare your result with the temperature increase computed on the assumption that the increase in air temperature is due to adiabatic compression of the air. Take $\Delta P = 770 - 600 = 170$ torr, and $T_1 = 273$ K.

33. Explain why the operation of an externally vented clothes dryer *decreases* the temperature of the inside air in a house when $T_{out} < T_{in}$ and the house air heater is off.

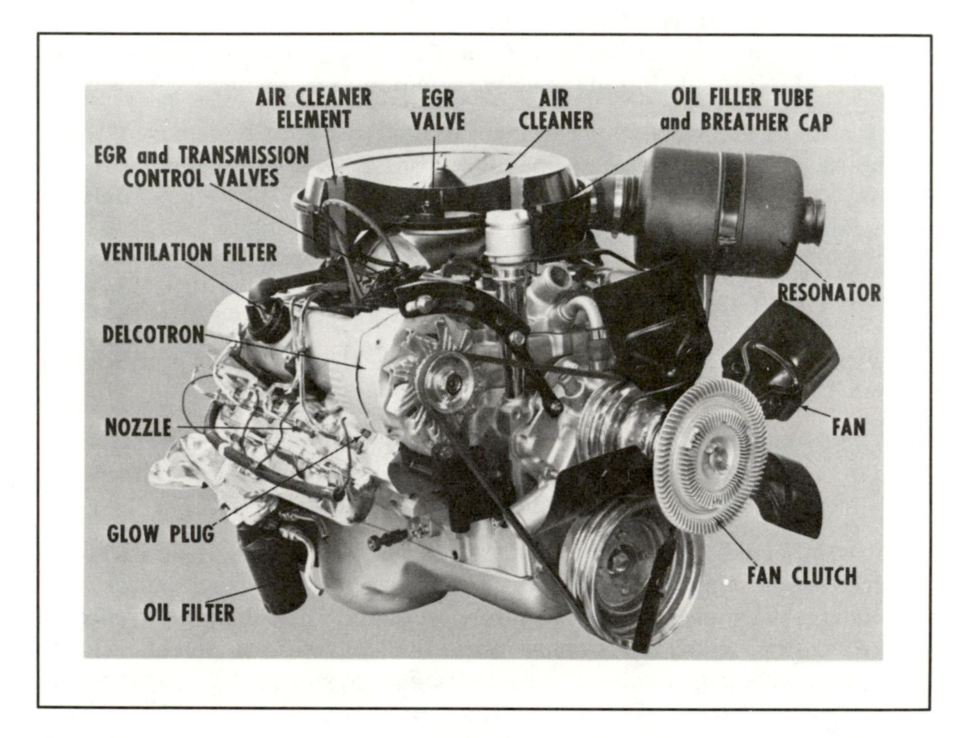

General Motors' 5.7 liter, 8 cylinder (90° V) diesel engine produced by Oldsmobile. The total engine mass, including attachments and fluids, is 330.2 kg. The pistons are aluminum alloy with a 0.78 kg mass. The bore is 103.05 mm and the stroke is 85.98 mm. (EGR denotes ***e****xhaust* ***g****as* ***r****ecirculation and the Delcotron is the alternator.)*

4

Entropy and the Second Law of Thermodynamics

Every physical or chemical process in nature takes place in such a way as to increase the sum of the entropies of all the bodies taking any part in the process. In the limit, that is for reversible processes, the sum of the entropies remains unchanged.

Max Planck. *Treatise on Thermodynamics.*, Translated by A. Ogg. New York: Dover Publications, Inc., n.d.

. . . and we thus come to Carnot's brilliant conclusion: that if an engine is reversible, it makes no difference how it is designed, because the amount of work one will obtain if the engine absorbs a given amount of heat at temperature T_2 and delivers heat at some other temperature T_1 [$T_2 > T_1$] does not depend on the design of the engine. It is a property of the world, not a property of a particular engine.

R. P. Feynman, R. B. Leighton, and M. Sands. *The Feynman Lectures on Physics*, Vol. I. Reading, Mass.: Addison-Wesley Publishing Company, Inc., 1965.

4-1 *Conservation Is Not the Only Restriction on Energy Transfers*

The First Law of Thermodynamics places no restrictions on energy transfers other than that of conservation. It is well known from experiments, however, that there are other restrictions on energy transfers. Some examples follow:

1. Heat always flows *spontaneously* from higher- to lower-temperature systems and never the reverse.
2. Work must be done to effect the transfer of energy as heat from a lower- to a higher-temperature system.
3. If any real (as opposed to hypothetical) system undergoes a change in state (thermodynamic process), then the system and its surroundings (i.e., everything but the system) cannot both be restored exactly to their original states. If the system is restored to its initial state, then it is impossible to restore the surroundings to their original state. All naturally occurring processes are in this sense irreversible.
4. Innumerable naturally occurring processes are *unidirectional*. Paper will burn spontaneously in oxygen if ignited, but the reverse process, that is, the spontaneous recombination of the combustion products to the piece of paper and oxygen gas, has never been observed in nature. There is no incantation that will cause a broken egg to reassemble itself. Ice around 1 atm pressure and any temperature above 0°C, always melts spontaneously. Certain occurrences are inevitable. Things are never exactly the same today as they were yesterday. Whenever an actual thermodynamic process occurs, the universe is irreversibly changed.

An irreversible thermodynamic process is a process for which it is impossible by any means whatsoever to restore everywhere the exact initial states once the process has taken place; all the other processes are termed *reversible*. No completely reversible thermodynamic change in state has ever been found in nature. A reversible process constitutes an ideal limiting case and is thus of theoretical importance.

The Second Law of Thermodynamics is fundamentally different from the first law in that the second law deals with the direction in which processes occur in nature. The implicit connection of the second law with the passage of time has led to characterization of *entropy*, the key concept embodied in the second law, as "time's arrow." For the transfer of energy as heat between two systems, the first law requires only that the amount of energy lost as heat by one system be exactly equal to the amount of energy taken up as heat by the other system. Whether the flow of heat takes place from the higher-temperature system to the lower-temperature system, or the reverse, cannot be deduced from the energy conservation principle. The energy conservation principle is completely silent on both the concept of temperature and the direction of spontaneous energy transfers.

4-2 *Entropy Is a Thermodynamic State Function*

An ideal gas is a gas with the equation of state:

$$PV = nRT \tag{4-1}$$

We also know from the kinetic theory of gases that the internal energy of a fixed mass of an ideal gas depends only on the absolute thermodynamic temperature; that is,

$$dU = C_V \, dT \tag{4-2}$$

where C_V is the heat capacity at constant volume of the ideal gas.

EXAMPLE 4-1

The translational kinetic energy of a gas composed of noninteracting molecules or atoms is given by the kinetic theory of gases as

$$U = \tfrac{3}{2}nRT$$

where n is the number of moles of the gas. Compute the change in the internal energy of 2.00 mol of He(g) when the temperature of the gas is increased from 10 K to 300 K.

Solution: The change in the internal energy is

$$\Delta U = U_2 - U_1 = \tfrac{3}{2}nR(T_2 - T_1)$$

thus

$$\Delta U = \tfrac{3}{2}(2.00 \text{ mol})(8.314 \text{ J}\cdot\text{K}^{-1}\cdot\text{mol}^{-1})(300 \text{ K} - 10 \text{ K})\left(\frac{1 \text{ kJ}}{1000 \text{ J}}\right)$$

and

$$\Delta U = 7.23 \text{ kJ}$$

Note that the value of ΔU is independent of the volume of the ideal gas. Thus, if we *isothermally* ($\Delta T = 0$) compress 2.00 mol of He(g) from V_1 to $V_2 = V_1/2$, then the value of U remains unchanged; that is, $\Delta U = 0$.

Consider the reversible, adiabatic expansion of an ideal gas. From the first law we have

$$dU = \delta q + \delta w = \delta q - P\,dV$$

but if the process is adiabatic, then $\delta q = 0$ and

$$dU = -P\,dV \tag{4-3}$$

Combination of Eqs. 4-3 and 4-2 yields

$$-P\,dV = C_V\,dT$$

or

$$C_V\,dT + P\,dV = 0 \tag{4-4}$$

Substitution of the equation of state, Eq. 4-1, into Eq. 4-4 yields

$$C_V\,dT + \frac{nRT}{V}\,dV = 0 \tag{4-5}$$

On dividing Eq. 4-5 through by T, we obtain

$$C_V \frac{dT}{T} + nR\frac{dV}{V} = 0 \tag{4-6}$$

Integration of Eq. 4-6 from the initial state (T_1, V_1) to the final state (T_2, V_2), with C_V equal to a constant, yields

$$\int_{T_1}^{T_2} C_V\,d\ln T + \int_{V_1}^{V_2} nR\,d\ln V = 0$$

or

$$C_V \ln\left(\frac{T_2}{T_1}\right) + nR\ln\left(\frac{V_2}{V_1}\right) = 0 \tag{4-7}$$

Equation 4-7 has an interesting consequence, namely, that given T_1 and V_1 we are free to pick T_2 *or* V_2 but not both; otherwise Eq. 4-7, which relates T_1, V_1 and T_2, V_2, would be violated. This is a general result for the reversible, adiabatic expansion of gases. Even if the gas is nonideal, even if U depends on both T and V, and even if C_V is not constant, the result is still the same—that is, when a fixed mass of gas undergoes a reversible, adiabatic expansion, we are free to pick the final value of only one of the thermodynamic state functions. The final values of all the other thermodynamic state functions are determined by Mother Nature.

We know from Chapter 3 that the number of independent thermodynamic state functions for a system of fixed mass is equal to the number of work terms plus one. If only pressure-volume work is involved, then there are only two independent state functions, say T and V. The fact that for a reversible, adiabatic expansion we are free to pick the value of either T or V but not both in the final state implies that there must be another thermodynamic state function of the system, call it S, that does not change in a reversible adiabatic expansion; that is,

$$\Delta S = S_2 - S_1 = 0 \qquad \text{(reversible, adiabatic process)} \tag{4-8}$$

Consider a reversible, isothermal expansion of an ideal gas. From the first law we have

$$dU = \delta q + \delta w = \delta q - P\,dV \tag{4-9}$$

Further, because the gas is ideal and the process is isothermal,

$$dU = C_V\,dT = 0$$

and thus

$$\delta q - P\,dV = 0 \tag{4-10}$$

Substitution of the ideal-gas equation of state into Eq. 4-10 yields

$$\delta q - \frac{nRT}{V}\,dV = 0$$

or

$$\frac{\delta q}{T} = nR\,d\ln V \qquad \text{(reversible, isothermal process)} \tag{4-11}$$

Integration of Eq. 4-11 yields

$$\int_1^2 \frac{\delta q}{T} = \int_{V_1}^{V_2} nR\,d\ln V$$

and

$$\frac{q}{T} = nR\ln\frac{V_2}{V_1} \tag{4-12}$$

Note that for the reversible, isothermal expansion of the ideal gas, from T, V_1 to T, V_2, we are free to pick the final value of the volume (i.e., V_2) even though T is fixed; this is a general result that is independent of the thermodynamic properties of ideal gases. Equation 4-12 means that the state function S changes in the reversible, isothermal expansion of the ideal gas; otherwise with both T and S fixed we would not be free to pick V_2. Thus we have

$$\Delta S \neq 0 \qquad \text{(reversible, isothermal process)} \tag{4-13}$$

Consider the general change in state:

$$\text{ideal gas at } T_1, V_1 \longrightarrow \text{ideal gas at } T_2, V_2$$

From the first law we have

$$\delta q = dU - \delta w = dU + P\,dV \tag{4-14}$$

Substituting Eqs. 4-1 and 4-2 into Eq. 4-14 yields

$$\delta q = C_V\,dT + \frac{nRT}{V}\,dV \tag{4-15}$$

Dividing Eq. 4-15 through by T, we obtain

$$\frac{\delta q}{T} = C_V\,d\ln T + nR\,d\ln V$$

or in integral form

$$\int_1^2 \frac{\delta q}{T} = \int_{T_1}^{T_2} C_V\,d\ln T + \int_{V_1}^{V_2} nR\,d\ln V \tag{4-16}$$

Assuming that C_V is constant we obtain from Eq. 4-16

$$\int_1^2 \frac{\delta q}{T} = C_V\ln\left(\frac{T_2}{T_1}\right) + nR\ln\left(\frac{V_2}{V_1}\right) \tag{4-17}$$

The value of the right-hand side of Eq. 4-17 depends only on the initial and final states (T_1, V_1 and T_2, V_2) and thus the left-hand side of Eq. 4-17 must be an exact differential. Note that if the process is adiabatic ($\delta q = 0$), then Eq. 4-7 results; whereas if the process is isothermal, then Eq. 4-12 results. The foregoing suggests that we can write

$$dS = \frac{\delta q}{T} \qquad \text{(reversible process)}$$

because

$$\int_1^2 \frac{\delta q}{T} = \int_1^2 dS = S_2 - S_1 = \Delta S$$

where dS is an exact differential and S is a thermodynamic state function, $S = S(T, V)$. The thermodynamic state function S is called the *entropy*. Notice that although δq is not an exact differential, the ratio $\delta q/T$ is an exact differential, the value of which is therefore independent of the path by which the system passes from the initial to the final state, provided that the process is reversible.

4-3 $dS_{sys} \geqslant \delta q_{sys}/T$

The preceding development of the entropy function involves a specific type of thermodynamic system, the ideal gas. The development was restricted to ideal gases in the interests of simplicity; however, the key result, namely, that $dS = \delta q/T$ for a reversible process, is perfectly general and constitutes part of the Second Law of Thermodynamics. The Second Law of Thermodynamics can be stated in several different ways. The second-law statement that we shall employ is the following:

For any change in defined thermodynamic states of a system,

$$dS_{sys} \geqslant \frac{\delta q_{sys}}{T} \qquad (4\text{-}18)$$

where the equality sign holds if the process is reversible, and the inequality sign holds if the process is irreversible. The function S, called the "entropy," is a thermodynamic state function.

Note that entropy has the units joule per kelvin, $J \cdot K^{-1}$.

When a system undergoes a change in state, the entropy change of the system is given by the second law (Eq. 4-18) as

$$dS_{sys} \geqslant \frac{\delta q_{sys}}{T}$$

If the process is adiabatic ($\delta q_{sys} = 0$), then

$$dS_{sys} \geqslant 0$$

and thus

$$\Delta S_{sys} \geqslant 0 \qquad \text{(adiabatic process)} \qquad (4\text{-}19)$$

When a system undergoes an adiabatic process, the entropy of the system either remains constant (reversible process) or increases (irreversible process). The entropy of the system cannot decrease in an adiabatic process.

EXAMPLE 4-2

Suppose n mol of an ideal gas at T_1, V_1 undergoes an irreversible adiabatic expansion into a vacuum to a final volume V_2. Obtain an expression for the entropy change of the gas.

Solution: From the first law we have

$$dU = \delta q + \delta w$$

The process is adiabatic and thus

$$dU = \delta w = -P_{ext}\, dV$$

The external pressure on the gas is $P_{ext} = 0$ (vacuum) and thus

$$dU = \delta w = 0$$

Because the gas is ideal,

$$dU = C_V\, dT$$

therefore

$$C_V\, dT = 0$$

or, in other words, the process is also isothermal. Entropy is a state function and thus the value of ΔS for the gas in the above process is independent of path. Therefore, the value of ΔS_{gas} for the process

$$\text{gas}(T, V_1) \xrightarrow[\text{expansion}]{\substack{\text{Irreversible} \\ \text{adiabatic}}} \text{gas}(T, V_2)$$

is the same as ΔS_{gas} for the process

$$\text{gas}(T, V_1) \xrightarrow[\text{expansion}]{\substack{\text{Reversible} \\ \text{isothermal}}} \text{gas}(T, V_2)$$

For the reversible process

$$dS_{gas} = \frac{\delta q_{gas}}{T}$$

Eliminating δq_{gas} using the first law, we obtain

$$dS_{gas} = \frac{dU_{gas} - \delta w_{gas}}{T} = \frac{-\delta w_{gas}}{T} = \frac{P\, dV}{T}$$

where we used $dU_{gas} = C_V\, dT = 0$ (isothermal process). Using the equation of state for the gas, we have

$$dS_{gas} = nR\frac{dV}{V} = nR\, d\ln V$$

and thus

$$\int_1^2 dS_{gas} = \int_{V_1}^{V_2} nR\,d\ln V$$

or

$$\Delta S_{gas} = nR\ln\left(\frac{V_2}{V_1}\right)$$

Because $V_2 > V_1$, $\Delta S_{gas} > 0$, as required by the second law. Note that we obtained ΔS_{gas} by devising a *reversible* path between the same initial and final states as in the irreversible process and then we equated ΔS_{gas} for the reversible path to ΔS_{gas} for the irreversible path. Entropy is a thermodynamic state function and thus the value of ΔS_{sys} is independent of path.

When a system undergoes an isothermal process, the entropy change of the system is given by the second law as

$$\int_1^2 dS_{sys} \geqslant \int_1^2 \frac{\delta q_{sys}}{T} = \frac{1}{T}\int_1^2 \delta q_{sys}$$

thus

$$\Delta S_{sys} \geqslant \frac{q_{sys}}{T} \qquad \text{(isothermal process)} \qquad (4\text{-}20)$$

Because q_{sys} can be either positive or negative depending on the process, the value of ΔS_{sys} can be either positive or negative, but in any case Eq. 4-20 holds, provided that the initial and final states for the process are defined.

EXAMPLE 4-3

When liquid water at 0°C and 1 atm is converted to ice at 0°C and 1 atm, 6.01 kJ·mol^{-1} of energy is released as heat to the surroundings. Given that the surroundings are at −2°C throughout the process, compute ΔS_{H_2O} and ΔS_{sur} when 1.00 mol of water freezes at 0°C.

Solution: Water at 1 atm has a melting point of 0°C; thus the conversion of water to ice at 0°C and 1 atm can be regarded as reversible. The entropy change of the H_2O is

$$\Delta S_{H_2O} = \frac{q_{H_2O}}{T} = \frac{-(6.01 \times 10^3\ \text{J}\cdot\text{mol}^{-1})(1.00\ \text{mol})}{273\ \text{K}} = -22.0\ \text{J}\cdot\text{K}^{-1}$$

The heat evolved when the water freezes is absorbed by the surroundings:

$$q_{sur} = -q_{H_2O}$$

The entropy change of the surroundings is

$$\Delta S_{sur} = \frac{q_{sur}}{T} = \frac{6.01 \times 10^3\ \text{J}}{271\ \text{K}} = +22.2\ \text{J}\cdot\text{K}^{-1}$$

The total entropy change for the process is

$$\Delta S_{tot} = \Delta S_{sys} + \Delta S_{sur} = \Delta S_{H_2O} + \Delta S_{sur}$$
$$\Delta S_{tot} = -22.0\ J \cdot K^{-1} + 22.2\ J \cdot K^{-1} = +0.2\ J \cdot K^{-1}$$

There is a net entropy production in the process, because $T_{sys} > T_{sur}$ and the system and the surroundings are therefore not in thermal equilibrium during the process. Whenever thermodynamic equilibrium is not maintained throughout a process, irreversibility is involved, and there is a net entropy production.

4-4 *There Is Always a Net Entropy Production in an Irreversible Process*

Suppose that we have a system and its surroundings between which energy can be transferred, and further suppose that the system and its surroundings are isolated from the rest of the universe. If the system undergoes a change in state, then the total entropy change is

$$\Delta S_{tot} = \Delta S_{sys} + \Delta S_{sur}$$

If the process is adiabatic, then

$$\delta q_{sys} = \delta q_{sur} = 0$$

and

$$\Delta S_{sys} \geqslant 0 \quad \text{and} \quad \Delta S_{sur} \geqslant 0$$

thus

$$\Delta S_{tot} \geqslant 0$$

If the adiabatic process is reversible, then the total entropy remains unchanged; whereas if the process is in any way irreversible, then there is a net increase in the total entropy.

Suppose now that the system undergoes an isothermal process with $T_{sur} = T_{sys} = T$. From the second law, we have

$$\Delta S_{sys} \geqslant \frac{q_{sys}}{T} \qquad \Delta S_{sur} \geqslant \frac{q_{sur}}{T}$$

and thus

$$\Delta S_{tot} = \Delta S_{sys} + \Delta S_{sur} \geqslant \frac{q_{sys}}{T} + \frac{q_{sur}}{T}$$

but $q_{sys} = -q_{sur}$ and thus

$$\Delta S_{tot} \geqslant 0$$

Note once again that there is a net entropy production if the process is irreversible, whereas the total entropy remains unchanged if the process is reversible.

All real processes involve some degree of irreversibility, and thus all real processes lead to an increase in the *total* entropy. *Entropy is not conserved,* except in the hypothetical limiting case of a reversible process. The total entropy can never decrease; it can only increase.

A *spontaneous thermodynamic process* is a process that occurs on its own without the need for any outside intervention. Examples of spontaneous processes are the melting of ice above 0°C, the diffusion of a chemical species until a uniform concentration is obtained, various metal corrosion processes, and innumerable chemical reactions. All spontaneous processes are naturally occurring processes and thus are thermodynamically irreversible. The occurrence of a spontaneous process always involves a net entropy production. A thermodynamic system is said to be in an *equilibrium state* when it can no longer undergo any spontaneous processes. Thus, *the entropy of a system in an equilibrium state is a maximum for the given energy and volume.*

4-5 *Entropy Is a Measure of the Disorder in a System*

Entropy, like energy, is a fundamental abstract thermodynamic concept. It is not difficult, as we have already seen, to state the essential mathematical properties of the *entropy function S*, but it is not a simple matter to answer the question "What is entropy?". The first point to recognize is that entropy is not a thing. There are no entropy meters. Entropy is not a directly measurable quantity, but rather entropy changes are calculated from measurable quantities such as temperature, pressure, volume, heat capacity, current, and magnetic field strength.

Entropy was described by the great thermodynamicist J. Willard Gibbs as a measure of the "mixed-up-ness" of a system; that is, *the more disordered or randomized a system is, the higher is its entropy*. Things left to themselves proceed to a state of maximum possible disorder (entropy), subject to certain constraints such as fixed energy and volume. Constructions fall apart, and living matter, upon death, decays. The existence of life, which spontaneously builds itself up into a more ordered state, appears to contradict the Second Law of Thermodynamics. However, all living systems build themselves up at the expense of their environment. For example, life could not exist on earth without the sun. The spontaneous radiation production by the sun is accompanied by tremendous entropy production. The second law does not say that the entropy of a system can never spontaneously decrease. What the second law does say is that the entropy of any real system (or collection of systems) in an adiabatic enclosure can never spontaneously decrease. There is a world of difference between a system that is *open* (i.e., a system that can exchange matter and energy with its environment) and a system that is *adiabatically enclosed* (cannot exchange matter or heat) or a system that is *isolated* (cannot exchange either matter or energy).

Compared at the same pressure and temperature, liquids are more disordered than solids, and gases are more disordered than liquids. This is because the molecules of a gas have more freedom to move around than the molecules of a liquid; the same holds for the molecules of a liquid as compared to those of a solid. The melting of a solid involves an entropy increase of $8\ \mathrm{J \cdot K^{-1} \cdot mol^{-1}}$ to $40\ \mathrm{J \cdot K^{-1} \cdot mol^{-1}}$; the vaporization of a liquid involves an entropy increase of $75\ \mathrm{J \cdot K^{-1} \cdot mol^{-1}}$ to $120\ \mathrm{J \cdot K^{-1} \cdot mol^{-1}}$.

An increase in the temperature of a system held at constant pressure or at constant volume always produces an increase in the entropy of the system. As the temperature increases, the particles increase the extent of their motion, and there is a corresponding loss of information as to the precise location of all the atoms at any instant; the substance is more disordered. The spontaneous process of temperature equalization that occurs when two systems with different temperatures are brought into contact produces an entropy increase that can be visualized as arising from the distribution of the total energy over a larger number of particles (increased *thermal randomness*). The entropy increase that arises when two different liquids are mixed can be attributed to an increase in *spatial randomness* that arises from the spreading of the particles over a greater volume of space.

The entropy change ΔS of an ideal gas that undergoes the change in state

$$\text{ideal gas}(T_1, V_1) \longrightarrow \text{ideal gas}(T_2, V_2)$$

is given by Eq. 4-17 as

$$\Delta S = C_V \ln\left(\frac{T_2}{T_1}\right) + nR \ln\left(\frac{V_2}{V_1}\right) \qquad (4\text{-}21)$$

Recall that Eq. 4-21 holds for reversible and irreversible processes. Consider the case in which the volume of the gas is increased isothermally; that is, $T_2 = T_1$. Then from Eq. 4-21 we have

$$\Delta S = nR \ln\left(\frac{V_2}{V_1}\right) \qquad \text{(isothermal)}$$

Note that if $V_2 > V_1$, then $\Delta S > 0$. The same number of gas molecules in a larger volume at the same temperature have a higher entropy than when the gas molecules are in a smaller volume. The isothermal expansion of a gas leads to an increase in the spatial disorder (randomness) of the gas molecules and thus the entropy of the gas increases. The *entropy is an extensive state function* because ΔS is proportional to the number of moles n of the gas.

Consider the case in which the temperature of the gas is increased while the volume of the gas is held constant. In such a case, we have from Eq. 4-21

$$\Delta S = C_V \ln\left(\frac{T_2}{T_1}\right) \qquad \text{(constant volume)}$$

Note that if $T_2 > T_1$, then $\Delta S > 0$, because $C_V > 0$. An increase in the temperature of a gas held at fixed volume produces an increase in the entropy of the gas. An increase in the temperature of the gas held at fixed volume is produced by an input of energy as heat to the gas:

$$q_V = C_V(T_2 - T_1)$$

The additional energy produces an increased thermal disorder (i.e., higher average velocity of the gas molecules) and thus the entropy of the gas increases.

The increase in the entropy of the gas produced by a given increase in the temperature of the gas held at constant volume is greater the larger the value of the heat capacity C_V of the gas. The total heat capacity is equal to

$$C_V = n\bar{C}_V$$

where n is the number of moles of the gas and $\bar{C}_V$ is the molar heat capacity at constant volume. We know from the kinetic theory of gases that there are various possible motions of the molecules that can make contributions to $\bar{C}_V$:

Molecular motion	*Contribution to* $\bar{C}_V$
Translation	$\frac{3}{2}R$
Rotation (nonlinear molecule)	$\frac{3}{2}R$
Rotation (linear molecule)	R
Vibration (N atom, nonlinear molecule)	$\leqslant(3N - 6)R$
Vibration (N atom, linear molecule)	$\leqslant(3N - 5)R$

The vibrational motions of most molecules do not make major contributions to $\bar{C}_V$ below 300 K, because such motions are not excited (active) at these temperatures. Thus we have the following data for the molar heat capacity at constant volume of the species $He(g)$, $N_2(g)$, and $H_2O(g)$:

	Contribution to $\bar{C}_V$ *at 298 K*			$\bar{C}_V/J\cdot K^{-1}\cdot mol^{-1}$	
Molecule	*Translation*	*Rotation*	*Vibration*	*Estimated*	*Observed*
$He(g)$	$\frac{3}{2}R$	0	0	12.47	12.47
$N_2(g)$	$\frac{3}{2}R$	R	~ 0	20.79	20.81
$H_2O(g)$	$\frac{3}{2}R$	$\frac{3}{2}R$	~ 0	24.94	25.26

A substance with a large molar heat capacity has more ways in which the energy can be absorbed by the molecule (greater potential thermal randomness) and thus such a substance exhibits a larger molar entropy increase for a given increase in temperature.

4-6 *The Work Done by a System Is a Maximum When the Process Is Reversible and Isothermal*

In Section 4-5 we discussed the connection between entropy changes and molecular disorder. The question that we now consider is the origin of the increase in the *total* entropy in an irreversible process.

The work done by a system on the surroundings is given by the first law as

$$-\delta w = \delta q - dU \tag{4-22}$$

From the second law we have

$$\delta q \leqslant T\,dS$$

and combination with Eq. 4-22 gives

$$-\delta w \leqslant T\,dS - dU \tag{4-23}$$

For an isothermal process, we obtain from Eq. 4-23

$$-w \leqslant T\Delta S - \Delta U \tag{4-24}$$

Because the right-hand side of Eq. 4-24 involves only state functions, its value *for a given change in state* is unaffected by whether or not the process is reversible. The equality sign in Eq. 4-24 applies if the process is reversible, whereas the inequality sign applies if the process is irreversible. Thus, the work done by the system in a reversible isothermal process ($-w_{\text{rev}}$) is greater than the work done by the system in an irreversible isothermal process ($-w_{\text{irr}}$); that is

$$-w_{\text{rev}} > -w_{\text{irr}} \qquad \text{(for a given isothermal change in state)} \tag{4-25}$$

EXAMPLE 4-4

Compare the work done by 2.00 mol of an ideal gas that expands isothermally ($T = 300$ K) and reversibly from an initial volume of V_1 to final volume of $V_2 = 2V_1$, with the work done when the same gas expands isothermally ($T = 300$ K) into an evacuated space such that $V_2 = 2V_1$.

Solution: The process is isothermal, and thus we have from Eq. 4-24

$$-w \leqslant T\Delta S - \Delta U$$

Because the gas is ideal, we can write

$$\Delta U = 0$$

and, from Equation 4-21,

$$\Delta S = nR\ln\left(\frac{V_2}{V_1}\right)$$

(ideal gas, isothermal process)

Thus

$$-w \leqslant nRT\ln\left(\frac{V_2}{V_1}\right) = (2.00\text{ mol})(8.314\text{ J}\cdot\text{K}^{-1}\cdot\text{mol}^{-1})(300\text{ K})\ln 2$$

$$= 3.46 \times 10^3\text{ J}$$

For the reversible process the equality sign applies, and

$$-w_{\text{rev}} = 3.46 \times 10^3\text{ J}$$

The work done by the gas in the irreversible process must be less than 3.46×10^3 J. In the irreversible process the gas expands without restraint ($P_{ext} = 0$), and thus

$$-w_{irr} = \int_1^2 P_{ext}\,dV = 0$$

Therefore

$$-w_{irr} < -w_{rev}$$

EXAMPLE 4-5

Compare the work done by n mol of an ideal gas that expands adiabatically and reversibly with the work done by the same gas when it expands irreversibly between the same initial and final states.

Solution: From Eq. 4-23 we have for the work done by the gas in the reversible adiabatic expansion

$$-w = \int_1^2 T\,dS - \int_1^2 dU$$

The process is adiabatic and reversible; therefore

$$dS = \frac{\delta q}{T} = 0$$

and using the first law

$$-w = -\Delta U$$

The gas is ideal, and thus (C_V = constant) we obtain

$$-w = -\Delta U = -C_V(T_2 - T_1) = -n\bar{C}_V(T_2 - T_1) > 0$$

where $-w > 0$, because $T_2 < T_1$(the gas cools on adiabatic expansion). From the first law, we have for the reversible adiabatic expansion

$$\Delta U_{rev} = q_{rev} + w_{rev} = 0 + w_{rev}$$

whereas for the irreversible expansion between the same initial and final states ($\Delta U_{rev} = \Delta U_{irr}$)

$$\Delta U_{irr} = q_{irr} + w_{irr} = w_{rev}$$

or

$$-w_{rev} = -w_{irr} - q_{irr}$$

Note that if $-q_{irr} > 0$ then

$$-w_{rev} > -w_{irr}$$

In other words, the work done is less in the irreversible expansion, because some of the energy decrease of the gas was transferred as heat to the surroundings rather than as work. There will thus be an entropy increase in the surroundings in the irreversible expansion, whereas the entropy of the surroundings remains unchanged in the reversible adiabatic expansion. Whenever a process occurs and less than the maximum possible amount of work is done by the system, there is an increase in the amount of heat transferred to the surroundings and a net entropy production.

Whenever a system undergoes a change in state and the amount of work done by the system is less than the maximum possible amount of work, then there is a net entropy production, $\Delta S_{tot} > 0$. The increase in the total entropy that occurs whenever an irreversible process takes place is associated with the failure to achieve the maximum possible transfer of energy as work in the process. In effect, the net entropy production in an irreversible process is a result of a failure to capitalize fully on an opportunity to perform work.

If a process involves only the isothermal transfer of energy as heat (*pure heat transfer*), then $\delta w = 0$ and from the first law

$$dU = \delta q = dq$$

where dq is an exact differential because dU is an exact differential. For a pure heat transfer the second law can be written as

$$dS_{sys} = \frac{dq_{sys}}{T} \qquad \text{(pure heat transfer)} \qquad (4\text{-}26)$$

Note the absence of the inequality sign when the second law is applied to a pure heat transfer. The inequality sign in the most general statement of the second law; that is,

$$dS_{sys} \geqslant \frac{\delta q_{sys}}{T}$$

is a consequence of the inexactness of δq_{sys}. When $\delta q_{sys} = dU = dq_{sys}$, then dS_{sys} can be computed directly from dq_{sys}/T, irrespective of whether or not the process is reversible, provided only that no energy is transferred by the system as work.

Suppose energy is transferred spontaneously as heat q from a system at a fixed temperature T_2 to a system at a fixed temperature T_1 without the performance of any work (pure heat transfer). What is the value of ΔS_{tot} for the process? The total entropy change is equal to

$$\Delta S_{tot} = \Delta S_2 + \Delta S_1$$

For isothermal pure heat transfers, $\Delta S = q/T$; thus

$$\Delta S_{tot} = \frac{q_2}{T_2} + \frac{q_1}{T_1}$$

where q_1 and q_2 are the heats for the systems at T_1 and T_2, respectively. But $q_1 = -q_2 = q > 0$, thus

$$\Delta S_{tot} = q\left(\frac{1}{T_1} - \frac{1}{T_2}\right) = q\left(\frac{T_2 - T_1}{T_1 T_2}\right) > 0$$

The value of q is positive and $\Delta S_{\text{tot}} > 0$ by the second law; therefore, $T_2 > T_1$. Thus we see that the spontaneous transfer of energy as heat is from the higher to the lower temperature. The reverse process is impossible because it would lead to $\Delta S_{\text{tot}} < 0$, which is a violation of the second law.

4-7 *The Entropy of a Substance Held at Constant Pressure Increases When the Temperature Increases*

The input of energy as heat to a system ($\delta q_{\text{sys}} > 0$) necessarily increases the entropy of the system:

$$dS_{\text{sys}} \geqslant \frac{\delta q_{\text{sys}}}{T} > 0 \qquad (\delta q_{\text{sys}} > 0)$$

The input of energy as heat to a system may or may not increase the temperature of the system; the temperature increases if the heat capacity of the system is finite, and the temperature remains constant if the heat capacity is infinite. We shall now show that an increase in the temperature of a system held at constant pressure or at constant volume increases the entropy of the system.

The heat capacity of a system at constant pressure is given by Eq. 3-38 as

$$C_P = \lim_{\Delta T \to 0} \left(\frac{q_P}{\Delta T} \right) \tag{4-27}$$

If a process involves only the isothermal transfer of energy as heat at constant pressure (pure heat transfer), then from Eq. 4-26 we have

$$q_P = T\Delta S \tag{4-28}$$

and thus

$$C_P = T \lim_{\Delta T \to 0} \left(\frac{\Delta S}{\Delta T} \right)_P = T \left(\frac{\partial S}{\partial T} \right)_P \tag{4-29}$$

or

$$\left(\frac{\partial S}{\partial T} \right)_P = \frac{C_P}{T} \tag{4-30}$$

Equation 4-30 tells us how the entropy of a substance changes with temperature at constant pressure. For an isothermal pure heat transfer at constant volume, we have, by reasoning analogous to the constant pressure case,

$$C_V = \lim_{\Delta T \to 0} \left(\frac{q_V}{\Delta T} \right) = T \lim_{\Delta T \to 0} \left(\frac{\Delta S}{\Delta T} \right)_V \tag{4-31}$$

or

$$\left(\frac{\partial S}{\partial T} \right)_V = \frac{C_V}{T} \tag{4-32}$$

Equation 4-32 tells us how the entropy of a substance changes with temperature at constant volume.

EXAMPLE 4-6

The molar heat capacity at constant pressure of $H_2O(g)$ at 1.00 atm is given approximately by the equation

$$\bar{C}_P = 30.54 + 1.03 \times 10^{-2}T$$

with T in kelvins; $\bar{C}_P$ has the units joules per kelvin per mole, $J \cdot K^{-1} \cdot mol^{-1}$. Compute the change in the entropy of 2.00 mol of water at 1.00 atm when the temperature of the $H_2O(g)$ is increased from 300 K to 1000 K.

Solution: The entropy change is computed using Eq. 4-30:

$$\left(\frac{\partial S}{\partial T}\right)_P = \frac{C_P}{T} = \frac{2.00(30.54 + 1.03 \times 10^{-2}T)}{T}$$

or

$$dS = \left(\frac{61.08}{T} + 2.06 \times 10^{-2}\right)dT$$

where we have written $(\partial S/\partial T)_P = dS/dT$, because the pressure is constant in the process. Thus, on integration, we obtain

$$\Delta S = 61.08 \ln\left(\frac{T_2}{T_1}\right) + 2.06 \times 10^{-2}(T_2 - T_1)$$

and

$$\Delta S_{H_2O} = \Delta S = 61.08 \ln\left(\frac{1000}{300}\right) + 2.06 \times 10^{-2}(1000 - 300) = 87.96\ J \cdot K^{-1}$$

Note that $\Delta S_{H_2O} > 0$, because T increases and P is constant.

The increase in the entropy of a substance that undergoes an increase in temperature at constant pressure is a consequence of the increased thermal randomness of the molecules of the substance. A greater amount of energy is distributed among the same number of particles, which results in an increased degree of molecular and atomic disorder.

4-8 *An Isothermal Change in the Phase of a Substance Produced by the Input of Energy as Heat Always Leads to an Increase in the Entropy of the Substance*

We shall define a phase in precise thermodynamic terms in Chapter 9. It is sufficient for our purposes at this point to recognize that, for example, ice, liquid water, and gaseous water are three different, and easily distinguished, phases of H_2O.

A two-phase equilibrium system (e.g., ice + liquid water at 0°C, 1 atm or liquid water + gaseous water at 100°C, 1 atm) has an infinite heat capacity, as long as both phases are present. The input of energy as heat does not change the temperature but rather leads to the melting of some of the ice or the vaporization of some of the water. The entropy change of a substance that undergoes a reversible phase transition under isothermal conditions is given by the second law ($dS_{tr} = \delta q_{tr}/T_{tr}$) as

$$\Delta S_{tr} = \frac{q_{tr}}{T_{tr}} \tag{4-33}$$

where the subscript tr denotes a phase transition. If the process occurs at constant pressure, then $q_{tr} = q_p = \Delta H_{tr}$ and

$$\Delta S_{tr} = \frac{\Delta H_{tr}}{T_{tr}} \tag{4-34}$$

The conversion of a low-temperature phase of a substance to a higher-temperature phase of a substance, for example, the melting of ice,

$$H_2O(s) \longrightarrow H_2O(\ell)$$

requires an input of energy as heat. In such a case $q_{tr} > 0$ and thus $\Delta S_{tr} > 0$, reflecting the increased molecular disorder in the higher-temperature phase.

EXAMPLE 4-7

The molar enthalpy (heat) of fusion of ice at 1 atm is $\Delta\bar{H}_{fus} = 6.01\ \mathrm{kJ\cdot mol^{-1}}$; the molar enthalpy (heat) of vaporization of liquid water at 1 atm is $\Delta\bar{H}_{vap} = 44.0\ \mathrm{kJ\cdot mol^{-1}}$. Compute (a) $\Delta\bar{S}_{fus}$ at the melting point of ice (0°C) and (b) $\Delta\bar{S}_{vap}$ at the normal boiling point of water (100°C).

Solution: From Eq. 4-34 we have

$$\text{(a)}\ \Delta\bar{S}_{fus} = \frac{\Delta\bar{H}_{fus}}{T_{fus}} = \left(\frac{6.01 \times 10^3\ \mathrm{J\cdot mol^{-1}}}{273\ \mathrm{K}}\right) = 22.0\ \mathrm{J\cdot K^{-1}\cdot mol^{-1}}$$

and

$$\text{(b)}\ \Delta\bar{S}_{vap} = \frac{\Delta\bar{H}_{vap}}{T_{vap}} = \left(\frac{44.0 \times 10^3\ \mathrm{J\cdot mol^{-1}}}{373\ \mathrm{K}}\right) = 118.0\ \mathrm{J\cdot K^{-1}\cdot mol^{-1}}$$

Note that $\Delta\bar{S}_{vap} > \Delta\bar{S}_{fus}$; this is a general result that is a consequence of the much greater disorder in a gas as compared to a liquid or a solid at the same pressure and temperature.

The increase in the molar entropy with temperature of oxygen is shown in Figure 4-1. The discontinuities in the $\bar{S}$ versus T curve for O_2 arise at phase transition temperatures. There are three different solid phases of O_2, which are

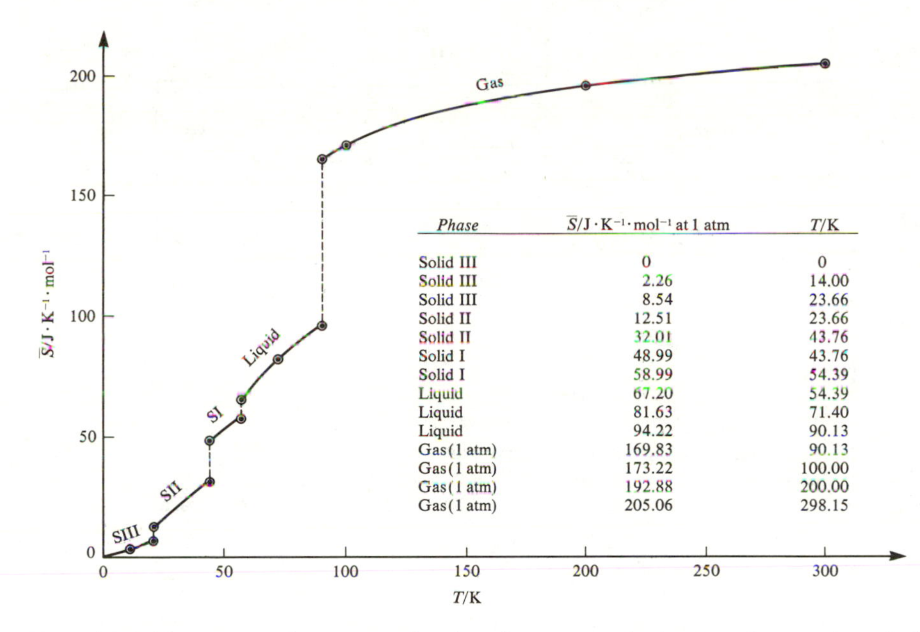

Phase	$\overline{S}$/J · K^{-1} · mol^{-1} at 1 atm	*T*/K
Solid III	0	0
Solid III	2.26	14.00
Solid III	8.54	23.66
Solid II	12.51	23.66
Solid II	32.01	43.76
Solid I	48.99	43.76
Solid I	58.99	54.39
Liquid	67.20	54.39
Liquid	81.63	71.40
Liquid	94.22	90.13
Gas(1 atm)	169.83	90.13
Gas(1 atm)	173.22	100.00
Gas(1 atm)	192.88	200.00
Gas(1 atm)	205.06	298.15

Figure 4-1. Molar entropy of oxygen as a function of temperature (at 1 atm total pressure); vertical jumps correspond to phase transitions. Note that the entropy increase on vaporization is much larger than the entropy increase on melting. Note also that for any particular phase the entropy increases with increasing temperature.

labeled SIII, SII, and SI in the figure. Note that, as required by the second law, the entropy of any particular phase increases with increasing temperature. Note also that $\overline{S} = 0$ at $T = 0$; this result is predicted by the Third Law of Thermodynamics, which we shall discuss in Chapter 6.

4-9 *All Heat Engines Require a Temperature Difference to Do Work*

The Second Law of Thermodynamics was discovered by the great French engineer Sadi Carnot about 1820. Carnot's tremendous discovery was accomplished before the First Law of Thermodynamics was known. Carnot deduced the second law by analyzing the thermodynamic behavior of an idealized (i.e., reversible) heat engine, which is now called a *Carnot engine*. We shall not outline Carnot's original line of reasoning here, because it would be unnecessarily complex without the first law. Rather, we shall briefly analyze a Carnot engine with the same original objective as Carnot. What is the maximum amount of work that can be obtained per cycle from a heat engine that, operating in a cyclic manner, takes in energy as heat and performs work on the surroundings? A schematic of a Carnot engine is shown in Figure 4-2.

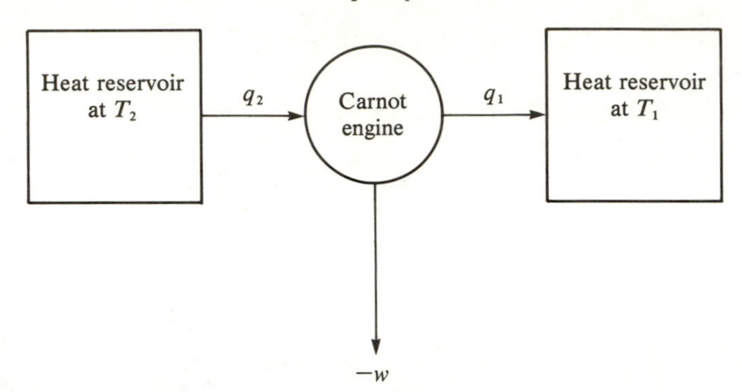

Figure 4-2. Schematic of a Carnot heat engine. The engine takes in energy as heat ($q_2 > 0$) from a high-temperature (T_2) "heat reservoir," performs work on the surroundings ($-w$), deposits energy as heat ($q_1 < 0$) in a low-temperature (T_1) heat reservoir, and then returns to the high-temperature heat reservoir to begin another cycle.

A reversible Carnot engine operates on the following four-step cycle ($T_2 > T_1$):

Step (reversible)		*Temperature*	*Work*	*Heat*	*Entropy change of engine*	*Energy change of engine*
(a) Isothermal (expansion)	$A \to B$	T_2	w_2	q_2	$\Delta S_2 = q_2/T_2$	$q_2 + w_2$
(b) Adiabatic (expansion)	$B \to C$	$T_2 \to T_1$	w_{21}	0	0	w_{21}
(c) Isothermal (compression)	$C \to D$	T_1	w_1	q_1	$\Delta S_1 = q_1/T_1$	$w_1 + q_1$
(d) Adiabatic (compression)	$D \to A$	$T_1 \to T_2$	w_{12}	0	0	w_{12}

The terms in parentheses in the steps above refer to the case of pressure-volume work for a gas. However, in general, a Carnot engine can operate with systems involving other types of work as well, for example, magnetic work.

The efficiency ε of the engine is defined as the ratio of the work done by the engine on the surroundings to the heat taken in at the high temperature; that is,

$$\varepsilon = \frac{-w}{q_2} \tag{4-35}$$

where

$$-w = -(w_2 + w_{21} + w_1 + w_{12})$$

The engine operates in a cyclic manner; that is, it returns to its original state at the end of each cycle. Application of the first law to one complete cycle of the engine yields

$$\Delta U_{eng} = 0 = w + q_2 + q_1$$

Thus the work done by the engine ($-w$) per cycle is

$$-w = q_2 + q_1 \tag{4-36}$$

Application of the second law to one complete cycle of the engine yields (see page 88)

$$\Delta S_{eng} = 0 = \Delta S_2 + \Delta S_1 = \frac{q_2}{T_2} + \frac{q_1}{T_1} \tag{4-37}$$

Given that the engine takes in energy as heat ($q_2 > 0$) from the high-temperature reservoir and therefore must undergo an entropy increase ΔS_2,

$$\Delta S_2 = \frac{q_2}{T_2} > 0$$

we conclude that to return to its initial state the engine must *discharge* energy as heat to produce an entropy *decrease* that will offset the entropy increase in the heat intake step. However, because the engine puts out energy as work, the amount of energy discharged as heat must be less than the amount of energy taken in as heat (Eq. 4-36). From Eq. 4-37 we have

$$\frac{q_2}{T_2} = \frac{-q_1}{T_1} \tag{4-38}$$

but $-q_1$ is less than q_2 and thus T_1 *must* be less than T_2. In other words, a heat engine requires a temperature difference to take in energy as heat and to do work on the surroundings.

The efficiency of the Carnot engine is given by

$$\varepsilon = \frac{-w}{q_2} = \frac{q_2 + q_1}{q_2} = 1 + \frac{q_1}{q_2} \tag{4-39}$$

but from Eq. 4-38

$$\frac{q_1}{q_2} = -\frac{T_1}{T_2}$$

and thus

$$\frac{-w}{q_2} = \frac{T_2 - T_1}{T_2} \tag{4-40}$$

Equation 4-40 is a truly remarkable result; it says that *the efficiency of a reversible Carnot engine depends only on the temperatures of the two heat reservoirs.* The efficiency is independent of the nature of the engine, the engine working substance, or the type of work performed. If any irreversibility enters into the operation of the engine, then the work done must be less than for the reversible case, and thus we can write Eq. 4-40 as

$$\frac{-w}{q_2} \leqslant \frac{T_2 - T_1}{T_2} \tag{4-41}$$

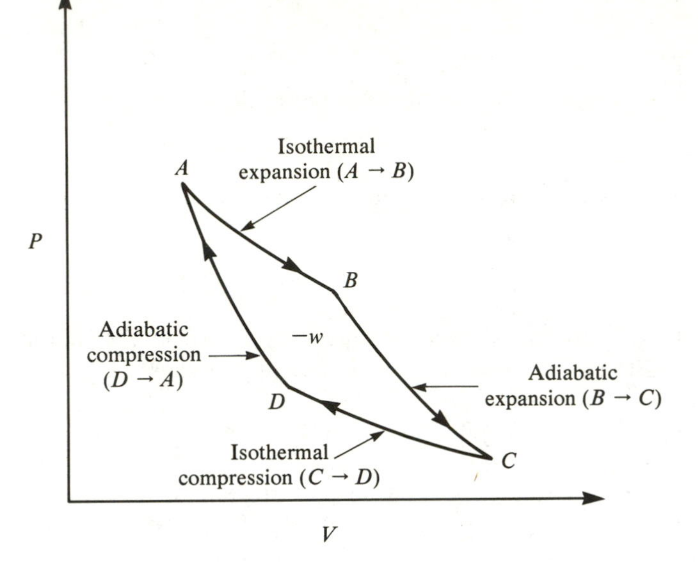

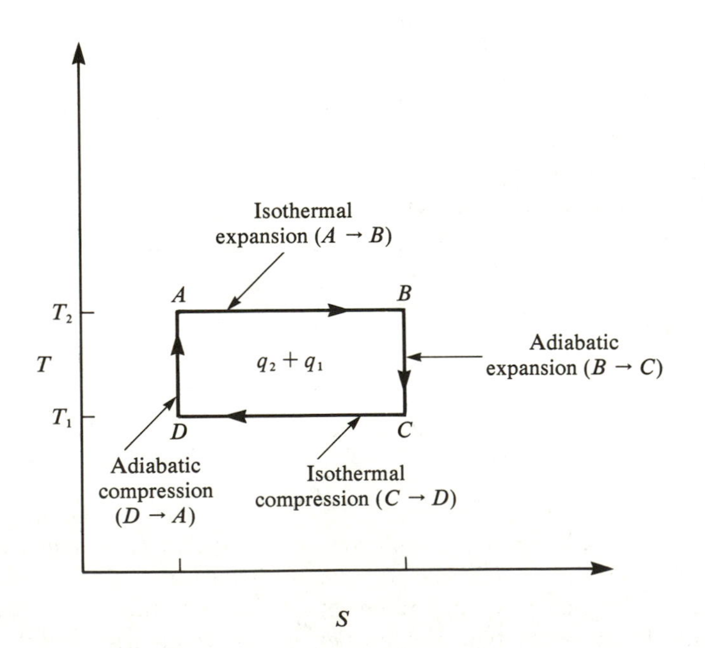

Figure 4-3. Reversible Carnot cycle on *P-V* and *T-S* coordinates for a gas as working fluid. The area enclosed by the *P-V* path is equal to the net work done on the surroundings by the gas in one cycle. The area enclosed by the *T*-S path is equal to the net intake of energy as heat by the gas in one cycle. Further, $-w = q_2 + q_1$.

where the inequality sign applies if any irreversibility is involved in the operation of the engine. Equation 4-41 can be obtained directly from Eq. 4-39 by noting that if any irreversibility is involved, then $q_2 < T_2 \Delta S_2$, and $q_1 < T_1 \Delta S_1$, but in any event $\Delta S_1 = -\Delta S_2$.

It is impossible to construct an engine that operates in a cycle that is more efficient than a reversible Carnot engine, because such a device would violate Eq. 4-41 and therefore would violate the laws of thermodynamics. A real engine necessarily involves some degree of irreversibility in its operation and therefore the efficiency of any real engine is less than that of a Carnot engine operating between the same two temperatures.

A reversible Carnot cycle for a Carnot engine with a gas as working substance is sketched in Figure 4-3 on P-V and T-S coordinate systems. The enclosed area on P-V coordinates is the net work output $(-w)$ of the engine, and the enclosed area on T-S coordinates is the net heat intake $(q_2 + q_1)$.

The maximum amount of work that can be obtained $(-w_{rev})$ from a given input of energy as heat (q_2) is given by Eq. 4-41 as

$$-w_{rev} = q_2\left(\frac{T_2 - T_1}{T_2}\right) \tag{4-42}$$

Thus for a given value of q_2, the value of $-w_{rev}$ is greater the greater the temperature difference $T_2 - T_1$; this conclusion has important practical consequences, as we shall soon see.

A Carnot engine can be run in reverse and used to extract energy as heat from a low-temperature reservoir and discharge energy as heat to a high-temperature reservoir (see Figure 4-4). In such a case the device is called a *heat pump*, if it is used as a heat source, or a *refrigerator*, if it is used to remove energy as heat. From the second law we know that work must be done on the engine to accomplish the transfer of energy as heat from the low- to the high-temperature reservoir (reverse Carnot cycle). If the process is cyclic ($\Delta U_{eng} = 0$),

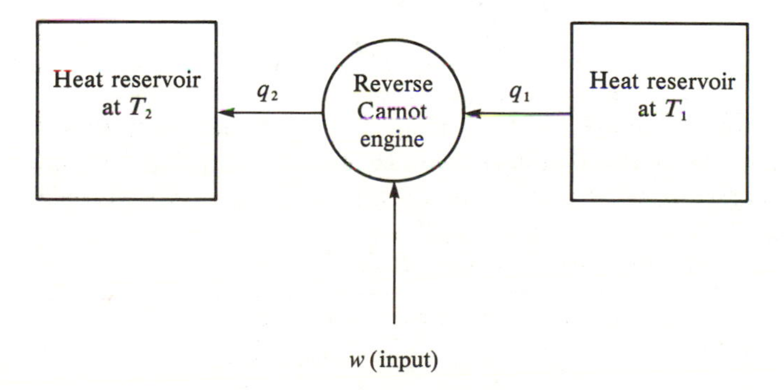

Figure 4-4. Schematic of a Carnot engine used as a refrigerator or a heat pump. Compare with Figure 4-2.

then the amount of energy deposited as heat in the high-temperature reservoir (q_2) will be equal to the amount of heat taken from the low-temperature reservoir (q_1) *plus* an amount of heat equal to the work done on the system.

Just as for a Carnot engine running in the forward direction we have for the engine running in the reverse direction

$$\Delta U_{\text{eng}} = q_1 + q_2 + w = 0$$

The *coefficient of performance* (cop) *of a heat pump is defined as*

$$\text{cop} = \frac{q_1}{w}$$

and thus

$$\text{cop} = \frac{q_1}{w} = -\frac{q_1}{q_1 + q_2}$$

Application of the second law to the cycle ($q_1 \leqslant T_1 \Delta S_1$ and $q_2 \leqslant T_2 \Delta S_2$) yields

$$\frac{q_1}{w} \leqslant -\frac{T_1 \Delta S_1}{T_1 \Delta S_1 + T_2 \Delta S_2}$$

but the process is cyclic and therefore $\Delta S_1 = -\Delta S_2$; hence

$$\text{cop} = \frac{q_1}{w} \leqslant \frac{T_1}{T_2 - T_1} \tag{4-43}$$

Note that q_1/w can be greater than unity, and q_1/w is greater the smaller the temperature difference, $T_2 - T_1$. In other words, the work required to pump a given amount of energy as heat out of your refrigerator box is greater the greater the difference in temperature between the refrigerator and the room into which the heat is pumped. It costs more to run your refrigerator in the summer.

4-10 *The Second-Law Efficiency Rates the Performance of a Device Relative to the Ideal*

A primary consideration in the thermodynamic analysis of energy utilization is the *efficiency* with which the energy is utilized. Unfortunately, there are several definitions of efficiency in use and this situation can lead to ambiguities, unless the definition in use is stated clearly. The definition of efficiency can be based on either the First Law or the Second Law of Thermodynamics.

The first-law efficiency ε_1 is defined as

$$\varepsilon_1 = \frac{\text{desired energy transfer achieved by the device}}{\text{energy input to the device}} \tag{4-44}$$

If the maximum possible value of $\varepsilon_1 = 1$, then ε_1 is called an *efficiency*; whereas if necessarily $\varepsilon_1 > 1$, then ε_1 is referred to as a *coefficient of performance*.

The second-law efficiency ε_2 is defined as

$$\varepsilon_2 = \frac{\text{minimum amount of energy required to perform a given task}}{\text{amount of energy actually used to perform the task}} < 1 \tag{4-45}$$

The value of ε_2 cannot exceed unity. The numerator in the definition of ε_2 is a *task minimum*, not a device minimum. The second-law efficiency rates the performance of a device relative to the ideal, and thus ε_2 is more useful than ε_1 in assessing the possibility for reducing the amount of energy used to perform a given task. The first-law efficiency measures how much of the energy used actually reaches where it is supposed to go; it ignores alternative ways to accomplish the same task and it is concerned only with how well a particular device works. The second-law efficiency assesses the task and compares the minimum amount of energy required to accomplish the task with the amount of energy actually used to accomplish the task.

A typical furnace providing hot air at 100°F (316 K) to a house when the outside air is 32°F (273 K) has a first-law efficiency of about 0.60, because about 40% of the heat is vented to the outdoors. The second-law efficiency is

$$\varepsilon_2 = \frac{q_2(T_2 - T_1)/T_2}{q} = \frac{(0.60)(316 - 273)}{316} = 0.082$$

where the numerator $q_2(T_2 - T_1)/T_2$ was obtained from Eq. 4-42 for $-w$, and where $q_2/q = 0.60$ (given). A well-designed space heater has $\varepsilon_1 \approx 0.75$ (25% more first-law efficient than the $q_2/q = 0.60$ case), because of the reduced amount of heat that is vented. Burning wood in a typical, decorative home fireplace to provide home heat has an $\varepsilon_1 \approx 0.2$; that is, 80% of the heated air goes up the chimney. Woodburning stoves have ε_1 values as high as 0.85.

The conversion of electricity to heat by resistance heating (I^2Rt or Joule heat) has a first-law efficiency of essentially 100%. The second-law efficiency for the 110°F/32°F temperature case is

$$\varepsilon_2 = \frac{1.0(316 - 273)}{316} = 0.14$$

If the electricity is generated in a power plant using fossil fuels, then the second-law efficiency of the conversion of chemical energy (fuel) to electricity at the plant is typically about 0.33, and therefore the *overall* second-law efficiency is $0.14 \times 0.33 = 0.046$. Actually the overall efficiency is even less than 0.046, because of energy losses in the transmission of the electricity from the power plant to the home. If the electricity is generated in a hydroelectric plant, then the value of ε_2 at the power plant is about 0.95 to 0.99, and the overall value of ε_2 is about 0.14 (neglecting transmission losses).

The use of electricity for space heating by passing current through resistors is much less efficient on a second-law basis than using the electricity to run a heat pump, provided the temperature of the low-temperature reservoir (T_1) is not too low relative to the desired room temperature (T_2). The value of T_1 can be held at, say, 285 K (12°C) by locating the low-temperature heat exchanger in a well. The minimum amount of energy w that must be expended to "pump" an amount of heat q_1 from $T_1 = 285$ K to $T_2 = 293$ K (68°F) is given by Eq. 4-43 as

$$w = \frac{q_1(T_2 - T_1)}{T_1} = 0.028q_1$$

The total amount of energy transferred as heat to the home is $w + q_1$, and equating this to the production of heat via resistance heating gives

$$w + q_1 = I^2Rt$$

or

$$w + \frac{w}{0.028} = I^2Rt$$

and thus

$$w = 0.027I^2Rt$$

where I is the current, R is the resistance, and t is the time. Even if the heat pump was only 50% efficient (actual value of w twice the above value), the amount of energy required by the heat pump would only be about 6% of that required by resistance heating. If the heat pump works directly off the outside air (rather than a well) and the outside air is 0°C, then the above 6% figure is increased to 14%. Although the possible energy savings in a particular case obtained by using heat pumps rather than resistance heaters are significant (space heating accounts for 68% of the total average residential energy use), the total possible savings are not large, because most homes are not electrically heated. Also, electrical heating has an advantage over central-forced-air heating in that individual rooms can be thermostated at different temperatures and the heating can be concentrated where it is most needed at any time.

A modern fossil-fuel power plant using steam turbines operates between 811 K (1000°F) (steam from the boiler) and 311 K (100°F) (ejection of heat into the cooling water). The maximum possible second-law efficiency ε_2 is

$$\varepsilon_2 = \frac{-w}{q_2} = \frac{T_2 - T_1}{T_2} = \frac{811 \text{ K} - 311 \text{ K}}{811 \text{ K}} = 0.62$$

However, the actual operational efficiency is only about 0.41 owing to irreversibility. The power output of a heat engine is given by

$$\text{power output} = p = \frac{-w}{\tau} \tag{4-46}$$

where $-w$ is the work done in one complete cycle of duration τ. For reversible operation, τ is infinite and p is zero. The major limiting factor on the value of p is the rate at which heat is transferred from the heat source (T_2) to the engine and from the engine to the heat sink (T_1). Frictional losses are of less importance.

To transfer a finite amount of energy as heat in a finite time requires a finite temperature difference. That is, at the high temperature T_2 where heat is absorbed, the temperature of the engine working fluid T'_2 must be less than T_2; whereas at the low temperature where heat is ejected, the temperature of the engine working fluid T'_1 must be greater than T_1. The transfer of energy as heat down a temperature gradient is an irreversible process that leads to entropy production and a lowering of the amount of work obtainable. The rate of nonradiative heat transfer is proportional to the temperature difference between the two bodies (Newton's law of cooling):

$$-\frac{dT}{dt} = k(T - T_0) \qquad T > T_0 \tag{4-47}$$

In such a case the *engine efficiency for maximum power output* $\varepsilon_2(p \max)$ can be shown to be given by (J. Ross)

$$\varepsilon_2(p \max) = 1 - \left(\frac{T_1}{T_2}\right)^{1/2} \tag{4-48}$$

For the 811 K/311 K power plant, $\varepsilon_2(p \max) = 0.38$. The practical operation of a power plant involves a compromise between maximum power and maximum efficiency.

4-11 *A Reversible Isothermal Change of State Always Involves the Transfer of Energy as Heat*

For any isothermal reversible process we have from the second law

$$q_{\text{sys}} = T\Delta S_{\text{sys}}$$

Because S is a thermodynamic state function, $S = S(T, P, \ldots)$; if we have a change of state with T constant, then $\Delta S_{\text{sys}} \neq 0$. These considerations enable us to define *absolute zero* on the Kelvin temperature scale: *If any system undergoes a reversible isothermal change of state without heat transfer, then the temperature at which this process occurs is called "absolute zero."*

It appears that the equation

$$\frac{-w}{q_2} = \frac{T_2 - T_1}{T_2}$$

implies that when $T_1 = 0$, $-w = q_2$, which contradicts Kelvin's statement of the second law, because it implies the conversion of a heat input completely into work in a cyclic process. For the complete Carnot cycle,

$$\Delta S_{\text{tot}} = \Delta S_1 + \Delta S_2 = \frac{q_1}{T_1} + \frac{q_2}{T_2}$$

but if $q_1 = 0$ and $T_1 = 0$, then $\Delta S_1 = 0/0$, which is undefined. The only way out of this dilemma involves the introduction of the postulate that absolute zero is unattainable. In other words, when any system undergoes a reversible isothermal change of state, energy must be transferred as heat between the system and the surroundings. We shall return to a further consideration of this matter when we discuss the Third Law of Thermodynamics in Chapter 6.

When a system with two independent thermodynamic coordinates undergoes a reversible adiabatic change of state, then we can specify one thermodynamic coordinate for the final state but not two, because $\Delta S = 0 = S_2 - S_1$; that is, S is fixed. Thus a reversible adiabatic process cannot also be an isothermal process, because such a process fixes both T and S and therefore is impossible.

4-12 *Entropy Is a Direct Measure of the Quality of an Energy Source*

The quality of an energy source can be characterized by the ratio of the entropy change to the energy change $\Delta S/\Delta U$ when energy is extracted from the source. The lower the value of $\Delta S/\Delta U$, the higher the quality of the energy. The flow of energy in a spontaneous process is always such that the total entropy increases while the total energy remains unchanged. The utilization of energy inevitably leads to a lowering of the quality of the energy that is used. The energy, although conserved, is thus said to be *degraded* (lowered in quality). Energy degradation is a fundamental aspect of nature that cannot be circumvented, and energy that has been degraded has a lowered capacity for utilization in the performance of tasks. The lower the value of $\Delta S/\Delta U$ of an energy source, the greater the fraction of the energy extracted from the energy source that can be converted to work.

The entropy produced in the transfer of a given amount of energy *as heat* varies inversely with the temperature of the energy source ($dS \geqslant \delta q/T$). The higher the temperature, the lower the amount of entropy produced for the transfer of a given amount of energy as heat. Thus sunlight with an effective temperature of 6000 K is a higher-quality energy source than most chemical reactions, because almost all chemical reactions occur with an effective temperature of less than 6000 K. For example, methane burns in oxygen with a flame temperature of less than 2300 K.

The highest possible quality energy source is one from which the energy can be transferred directly *as work.* The primary example is gravitational energy. For example, falling water can be used directly to produce mechanical work without the need for a heat engine. Table 4-1 gives the $\Delta S/\Delta U$ values for several major types of energy sources. The natural flow of energy in the universe is from forms of higher quality to forms of lower quality. Cosmic microwave radiation with an effective temperature of about 2.9 K is apparently the ultimate energy sink in the universe. No way is known in which this form of energy can be degraded further.

TABLE 4-1
Quality of Energy Sources[a]

Form of energy	*Associated temperature*	*Entropy per unit energy, $\Delta S/\Delta U$*
Gravitation	—	0
Nuclear reactions	10^{10} K	10^{-6}
Internal energy of stars	10^{7} K	10^{-3}
Sunlight	6000 K	1
Chemical reactions	600 K–6000 K	1–10
Terrestrial waste heat	300 K–600 K	10–100
Cosmic microwave radiation	~3 K	10^{4}

[a] Adapted from F. Dyson, 1971.

4-13 *Entropy and Probability*

The discussion of entropy that we have presented up to this juncture is primarily classical, that is, nonmolecular and nonquantum mechanical. The fundamental link between the macroscopic world of classical thermodynamics and the microscopic world of quantum mechanics was discovered by the Austrian physicist Ludwig Boltzmann, on whose gravestone in a Vienna cemetery is inscribed the following epitaph:

$$S = k \ln W \tag{4-49}$$

In the Boltzmann equation, Eq. 4-49, S is the entropy; k is Boltzmann's constant, that is, the gas constant per molecule,

$$k = \frac{R}{N_0} = \frac{8.314\ \mathrm{J \cdot K^{-1} \cdot mol^{-1}}}{6.022 \times 10^{23}\ \mathrm{mol^{-1}}} = 1.381 \times 10^{-23}\ \mathrm{J \cdot K^{-1}}$$

and W is the probability of a thermodynamic state. More precisely, W is the number of quantum states that correspond to a given thermodynamic state. In Chapter 14 we shall describe how to compute W for some systems. However, we note here that W is proportional to the probability of observing a given thermodynamic state or, in other words, that entropy and probability are interrelated. As a simple example, consider N molecules of a gas that are distributed between two chambers of equal volume, which are connected via a stopcock (Figure 4-5). What is the probability that all the gas molecules will be found in only one of the two chambers when the stopcock is open? For one molecule, the probability that the molecule will be found in the left chamber is $\frac{1}{2}$. The probability that 100 molecules all will be found in the left chamber

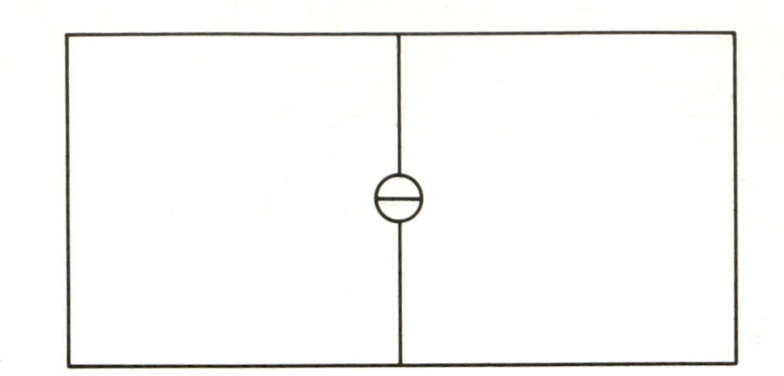

Figure 4-5. Two chambers of equal volume connected by a stopcock.

is $(\frac{1}{2})^{100}$ or about 1 in 10^{30}. The probability that N molecules will all be found in the left chamber is $(\frac{1}{2})^N$. When N is of the order of Avogadro's number (6.02×10^{23}), $(\frac{1}{2})^N$ is an incredibly small number. It is not difficult to show in the general case that if the volume of the left chamber is V_1 and if the total volume of the two chambers combined is V_2, then the probability of finding all N molecules in the left chamber is $(V_1/V_2)^N$. When the stopcock is open, the probability of finding the molecules in the total volume V_2 is unity, provided there are no leaks. Thus the change in entropy for the process in which all the molecules are initially in the left chamber of volume V_1 and then the stopcock is opened (thereby allowing the molecules access to the entire volume V_2) is

$$\Delta S = S_2 - S_1 = k \ln W_2 - k \ln W_1$$

and thus

$$\Delta S = k \ln 1 - k \ln \left(\frac{V_1}{V_2}\right)^N$$

or

$$\Delta S = Nk \ln \left(\frac{V_2}{V_1}\right)$$

If $N = N_0$ (Avogadro's number), then we have ($N_0 k = R$):

$$\Delta S = R \ln \left(\frac{V_2}{V_1}\right) \qquad (4\text{-}50)$$

Equation 4-50 is the same result that we obtained for the isothermal expansion of an ideal gas (see Eq. 4-21). There is, however, a crucial difference between the statistical thermodynamic (i.e., probabilistic) interpretation of ΔS values and the second-law interpretation of ΔS values. The difference is that the spontaneous process

$$\text{gas}(T, V_2) \longrightarrow \text{gas}(T, V_1) \qquad V_1 < V_2$$

for which $\Delta S_{tot} < 0$ is not ruled out but rather is only highly improbable. The probability of observing such a process is given by

$$W = \left(\frac{V_1}{V_2}\right)^N = e^{\Delta S/k} \tag{4-51}$$

In statistical thermodynamic terms, the impossibility of observing a spontaneous process with $\Delta S_{tot} < 0$ is reduced to an improbability. For N of the order of Avogadro's number, ΔS is of the order of $-R$, and thus

$$W = \left(\frac{V_1}{V_2}\right)^N \approx e^{-R/k} = e^{-N} = e^{-6.02 \times 10^{23}}$$

which is a vanishingly small number. Thus it is *extremely unlikely* that, with the stopcock open, all the molecules will be found in the left chamber.

Thermodynamic states of high probability have a high entropy; thermodynamic states of low probability have a low entropy. The energy levels of molecules are quantized and thus there are, in general, many different ways in which the available energy can be distributed among the accessible energy levels of the molecules. The value of W for a system is, in effect, equal to the total number of ways in which the available energy can be distributed among the accessible energy levels of the molecules. The amount of energy to be distributed among the energy levels and the number of accessible energy levels decreases with decreasing temperature, and therefore the entropy also decreases with decreasing temperature. In the special case for which $W = 1$, that is, the case in which there is only one way in which the energy can be distributed, we have

$$S = k \ln W = k \ln(1) = 0$$

As we shall see in Chapter 6, $S = 0$, and therefore $W = 1$ at the absolute zero of temperature. At the absolute zero of temperature, all the molecules are in the lowest possible quantum state; there is no ambiguity about how the energy is distributed and thus there is no entropy at absolute zero.

Problems

1. For $CH_4(g)$ at 1 atm,

$$\bar{C}_P = 23.64 + 4.79 \times 10^{-2}T - 1.93 \times 10^{-5}T^2 \qquad (\text{J}\cdot\text{K}^{-1}\cdot\text{mol}^{-1})$$

over the range 298 K to 2000 K. Compute ΔS for 3.00 mol of $CH_4(g)$ that is heated from 25.0°C to 825°C at 1 atm.

2. The molar enthalpy change on sublimation for dry ice is

$$CO_2(s) \longrightarrow CO_2(g) \qquad \Delta\bar{H}_{sub} = 25.2 \text{ kJ}\cdot\text{mol}^{-1}$$

The normal (1 atm) sublimation temperature of dry ice is 195 K. Compute q_P and ΔS for CO_2 when 1.00 kg of $CO_2(s)$ is sublimed at 195 K.

3. Compute the change in entropy of water when 55.5 mol of water is converted from ice at $-10°C$ and 1 atm to water vapor at 110°C and 1 atm.

For $H_2O(s)$: $\Delta\bar{H}_{fus} = 6.01\ kJ\cdot mol^{-1}$ $\quad T_{fus} = 273.15\ K$
$\bar{C}_P = 37.66\ J\cdot K^{-1}\cdot mol^{-1}$
For $H_2O(\ell)$: $\Delta\bar{H}_{vap} = 44.0\ kJ\cdot mol^{-1}$ $\quad T_{vap} = 373.15\ K$
$\bar{C}_P = 75.32\ J\cdot K^{-1}\cdot mol^{-1}$
For $H_2O(g)$: $\bar{C}_P = 30.54 + 1.03 \times 10^{-2}T \quad (J\cdot K^{-1}\cdot mol^{-1})$

4. A 100-Ω resistor of 2.0-g mass with a $C_P = 7.55\ J\cdot K^{-1}\cdot g^{-1}$ is placed in an ice-water mixture of 0°C in a Dewar flask and a current of 300 mA is passed through the resistor for 2.00 min.
(a) Compute the entropy change of the resistor.
(b) Compute the entropy change of the ice-water bath.
(c) Compute the number of grams of ice that is melted in the process.
(d) Compute the volume change of the ice-water mixture. (The densities of ice and water are $0.917\ g\cdot cm^{-3}$ and $1.000\ g\cdot cm^{-3}$, respectively.)
(e) Suppose the resistor is thermally insulated and the experiment is repeated. Compute ΔS for the resistor and ΔS for the surroundings.

5. Explain why it is more expensive to operate a given refrigerator in a room at 100°F than in a room at 75°F.

6. The molar heat capacity at constant pressure of $H_2O(g)$ over the temperature range 100°C to 200°C is given by

$$\bar{C}_P = 30.54 + 1.03 \times 10^{-2}T \qquad (J\cdot K^{-1}\cdot mol^{-1})$$

Compute ΔS, ΔH, and ΔU when 100 g of $H_2O(g)$ at 100°C, 1.00 atm is heated to 200°C, 1.00 atm. Assume ideal gas behavior.

7. Can you devise an isothermal process for which (a) $\Delta S_{sys} < 0$ or (b) $\Delta S_{sys} < q_{sys}/T$? Explain.

8. Air conditioners for home, office, and industrial use are rated in *tons*. A 1.00-ton air conditioner is capable of removing 12,000 Btu/hr (1 Btu = 1.05 kJ). Given that the heat of fusion of ice is $6.01\ kJ\cdot mol^{-1}$, compute the number of pounds of water at 0°C that can be converted to ice by a 1.00-ton unit operating continuously for 24 hr.

9. From a practical (i.e., operational) point of view, does the concept of thermodynamic equilibrium require that the system be at equilibrium with respect to all *possible* processes? Explain.

10. Does the second law *require* that all naturally occurring processes be irreversible? Explain.

11. Is it possible for a system to have a negative heat capacity? Explain.

12. Is it possible to deduce unequivocally from the second law that the universe is "running down" owing to the continual occurrence of spontaneous processes (all the way from nuclear disintegration to supernovae explosions) in the universe? Is the universe an isolated system?

13. Discuss the connection between entropy and time. Can the second law be used to decide which of two events preceded the other? If so, under what conditions? Does the second law require that the universe must have had a beginning?

14. Suppose a low-boiling liquid like ammonia is placed in a heat reservoir and part of the energy of the surroundings is used to vaporize the liquid. The adiabatically expanding gas does work by driving a piston and during the expansion the gas cools until it condenses to a liquid, the liquid is once again vaporized using energy available from the surroundings, and the cycle is repeated. Can such an engine be made operational?

15. Suppose we have a wire loop that is a superconductor (specific resistance of zero ohm centimeters, $\Omega \cdot \text{cm}$). If such a loop is subjected to a magnetic field, a current is induced that in a superconductor will evidently flow indefinitely. Does such a system constitute a perpetual motion machine? Does it violate the second law? Explain.

16. Prove, using the second law, that heat always flows spontaneously from a higher to a lower temperature and never the reverse.

17. Prove that a hypothetical engine that is more efficient than a Carnot engine would violate the laws of thermodynamics. (*Hint:* Use the hypothetical engine to run a Carnot engine in reverse.)

18. The manufacturers of air conditioners have defined an energy efficiency rating (EER) for air-conditioning units. The EER, which is actually a coefficient of performance in hybrid units, is defined as follows:

$$\text{EER} = \frac{\text{cooling capacity in Btu per hour}}{\text{power used in watts}}$$

For example, if the unit can remove 30,000 $\text{Btu} \cdot \text{hr}^{-1}$ with an energy input of 3750 W, then

$$\text{EER} = \frac{30{,}000 \text{ Btu} \cdot \text{hr}^{-1}}{3750 \text{ W}} = 8.0$$

where the EER is stated without units (actually British thermal units per hour per watt, $\text{Btu} \cdot \text{hr}^{-1} \cdot \text{W}^{-1}$). An EER of 8 to 10 is rated as very good, more than 10 is excellent, and less than 7 is poor. Calculate the second-law efficiency of an air conditioner with an EER = 8.0, which is used to maintain an inside temperature of 78°F when the outside air temperature is 95°F.

19. Conventional modern fossil-fuel power plants operate their steam turbines between about 810 K (steam from boiler) and 310 K (cooling water). Nuclear power plants cannot be run at as high a temperature as fossil-fuel plants because of possible thermal damage to the fuel rod assemblies. A typical nuclear power plant operates between 620 K and 310 K. Compare the efficiencies of the heat engine portions of the two types of power plants. Other factors that lower the overall efficiencies of power plants are as follows:

	Efficiency factor
Steam cycle temperature variations	0.85
Turbine efficiency	0.90
Boiler losses	0.88
Generator	0.99

Compute the overall efficiency of the two types of power plants and also compare the overall efficiencies with the efficiency calculated from the expression

$$\varepsilon_2(p\ \max) = 1 - \left(\frac{T_1}{T_2}\right)^{1/2}$$

20. Consider two identical Hooke's law springs, one at length L_0 (and, therefore, not under tension) and the other at length $3L_0$. If these two springs are separately dissolved in equal volumes of an aqueous acid solution at T, then will there be any difference in the heats of solution for the two springs and, if so, by how much will they differ?

21. Is it possible for a system to output an amount of energy as work that is equal to the amount of energy taken in as heat by the system in the process? Explain by specific example(s).

22. The following "capillary engine" has been proposed. A bundle of capillary tubes in which a certain liquid rises spontaneously to a height h are cut off at a height less than h and the ends are bent over (like a cane). It is now proposed to use the liquid issuing from the tube ends, when the straight ends of the tubes are placed in the liquid, to turn a paddle wheel. The liquid falling off the paddle wheel is returned to the main liquid reservoir. Can such an engine be made operational? If in doubt, then try it out!

23. Suppose a 4×10^3 ton meteorite strikes the earth and is buried beneath the surface. If the temperature of the meteorite is 2000°C and its heat capacity at constant pressure is $(0.42 + 4.2 \times 10^{-5}T)\ \mathrm{J \cdot K^{-1} \cdot g^{-1}}$, compute the maximum work obtainable using the meteorite as a high-temperature reservoir and the atmospheric surroundings (30°C) as a low-temperature reservoir. (*Hint:* Use the expression

$$\frac{-\delta w}{\delta q_2} = \frac{T_2 - T_1}{T_2}$$

which is the differential form of Eq. 4-40.)

24. The internal energy of 1 mol of a van der Waals gas is given by

$$\bar{U} = \bar{C}_V T - \frac{a}{\bar{V}} + c$$

where $\bar{C}_V$ a and c are constants.

(a) Obtain expressions for $\Delta\bar{U}$ and $\Delta\bar{S}$ for the reversible adiabatic expansion of a van der Waals gas from T_1, V_1 to T_2, V_2.

(b) Show that

$$T(\bar{V} - b)^{R/\bar{C}_V} = \text{constant}$$

for the reversible adiabatic expansion of a van der Waals gas.

25. Show that if a system undergoes an irreversible isothermal process from state a to state b without doing any work, then the total entropy production is given by

$$\Delta S_{\text{tot}} = \frac{w_{ba}}{T}$$

where w_{ba} is the work that must be done in a reversible process that restores the system to its initial state.

26. Calculate the total entropy change when a 2.00-kg mass at 300 K falls freely a distance of 25.0 m into a heat bath at 300 K. Take $g = 9.81\ \mathrm{m \cdot s^{-2}}$.

27. Consider two energy reservoirs initially at T_2 and $T_1 (T_2 > T_1)$ both with a finite constant heat capacity at constant pressure of C_P. Calculate the final common temperature of the two reservoirs for two cases:

(a) The reservoirs are used in conjunction with an engine to provide the maximum work possible.

(b) The reservoirs are connected with a thin copper rod and no work at all is done.

(c) Show that the work done in case (a) is given by

$$-w = C_P[(T_2 + T_1) - 2(T_1 T_2)^{1/2}]$$

(d) Show that the total entropy production in case (b) is given by

$$\Delta S_{\text{tot}} = 2C_P \ln\left[\frac{T_1 + T_2}{2(T_1 T_2)^{1/2}}\right] > 0$$

(e) What is ΔS_{tot} equal to for case (a)?

28. Only about 4% of the energy input to an incandescent light bulb is converted to light, whereas about 20% of the energy input to a fluorescent light bulb is converted to light, the rest of the energy is transferred as heat to the surrounding air. The light energy is ultimately absorbed by the walls, furnishings, etc. Compute the entropy production per second for an energized 40-W incandescent bulb (80°C) and a 40-W fluorescent bulb (40°C). Take the air temperature as 15°C.

29. Make a sketch of the various components of a typical home refrigerator unit. Explain the function of each component and trace the flow of the refrigerant through the unit.

30. Devise a heat engine that runs on solar energy.

31. Prove that a reversible adiabatic process cannot also be isothermal for $T > 0$.

32. Pure water that is free of dust can be supercooled, that is, cooled below the freezing point. Under carefully controlled conditions water can be supercooled to -45°C. Consider an adiabatically isolated 100-g sample of supercooled water at -10°C that spontaneously begins to freeze. Ice is more ordered than liquid water at the same temperature and thus ice has a lower entropy than liquid water at the same temperature. How do you reconcile the above spontaneous process with the second law? [*Hint:* What is the final temperature after $H_2O(s)$ formation ceases? What is the composition of the system at equilibrium assuming a negligible vapor space?] Calculate the total entropy change for the process. See Problem 3 for the necessary data.

33. The *Clausius inequality* is usually stated in the form $\oint \delta q_{\text{sys}}/T \leqslant 0$ wherein the equality sign is said to hold only if the cyclic process is reversible. Consider a system that is placed in an adiabatic enclosure at temperature T_1 with two heat baths at temperatures T_2 and T_3 ($T_3 > T_1 > T_2$). The system is brought in contact with the heat bath at T_2 until its temperature becomes equal to T_2 (the system volume being held constant) and then the system is removed from contact with the heat bath at T_2 and placed in contact with the heat bath at T_3 until its temperature reaches T_1 (the system volume again being held constant). First calculate the entropy changes in the system and the heat baths for the processes described. Then derive the Clausius inequality. How do you reconcile your results with the above form of the Clausius inequality?

How a Steam Electric Power Plant Works

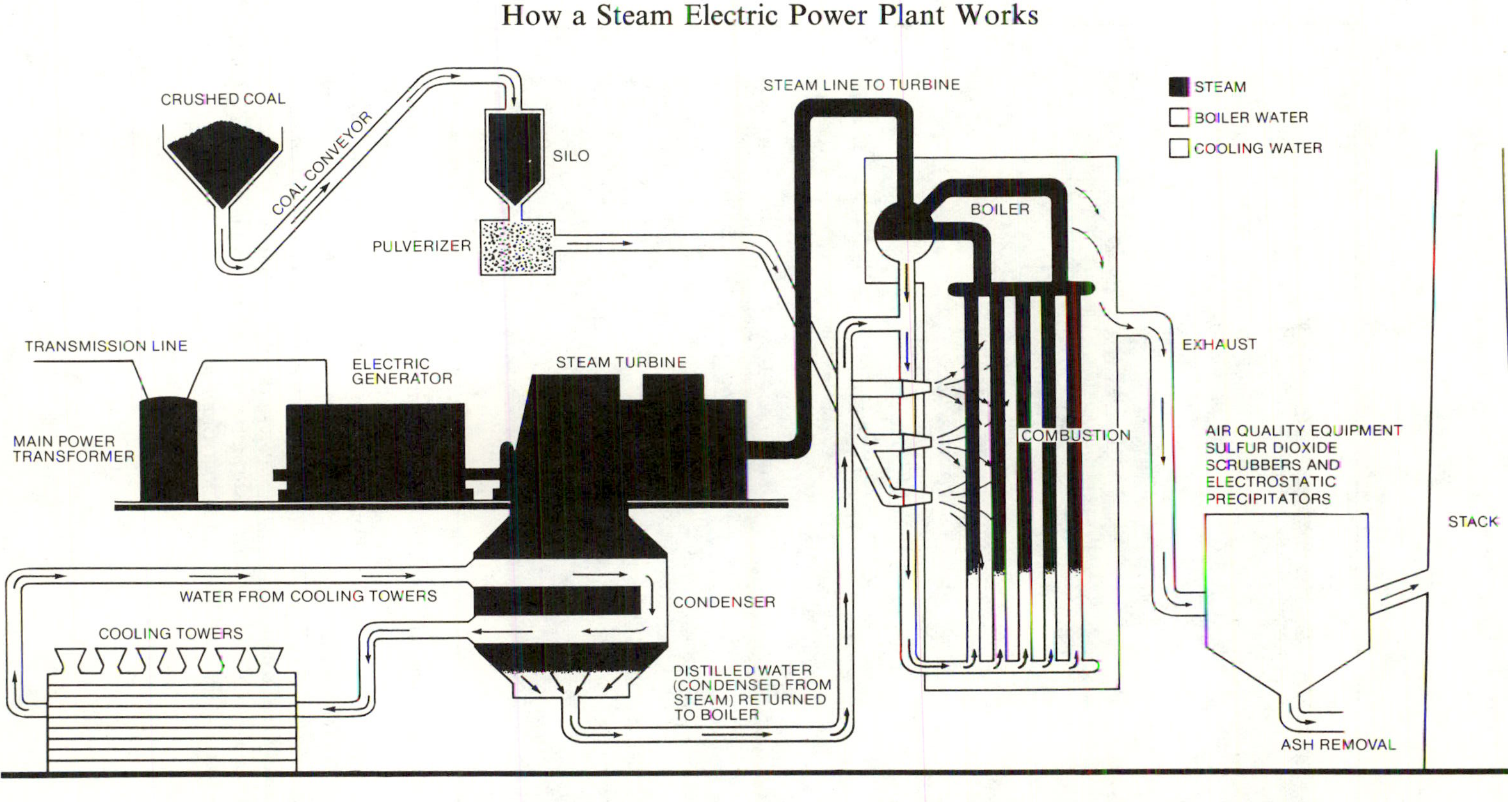

A schematic diagram for the operation of the coal-fired power plant shown in the photograph on the following page. The coal is washed to remove dirt, crushed, and then pulverized to a talcumlike powder. The powdered coal is blown through 64 large nozzles into large boilers where the coal is consumed in two mammoth fireballs. The boilers are each equal in size to a 16-story building. The fireball turns the water running through 162 miles of pipes into steam at 540°C with a pressure of 2400 psi (163 atm). The high-pressure steam turns the two 665,000-kW generators at 3600 rpm. The electricity leaves the plant through 500,000-V transmission lines (Data and figure courtesy of Pacific Power and Light Company)

This coal-fired power plant in Centralia, Washington, has two 665-megawatt units. Low sulfur (0.5 to 0.75%) sub-bituminous coal (7500 to 8100 Btu per pound) is the primary fuel for the plant. There are an estimated 500 million tons of coal in the nearby Centralia Deposit, where the coal seams are 8 to 50 feet thick. Each boiler unit can produce 5.27 million pounds of 2400-psi steam per hour with a total coal input of 15 tons per minute. The plant uses about 4.8 million tons of coal a year. The plant chimneys are 470 feet in height. Electrostatic precipitators remove 99.8% of the particulate matter (fly ash) from the flue gases. (Data and photo courtesy of Pacific Power and Light Company)

5

Thermodynamic Functions

Mathematics is a language.

J. Willard Gibbs

In doing a problem involving a given mass of some substance, the condition of the substance at any moment can be described by telling what its temperature is and what its volume is. If we know the temperature and volume of a substance, and that the pressure is some function of the temperature and the volume, then we know the internal energy. One could say, "I do not want to do it that way. Tell me the temperature and the pressure and I will tell you the volume. I can think of the volume as a function of temperature and pressure, and the internal energy as a function of temperature and pressure, and so on." That is why thermodynamics is hard, because everyone uses a different approach. If we could only sit down once and decide on our variables, and stick to them, it would be fairly easy.

R. P. Feynman, R. B. Leighton, and M. Sands. *The Feynman Lectures on Physics*, Vol. I. Reading, Mass.: Addison-Wesley Publishing Company, Inc., 1965.

5-1 *Partial Derivatives*

The functional relation

$$U = U(S, V) \tag{5-1}$$

which expresses the fact that the value of U is determined by the values of the *independent* variables S and V, may be represented by a surface in three-dimensional space (Figure 5-1). If one of the independent variables is restricted to a constant value, then U can be considered as a function of only the other variable. In such a case *partial derivatives* can be defined. We define the partial derivative of the internal energy U with respect to the entropy S at constant volume V as

$$\left(\frac{\partial U}{\partial S}\right)_V \equiv \lim_{\Delta S \to 0} \left[\frac{U(S + \Delta S, V) - U(S, V)}{\Delta S}\right] = f(S, V) \tag{5-2}$$

The function $(\partial U/\partial S)_V$ is called the *partial of U with respect to S at constant V.* Note that $(\partial U/\partial S)_V$ is, in general, also a function of S and V; this is noted in Eq. 5-2 by the function $f(S, V)$. We define the partial derivative of U with respect to V at constant S as

$$\left(\frac{\partial U}{\partial V}\right)_S \equiv \lim_{\Delta V \to 0} \left[\frac{U(S, V + \Delta V) - U(S, V)}{\Delta V}\right] = g(S, V) \tag{5-3}$$

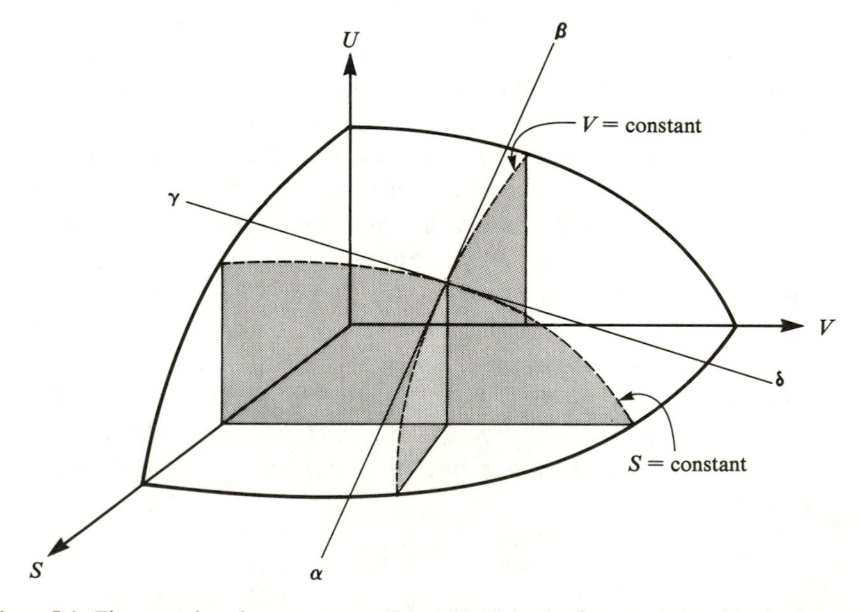

Figure 5-1. The curved surface represents $U = U(S, V)$ in the first quadrant. The horizontally and vertically lined planes represent particular values of V = constant and S = constant, respectively. The lines $\alpha\beta$ and $\gamma\varepsilon$ are tangents to the intersection curves. At the intersection of $\alpha\beta$ and $\gamma\varepsilon$, the partial derivative $(\partial U/\partial S)_V$ is equal to the slope of $\alpha\beta$, and $(\partial U/\partial V)_S$ is equal to the slope of $\gamma\varepsilon$ (adapted from Blinder, 1966).

A partial derivative, just like an ordinary derivative, is the slope of a curve. Thus, the partial derivative $(\partial U/\partial S)_V$ is the slope of the curve defined by the intersection of a V = constant plane with the $U = U(S, V)$ surface (Figure 5-1). Similarly, the partial derivative $(\partial U/\partial V)_S$ is the slope of the curve defined by the intersection of an S = constant plane with the $U(S, V)$ surface.

Partial derivatives are obtained by the usual rules of differentiation except that the variable(s) held constant are treated as constants in the differentiation. For example, the partial derivative $(\partial P/\partial T)_V$ for a gas with the equation of state (van der Waals gas)

$$P = \frac{nRT}{V - nb} - \frac{n^2 a}{V^2}$$ (a, b, and R are constants; and n, the moles of gas, is constant for a fixed mass of the gas)

is

$$\left(\frac{\partial P}{\partial T}\right)_V = \frac{nR}{V - nb}$$

where we differentiated P with respect to T holding V constant. Note that the second term on the right-hand side of the equation of state makes no contribution to $(\partial P/\partial T)_V$ because n, a, and V are all constants in the differentiation.

EXAMPLE 5-1

Obtain the partial derivative $(\partial P/\partial V)_T$ for a van der Waals gas.

Solution: From the van der Waals equation (see above) we have

$$\left(\frac{\partial P}{\partial V}\right)_T = \frac{(V - nb)(0) - nRT(1)}{(V - nb)^2} - \frac{[V^2(0) - n^2a(2V)]}{V^4}$$

or

$$\left(\frac{\partial P}{\partial V}\right)_T = -\frac{nRT}{(V - nb)^2} + \frac{2n^2 a}{V^3}$$

Note that $(\partial P/\partial V)_T$ is a function of V and T.

5-2 *Exact Differentials*

Consider the differential quantity

$$M(X, Y)\,dX + N(X, Y)\,dY$$

We say that $M(X, Y)\,dX + N(X, Y)\,dY$ is an *exact differential* if, and only if, there exists a function $F(X, Y)$ such that

$$dF = M(X, Y)\,dX + N(X, Y)\,dY$$

Because it is true for any well-behaved function $F(X, Y)$ that

$$dF = \left(\frac{\partial F}{\partial X}\right)_Y dX + \left(\frac{\partial F}{\partial Y}\right)_X dY \tag{5-4}$$

we see that $M(X, Y)\,dX + N(X, Y)\,dY$ is an exact differential if $M(X, Y)$ and $N(X, Y)$ are derivatives of the same function $F(X, Y)$:

$$M(X, Y) = \left(\frac{\partial F}{\partial X}\right)_Y \qquad N(X, Y) = \left(\frac{\partial F}{\partial Y}\right)_X \tag{5-5}$$

If there is no function $F(X, Y)$ such that Eqs. 5-5 are satisfied, then $M(X, Y)\,dX + N(X, Y)\,dY$ is called an *inexact differential.*

To test if a differential quantity is exact, we use the fact that

$$\left[\frac{\partial}{\partial Y}\left(\frac{\partial F}{\partial X}\right)_Y\right]_X = \left[\frac{\partial}{\partial X}\left(\frac{\partial F}{\partial Y}\right)_X\right]_Y$$

provided that all the indicated partial derivatives are continuous. If we apply this condition to Eqs. 5-5, then we see that $M(X, Y)\,dX + N(X, Y)\,dY$ is exact if, and only if,

$$\left(\frac{\partial M}{\partial Y}\right)_X = \left(\frac{\partial N}{\partial X}\right)_Y \tag{5-6}$$

Equation 5-6 is both a necessary and a sufficient condition that $M(X, Y)\,dX + N(X, Y)\,dY$ be exact. In other words, *if a differential is exact, then the second-cross-partial derivatives* (Eq. 5-6) *are equal.* If the second-cross-partial derivatives are equal, then the differential is exact. Equations of the type of Eq. 5-6 involving thermodynamic state functions often are called *Maxwell relations* in thermodynamics. If a differential is exact, then the integral of the differential depends only on the endpoints of the integration.

EXAMPLE 5-2

Determine if the following differentials are exact or inexact.

(a) $(3X + Y^2)\,dX + 2XY\,dY$

(b) $(X^2 + X + 1)\,dX + (X + Y - 3)\,dY$

Solution:

(a) In this case $M(X, Y) = 3X + Y^2$ and $N(X, Y) = 2XY$. Consequently,

$$\left(\frac{\partial M}{\partial Y}\right)_X = 2Y = \left(\frac{\partial N}{\partial X}\right)_Y$$

and so $(3X + Y^2)\,dX + 2XY\,dY$ is an exact differential.

(b) We have $M(X, Y) = X^2 + X + 1$ and $N(X, Y) = X + Y - 3$, and so

$$\left(\frac{\partial M}{\partial Y}\right)_X = 0 \neq \left(\frac{\partial N}{\partial X}\right)_Y = 1$$

Therefore $(X^2 + X + 1)\,dX + (X + Y - 3)\,dY$ is an inexact differential.

EXAMPLE 5-3

Show that $dw = -P\,dV$ is an inexact differential for 1 mol of an ideal gas undergoing a reversible process.

Solution: Let the work w be a function of T and V so that

$$dw = M(T, V)\,dT + N(T, V)\,dV$$

Comparing this differential to $dw = -P\,dV$, we see that

$$M(T, V) = 0 \qquad N(T, V) = -P = -\frac{RT}{V}$$

and that

$$\left(\frac{\partial M}{\partial V}\right)_T = 0 \neq \left(\frac{\partial N}{\partial T}\right)_V = -\frac{R}{V}$$

Therefore the work in this case is an inexact differential. Exact differentials play a central role in thermodynamics because an integral of an exact differential depends only on the limits of the integral

$$\int_1^2 dF = F_2 - F_1 = \Delta F$$

The integrals of inexact differentials, on the other hand, depend on the path of the integral in addition to the endpoints. Thus we have that

$$\int_1^2 dw \neq w_2 - w_1 = \Delta w$$

Consider the general case of a variation in U produced by *simultaneous* variations in S and V. In such a case, because dU is an exact differential (first law), the total differential dU is given by (compare with Eq. 5-4)

$$dU = \left(\frac{\partial U}{\partial S}\right)_V dS + \left(\frac{\partial U}{\partial V}\right)_S dV \tag{5-7}$$

When a function and its derivatives are single valued, finite, and (at least piecewise) continuous, then the total differential of the function is exact. *All thermodynamic state functions have exact differentials* by definition, and equations analogous to Eq. 5-4 can be written for any thermodynamic state function. For example, if $G = G(T, P)$, where G is a state function, then dG must be an exact differential, and thus from Eq. 5-4

$$dG = \left(\frac{\partial G}{\partial T}\right)_P dT + \left(\frac{\partial G}{\partial P}\right)_T dP \tag{5-8}$$

Further, from Eq. 5-6, we have the equality of the second-cross-partial derivatives in Eq. 5-8:

$$\left[\frac{\partial}{\partial P}\left(\frac{\partial G}{\partial T}\right)_P\right]_T = \left[\frac{\partial}{\partial T}\left(\frac{\partial G}{\partial P}\right)_T\right]_P \tag{5-9}$$

From the first law we have

$$dU = \delta q + \delta w \tag{5-10}$$

For a reversible process involving only pressure-volume work, we have $\delta w = -P\,dV$. From the second law, $\delta q = T\,dS$; thus Eq. 5-10 can be written as

$$dU = T\,dS - P\,dV \tag{5-11}$$

Equation 5-11 is the differential equation that combines the First and Second Laws of Thermodynamics. Comparison of Eq. 5-11 with Eq. 5-7 yields the results

$$T = \left(\frac{\partial U}{\partial S}\right)_V \tag{5-12}$$

and

$$-P = \left(\frac{\partial U}{\partial V}\right)_S \tag{5-13}$$

Note that the thermodynamic state functions T and P can be expressed as partial derivatives of the internal energy function.

Equation 5-11 expresses the change in the internal energy U of a thermodynamic system in terms of simultaneous changes in the entropy and the volume of the system. That is, given U as a function of S and V,

$$U = U(S, V)$$

The differential equation

$$dU = T\,dS - P\,dV$$

governs the simultaneous changes in the variables U, S, and V. Because Eq. 5-11 is obtained directly from the First and Second Laws of Thermodynamics, the variables S and V are said to be the *natural independent* variables for U.

Because U is a state function, if the actual process is irreversible, then Eq. 5-11 can still be used, provided that it is applied to a reversible process between the same initial and final states as in the irreversible process.

EXAMPLE 5-4

Show that

$$dU = C_V\,dT$$

for a constant volume process.

Solution: The thermodynamic variables of interest are U, T, and V. We begin by writing $U = U(T, V)$, and then from Eq. 5-4 we have

$$dU = \left(\frac{\partial U}{\partial T}\right)_V dT + \left(\frac{\partial U}{\partial V}\right)_T dV \tag{5-14}$$

By definition, $C_V = (\partial U/\partial T)_V$; thus we can write Eq. 5-14 as

$$dU = C_V\,dT + \left(\frac{\partial U}{\partial V}\right)_T dV \tag{5-15}$$

If the volume is constant, then $dV = 0$, and thus

$$dU = C_V\,dT$$

5-3 *The Helmholtz Energy A and the Gibbs Energy G Are Thermodynamic State Functions*

We have already seen in Chapter 3 that it is convenient for thermodynamic purposes to define the enthalpy function H as

$$H = U + PV \tag{5-16}$$

Differentiation of Eq. 5-16 yields

$$dH = dU + P\,dV + V\,dP \tag{5-17}$$

Substitution of Eq. 5-11 into Eq. 5-17 yields

$$dH = T\,dS - P\,dV + P\,dV + V\,dP$$

or

$$dH = T\,dS + V\,dP \tag{5-18}$$

From Eq. 5-18 we note that the natural independent variables for the enthalpy are $H = H(S, P)$. Furthermore, because

$$dH = \left(\frac{\partial H}{\partial S}\right)_P dS + \left(\frac{\partial H}{\partial P}\right)_S dP \tag{5-19}$$

we have

$$T = \left(\frac{\partial H}{\partial S}\right)_P \tag{5-20}$$

and

$$V = \left(\frac{\partial H}{\partial P}\right)_S \tag{5-21}$$

EXAMPLE 5-5

Show that

$$\left(\frac{\partial T}{\partial P}\right)_S = \left(\frac{\partial V}{\partial S}\right)_P \tag{5-22}$$

Solution: The enthalpy is a state function and dH is an exact differential; thus the second-cross-partial derivatives obtained from Eq. 5-18 must be equal; that is

$$\left(\frac{\partial T}{\partial P}\right)_S = \left(\frac{\partial V}{\partial S}\right)_P$$

It is convenient for the thermodynamic analysis of processes occurring at constant temperature and constant volume to define a thermodynamic energy state function with T and V as the natural independent variables. The appropriate function is obtained from Eq. 5-11 as follows. We first rewrite Eq. 5-11 as

$$dU - T\,dS = -P\,dV \tag{5-23}$$

Equation 5-23 has V as a natural variable on the right-hand side. If we now add $-S\,dT$ to both sides of Eq. 5-23, then we obtain

$$dU - T\,dS - S\,dT = -S\,dT - P\,dV$$

or

$$d(U - TS) = -S\,dT - P\,dV \tag{5-24}$$

Equation 5-24 shows that the function $U - TS$ has T and V as natural independent variables, and thus the function $U - TS$ is the desired thermodynamic energy state function. We define the *Helmholtz energy function A* as

$$A \equiv U - TS \tag{5-25}$$

From Eq. 5-24 we have

$$dA = -S\,dT - P\,dV \tag{5-26}$$

EXAMPLE 5-6

Show that

$$-S = \left(\frac{\partial A}{\partial T}\right)_V \quad \text{and} \quad -P = \left(\frac{\partial A}{\partial V}\right)_T$$

Solution: The presence of the variables A, T, and V in the partial derivatives suggests that we begin with $A = A(T, V)$, from which we have

$$dA = \left(\frac{\partial A}{\partial T}\right)_V dT + \left(\frac{\partial A}{\partial V}\right)_T dV \tag{5-27}$$

A comparison of Eqs. 5-27 and 5-26 yields

$$-S = \left(\frac{\partial A}{\partial T}\right)_V \quad \text{and} \quad -P = \left(\frac{\partial A}{\partial V}\right)_T \tag{5-28}$$

Because dA is an exact differential, the second-cross-partial derivatives obtained from Eq. 5-26 must be equal; that is,

$$\left[\frac{\partial(-S)}{\partial V}\right]_T = \left[\frac{\partial(-P)}{\partial T}\right]_V$$

or

$$\left(\frac{\partial S}{\partial V}\right)_T = \left(\frac{\partial P}{\partial T}\right)_V \tag{5-29}$$

The Maxwell relation Eq. 5-29 is especially useful for calculating the change in entropy with volume under isothermal conditions, given the equation of state of a substance in terms of P, V, and T. The procedure is to obtain $(\partial P/\partial T)_V$ from the equation of state, substitute the result into Eq. 5-29, and then use the result to obtain ΔS.

EXAMPLE 5-7

Obtain an expression for the change in the entropy of a van der Waals gas that undergoes an isothermal change in volume from V_1 to V_2.

Solution: We have already shown in Section 5-1 that for a van der Waals gas

$$\left(\frac{\partial P}{\partial T}\right)_V = \frac{nR}{V - nb}$$

and thus using Eq. 5-29 we have

$$\left(\frac{\partial S}{\partial V}\right)_T = \frac{nR}{V - nb}$$

The objective is to find ΔS for a process in which V changes while T is held constant. We begin by writing $S = S(T, V)$ and thus

$$dS = \left(\frac{\partial S}{\partial V}\right)_T dV + \left(\frac{\partial S}{\partial T}\right)_V dT = \left(\frac{\partial S}{\partial V}\right)_T dV$$

where the second equality follows because $dT = 0$. We now have

$$dS = \left(\frac{nR}{V - nb}\right) dV$$

which yields on integration

$$\int_{S_1}^{S_2} dS = \int_{V_1}^{V_2} \left(\frac{nR}{V - nb}\right) dV$$

$$\Delta S = nR \ln\left(\frac{V_2 - nb}{V_1 - nb}\right)$$

Note that the result for ΔS of the gas is independent of whether or not the process is reversible because the change in the entropy of the gas is independent of path.

It is convenient for the thermodynamic analysis of processes occurring at constant temperature and constant pressure to define a thermodynamic energy function with T and P as natural independent variables. The appropriate function is obtained from Eq. 5-18 as follows. We first rewrite Eq. 5-18 as

$$dH - T\,dS = V\,dP \tag{5-30}$$

Equation 5-30 has P as a natural variable on the right-hand side. If we now add $-S\,dT$ to both sides of Eq. 5-30, then we obtain

$$dH - T\,dS - S\,dT = -S\,dT + V\,dP$$

or

$$d(H - TS) = -S\,dT + V\,dP \tag{5-31}$$

Equation 5-31 shows that the function $H - TS$ has T and P as natural independent variables, and thus $H - TS$ is the desired thermodynamic energy state function. We define the *Gibbs energy function* G as

$$G \equiv H - TS \tag{5-32}$$

Substitution of Eq. 5-32 into Eq. 5-31 yields

$$dG = -S\,dT + VdP \tag{5-33}$$

From Eq. 5-33 we obtain

$$-S = \left(\frac{\partial G}{\partial T}\right)_P \quad \text{and} \quad V = \left(\frac{\partial G}{\partial P}\right)_T$$

and also the Maxwell relation

$$-\left(\frac{\partial S}{\partial P}\right)_T = \left(\frac{\partial V}{\partial T}\right)_P \tag{5-34}$$

Equation 5-34 is used in a manner analogous to that illustrated for Eq. 5-29.

EXAMPLE 5-8

Obtain an expression for $(\partial S/\partial P)_T$ for a gas with the equation of state

$$P(V - nB) = nRT \qquad (B \text{ is a constant})$$

Solution: From Eq. 5-34 we have

$$\left(\frac{\partial S}{\partial P}\right)_T = -\left(\frac{\partial V}{\partial T}\right)_P$$

We obtain $(\partial V/\partial T)_P$ from the equation of state:

$$\left(\frac{\partial V}{\partial T}\right)_P = \left[\frac{\partial}{\partial T}\left(\frac{nRT}{P} + nB\right)\right]_P = \frac{nR}{P}$$

thus

$$-\left(\frac{\partial S}{\partial P}\right)_T = \frac{nR}{P}$$

EXAMPLE 5-9

A gas with the equation of state

$$PV = n(RT + BP) \qquad B \text{ is a constant}$$

undergoes an isothermal expansion from an initial pressure P_1 to a final pressure P_2. Obtain expressions for ΔG and ΔA for the gas.

Solution: Note that we do not know from the information given whether or not the process is reversible. However, both G and A are thermodynamic state functions, and therefore the values of ΔG and ΔA are independent of path. Thus, we shall obtain ΔG and ΔA assuming a reversible expansion.

From Eq. 5-33 we have

$$dG = -S\,dT + V\,dP = V\,dP$$

where the second equality follows because $dT = 0$. Substitution of the equation of state into the above equation yields

$$dG = n\left(\frac{RT}{P} + B\right)dP$$

Integration with T fixed from P_1 to P_2 yields

$$\int_1^2 dG = \int_{P_1}^{P_2} n\left(\frac{RT}{P} + B\right)dP$$

$$\Delta G = nRT\ln\left(\frac{P_2}{P_1}\right) + nB(P_2 - P_1)$$

From Eq. 5-26 we have ($dT = 0$)

$$dA = -S\,dT - P\,dV = -P\,dV$$

and from the equation of state

$$P = \frac{nRT}{V - nB}$$

and thus

$$\int_1^2 dA = -\int_{V_1}^{V_2}\left(\frac{nRT}{V - nB}\right)dV$$

$$\Delta A = -nRT\ln\left(\frac{V_2 - nB}{V_1 - nB}\right)$$

From the equation of state with T constant, we have

$$\frac{V_2 - nB}{V_1 - nB} = \frac{(nRT/P_2)}{(nRT/P_1)} = \frac{P_1}{P_2}$$

and thus

$$\Delta A = nRT\ln\left(\frac{P_2}{P_1}\right)$$

From the definitions of G, A, and H we have

$$G = H - TS = U + PV - TS$$

and

$$A = U - TS$$

Therefore

$$G = A + PV$$

For any change in thermodynamic state the relationship between ΔG and ΔA is thus given by the equation

$$\Delta G = \Delta A + \Delta(PV) = \Delta A + P_2V_2 - P_1V_1$$

We could have used the above equation to obtain the ΔA expression from the ΔG expression in the above instance, as shown below:

$$\Delta A = \Delta G - (P_2V_2 - P_1V_1)$$

Substituting in the ΔG expression and using the equation of state for P_2V_2 and P_1V_1, we have

$$\Delta A = nRT\ln\left(\frac{P_2}{P_1}\right) + nB(P_2 - P_1) - [n(RT + BP_2) - n(RT + BP_1)]$$

thus

$$\Delta A = nRT\ln\left(\frac{P_2}{P_1}\right)$$

5-4 *The Partial Derivatives $(\partial U/\partial V)_T$ and $(\partial H/\partial P)_T$ Can Be Obtained from the P, V, T Equation of State for a Substance*

A general expression for $(\partial U/\partial V)_T$, that is, for the change in internal energy with change in volume under isothermal conditions, is obtained from Eq. 5-11:

$$dU = T\,dS - P\,dV \tag{5-11}$$

Division of Eq. 5-11 by dV, followed by the imposition of the constant T condition, yields*

$$\left(\frac{dU}{dV}\right)_T = T\left(\frac{dS}{dV}\right)_T - P$$

or more properly

$$\left(\frac{\partial U}{\partial V}\right)_T = T\left(\frac{\partial S}{\partial V}\right)_T - P \tag{5-35}$$

* See Problems 9 and 12 for a discussion of the restrictions on division by total differentials.

Substitution of Eq. 5-29 into Eq. 5-35 yields

$$\left(\frac{\partial U}{\partial V}\right)_T = T\left(\frac{\partial P}{\partial T}\right)_V - P \tag{5-36}$$

Equation 5-36 is used, in conjunction with a P, V, T equation of state, to obtain an expression for $(\partial U/\partial V)_T$. The resulting expression for $(\partial U/\partial V)_T$ can be used to calculate the change in internal energy ΔU that results from an isothermal change in volume. Note that for an ideal gas

$$PV = nRT$$

and

$$\left(\frac{\partial U}{\partial V}\right)_T = T\left(\frac{nR}{V}\right) - \frac{nRT}{V} = 0$$

Thus the internal energy of an ideal gas is independent of volume. In the general case, $U = U(T, V)$ and hence (Eq. 5-15)

$$dU = C_V\,dT + \left(\frac{\partial U}{\partial V}\right)_T dV$$

but for an ideal gas $(\partial U/\partial V)_T = 0$, and therefore we have

$$dU = C_V\,dT$$

for an ideal gas, irrespective of whether or not the volume of the gas changes in the process.

EXAMPLE 5-10

Derive an expression for $(\partial U/\partial V)_T$ for a van der Waals gas.

Solution: From the van der Waals equation of state,

$$P = \frac{nRT}{V - nb} - \frac{n^2 a}{V^2}$$

we obtain

$$\left(\frac{\partial P}{\partial T}\right)_V = \frac{nR}{V - nb}$$

Substitution of the above expression for $(\partial P/\partial T)_V$ into Eq. 5-36 yields

$$\left(\frac{\partial U}{\partial V}\right)_T = \frac{nRT}{V - nb} - \frac{nRT}{V - nb} + \frac{n^2 a}{V^2} = \frac{n^2 a}{V^2}$$

If a van der Waals gas undergoes an isothermal expansion, then

$$dU = \left(\frac{\partial U}{\partial V}\right)_T dV = \frac{n^2 a}{V^2}\,dV$$

and

$$\Delta U = \int_{V_1}^{V_2} \frac{n^2 a}{V^2}\,dV = -n^2 a\left(\frac{1}{V_2} - \frac{1}{V_1}\right) = n^2 a\left(\frac{1}{V_1} - \frac{1}{V_2}\right)$$

Note that for $V_2 > V_1$ (expansion) $\Delta U > 0$, because $a > 0$. The increase in U on isothermal expansion of a van der Waals gas is a result of an energy input to overcome the attractive forces between the gas molecules.

A general expression for the isothermal pressure dependence of the enthalpy $(\partial H/\partial P)_T$ is obtained from Eq. 5-18:

$$dH = T\,dS + V\,dP \tag{5-18}$$

Division of both sides of Eq. 5-18 by dP, followed by the imposition of the constant temperature condition, yields

$$\left(\frac{dH}{dP}\right)_T = T\left(\frac{dS}{dP}\right)_T + V$$

or more precisely

$$\left(\frac{\partial H}{\partial P}\right)_T = T\left(\frac{\partial S}{\partial P}\right)_T + V \tag{5-37}$$

Using Eq. 5-34 we can eliminate $(\partial S/\partial P)_T$ from Eq. 5-37 to obtain the desired result:

$$\left(\frac{\partial H}{\partial P}\right)_T = V - T\left(\frac{\partial V}{\partial T}\right)_P \tag{5-38}$$

Equation 5-38 is used, in conjunction with an equation of state $V = V(T, P)$, to obtain an expression for $(\partial H/\partial P)_T$ for a substance.

EXAMPLE 5-11

A Berthelot gas is a gas whose P, V, T behavior is described by the equation of state

$$V = \frac{nRT}{P} + na\left(1 + \frac{b}{T^2}\right)$$

where a and b are constants characteristic of a particular gas. Obtain an expression for the change in enthalpy ΔH of n mol of a Berthelot gas that undergoes the change in state P_1, T to P_2, T.

Solution: Because we want ΔH arising from a change in P at constant T, we begin by writing

$$H = H(T, P)$$

from which we obtain

$$dH = \left(\frac{\partial H}{\partial T}\right)_P dT + \left(\frac{\partial H}{\partial P}\right)_T dP$$

But $dT = 0$, and therefore

$$dH = \left(\frac{\partial H}{\partial P}\right)_T dP$$

We obtain an expression for $(\partial H/\partial P)_T$ from Eq. 5-38 applied to the equation of state:

$$\left(\frac{\partial H}{\partial P}\right)_T = V - T\left(\frac{\partial V}{\partial T}\right)_P = \frac{nRT}{P} + na\left(1 + \frac{b}{T^2}\right) - T\left(\frac{nR}{P} - \frac{2nab}{T^3}\right)$$

or

$$\left(\frac{\partial H}{\partial P}\right)_T = na\left(1 + \frac{3b}{T^2}\right)$$

Thus

$$dH = na\left(1 + \frac{3b}{T^2}\right)dP$$

Integration with T held constant yields the desired result:

$$\Delta H = na\left(1 + \frac{3b}{T^2}\right)\Delta P$$

Note from the equation of state that if $a = b = 0$, then the gas is ideal, and thus $(\partial H/\partial P)_T = 0$ for an ideal gas.

5-5 *Transformations of Partial Derivatives*

The *constant pressure coefficient of thermal expansion* α of a substance is defined by the equation

$$\alpha = \frac{1}{V}\left(\frac{\partial V}{\partial T}\right)_P \tag{5-39}$$

Note that α is the fractional change in volume $(\Delta V/V)$ divided by the corresponding change in temperature ΔT, and thus α has the units of per kelvin K^{-1}; α is almost always a positive quantity—most substances expand on heating. The isothermal compressibility β of a substance is defined by the equation

$$\beta = -\frac{1}{V}\left(\frac{\partial V}{\partial P}\right)_T \tag{5-40}$$

where the minus sign is inserted to make β a positive quantity—all substances decrease in volume on compression (assuming no chemical reactions occur as in a compression-induced explosion). Note that β is the fractional change in volume divided by the corresponding change in pressure, and thus the units of β are, for example, per atmosphere, atm^{-1}, or per pascal, Pa^{-1}. The quantities α and β are related by the equation

$$\left(\frac{\partial P}{\partial T}\right)_V = \frac{\alpha}{\beta} \tag{5-41}$$

The problem of interest here is the derivation of Eq. 5-41. A general procedure that is often useful in derivations involving partial derivatives is as follows:

1. Express the variable held constant in the partial derivative as a function of the other two variables in the derivative, and then write the expression for the total differential:

$$V = V(P, T)$$

$$dV = \left(\frac{\partial V}{\partial P}\right)_T dP + \left(\frac{\partial V}{\partial T}\right)_P dT$$

2. Set the total differential equal to zero and solve for the ratio of the differentials of the other two variables:

$$0 = \left(\frac{\partial V}{\partial P}\right)_T dP_V + \left(\frac{\partial V}{\partial T}\right)_P dT_V$$

Thus, writing $(dP/dT)_V = (\partial P/\partial T)_V$, we have

$$\left(\frac{\partial P}{\partial T}\right)_V = -\frac{(\partial V/\partial T)_P}{(\partial V/\partial P)_T} \tag{5-42}$$

If we now multiply the numerator and the denominator on the right-hand side of Eq. 5-42 by $1/V$, then we obtain

$$\left(\frac{\partial P}{\partial T}\right)_V = -\frac{(1/V)(\partial V/\partial T)_P}{(1/V)(\partial V/\partial P)_T} \tag{5-43}$$

Comparison of Eq. 5-43 with Eqs. 5-39 and 5-40 leads to the desired result:

$$\left(\frac{\partial P}{\partial T}\right)_V = \frac{\alpha}{\beta}$$

EXAMPLE 5-12

Derive the relationship

$$\left(\frac{\partial T}{\partial P}\right)_H = -\frac{1}{C_P}\left(\frac{\partial H}{\partial P}\right)_T$$

Solution: Following the procedure outlined in the text, we take $H = H(T, P)$, write the expression for dH, and then set $dH = 0$:

$$dH = \left(\frac{\partial H}{\partial T}\right)_P dT + \left(\frac{\partial H}{\partial P}\right)_T dP = 0$$

Solving for $(dT/dP)_H = (\partial T/\partial P)_H$, we obtain

$$\left(\frac{\partial T}{\partial P}\right)_H = -\frac{(\partial H/\partial P)_T}{(\partial H/\partial T)_P}$$

but $(\partial H/\partial T)_P = C_P$; thus

$$\left(\frac{\partial T}{\partial P}\right)_H = -\frac{1}{C_P}\left(\frac{\partial H}{\partial P}\right)_T \tag{5-44}$$

Note that if we have a P, V, T equation of state for a substance, then we can use Eq. 5-38 to obtain an expression for $(\partial H/\partial P)_T$, which we can then substitute into Eq. 5-44 to obtain an expression for $(\partial T/\partial P)_H$ for the substance. The partial derivative $(\partial T/\partial P)_H$ for a gas is called the *Joule-Thomson coefficient* of the gas. The Joule-Thomson coefficient is a key quantity in the liquefaction of gases by adiabatic expansion.

As a second example of a thermodynamic derivation involving manipulation of partial derivatives, consider the problem of deriving a relationship between the two types of heat capacities C_P and C_V. The heat capacities at constant pressure C_P and constant volume C_V are given by (Chapter 3)

$$C_P = \left(\frac{\partial H}{\partial T}\right)_P \qquad C_V = \left(\frac{\partial U}{\partial T}\right)_V \tag{5-45}$$

We shall now show that thermodynamics provides a general relationship between C_P and C_V for a system. The enthalpy is defined as $H = U + PV$. Differentiation of H with respect to T at constant P yields

$$C_P = \left(\frac{\partial H}{\partial T}\right)_P = \left(\frac{\partial U}{\partial T}\right)_P + P\left(\frac{\partial V}{\partial T}\right)_P \tag{5-46}$$

Equation 5-46 contains C_P; we introduce C_V into Eq. 5-46 via the partial derivative $(\partial U/\partial T)_P$. Taking V and T as independent variables,* we have

$$U = U(V, T)$$

and thus

$$dU = \left(\frac{\partial U}{\partial V}\right)_T dV + \left(\frac{\partial U}{\partial T}\right)_V dT = \left(\frac{\partial U}{\partial V}\right)_T dV + C_V\,dT \tag{5-47}$$

If we now divide Eq. 5-47 through by dT and impose the condition of constant pressure on the result, then we obtain

$$\left(\frac{\partial U}{\partial T}\right)_P = \left(\frac{\partial U}{\partial V}\right)_T\left(\frac{\partial V}{\partial T}\right)_P + C_V \tag{5-48}$$

Substitution of Eq. 5-48 into Eq. 5-46 yields

$$C_P = C_V + \left[P + \left(\frac{\partial U}{\partial V}\right)_T\right]\left(\frac{\partial V}{\partial T}\right)_P \tag{5-49}$$

In Eq. 5-49, $P(\partial V/\partial T)_P$ represents the contribution to C_P arising from the expansion of the system against the external pressure P, and $(\partial U/\partial V)_T(\partial V/\partial T)_P$

* Note that of the variables T, V, P, U, S, H, A, G, C_P, and C_V only two are independent.

represents the contribution to C_P due to intermolecular forces. The so-called *internal pressure* $(\partial U/\partial V)_T$ is large for condensed phases; for example, $(\partial U/\partial V)_T$ is equal to 2.4×10^3 atm and 2.0×10^4 atm for diethyl ether and water, respectively, but $(\partial U/\partial V)_T$ is small for gases (<0.1 atm). Substitution of Eq. 5-36 into Eq. 5-49 yields

$$C_P = C_V + T\left(\frac{\partial P}{\partial T}\right)_V\left(\frac{\partial V}{\partial T}\right)_P \tag{5-50}$$

Substitution of Eq. 5-39 and 5-41 into Eq. 5-50 yields

$$C_P = C_V + \frac{\alpha^2 TV}{\beta} \tag{5-51}$$

EXAMPLE 5-13

Obtain an expression relating C_P and C_V for an ideal gas.

Solution: From the equation of state $PV = nRT$, we have

$$\left(\frac{\partial P}{\partial T}\right)_V = \frac{nR}{V} \quad \text{and} \quad \left(\frac{\partial V}{\partial T}\right)_P = \frac{nR}{P}$$

Substitution of the above expressions into Eq. 5-50 yields

$$C_P = C_V + T\left(\frac{nR}{V}\right)\left(\frac{nR}{P}\right)$$

and thus

$$C_P = C_V + nR \qquad \text{(ideal gas)} \tag{5-52}$$

From the kinetic theory of gases we know that, for gases with noninteracting molecules, the contributions of various fully excited (classical motion) "degrees of freedom" to the molar heat capacity at constant volume $\bar{C}_V$ are as follows:

Degree of freedom	*Contribution to* $\bar{C}_V$
One-dimensional translation	$R/2$
One-dimensional rotation	$R/2$
Each vibrational mode	R

All gas-phase molecules have three translational degrees of freedom and thus the contribution of translational motion to the total heat capacity per mole is $\bar{C}_{V,\text{trans}} = 3R/2$. Linear gas-phase molecules have two rotational degrees of freedom (two angles must be specified to define the spatial orientation of the molecule), and the contribution of rotational motion to the total heat capacity per mole is $\overline{C}_{V,\text{rot}} = R$. Nonlinear molecules have three rotational degrees of freedom (three angles must be specified to define the spatial orientation of a nonlinear molecule), and thus the molar rotational heat capacity is $\overline{C}_{V,\text{rot}} = 3R/2$. The vibrational contributions to $\overline{C}_V$ for a molecule with N atoms are

$$\overline{C}_{V,\text{vib}} \leqslant (3N - 6)R \qquad \text{(nonlinear)}$$
$$\overline{C}_{V,\text{vib}} \leqslant (3N - 5)R \qquad \text{(linear)}$$

The translational and rotational motions of gas-phase molecules are effectively classical at all T, whereas quantum-mechanical effects are important for vibrations except at very high temperatures. For molecules like H_2 and N_2 with strong bonds there is essentially no contribution to $\overline{C}_V$ from vibration around room temperature (i.e., almost all the molecules are in the lowest vibrational

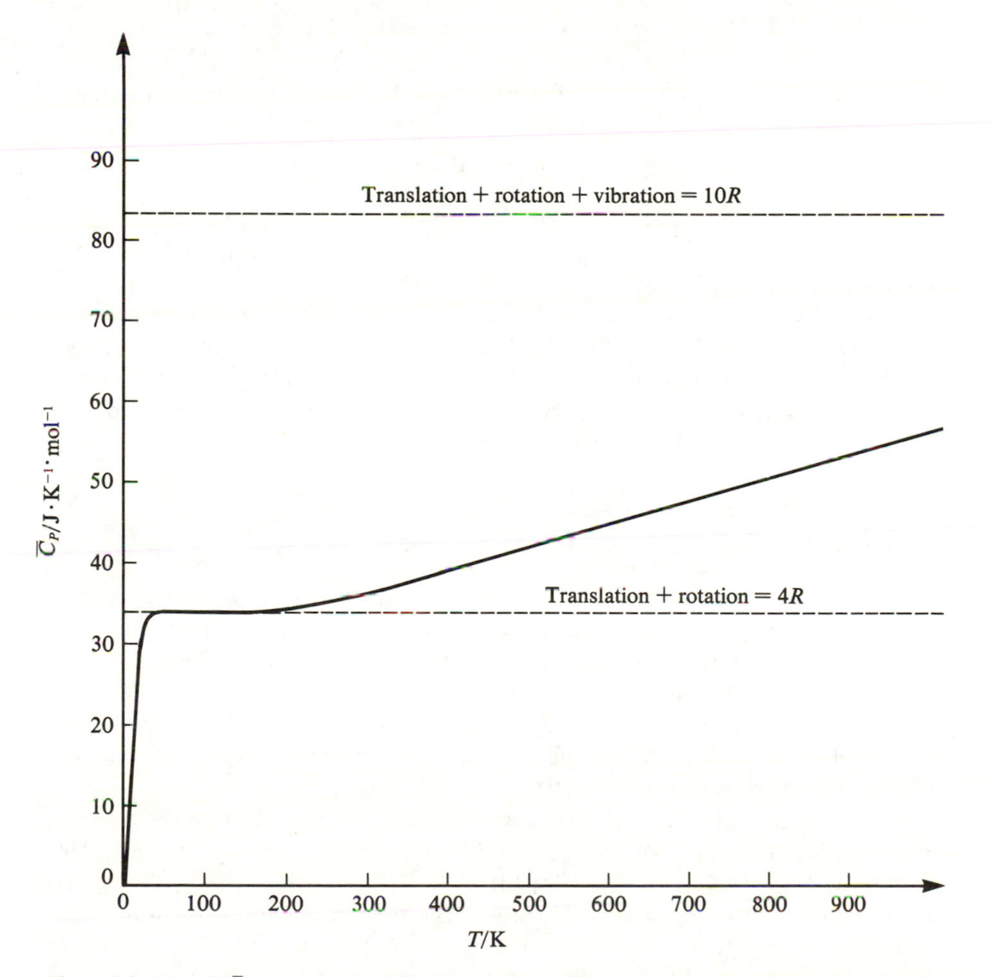

Figure 5-2. Plot of $\overline{C}_P$ versus T for $NH_3(g)$ at 1 atm over the temperature range 0 K to 900 K. Note that $\overline{C}_P = \overline{C}_V + R$ for an ideal gas. We have neglected the contribution to $\overline{C}_P$ arising from nonideality.

level). Thus for $N_2(g)$ at 300 K, we estimate

$$\bar{C}_P = \bar{C}_V + R \approx \tfrac{3}{2}R_{\text{trans}} + R_{\text{rot}} + 0_{\text{vib}} + R = \tfrac{7}{2}R = 29.10\ \text{J}\cdot\text{K}^{-1}\cdot\text{mol}^{-1}$$

whereas the observed value of $\bar{C}_P$ for $N_2(g)$ is 29.12 $\text{J}\cdot\text{K}^{-1}\cdot\text{mol}^{-1}$. For $H_2O(g)$ at 300 K, we estimate

$$\bar{C}_P = \bar{C}_V + R \approx \tfrac{3}{2}R_{\text{trans}} + \tfrac{3}{2}R_{\text{rot}} + 0_{\text{vib}} + R = 4R = 33.26\ \text{J}\cdot\text{K}^{-1}\cdot\text{mol}^{-1}$$

whereas the observed value of $\bar{C}_P$ for $H_2O(g)$ at 300 K is 33.58 $\text{J}\cdot\text{K}^{-1}\cdot\text{mol}^{-1}$.

The temperature-dependent expressions for $\bar{C}_V$ and $\bar{C}_P$ reflect the increasing importance of vibrational motions as T increases. A plot of $\bar{C}_P$ for $NH_3(g)$ as a function of T is given in Figure 5-2. The data in the figure from 300 K to 900 K are based on the expression

$$\bar{C}_P/\text{J}\cdot\text{K}^{-1}\cdot\text{mol}^{-1} = \left(29.75 + 2.51 \times 10^{-2}T - \frac{1.55 \times 10^5}{T^2}\right)$$

which at 300 K gives the value $\bar{C}_P = 35.56\ \text{J}\cdot\text{K}^{-1}\cdot\text{mol}^{-1}$. For comparison, we estimate

$$\bar{C}_P = \bar{C}_V + R = \tfrac{3}{2}R_{\text{trans}} + \tfrac{3}{2}R_{\text{rot}} + R = 4R = 33.26\ \text{J}\cdot\text{K}^{-1}\cdot\text{mol}^{-1}$$

The difference $35.56 - 33.26 = 2.30\ \text{J}\cdot\text{K}^{-1}\cdot\text{mol}^{-1}$ ($\sim 7\%$) arises primarily from vibrational motion. The maximum possible vibrational contribution to $\bar{C}_P$ for $NH_3(g)$ is $(3N - 6)R = (3 \times 4 - 6)R = 6R = 6 \times 8.314 = 49.88\ \text{J}\cdot\text{K}^{-1}\cdot\text{mol}^{-1}$. The maximum possible vibrational contribution is never reached because the NH_3 molecule dissociates before all the vibrational motions can be regarded as classical (i.e., nonquantized). At high temperatures, electronic excitations may contribute to $\bar{C}_P$; this is especially true for molecules with unpaired electrons, for example, NO.

5-6 *The Thermodynamic Square*

The thermodynamic square shown in Figure 5-3 is a mnemonic diagram that can be used to obtain a number of the more useful thermodynamic relations (Max Born and F. O. Koenig). There are several different types of these diagrams applicable to thermodynamic systems of various types. The one presented here is the best known and most generally useful of these diagrams. It is applicable to thermodynamic systems of fixed mass and composition involving only pressure volume work.

In the diagram, each of the thermodynamic energy functions U, H, G, and A is placed on a side of the square and is flanked by its natural independent variables: $U = U(S, V)$; $H = H(S, P)$; $G = G(P, T)$; and $A = A(V, T)$. To write the differential expression for any of these functions in terms of differentials of its natural independent variables, we need only note that the coefficient of the differential of the independent variable is placed diagonally opposite this variable on the square. An arrow pointing away from a natural variable

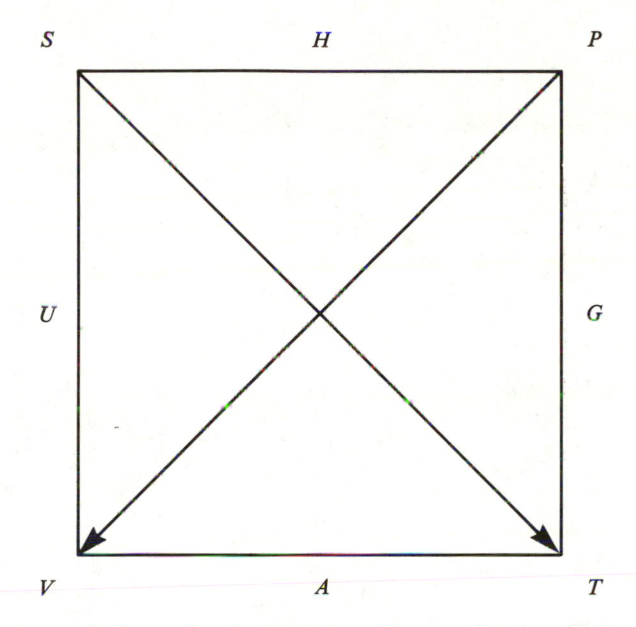

Figure 5-3. A mnemonic diagram for the thermodynamic energy functions U, H, G, and A. Each energy function is flanked by its natural independent variables; thus, $H = H(S, P)$; $U = U(S, V)$; and so forth.

indicates a positive sign. An arrow pointing toward a natural variable indicates a negative sign. Thus, for example, we obtain the dG equation as follows:

1. Write the natural variables of the energy function of interest in differential form:

$$dG \qquad dT \qquad dP$$

2. Read the coefficients of dT and dP with appropriate signs from the thermodynamic square and insert the equality sign in the result:

$$dG = -S\,dT + V\,dP$$

For the dU expression we find

$$\text{(a)}\quad dU \qquad dS \qquad dV$$

$$\text{(b)}\quad dU = T\,dS - P\,dV$$

and so forth. The Maxwell relations can also be read off the diagram using the corners of the square with the signs being determined by the placement of the arrows. If the arrows are symmetrically placed relative to the corners of interest, then the sign is plus; if the arrows are asymmetrically placed, then the sign is minus. For example, from the orientation

$$\begin{matrix} S & & & & P \\ V & T & & V & T \end{matrix}$$

we obtain

$$\left(\frac{\partial S}{\partial V}\right)_T = \left(\frac{\partial P}{\partial T}\right)_V$$

Whereas from the orientation

$$\begin{matrix} V & & & S \\ T & P & T & P \end{matrix}$$

we obtain

$$\left(\frac{\partial V}{\partial T}\right)_P = -\left(\frac{\partial S}{\partial P}\right)_T$$

The mnemonic diagram can be remembered conveniently by the nonsense phrase

"*SHiP*
UG
VAT"

The thermodynamic energy state functions U, H, A, and G are all *extensive* state functions and they all have the dimensions of energy. The SI unit of energy, as we have already noted, is the joule, J. In Section 5-7 we shall describe the use of the functions G and A to obtain additional criteria for thermodynamic equilibrium. In Table 5-1 we have summarized the various key relationships involving the energy functions U, H, A, and G.

5-7 *The Gibbs and Helmholtz Energies Provide Criteria for Equilibrium States*

Consider a system for which

$$\delta w_{\text{tot}} = -P\,dV + \delta w_{npv} \tag{5-53}$$

where δw_{npv} (*n*on*p*ressure-*v*olume) represents the work done on the system exclusive of pressure-volume work. From the first and second laws, assuming that the change in state takes place reversibly, we have

$$dU = T\,dS + \delta w_{\text{tot}} = T\,dS - P\,dV + \delta w_{npv} \tag{5-54}$$

Taking $A = U - TS$ and $G = U + PV - TS$, we have

$$dA = -S\,dT + \delta w_{\text{tot}} = -S\,dT - P\,dV + \delta w_{npv} \tag{5-55}$$

and

$$dG = -S\,dT + V\,dP + \delta w_{npv} \tag{5-56}$$

From Eq. 5-55, we note that for an isothermal, reversible process the change in the Helmholtz energy of the system is equal to the total work done on the system, $\Delta A = w_{\text{tot}}$. For any given isothermal change in state of the system, ΔA has a definite value that is equal to the minimum amount of work that would have to be done on the system to bring about this change in state. If however any irreversibility enters into the process, then, by the second law,

TABLE 5-1
The Thermodynamic Energy State Functions: U, H, A, and G

Symbol	Name	Defining equation	Differential equation involving the natural variables[a]	Coefficients in the Differential equation		Corresponding Maxwell relation
U	Internal energy	(First law)	$dU = T\,dS - P\,dV$	$T = (\partial U/\partial S)_V$	$-P = (\partial U/\partial V)_S$	$(\partial T/\partial V)_S = -(\partial P/\partial S)_V$
H	Enthalpy ("heat content")	$H = U + PV$	$dH = T\,dS + V\,dP$	$T = (\partial H/\partial S)_P$	$V = (\partial H/\partial P)_S$	$(\partial T/\partial P)_S = (\partial V/\partial S)_P$
A	Helmholtz energy	$A = U - TS$	$dA = -S\,dT - P\,dV$	$-S = (\partial A/\partial T)_V$	$-P = (\partial A/\partial V)_T$	$(\partial S/\partial V)_T = (\partial P/\partial T)_V$
G	Gibbs energy ("free energy")	$G = H - TS$ ($G = A + PV$)	$dG = -S\,dT + V\,dP$	$-S = (\partial G/\partial T)_P$	$V = (\partial G/\partial P)_T$	$(\partial S/\partial P)_T = -(\partial V/\partial T)_P$

[a] These equations apply to a system of fixed mass and chemical composition for which only pressure-volume work is possible. If other work terms are involved, or if the composition or mass can change, then additional terms will appear in the equations.

$w_{tot} > \Delta A$, and therefore we have

$$w_{tot} \geqslant \Delta A \qquad \text{(isothermal)} \tag{5-57}$$

From Eq. 5-56, we conclude that, for a reversible process taking place at constant T and P, $\Delta G = w_{npv}$, where w_{npv} represents the work exclusive of pressure-volume work that must be done on the system to bring about the particular change in state. If any irreversibility enters into the process, then by the second law, $w_{npv} \geqslant \Delta G$, and therefore we have

$$w_{npv} \geqslant \Delta G \qquad \text{(isothermal, constant pressure)} \tag{5-58}$$

We also note from Eq. 5-55 that at constant T and V

$$w_{npv} \geqslant \Delta A \qquad \text{(isothermal, constant volume)} \tag{5-59}$$

If we consider, as implicitly assumed above, that the system is in thermal contact with a heat bath and the system and heat bath are placed within an adiabatic enclosure, then

$$w_{sys} = -w_{sur} \tag{5-60}$$

where the surroundings consist of everything outside the adiabatic enclosure. Combining Eq. 5-60 with Eqs. 5-57, 5-58, and 5-59 yields

$$-\Delta A \geqslant (w_{tot})_{sur} \qquad \text{(isothermal)} \tag{5-61}$$

$$-\Delta G \geqslant (w_{npv})_{sur} \qquad \text{(isothermal, constant pressure)} \tag{5-62}$$

$$-\Delta A \geqslant (w_{npv})_{sur} \qquad \text{(isothermal, constant volume)} \tag{5-63}$$

The rationale for obtaining expressions for the nonpressure-volume (*npv*) done on or by the system in a process is as follows. Suppose that we have an electrochemical cell operating at constant T and P. The cell reaction will, in general, involve some volume change ΔV, and thus the performance of work $P\Delta V$ on the surroundings, as well as the performance of a quantity of electrical work, which at constant cell voltage is given by $\mathscr{E}Z$. The design of the cell is invariably such, however, that (assuming $\Delta V > 0$) the expansion work done by the system on the surroundings is not utilized and, therefore, is said to be "useless." In other words, only the electrical work done by the system is "useful" work. Clearly, in the case where $\Delta V < 0$ for the cell reaction, it is the surroundings that do work on the system, and thus this exchange of energy as work is not available to, say, lift an external weight and hence is not useful work.

We have already seen that the second law requires that for any change in state of a system

$$dS_{sys} \geqslant \frac{\delta q_{sys}}{T} \tag{5-64}$$

From the first law,

$$dU = \delta q + \delta w_{tot} \tag{5-65}$$

and combination of Eq. 5-65 with Eq. 5-64 yields

$$dU \leqslant T\,dS + \delta w_{tot} \tag{5-66}$$

From Eq. 5-66, together with the definitions

$$A = U - TS \quad \text{and} \quad G = A + PV \tag{5-67}$$

we obtain

$$dA \leqslant -S\,dT - P\,dV + \delta w_{npv} \tag{5-68}$$

and

$$dG \leqslant -S\,dT + V\,dP + \delta w_{npv} \tag{5-69}$$

Therefore, if only pressure-volume work is involved (i.e., if $\delta w_{npv} = 0$), then we have as a criterion for a *spontaneous* (*irreversible*) *process* at constant T and V, $\Delta A < 0$; at constant T and P, $\Delta G < 0$. In addition, *our equilibrium criteria for a system in which only pressure-volume work is possible are as follows:*

(a) At constant T and V, A is a minimum and $dA = 0$ (5-70)

(b) At constant T and P, G is a minimum and $dG = 0$ (5-71)

As an example of the application of Eq. 5-69 to a case for which $\delta w_{npv} \neq 0$, consider an electrochemical cell ($\delta w_{npv} = -\mathscr{E}\,dZ$)

$$dG \leqslant -\mathscr{E}\,dZ$$

The criterion for spontaneity *in the cell* is

$$dG + \mathscr{E}\,dZ < 0$$

On the other hand, if we determine $\Delta G = -nF\mathscr{E}$ under equilibrium conditions in the cell, then we can use the measured ΔG as a criterion of whether or not the process will take place spontaneously outside the cell, when the substances at the same partial pressures, concentrations, and so forth, as prevailed in the cell, are simply mixed outside the cell. If $\Delta G < 0$, then the reaction will take place spontaneously until equilibrium is attained. The mixing of the reactants and products external to the cell is equivalent to short-circuiting the electrochemical cell. In such a case, current will flow in the external circuit until $\mathscr{E} = 0$, that is, until $\Delta G = 0$.

5-8 *Gases Can Be Liquefied by a Joule-Thomson Expansion*

In the Joule-Thomson experiment (Figure 5-4), part of a mass of gas maintained at pressure P_1 (by means of a piston) and temperature T_1 is allowed to escape through a *rigid porous plug* into a second chamber where the pressure is maintained at P_2 ($P_2 < P_1$) by means of a second piston. The temperatures T_1 and T_2 of the gas are measured by thermometers placed in the two chambers. The entire apparatus is adiabatically isolated and therefore $\delta q = 0$. Although the expansion is irreversible, w is calculable because $\delta q = 0$ and therefore

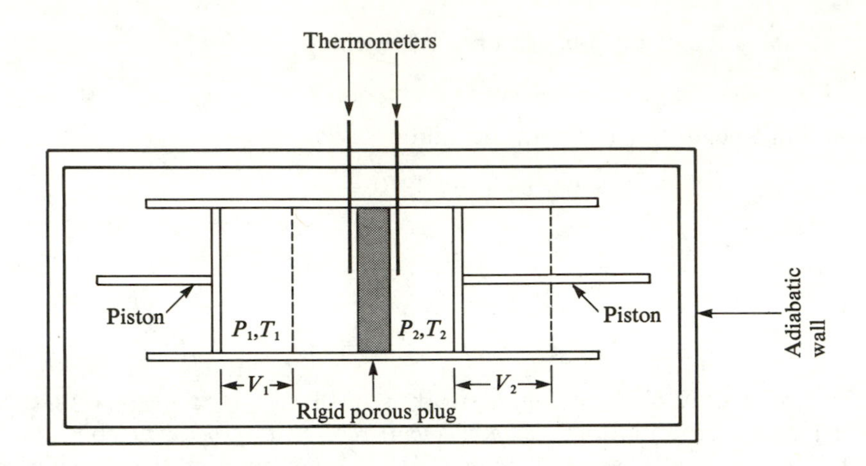

Figure 5-4. A Joule-Thomson expansion of a gas ($P_1 > P_2$). The gas is forced through a rigid porous plug while the pressures P_1 and P_2 are maintained at constant values by the movable pistons.

δw ($=dU$) is exact; in other words, δw is calculable because the path is specifiable, even though the expansion is irreversible. The work done on the system is the sum of the work done on the gas in the two chambers:

$$w = -\int_{V_1}^{0} P_1\, dV - \int_{0}^{V_2} P_2\, dV = P_1V_1 - P_2V_2$$

Because $\delta q = 0$, we have from the first law

$$\Delta U = U_2 - U_1 = w$$

and thus

$$U_2 - U_1 = P_1V_1 - P_2V_2$$

Rearrangement of the above equation yields

$$U_1 + P_1V_1 = U_2 + P_2V_2$$

or

$$H_1 = H_2$$

that is, in a Joule-Thomson expansion the enthalpy of the gas is constant:

$$\Delta H = 0$$

Therefore, measurements of P_1, T_1 and P_2, T_2 for a Joule-Thomson expansion experiment yields the Joule-Thomson coefficient $(\partial T/\partial P)_H$:

$$\left(\frac{\partial T}{\partial P}\right)_H = \lim_{\Delta P \to 0} \left(\frac{\Delta T}{\Delta P}\right)_H$$

From Eq. 5-44 we have

$$\left(\frac{\partial T}{\partial P}\right)_H = -\frac{(\partial H/\partial P)_T}{C_P}$$

Because C_P is finite for all known gases, a zero value for the Joule-Thomson coefficient requires that $(\partial H/\partial P)_T = 0$.

The Joule-Thomson coefficient is a function of pressure and temperature. Fixed-composition gases around 300 K and moderate pressures, except hydrogen and helium, have a positive Joule-Thomson coefficient; that is, they cool on expansion against a fixed external pressure. At low temperatures $(\partial T/\partial P)_H$ becomes greater than zero for hydrogen and helium, whereas for other gases $(\partial T/\partial P)_H$ becomes less than zero at high temperatures. The Joule-Thomson coefficient for air follows the equation

$$\left(\frac{\partial T}{\partial P}\right)_H = -0.1975 + \frac{138}{T} - \frac{319P}{T^2} \qquad (\text{K}\cdot\text{atm}^{-1})$$

from which we estimate that for $\Delta P = 50$ atm; that is, for an average pressure of 25 atm, $(\partial T/\partial P)_H = 0.17\ \text{K}\cdot\text{atm}^{-1}$ at 300 K and 0.29 at 200 K. Thus when air undergoes a Joule-Thomson expansion from 50 atm to some low pressure ($\sim$1 atm), we compute $\Delta T \approx -9$, if the gas was initially at 300 K, whereas we obtain $\Delta T \approx -14$, if the gas was initially at 200 K.

The Joule-Thomson effect is used in the liquefaction of gases with low boiling points, for example, nitrogen and helium. The gas to be liquefied is compressed and part of the compressed gas is allowed to undergo an adiabatic expansion which decreases the temperature. The low-temperature expanded gas is used to cool the remaining high-pressure gas, a part of which is then expanded and the process repeated. In this manner part of the gas can be liquefied, more gas can then be added, and the process repeated. The Joule-Thomson effect is also utilized in refrigeration devices.

5-9 *Partial Molar Quantities Give the Dependence of the Extensive Thermodynamic State Functions on the Composition*

The thermodynamic analysis of solutions is facilitated by the introduction of quantities that measure how the extensive thermodynamic quantities (V, U, H, S, G, A, C_X, and so forth) of a system as a whole vary on addition of the various components. If we let Y be any extensive thermodynamic coordinate of the system (e.g., the volume V), then *the partial molar value of Y for the ith component* is defined as

$$\bar{Y}_i \equiv \left(\frac{\partial Y}{\partial n_i}\right)_{T,P,n_j} = \bar{Y}_i(T, P, n_1, \ldots, n_c) \tag{5-72}$$

when n_1 is the moles of component 1 and so forth, n_j stands for all the mole numbers of the various components except n_i, and c is the number of components.

In general, the partial molar quantities are functions of temperature, pressure, and composition (and conceivably other variables as well, if work terms in addition to pressure-volume work are involved). Although the partial molar quantities defined in Eq. 5-72 pertain to the individual components of the system, *they are properties of the system as a whole*—that is to say, the values of the partial molar quantities depend not only on the nature of the particular substance in question but also on the nature and relative amounts of the other components present as well. *Partial molar quantities are intensive thermodynamic coordinates* (units of Y per mole), and thus their values are independent of the size of the system.

For a system of variable composition at fixed temperature and pressure we can write

$$Y = Y(n_1, n_2, \ldots, n_c)$$

The variable Y is by definition an extensive thermodynamic state function, and thus we can also write

$$dY = \left(\frac{\partial Y}{\partial n_1}\right)_{T,P,n_j} dn_1 + \left(\frac{\partial Y}{\partial n_2}\right)_{T,P,n_j} dn_2 + \cdots + \left(\frac{\partial Y}{\partial n_c}\right)_{T,P,n_j} dn_c$$

or

$$dY = \bar{Y}_1\, dn_1 + \bar{Y}_2\, dn_2 + \cdots + \bar{Y}_c\, dn_c = \sum_{i=1}^{c} \bar{Y}_i\, dn_i \qquad (5\text{-}73)$$

If we now enlarge the system, holding all the intensive variables constant, to k times its original mass, then integration of Eq. 5-73 yields

$$(k-1)Y = \bar{Y}_1(k-1)n_1 + \bar{Y}_2(k-1)n_2 + \cdots + \bar{Y}_c(k-1)n_c$$

or

$$Y = \bar{Y}_1 n_1 + \bar{Y}_2 n_2 + \cdots + \bar{Y}_c n_c = \sum_i \bar{Y}_i n_i \qquad (5\text{-}74)$$

For some change in state ($A \rightarrow B$), taking place without change in the mole numbers ($n_1, n_2, \ldots, n_c$) at fixed T and P, Eq. 5-74 yields

$$\Delta Y = Y_B - Y_A = n_1(\bar{Y}_{1B} - \bar{Y}_{1A}) + n_2(\bar{Y}_{2B} - \bar{Y}_{2A}) + \cdots + n_c(\bar{Y}_{cB} - \bar{Y}_{cA})$$

or

$$\Delta Y = \sum_i n_i\, \Delta \bar{Y}_i \qquad (5\text{-}75)$$

Equation 5-75 is useful in dealing with thermodynamic coordinates like enthalpy or Gibbs energy for which only differences are measurable. We note here that, for any pure substance, $\bar{Y}_i$ is equal to Y/n_i, and thus $\Delta\bar{Y}_i$ is equal to $\Delta Y/n_i$ for a pure substance.

Differentiation of Eq. 5-74 yields

$$dY = n_1\, d\bar{Y}_1 + \bar{Y}_1\, dn_1 + n_2\, d\bar{Y}_2 + \bar{Y}_2\, dn_2 + \cdots + n_c\, d\bar{Y}_c + \bar{Y}_c\, dn_c \qquad (5\text{-}76)$$

Comparison of Eq. 5-76 with Eq. 5-73 yields the result

$$n_1\, d\bar{Y}_1 + n_2 d\bar{Y}_2 + \cdots + n_c\, d\bar{Y}_c = \sum_i n_i\, d\bar{Y}_i = 0 \qquad (5\text{-}77)$$

Equation 5-77, like Eq. 5-74, is valid at fixed T and P. Equation 5-77 is particularly interesting, because it tells us that there is a necessary relationship among the variations in the partial molar quantities of a system. For example, with a two-component system we have

$$n_1\, d\bar{Y}_1 + n_2\, d\bar{Y}_2 = 0 \tag{5-78}$$

and thus if we have determined, say,

$$\bar{Y}_1 = \bar{Y}_1(n_1, n_2) \tag{5-79}$$

then Eq. 5-78 fixes the form of the function $\bar{Y}_2(n_1, n_2) = \bar{Y}_2$.

It is helpful at this point to state explicitly the definitions of several of the more useful partial molar quantities. For example, we have for the partial molar entropy, volume, enthalpy, and Gibbs energy of the ith component

$$\bar{S}_i = \left(\frac{\partial S}{\partial n_i}\right)_{T,P,n_j} \tag{5-80}$$

$$\bar{V}_i = \left(\frac{\partial V}{\partial n_i}\right)_{T,P,n_j} \tag{5-81}$$

$$\bar{H}_i = \left(\frac{\partial H}{\partial n_i}\right)_{T,P,n_j} \tag{5-82}$$

$$\bar{G}_i = \left(\frac{\partial G}{\partial n_i}\right)_{T,P,n_j} \tag{5-83}$$

respectively.

Fortunately, all the general thermodynamic relations obtained in this chapter can be applied with minor symbolic modifications to the partial molar quantities. For example, consider the expression

$$G = H - TS$$

Differentiation of G with respect to n_i, at constant T, P, n_j, yields

$$\left(\frac{\partial G}{\partial n_i}\right)_{T,P,n_j} = \left(\frac{\partial H}{\partial n_i}\right)_{T,P,n_j} - T\left(\frac{\partial S}{\partial n_i}\right)_{T,P,n_j}$$

or

$$\bar{G}_i = \bar{H}_i - T\bar{S}_i \tag{5-84}$$

Similarly, from the definition $H = U + PV$, we obtain

$$\bar{H}_i = \bar{U}_i + P\bar{V}_i \tag{5-85}$$

From the definition for C_P, that is,

$$C_P = \left(\frac{\partial H}{\partial T}\right)_P$$

we obtain

$$\left(\frac{\partial C_P}{\partial n_i}\right)_{T,P,n_j} = \left[\frac{\partial}{\partial n_i}\left(\frac{\partial H}{\partial T}\right)_P\right]_{T,P,n_j} = \left[\frac{\partial}{\partial T}\left(\frac{\partial H}{\partial n_i}\right)_{T,P,n_j}\right]_P$$

or

$$\bar{C}_{P_i} = \left(\frac{\partial \bar{H}_i}{\partial T}\right)_P \tag{5-86}$$

In obtaining Eq. 5-86 we have utilized the property that the order of differentiation is irrelevant. Proceeding in a fashion analogous to the development of Eq. 5-86, we obtain from the relations $V = (\partial G/\partial P)_T$ and $-S = (\partial G/\partial T)_P$,

$$\bar{V}_i = \left(\frac{\partial \bar{G}_i}{\partial P}\right)_T \tag{5-87}$$

and

$$-\bar{S}_i = \left(\frac{\partial \bar{G}_i}{\partial T}\right)_P \tag{5-88}$$

From the relation (Eq. 5-33)

$$dG = -S\,dT + VdP$$

we obtain

$$\left[\frac{\partial}{\partial n_i}(dG)\right]_{T,P,n_j} = -\left(\frac{\partial S}{\partial n_i}\right)_{T,P,n_j} dT + \left(\frac{\partial V}{\partial n_i}\right)_{T,P,n_j} dP \tag{5-89}$$

or

$$d\bar{G}_i = -\bar{S}_i\,dT + \bar{V}_i\,dP$$

From the relation

$$dH = TdS + V\,dP$$

we obtain

$$d\left(\frac{\partial H}{\partial n_i}\right)_{T,P,n_j} = Td\left(\frac{\partial S}{\partial n_i}\right)_{T,P,n_j} + \left(\frac{\partial V}{\partial n_i}\right)_{T,P,n_j} dP$$

or

$$d\bar{H}_i = Td\bar{S}_i + \bar{V}_i\,dP \tag{5-90}$$

Partial molar quantities are essential for analyzing the thermodynamics of solutions. Fortunately, most of the thermodynamic expressions that we have developed for systems of fixed chemical composition can be transformed into the corresponding equations for the various components of a system with variable composition. In essence, only Eqs. 5-74 and 5-77 are "new" equations.

Problems

1. Given the function

$$y = xz^2$$

find $(\partial y/\partial x)_z$ and $(\partial y/\partial z)_x$.

2. Given

$$dy = z^2\,dx + 2xz\,dz$$

is dy an exact differential?

3. The equation of state for a Berthelot gas is

$$V = \frac{nRT}{P} + na\left(1 + \frac{b}{T^2}\right)$$

where a and b are constants. Obtain $(\partial V/\partial T)_P$ and $(\partial V/\partial P)_T$ for a Berthelot gas.

4. For a thermodynamic system of fixed mass and composition, the internal energy can be expressed as a function of any two thermodynamic state functions; thus we can write

(a) $U = U(S, V)$ (b) $U = U(T, V)$

(c) $U = U(T, P)$ (d) $U = U(S, T)$

as well as others. For each of the above, write out the expression for dU.

5. The coefficient of the thermal expansion of ethanol is given by

$$\alpha/{}^\circ\text{C}^{-1} = 1.0414 \times 10^{-3} + 1.5672 \times 10^{-6}t + 5.148 \times 10^{-8}t^2$$

Calculate the percentage change in volume when ethanol is heated from 0°C to 100°C at constant pressure.

6. Derive the expressions

(a) $\left(\frac{\partial \alpha}{\partial P}\right)_T = -\left(\frac{\partial \beta}{\partial T}\right)_P$ (b) $\left(\frac{\partial U}{\partial P}\right)_V = \frac{\beta C_V}{\alpha}$

(c) $\left(\frac{\partial S}{\partial P}\right)_T = -\alpha V$ (d) $\left(\frac{\partial S}{\partial V}\right)_T = \frac{\alpha}{\beta}$

7. A 2.00-mol sample of an ideal gas with $\bar{C}_V = 5R/2$ expands adiabatically and reversibly from $P_1 = 1.00$ atm and $V_1 = 20.0$ L to a final volume $V_2 = 2V_1$. Compute q, w, ΔU, ΔH, and ΔS for the gas, and also compute ΔS_{sur} and ΔS_{tot}.

8. Derive the following expressions for an ideal gas ($PV = nRT$):

(a) $\left(\frac{\partial U}{\partial V}\right)_T = 0$ (b) $\left(\frac{\partial H}{\partial P}\right)_T = 0$

(c) $dU = C_V\,dT$ (d) $dH = C_P\,dT$

(e) $\alpha = \frac{1}{T}$ (f) $\beta = \frac{1}{P}$

(g) $\left(\frac{\partial C_V}{\partial V}\right)_T = 0$ (h) $\left(\frac{\partial C_P}{\partial P}\right)_T = 0$

9. The partial derivative $(\partial y/\partial x)_z$ tells us that y is a function of x and z; that is, $y = y(x, z)$. The total derivative dy/dx tells us that y is a function of x alone; that is, $y = y(x)$. If we have $y = y(x, z)$, then the total differential

dy is given by

$$dy = \left(\frac{\partial y}{\partial x}\right)_z dx + \left(\frac{\partial y}{\partial z}\right)_x dz$$

The operation of dividing the above equation through by, say, dx to obtain

$$\frac{dy}{dx} = \left(\frac{\partial y}{\partial x}\right)_z + \left(\frac{\partial y}{\partial z}\right)_x \frac{dz}{dx}$$

is not mathematically defined because we cannot convert y from a function of x and z to a function of x alone (dy/dx) by a simple act of division. However, if x and z are also functions of, say, t, then the expression

$$\left(\frac{\partial y}{\partial x}\right)_t = \left(\frac{\partial y}{\partial x}\right)_z + \left(\frac{\partial y}{\partial z}\right)_x \left(\frac{\partial z}{\partial x}\right)_t$$

is a valid mathematical expression, provided only that the functions y, x, z, and t (only two of which are independent) and the various partial derivatives are mathematically well-behaved functions (finite, piecewise continuous, and differentiable). Thus, Eq. 5-35 is valid but the "derivation" given in the text is not really a derivation but a mnemonic device that gives us a desired valid result with a minimum of effort.

The rigorous derivation of Eq. 5-35 proceeds as follows:

(a) Take $U = U(V, T)$ and write out dU.

(b) Eliminate dS from

$$dU = T\,dS - P\,dV$$

by expressing $S = S(T, V)$ and substituting the dS expression into the dU expression.

(c) Rearrange, by collecting terms, your result in part (b) to the form

$$dU = M\,dV + N\,dT$$

(d) Now equate the coefficients of the differentials dV and dT in the dU expressions for parts c and a. Note that in addition to Eq. 5-35 we also obtain

$$\left(\frac{\partial S}{\partial T}\right)_V = \frac{C_V}{T}$$

which we derived in Chapter 4 by a very different method.

10. Derive Eq. 5-38 by the method outlined in Problem 9.

11. (a) Take $S = S(T, P)$ and show that

$$dS = \frac{C_P}{T}\,dT - \left(\frac{\partial V}{\partial T}\right)_P dP$$

(b) Show that

$$dS = \frac{C_V}{T}\,dT + \left(\frac{\partial P}{\partial T}\right)_V dV$$

12. Eliminate dV in Problem 11(b) by expressing $V = V(T, P)$. Now rearrange the resulting dS expression to the form

$$dS = M\,dT + N\,dP$$

Equate M and N to the corresponding coefficients of dT and dP, respectively, in the dS expression in Problem 11(a). Do you recognize the resulting expressions? Compare your derivations of these expressions with those given in the text. Which derivations are mathematically more rigorous?

13. Given that

$$G = G(T, P, \mathscr{A})$$

where $\mathscr{A}$ is the surface area of the substance, write the total differential dG. Note that

$$\sigma = \left(\frac{\partial G}{\partial \mathscr{A}}\right)_{T,P}$$

is the surface tension. Also write the three Maxwell relations obtainable from the dG expression.

14. Consider a thermodynamic system whose mass and/or composition can change. For such a system the extensive thermodynamic functions depend on the mole numbers $(n_1, n_2, \ldots)$ of the various components of the system. Given

$$U = U(S, V, n_1, n_2) \quad \text{and} \quad G = G(T, P, n_1, n_2)$$

Write the expressions for dU and dG.

15. Given that (see problem 13)

$$U = U(S, V, \mathscr{A}, n_1, n_2)$$

give an expression for dU.

16. Compute the values (in joules) of w, q, ΔU, ΔH, ΔS, ΔA, and ΔG for 2.00 mol of an ideal gas that undergoes a reversible isothermal expansion from $P_1 =$ 5.00 atm to $P_2 =$ 0.500 atm at 300 K. Also compute ΔS_{tot} for the process.

17. Compute the values of q, w, ΔU, ΔH, and ΔS when 2.00 mol of an ideal gas with $\bar{C}_V = 3R$ undergoes a reversible adiabatic expansion from $P_1 =$ 5.00 atm, $T_1 =$ 300 K to $P_2 =$ 0.500 atm. Is it possible to compute the values of ΔA and ΔG for the gas? Explain.

18. Consider an isothermal reversible expansion of 2.50 mol of $N_2(g)$ at 50.0 atm, 500 K to 1.00 atm. Assume that $N_2(g)$ is a van der Waals gas with $a = 1.39\ \text{L}^2\cdot\text{atm}\cdot\text{mol}^{-2}$ and $b = 39.1\ \text{cm}^3\cdot\text{mol}^{-1}$. Compute q, w, ΔU, ΔH, ΔS, ΔG, and ΔA for the gas in this process. Compare your result with the same quantities calculated assuming that $N_2(g)$ is an ideal gas.

19. Using the data given in Problem 18, compute q, w, ΔU, ΔH, and ΔS for the reversible adiabatic expansion of 2.50 mol $N_2(g)$ from 50.0 atm, 500 K to 1.00 atm. Take $\bar{C}_V = 5R/2$. Compare results for the van der Waals and ideal gas cases.

20. Generally, the coefficient of expansion α is found to be independent of pressure, and the compressibility β is essentially independent of temperature. Given these results, start from $V = V(T, P)$ and show that

$$\ln\left(\frac{V_2}{V_1}\right) \approx \int_{T_1}^{T_2} \alpha(T)\,dT - \int_{P_1}^{P_2} \beta(P)\,dP$$

21. A sample of mercury is placed in a sealed container at 20°C. Assume the mercury completely fills the container (i.e., there is no vapor space). Calculate the increase in pressure of the mercury when its temperature is raised 20 K. Take $\alpha = 0.182 \times 10^{-3}\ \mathrm{K^{-1}}$ and $\beta = 3.9 \times 10^{-6}\ \mathrm{atm^{-1}}$ for mercury. Comment on the significance of your results concerning the subjection of mercury-in-glass thermometers to a sudden large temperature increase.

22. Show that

$$S(T_2, P_2) - S(T_1, P_1) = n\left[a\ln\left(\frac{T_2}{T_1}\right) + b(T_2 - T_1) - R\ln\left(\frac{P_2}{P_1}\right)\right]$$

$$H(T_2, P_2) - H(T_1, P_1) = nB(P_2 - P_1) + n\left[a(T_2 - T_1) + \frac{b}{2}(T_2^2 - T_1^2)\right]$$

for a gas with $\bar{C}_P = a + bT$ and $PV = n(RT + BP)$, where a, b, and B are constants.

23. Show that for a van der Waals gas

$$\alpha = \frac{R}{P\bar{V} - (a/\bar{V}) + (2ab/\bar{V}^2)} \qquad \beta = \frac{\bar{V} - b}{P\bar{V} - (a/\bar{V}) + (2ab/\bar{V}^2)}$$

where $\bar{V} = V/n$. Evaluate α and β for $N_2(g)$ at 25.00°C, 1.00 atm, given $a = 1.39\ \mathrm{L^2 \cdot atm \cdot mol^{-2}}$ and $b = 39.1\ \mathrm{cm^3 \cdot mol^{-1}}$. Compare these results with those calculated assuming N_2 is an ideal gas.

24. (a) Compute $\bar{C}_P - \bar{C}_V$ for $N_2(g)$ at 10.0 atm and 25°C assuming that $N_2(g)$ is a van der Waals gas with $a = 1.39\ \mathrm{L^2 \cdot atm \cdot mol^{-2}}$ and $b = 39.1\ \mathrm{cm^3 \cdot mol^{-1}}$. Compare your result with $\bar{C}_P - \bar{C}_V = R$ and interpret the discrepancy.

(b) Compute $\bar{C}_P - \bar{C}_V$ for $H_2O(\ell)$ at 25°C. Take $\alpha = 2.8 \times 10^{-4}\ \mathrm{K^{-1}}$ and $\beta = 4.5 \times 10^{-5}\ \mathrm{atm^{-1}}$ for $H_2O(\ell)$.

25. The coefficient of linear expansion of Pyrex glass in the range 0°C–100°C is $3.6 \times 10^{-6}\ \mathrm{°C^{-1}}$ and for mercury

$$V_t = V_0(1 + 0.18182 \times 10^{-3}t + 0.0078 \times 10^{-6}t^2)$$

If 0°C and 100°C are taken as fixed points on a uniform bore mercury-in-glass thermometer, what will this thermometer read when an ideal gas thermometer reads 50.00°C? At what ideal gas temperature will this thermometer read 50.00°C?

26. Take an ideal gas with constant heat capacity around a reversible Carnot cycle. Obtain q, w, ΔU, and ΔS for each of the four steps. Show that $\Delta U_{tot} = \Delta S_{tot} = 0$, and also show that

$$\frac{-w_{tot}}{q_2} = \frac{T_2 - T_1}{T_2} \qquad (T_2 > T_1)$$

27. Show that

$$\left(\frac{\partial T}{\partial V}\right)_U = -\frac{1}{C_V}\left(\frac{\partial U}{\partial V}\right)_T$$

28. Consider an electrochemical cell of fixed mass and composition, the reversible voltage of which (as a function of temperature and pressure) is given by

$$\mathscr{E} = a + bT + CT^2 + hP + kP^2$$

where a, b, c, h, and k are constants. Starting from the laws of thermodynamics and noting that

$$\delta w_{elec} = -\mathscr{E}\, dZ$$

show that

(a) $\Delta G = -nF\mathscr{E}$ (at constant T and P)
(where F is the faraday (96,495 $C \cdot V^{-1} \cdot mol^{-1}$) and n is the number of moles of electrons transferred for the reaction as written)

(b) $\Delta V = -nF(h + 2kP)$

(c) $\Delta S = nF(b + 2cT)$

(d) $\Delta H = -nF(a - cT^2 + hP + kP^2)$

(e) $\Delta U = -nF(a - cT^2 - kP^2)$

Also obtain expressions for [q, w, w_{elec} (reversible)], ΔA, ΔC_P, and ΔC_V.

29. (a) Consider a thermodynamic system for which surface effects are important. If the work required to expand the surface area of the system is $\delta w = \sigma\, d\mathscr{A}$, show that

$$\sigma = \left(\frac{\partial G}{\partial \mathscr{A}}\right)_{T,P} \quad \text{and} \quad \left(\frac{\partial \sigma}{\partial T}\right)_{\mathscr{A},P} = -\left(\frac{\partial S}{\partial \mathscr{A}}\right)_{T,P}$$

(b) Consider a spherical drop of radius r subject to an external pressure P_0. Take the pressure P inside the particle to be directed outward from the center and [utilizing the fact that the drop is held together by surface tension; that is, $\sigma\, d\mathscr{A} = (P - P_0)\, dV$] show that $P - P_0 = 2\sigma/r$. Also show that, for a spherical bubble, $P - P_0 = 4\sigma/r$. Is the pressure inside a small bubble greater or less than that inside a larger bubble?

30. (a) Suppose that n mol of a gas with the equation of state

$$PV = n(RT + BP)$$

are compressed isothermally and reversibly from an initial volume V_1 to a final volume V_2. Obtain expressions for w, q, ΔU, ΔH, ΔA, ΔG, and ΔS for the gas and also ΔS for the surroundings for the process.

(b) Suppose the gas in the final state in part (a) is allowed to return to the initial state in part (a) by expanding isothermally into a vacuum. Obtain expressions for w, q, ΔU, ΔH, ΔA, ΔG, and ΔS for the gas and also ΔS for the surroundings in the process.

31. One mole of a gas at 100 atm and 25°C was slowly passed through a tube containing a porous plug in an arrangement similar to the Joule-Thomson experiment. The gas left the system at 1 atm pressure and the whole apparatus was kept at 25°C in a large thermostat. During the passage of the mole of gas, the thermostat absorbed 202 J of heat. If this gas follows the equation of state $P(\bar{V} - b) = RT$ and b is 20 $\text{cm}^3\cdot\text{mol}^{-1}$, calculate w, ΔU, ΔH, and ΔS for the gas in the process.

32. Obtain expressions for dU, dH, dA, and dG for (a) a van der Waals gas and (b) a Berthelot gas.

33. Show that

$$\left(\frac{\partial T}{\partial P}\right)_S = \frac{T}{C_P}\left(\frac{\partial V}{\partial T}\right)_P$$

34. Derive an expression for $(\partial T/\partial P)_H$ for a Berthelot gas.

35. A 3.00-mol sample of a gas with the equation of state $PV = n(RT + BP)$, where $B = 30\ \text{cm}^3\cdot\text{mol}^{-1}$ and for which $\bar{C}_P = (27.20 + 4.81 \times 10^{-3}T)$ in units of $\text{J}\cdot\text{K}^{-1}\cdot\text{mol}^{-1}$, undergoes the irreversible change in state (600 K, 10.0 atm) → (300 K, 5.0 atm). Calculate ΔS, ΔH, and ΔU for the gas. Can ΔG and ΔA be calculated from the data given? How do you reconcile the sign of ΔS_{gas} that you calculated with the second law?

36. Consider a 2.50-mol sample of a gas with $\bar{C}_P = 5R/2$ and for which

$$PV = n(RT + BP) \qquad B = 82\ \text{cm}^3\cdot\text{mol}^{-1}$$

Starting from first principles, together with any necessary definitions, compute the following:

(a) The final temperature of the gas when it undergoes a Joule-Thomson expansion from 500 K, 50 atm to a final pressure of 1.0 atm.

(b) ΔH and w for part (a).

(c) ΔS_{gas} and ΔS_{sur} for part (a).

37. (a) Suppose that we allow 3.50 mol of an ideal gas with $\bar{C}_V = 5R/2$ to expand isothermally and reversibly from 100 atm, 10 L to 10.0 atm, and then the gas is allowed to expand adiabatically and reversibly to a final pressure of 1.00 atm. Calculate q, w, ΔU, and ΔH for each step and the total values for the two steps. Suppose now that the processes are carried out irreversibly with P dropping discontinuously from 100 atm to 10.0 atm in the isothermal expansion and from 10.0 atm to 1.0 atm in the adiabatic expansion. Compute q, w, ΔH, and ΔU for each of the two steps. Compare with the reversible case.

38. Consider the adiabatic reversible compression of n mol of a gas with the equation of state

$$PV = n(RT + BP) \qquad B \text{ is a constant}$$

and for which

$$C_V = n(a + bT) \qquad a \text{ and } b \text{ are constants}$$

from an initial state P_1, T_1 to a final state with pressure P_2. First show that $(\partial U/\partial V)_T = 0$ for the gas and, using this result, verify the following results:

(a) $a \ln\left(\dfrac{T_2}{T_1}\right) + b(T_2 - T_1) = R \ln\left(\dfrac{P_2 T_1}{P_1 T_2}\right)$

(b) $\Delta U = n\left[a(T_2 - T_1) + \dfrac{b}{2}(T_2^2 - T_1^2)\right]$

(c) $\Delta H = n\left[(a + R)(T_2 - T_1) + B(P_2 - P_1) + \dfrac{b}{2}(T_2^2 - T_1^2)\right]$

(d) $\Delta G = \Delta H - S_1(T_2 - T_1)$

(e) Given n, a, b, and B, together with P_1, T_1, and P_2, can a value be calculated for ΔG? Explain.

(f) Given n, P_1, and T_1, are we free to specify both P_2 and T_2? Explain.

39. Consider the gas described in Problem 38 in the state P_2, T_2; suppose that the gas is allowed to expand adiabatically into a vacuum to a final pressure P_1 that is the same as the initial pressure of the gas in Problem 38. Obtain expressions for q, w, ΔU, ΔH, ΔA, ΔG, and ΔS for the gas. Also obtain expressions for ΔS_{sur} and ΔS_{tot} for the process. Is the final temperature of the gas equal to the initial temperature of the gas in Problem 38?

40. A 3.00-mol sample of an ideal gas with $\bar{C}_V = 3R$ undergoes, in an irreversible process, the change in state

$$600 \text{ K, } 10 \text{ atm} \longrightarrow 300 \text{ K, } 5.0 \text{ atm}$$

(a) Calculate ΔU, ΔH, and ΔS for the gas.

(b) How do you reconcile the sign of ΔS_{gas} with the second law?

(c) Is it possible to compute numerical values for ΔG and ΔA for the gas? Explain.

41. Given that $\Delta S = 40.0 \text{ J} \cdot \text{K}^{-1} \cdot \text{mol}^{-1}$ for the following change in state of a gas with $\bar{C}_P = 5R/2$:

$$\text{gas}(50 \text{ atm, } 400 \text{ K}) \longrightarrow \text{gas}(0.10 \text{ atm, } 400 \text{ K})$$

Compute the final temperature when the gas expands adiabatically and reversibly from 50 atm, 400 K to 0.10 atm. [*Hint:* Set up a closed cycle involving the three states (50 atm, 400 K; 0.10 atm, 400 K; and 0.10 atm, T) and use the state function property of S.]

42. Consider a gas with the equation of state

$$PV = n\left(RT - \frac{aP}{T}\right) \qquad a \text{ is a constant}$$

Suppose that n mol of this gas undergoes an isothermal expansion into a vacuum to a final volume $V_2 = 10V_1$. Obtain expressions for q, w, ΔU, ΔS, ΔG, and ΔS_{tot} for this process.

43. Show that for an adiabatic irreversible expansion of an ideal gas for which the external pressure drops instantaneously from P_1 to P_2 that

$$T_2 = \frac{T_1}{\bar{C}_P}\left[\bar{C}_V + R\left(\frac{P_2}{P_1}\right)\right]$$

Given $T_1 = 300$ K, $P_1 = 30$ atm, and $P_2 = 10$ atm, compute T_2, w, ΔU, ΔH, ΔS, and ΔS_{tot}. Take $\bar{C}_P = \bar{C}_V + R = 7R/2$, and one mole.

44. (a) Prove that $C_P \geqslant C_V$.
(b) What is the maximum possible value of $\gamma = C_P/C_V$?

45. Show that for an ideal gas

(a) $$\left(\frac{\partial P}{\partial V}\right)_T = -\frac{nRT}{V^2}$$

(b) $$\left(\frac{\partial P}{\partial V}\right)_S = \frac{nR}{V}\left[\left(\frac{\partial T}{\partial V}\right)_S - \frac{T}{V}\right]$$

46. In magnetic systems with $\delta w_{\text{tot}} = -P\,dV + \mathscr{H}\,dM$, it is convenient to *define enthalpy as*

$$H_m \equiv U + PV - \mathscr{H}M$$

Show that for such a definition of H_m we have

$$C_{P,\mathscr{H}} = \left(\frac{\partial H_m}{\partial T}\right)_{P,\mathscr{H}}$$

and derive a relationship between $C_{P,\mathscr{H}}$ and $C_{P,M}$. Also derive expressions for dA and dG and write all the Maxwell relations arising from the dA and dG expressions.

47. (a) Consider a column of gas of height h held at temperature T along its entire length in a gravitational field. Assume ideal-gas behavior and show that at equilibrium the ratio of gas pressures at heights h and 0 is given by

$$(P_h/P_0) = e^{-Mgh/RT}$$

where M is the molecular weight of the gas and g is the gravitational acceleration. Calculate the value of the ratio P_h/P_0 for a 10-km high column of N_2 at 300 K.

(b) Show that the pressure at depth d in a condensed phase exposed to the atmosphere is given by

$$P_d = P_0 + \rho g d$$

where ρ is the density of the phase and P_0 is the atmospheric pressure.

48. Show that when a gas with $(\partial U/\partial V)_T > 0$ expands adiabatically into a vacuum the gas temperature must decrease.

49. Derive the equation

$$\left(\frac{\partial C_P}{\partial P}\right)_T = -T\left(\frac{\partial^2 V}{\partial T^2}\right)_P$$

Also derive an expression for $(\partial C_V/\partial V)_T$.

50. Given the equations of state
(a) $PV = n(RT + BP)$

(b) $PV = n\left(RT + \dfrac{aP}{T}\right)$

Obtain expressions for $(\partial U/\partial V)_T$, $(\partial H/\partial P)_T$, $(\partial S/\partial V)_T$, $(\partial S/\partial P)_T$, $(\partial T/\partial V)_U$, and $(\partial T/\partial P)_H$.

51. (a) Show that $(\partial U/\partial V)_T = 0$ for any gas whose equation of state is of the form $Pf(V) = nRT$, where $f(V)$ is a function only of the volume.
(b) Show that $(\partial U/\partial V)_T = f_2(V)$ for any gas whose equation of state is of the form

$$[P + f_2(V)]f_1(V) = nRT$$

Give two examples considered in the text of gases with equations of state in class a and one example in class b.

52. Show that, for an electrochemical system ($\delta w_{\text{elec}} = -\mathscr{E}\,dZ$),

$$C_{V,Z} = \left(\frac{\partial U}{\partial T}\right)_{V,Z} \quad \text{and} \quad C_{P,Z} = \left(\frac{\partial H}{\partial T}\right)_{P,Z}$$

53. Consider a Hooke's law spring

$$\mathscr{T} = (a - bT)(L - L_0) \qquad (a, b \text{ positive})$$

where a, b, and L_0 (the length at zero tension) are constants. Take $G = U + PV - TS$ and show that
(a) $dG = -S\,dT + V\,dP + \mathscr{T}\,dL$

(b) $\left(\dfrac{\partial S}{\partial L}\right)_{T,P} = -\left(\dfrac{\partial \mathscr{T}}{\partial T}\right)_{L,P}$

If the spring is stretched from L_0 to $2L_0$ with T and P fixed, then show that

(with $\Delta V = 0$)

$$\Delta S = \frac{bL_0^2}{2}$$

$$\Delta G = \tfrac{1}{2}(a - bT)L_0^2$$

$$\Delta U \approx \Delta H = \frac{aL_0^2}{2}$$

Also obtain expressions for q_{sys}, w_{sys}, and ΔS_{sur}. Does a Hooke's law spring become stronger or weaker as T increases? Obtain an expression for the difference in the ΔH of solution in $HNO_3(aq)$ for identical stretched ($L = 2L_0$) and unstretched ($L = L_0$) Hooke's law springs.

54. Derive a general relationship between C_L and $C_{\mathcal{T}}$ for a system with $\delta w_{tot} = \mathcal{T}\, dL$. Which of these two heat capacities would be the easier to determine experimentally for an elastic cylinder?

55. Repeat Problem 53 for an elastic cylinder for which

$$\mathcal{T} = AT\left(\frac{L}{L_0} - \frac{L_0^2}{L^2}\right) \qquad (A > 0)$$

Also show that

$$\left(\frac{\partial U}{\partial L}\right)_{T,V} = \mathcal{T} - T\left(\frac{\partial \mathcal{T}}{\partial T}\right)_{L,V}$$

An ideal elastic cylinder has $(\partial U/\partial L)_{T,V} = 0$: Is this elastic cylinder ideal? Suppose an ideal elastic cylinder and a Hooke's law spring are stretched from $L_0 \to 2L_0$ adiabatically: Will the temperature of (a) the spring and (b) the cylinder increase or decrease? Suppose that two 100-g masses are separately suspended from an ideal elastic cylinder and a Hooke's law spring, respectively. Both hanging masses are exactly at the same height. If the temperature is increased, will the 100-g masses move up or down or remain unchanged in height in each case?

56. Show that

$$\left(\frac{\partial P}{\partial V}\right)_S = \left(\frac{\partial P}{\partial V}\right)_T \left(\frac{\partial S}{\partial T}\right)_P \left(\frac{\partial T}{\partial S}\right)_V$$

[*Hint:* Take $S = S(P, V)$ and $T = T(P, V)$; write out expressions for dS and dT and proceed.]

57. Derive the expression

$$C_P = \alpha TV\left(\frac{\partial P}{\partial T}\right)_S$$

58. Show that the isothermal compressibility β and the adiabatic compressibility β_S, where

$$\beta_S = -\frac{1}{V}\left(\frac{\partial V}{\partial P}\right)_S$$

are related by the expression

$$\beta = \beta_S + \frac{\alpha^2 T}{\rho C_P^*}$$

where α is the isobaric expansion coefficient, ρ is the density, and C_P^* is the specific heat (i.e., the heat capacity per gram) at constant pressure.

59. The velocity of sound v in a gas is given by

$$v^2 = -\frac{\bar{V}^2}{M}\left(\frac{\partial P}{\partial \bar{V}}\right)_S$$

where M is the molecular weight. Show that for an ideal gas

$$v = \left(\frac{\gamma RT}{M}\right)^{1/2} \qquad \gamma = \frac{C_P}{C_V}$$

Show that for a van der Waals gas

$$v^2 = \frac{\gamma}{M}\left[\frac{RTV^2}{(V - nb)^2} - \frac{2na}{V}\right]$$

60. A sample of gas is placed in a large heavy-walled glass bottle fitted with a stopcock and a pressure gauge. The pressure of gas in the bottle is initially 1.500 atm and the gas temperature is T_1. Some of the gas is then allowed to expand adiabatically and slowly through the stopcock until the gas pressure in the bottle reaches 1.000 atm. The remaining gas is now allowed to return to its initial temperature T_1; at this temperature the gas pressure was observed to be 1.176 atm. Calculate $\gamma = C_P/C_V$ for the gas.

61. Write the appropriate analogs of Eqs. 5-74, 5-75, and 5-77 for the cases in which Y is equal to V and G, respectively.

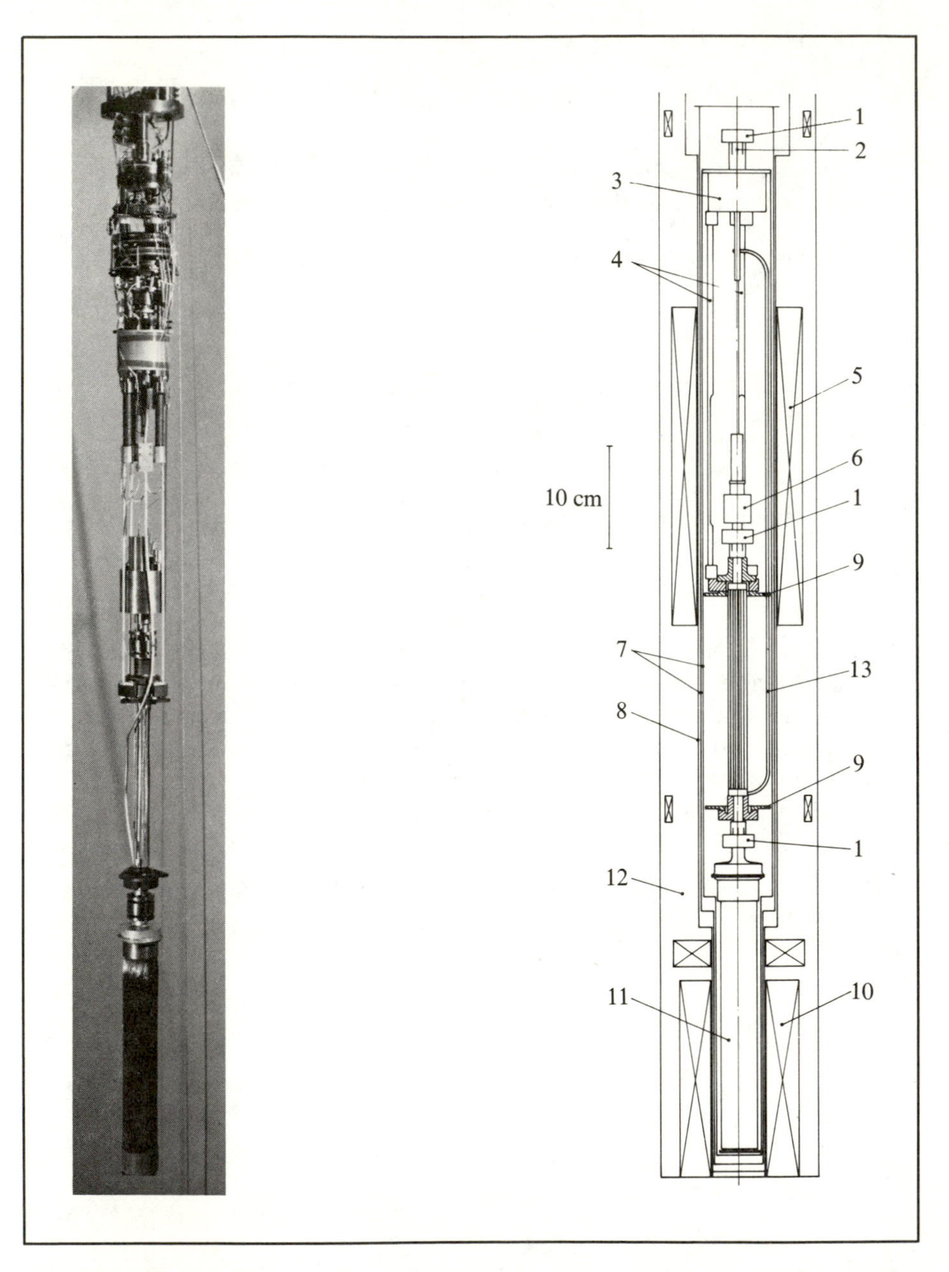

Copper Nuclear Adiabatic Demagnetization Cryostat. *The heat leak into the nuclear stage is 1.2 nW and the nuclear stage will cool 20 cm^3 of ^{3}He(l) to 0.16 mK. Temperatures below 1 mK can be maintained for several days. The key to the schematic is as follows: (1) demountable thermal connector; (2) location of superconducting heat switch; (3) mixing chamber with annular hole along central axis; (4) three phenol fiber support struts; (5) NMR magnet; (6) one of many interchangeable experimental cells; (7) lK and MC thermal radiation shields; (8) vacuum can; (9) graphite positioning flange; (10) demagnetization magnet; (11) copper wire bundle; (12) ^{4}He bath; (13) 3 mm dia silver rod for thermal contact with heat switch.*

6

The Third Law of Thermodynamics and Absolute Entropies

For any isothermal process

$$\lim_{T \to 0} \Delta S_T = 0.$$

W. Nernst, 1906

Absolute zero is unattainable.

W. Nernst, 1912

The entropy of a solid or liquid chemically homogeneous substance has the value of zero at the absolute zero of temperature.

M. Planck, 1912

If the entropy of each element in some crystalline state be taken as zero at the absolute zero of temperature: every substance has a finite positive entropy, but at the absolute zero temperature the entropy may become zero, and does so become in the case of perfect crystalline substances.

G. N. Lewis and M. Randall,
Thermodynamics. New York:
McGraw-Hill, 1923.

6-1 *The Entropy Change for an Isothermal Process Is Zero in the Limit of Zero Temperature*

It was first observed by T. W. Richards around the turn of the century that if ΔG and ΔH for a chemical reaction are plotted as a function of the absolute temperature on the same graph, then results like those shown in Figure 6-1 are obtained. The nature of the plots shown in Figure 6-1 suggest that as the temperature decreases, ΔG and ΔH approach equality. From the definition of G, we have

$$G = H - TS$$

and thus for some isothermal process

$$\Delta G = \Delta H - T\Delta S \tag{6-1}$$

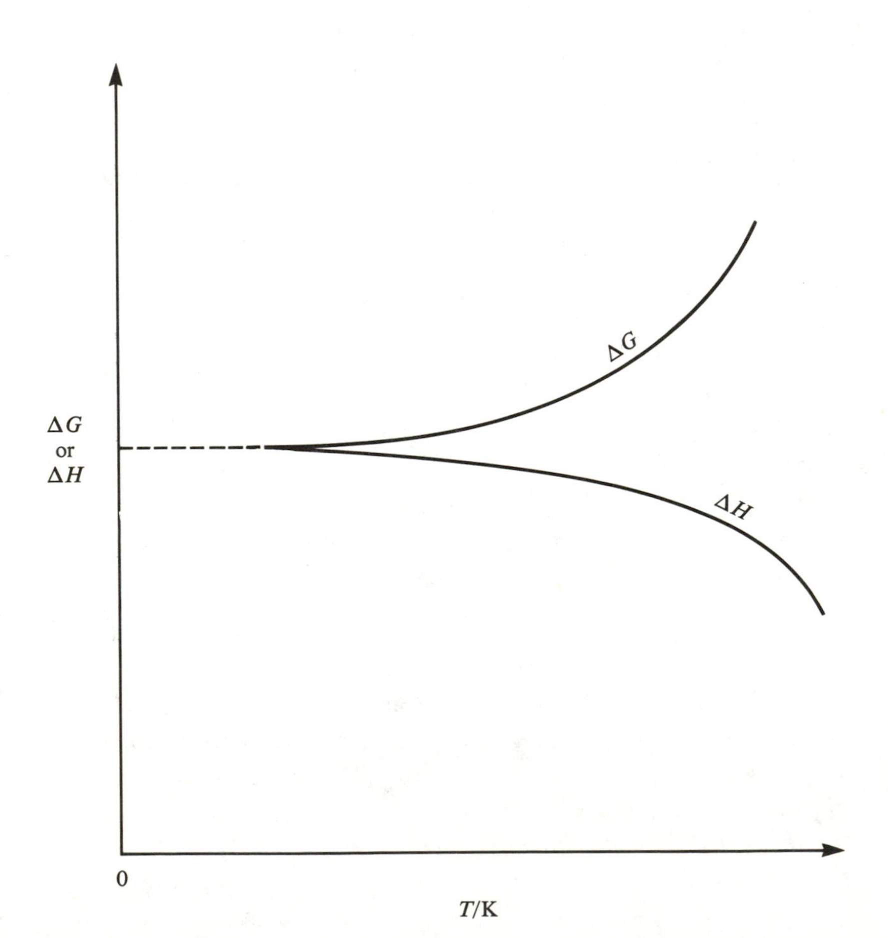

Figure 6-1. Plot of ΔG and ΔH for a process versus the absolute temperature. Note that as T decreases, $\Delta G \rightarrow \Delta H$ for the process. In some cases the ΔG and ΔH curves are interchanged relative to that shown above, but the curves still become coincident as $T \rightarrow 0$.

The observation that ΔG approaches ΔH in magnitude with decreasing T at a much faster rate than T approaches zero, led W. Nernst to the hypothesis that ΔS for an isothermal process approaches zero as T approaches zero; that is,

$$\lim_{T\to 0} \Delta S_T = 0 \tag{6-2}$$

Further studies and analysis led to a modification of Nernst's hypothesis, namely, that Eq. 6-2 holds, in general, only for a process involving phases in internal equilibrium. A phase in *internal equilibrium* is a phase at its minimum possible Gibbs energy at each T and P. Equation 6-1 also holds for substances that have internal *metastability* (e.g., strains, configurational disorder, or very finely divided particles), provided only that the process does not remove the metastability.

The *process* may be, for example, a chemical reaction, a change of phase, or a change of applied magnetic or electric field:

$$2\,\mathrm{Ag}(s) + \mathrm{PbSO_4}(s) \rightleftharpoons \mathrm{Ag_2SO_4}(s) + \mathrm{Pb}(s)$$

$$\mathrm{CaCO_3(aragonite)} \rightleftharpoons \mathrm{CaCO_3(calcite)}$$

$$\mathrm{CuSO_4}(s, \mathscr{H} > 0) \rightleftharpoons \mathrm{CuSO_4}(s, \mathscr{H} = 0)$$ ($\mathscr{H}$ represents an applied magnetic field)

Equation 6-2 is known as the *Third Law of Thermodynamics*. The third law is different from the first two laws in two significant respects. First, the third law does not give rise to a new thermodynamic state function (the first law gives rise to U, and the second law gives rise to S). Second, the third law is applicable only in the limit of the absolute zero of temperature, whereas the first and second laws hold at all temperatures. What the third law does say is that the entropy change ΔS_T for any isothermal thermodynamic process becomes equal to zero in the limit of zero temperature. For example, consider the phase change involving two different solid modifications of metallic tin:

$$\mathrm{Sn}(s, \text{gray}) \rightleftharpoons \mathrm{Sn}(s, \text{white})$$

These two phases of tin differ in their crystal structures. At 13.2°C, the temperature at which the two phases are in equilibrium at 1 atm, the value of ΔS for the conversion of gray to white tin is $+7.82\ \mathrm{J\cdot K^{-1}\cdot mol^{-1}}$ or, in other words,

$$\bar{S}(\text{white tin}) - \bar{S}(\text{gray tin}) = 7.82\ \mathrm{J\cdot K^{-1}\cdot mol^{-1}} \qquad (\text{at } 13.2^\circ\mathrm{C})$$

What the third law tells us is that in the limit of absolute zero there is no entropy difference between white and gray tin. The value of the entropy difference between white and gray tin decreases monotonically from $7.82\ \mathrm{J\cdot K^{-1}\cdot mol^{-1}}$ to zero as T decreases from 286.4 K to 0 K. For the reaction

$$2\,\mathrm{Ag}(s) + \mathrm{PbSO_4}(s) \rightleftharpoons \mathrm{Pb}(s) + \mathrm{Ag_2SO_4}(s)$$

at 298.15 K and 1 atm we find $\Delta S_{rxn} = 7.48\ \mathrm{J\cdot K^{-1}}$. The third law says that the value of ΔS_{rxn} approaches zero as T approaches 0 K and, in the limit at $T = 0$, $\Delta S_{rxn} = 0$.

Max Planck, who was the first to write the equation (see Chapter 4)

$$S = k \ln W \tag{6-3}$$

and who was also familiar with the new ideas in quantum theory that arose in the 1910s, reasoned that at absolute zero there was only one way in which the energy could be distributed in a system; that is, the atoms (or molecules) and the electrons were all in the lowest available quantum states and therefore $W = 1$ and $S = 0$ at absolute zero. Thus Planck offered a different version of the third law, namely, that

$$\lim_{T \to 0} S_T = 0 \tag{6-4}$$

Planck's hypothesis went beyond Nernst's hypothesis in that Planck stated that the entropy of a substance becomes equal to zero at absolute zero. Nernst's statement of the third law (Eq. 6-2) follows directly from Planck's statement of the third law (Eq. 6-4), but the reverse is not true; that is, Eq. 6-4 does not follow from Eq. 6-2. Furthermore, Planck's statement of the third law suggests that we can set up an *absolute entropy scale.* The entropy of a substance is simply set equal to zero (exactly) at absolute zero; then the entropy increase from 0 K to some temperature of interest is determined experimentally, and the result is set equal to the absolute entropy of the substance at that temperature.

The reason that the Nernst statement of the third law does not necessarily imply the Planck statement of the third law is that the atoms are conserved in all nonnuclear chemical transformations and reactions, and thus contributions to the entropy arising solely from the nuclei cancel out in any chemical reaction. For example, consider the general chemical reaction

$$a\text{A} + b\text{B} + \cdots \rightleftharpoons x\text{X} + y\text{Y} + \cdots \tag{6-5}$$

where a is the number of moles of A, b is the number of moles of B, and so forth. Let $\overline{S}_{\text{A},T}$ be the molar entropy of A at T, $\overline{S}_{\text{B},T}$ be the molar entropy of B at T, and so forth. The entropy change for reaction 6-5 under isothermal conditions is then

$$\Delta S_{\text{rxn},T} = (x\overline{S}_{X,T} + y\overline{S}_{Y,T} + \cdots) - (a\overline{S}_{\text{A},T} + b\overline{S}_{\text{B},T} + \cdots) \tag{6-6}$$

The entropy change of the reaction at 0 K is

$$\Delta S_{\text{rxn},0} = (x\overline{S}_{X,0} + y\overline{S}_{Y,0} + \cdots) - (a\overline{S}_{\text{A},0} + b\overline{S}_{\text{B},0} + \cdots) \tag{6-7}$$

According to Nernst, $\Delta S_{\text{rxn},0} = 0$; whereas, according to Planck, $\overline{S}_{X,0} = \overline{S}_{Y,0} = \overline{S}_{\text{A},0} = \overline{S}_{\text{B},0} = 0$. Because the nuclei are conserved in the balanced chemical reaction, the Nernst version of the third law could still hold even though the Planck version of the third law did not hold, provided that the source of the failure of the Planck version was associated with entropy contributions arising from the nuclei.

It is known from quantum mechanics that there is a contribution to the entropy arising from the existence of nuclear spin. The maximum nuclear spin entropy per mole of a compound is given by

$$\bar{S} = k \ln W = R \ln \left[\prod_j (2I_j + 1) \right] \tag{6-8}$$

where I_j is the nuclear spin of the ith nucleus in the compound, and the symbol $\prod_j$ denotes a repeat product:

$$\prod_{j=1}^{n} (2I_j + 1) = (2I_1 + 1)(2I_2 + 1)(2I_3 + 1) \cdots (2I_n + 1)$$

EXAMPLE 6-1

The nuclear spin of the ^1H is $\frac{1}{2}$, whereas the nuclear spin of ^{16}O is 0. Calculate the nuclear spin entropy of a mole of $^1H_2{}^{16}O$.

Solution: From Eq. 6-8 we have

$$\bar{S} = R \ln [2(\tfrac{1}{2}) + 1]^2[2(0) + 1] = R \ln 4 = 11.53 \text{ J}\cdot\text{K}^{-1}\cdot\text{mol}^{-1}$$

It is found experimentally that the nuclear spin entropy is removed (i.e., all nuclei go into their lowest nuclear spin state) at very low temperatures ($T < 1$ K), as predicted by the Planck version of the second law. However, it is also known from experiment that the entropy associated with the existence of mixtures of isotopes in chemical compounds is not removed at low temperatures. For example, ice contains the distinguishable species $^1H_2{}^{16}O$, $^1H_2{}^{17}O$, $^1H_2{}^{18}O$, $^2H_2{}^{16}O$, $^2H_2{}^{17}O$, $^2H_2{}^{18}O$, $^1H^2H^{16}O$, $^1H^2H^{17}O$, and $^1H^2H^{18}O$. There is an entropy term called the *entropy of mixing*,

$$\Delta\bar{S}_{\text{mix}} = -R \sum_i X_i \ln X_i \qquad (X_i \text{ is the mole fraction of species } i) \tag{6-9}$$

arising from the fact that there is a configurational disorder associated with the mixture of isotopic species. We shall discuss ΔS_{mix} in detail in Chapter 11; however, we note here that if the Planck version of the third law were correct, then we would expect that as $T \to 0$, the various isotopic species would unmix and thereby become segregated in separate phases. Isotopic segregation does not happen for solids because of the large energy barriers associated with moving molecules in the solid phase. For this reason we shall adopt the Nernst version of the third law. Furthermore, the Nernst version of the third law avoids uncertainties as to contributions to the entropy arising from possibly undiscovered nuclear properties. It is interesting to note, however, that an isotopic mixture of the helium isotopes ^3He and ^4He in liquid helium, which is the only substance capable of existence as a liquid as $T \to 0$, does indeed spontaneously segregate into a ^3He phase and a ^4He phase at $T < 1$ K. The Planck version of the third law offers some significant insights into the entropy function that we shall continue to call upon.

Although the third law stated in the form

$$\lim_{T \to 0} \Delta S_T = 0$$

does not imply the existence of an absolute entropy scale, it is desirable, for practical computational purposes, to be able to specify a number for the entropy S of a substance at temperature T. We have already seen in Chapter 4 that the temperature dependence of the entropy at constant pressure is given by

$$\left(\frac{\partial S}{\partial T}\right)_P = \frac{C_P}{T} \tag{6-10}$$

The change in entropy associated with a change in phase is given by

$$\Delta S_{\text{tr}} = \frac{\Delta H_{\text{tr}}}{T_{\text{tr}}} \tag{6-11}$$

Thus we have for the entropy of some substances at temperature T, assuming no phase transitions between 0 and T,

$$S_T = S_0 + \int_0^T \frac{C_P}{T}\, dT \tag{6-12}$$

where S_0 is the entropy at absolute zero. The entropy of a substance at T that undergoes a phase transition (α phase to the β phase) at $T_{\text{tr}} < T$ is given by

$$S_T = S_0 + \int_0^{T_{\text{tr}}} \frac{C_P^\alpha}{T}\, dT + \frac{\Delta H_{\text{tr}}}{T_{\text{tr}}} + \int_{T_{\text{tr}}}^T \frac{C_P^\beta}{T}\, dT \tag{6-13}$$

A graph of S_T versus T for oxygen is given in Figure 4-1. Whenever we have a phase transition for which S undergoes a discontinuity ($C_P = \infty$ at T_{tr}), then the integration specified in Eq. 6-12 must be stopped at the transition temperature; ΔS_{tr} is then added in, and the integration is continued above the transition temperature to the next transition or to the desired T, whichever comes first. Because C_P and ΔH_{tr} are calorimetric quantities, we can obtain $S_T - S_0$ from calorimetric data alone.

All that is required to establish a *conventional* (i.e., practical) *scale of absolute entropies* for tabulational purposes is to set $S_0 = 0$ *arbitrarily* for each substance. It is important to note that the practical "absolute" entropy S_T that we tabulate does not (by convention) contain the entropy contributions from the individual nuclei (nuclear spin entropy) nor does it include the entropy of isotope mixing. We can establish such a *practical absolute entropy scale* because it is known that a substance does not segregate itself into isotopically pure phases as its temperature is lowered nor is the temperature of third-law calorimetric studies usually taken to a low enough value to remove the nuclear spin entropy. In other words, for any chemical process in which nuclei are conserved, $\Delta S_0 = 0$ (given the qualification stated in the third law), even though the individual S_0 values may not be zero.

6-2 *The Third Law Predicts That* $\lim_{T\to 0} C_P = 0$

The value of ΔS_T for any process is finite at all temperatures. In addition, the value of ΔS_{tr} for a phase transition (low- to the high-temperature form) involving phase separation is always positive. The second law requires that the integrals in Eq. 6-13 must be finite and positive, and thus C_P must remain finite and positive at all T other than that of a phase transition. The third law requires that C_P must go to zero as T goes to zero. The proof that $\lim_{T\to 0} C_P = 0$ follows. The Gibbs energy is defined as

$$G = H - TS$$

Thus from the third law

$$\lim_{T\to 0} S_T = \lim_{T\to 0}\left(\frac{H - G}{T}\right) = S_0$$

Because of the value $\lim_{T\to 0}[(H - G)/T] = 0/0$, we must use l'Hôpital's rule to find the limit:

$$\lim_{T\to 0} S_T = \lim_{T\to 0}\left[\frac{(\partial H/\partial T)_P - (\partial G/\partial T)_P}{1}\right] = \lim_{T\to 0}(C_P + S_T) = S_0$$

but $\lim_{T\to 0} S_T = S_0$; therefore,

$$\lim_{T\to 0} C_P = 0 \tag{6-14}$$

All crystalline substances investigated have low-temperature C_P behavior consistent with Eq. 6-14. For example, as $T \to 0$, the following dependencies of C_P on T have been found:

$C_P \longrightarrow aT^3$ (most nonmagnetic, nonmetallic crystals)

$C_P \longrightarrow bT^2$ (layer lattice crystals, like graphite and boron nitride, and surface heat capacity)

$C_P \longrightarrow \gamma T + aT^3$ (metals)

$C_P \longrightarrow jT^{3/2} + aT^3$ (ferrimagnetic crystals below the magnetic transition temperature, if it occurs at low T)

$C_P \longrightarrow mT^3$ (antiferromagnetic crystals below the magnetic transition temperature, if it occurs at low T)

(a, b, γ, j, and m are constants for a given substance.)

In addition to the low-temperature behavior of C_P, the third law also predicts the low-temperature behavior of some other thermodynamic quantities. Consider the derivative $(\partial S/\partial X)_T$, where X is any thermodynamic state function. We know from the third law that $\lim_{T\to 0} S_T = S_0 =$ constant. Therefore,

$$\lim_{T\to 0}\left(\frac{\partial S}{\partial X}\right)_T = \left[\frac{\partial}{\partial X}\left(\lim_{T\to 0} S_T\right)_T\right] = \left(\frac{\partial S_0}{\partial X}\right)_T = 0 \tag{6-15}$$

The fact that T is held constant in the differentiation enables us to interchange the $\lim_{T\to 0}$ operation with $\partial/\partial X$.

As an example of the application of Eq. 6-15, we have (using Eqs. 5-34 and 5-39)

$$\left(\frac{\partial S}{\partial P}\right)_T = -\left(\frac{\partial V}{\partial T}\right)_P = -\alpha V$$

From Eq. 6-15 we have

$$\lim_{T\to 0}\left(\frac{\partial S}{\partial P}\right)_T = 0 = \lim_{T\to 0}(-\alpha V)$$

Because V remains positive and nonzero as $T \to 0$, we conclude that the coefficient of isothermal expansion goes to zero as T goes to zero; that is,

$$\lim_{T\to 0} \alpha = 0$$

which is observed experimentally.

EXAMPLE 6-2

Consider a system for which the equation of state

$$\left(\frac{\partial U}{\partial V}\right)_T = -P + kTP^2$$

(where k = constant) holds around room temperature. Can this equation be valid as $T \to 0$?

Solution: On comparing the equation of state with the general expression (Eq. 5-35)

$$\left(\frac{\partial U}{\partial V}\right)_T = T\left(\frac{\partial S}{\partial V}\right)_T - P$$

we note that $(\partial S/\partial V)_T = kP^2$. From the third law we know that $\lim_{T\to 0}(\partial S/\partial V)_T = 0$. Because T and P are independent variables, the $\lim_{T\to 0}(kP^2) \neq 0$, and hence the equation of state cannot be valid at low T.

6-3 *Absolute Zero Is Unattainable*

We shall now show that the Nernst statement of the Third Law is equivalent to the statement that absolute zero is unattainable. Consider the process

$$A(\mathscr{H} > 0) \longrightarrow B(\mathscr{H} = 0)$$

that is, a change in the externally applied magnetic field involving a phase in internal equilibrium in the states A and B. From the third law we have, at any temperature T,

$$\bar{S}_T^A = \bar{S}_0^A + \int_0^T \bar{C}_X^A \, d \ln T$$

$$\bar{S}_T^B = \bar{S}_0^B + \int_0^T \bar{C}_X^B \, d \ln T$$

where $\bar{S}_T^A$ and $\bar{S}_T^B$ indicate the entropy of some substance at T in the presence and absence of an external magnetic field, respectively. Suppose now that we put the system in the state denoted by A at the temperature T' and cause the process $A \rightarrow B$ to take place *adiabatically*. We shall let the final temperature be T'' and inquire as to whether or not it is possible for $T'' = 0$. From the second law, we know that for any process taking place within an adiabatic enclosure $\Delta S \geqslant 0$; thus T'' will be lowest when the process $A \rightarrow B$ is a reversible adiabatic one, $[\Delta S(A \rightarrow B) = 0]$, because $\int_0^{T''} \bar{C}_X^B \, d \ln T$ will then be smallest. In other words, we have the best possibility of reaching absolute zero when the process is reversible. For the reversible change in state,

$$\bar{S}_0^A + \int_0^{T'} \bar{C}_X^A \, d \ln T = \bar{S}_0^B + \int_0^{T''} \bar{C}_X^B \, d \ln T$$

and if $T'' = 0$, then

$$\bar{S}_0^B - \bar{S}_0^A = \int_0^{T'} \bar{C}_X^A \, d \ln T \tag{6-16}$$

Because $\bar{C}_X^A > 0$ for any nonzero T, and $\bar{S}_0^B - \bar{S}_0^A = 0$ by the third law, there is no nonzero T' that will satisfy Eq. 6-16, and thus absolute zero is unattainable using internally stable phases. Minor modifications of the above arguments show that absolute zero is also unattainable using internally metastable phases (see Problem 29). Thus the Third Law of Thermodynamics is equivalent to the statement that absolute zero is unattainable.

Using adiabatic demagnetization techniques, crystal temperatures as low as 10^{-3} K have been obtained. The predictions of the third law have been verified in a sufficiently large number of cases that experimental attempts to reach absolute zero are now placed in the same class as attempts to devise perpetual motion machines—which is to say there are much more productive ways to spend one's time. Much experimental work is carried out, however, at very low temperatures, because the behavior of matter under these conditions has produced many surprises and led to the uncovering of a great deal of new knowledge and the development of useful new devices, such as superconducting (zero resistance) magnets.

6-4 *Practical Absolute Entropies Can Be Obtained*

Equation 6-13 gives the prescription for obtaining the third-law entropy of a substance. The determination of $C_P = f(T)$ over the range 0 to T, together with the values of $\Delta \bar{H}_{tr}$ and T_{tr} for any phase changes that occur in the range 0 to T, enables us to calculate $\overline{S}_T - \overline{S}_0$. The contribution to $\overline{S}_T$ of each phase of the substance in the range 0 to T is obtained by graphical integration (over the appropriate range of T for that phase) of a plot of C_P/T versus T or C_P versus $\ln T$. Usually the experiments are not carried out below about 15 K, unless there is reason to believe that, for example, magnetic dipole ordering or the cessation of rotation of molecules or ions in the crystal occurs below 10 K; these effects, if present, can make large contributions to the entropy. For example, the magnetic contribution to the entropy per mole can be as large as

$$\overline{S}_{\text{mag}} = R \ln (2J + 1) \tag{6-17}$$

where J is the total angular-momentum (spin plus orbital) quantum number of the unpaired electrons. Thus for $J = \frac{7}{2}$, $\overline{S}_{\text{mag}} = R \ln 8$. If it is assumed that no special disorder exists in the crystal between 0 K and the lowest temperature of the experiments, then the contribution to $\overline{S}$ from the crystal lattice over this range is obtained by utilizing the Debye theory for the heat capacity of solids. Debye theory predicts (assuming a nonmetallic, nonlayer lattice solid) a T^3 dependence for C_P at low T. Thus, if it can be established that the measurements at the lowest temperatures of the experiment (say, 15 K to 30 K) are in the T^3 region, for example, by plotting C_P/T^3 versus T in the experimentally attained low-T region, then the constant of proportionality $a = C_P/T^3$ can be obtained. The T^3 region usually extends from 0 K to about 30 K. Once the T^3 law is established for a substance, then the entropy between, say, 0 K to 15 K can be obtained analytically as follows (taking $\overline{S}_0 = 0$):

$$\overline{S}_{15} = \int_0^{15} \frac{aT^3}{T}\, dT = a \int_0^{15} T^2\, dT = \frac{a}{3} T^3 \bigg]_0^{15} = \frac{a(15)^3}{3}$$

where a is known from experimental data in the 15 K to 30 K range.

In Table 6-1 we have summarized the data on HCl that were used to compute the third-law entropy of $HCl(g)$ at 25°C and 1 atm. Note that the contribution to $\overline{S}$ over the range 0 K to 16 K is not large; this is because $\overline{S}_T - \overline{S}_0$ approaches zero as $T \to 0$. Analogous data for O_2 are plotted in Figure 4-1.

Proceeding in the manner outlined above for $HCl(g)$ and $O_2(g)$, we can obtain the third-law entropies of various substances, and these entropies can in turn be combined to yield ΔS values for chemical reactions. For example, if the practical absolute entropies of all the reactants and products are known, then ΔS_{rxn} is calculated from the expression

$$\Delta S_{\text{rxn}} = \sum_i v_i \overline{S}_i \tag{6-18}$$

The v_i values in Eq. 6-18 are the reaction-balancing coefficients, which are positive for product species and negative for reactant species. Thus for the reaction

$$2\,H_2(g) + O_2(g) \longrightarrow 2\,H_2O(\ell)$$

we have

$$\Delta S_{\text{rxn}} = 2\overline{S}_{H_2O(\ell)} - 2\overline{S}_{H_2(g)} - \overline{S}_{O_2(g)}$$

TABLE 6-1

Contributions to the Third-Law Entropy of HCl(g) at 25°C, 1 atm[a]

T range/K		$J \cdot K^{-1} \cdot mol^{-1}$
0–16.00	Debye T^3 extrapolation for HCl(*s*, I)	1.26
16.00–98.36	Graphical integration of C_P vs. ln T for HCl(*s*, I)	29.54
98.36	Phase transition (*s*, I) → (*s*, II) 1189.2/98.36	12.09
98.36–158.91	Graphical integration of C_P vs. ln T for HCl(*s*, II)	21.13
158.91	Phase transition (*s*, II) → (ℓ) 1994.3/158.91	12.55
158.91–188.07	Graphical integration of C_P vs. ln T for HCl(ℓ)	9.87
188.07	Phase transition (ℓ) → (*g*) 16,147/188.07	85.86
188.07–298.15	Graphical integration of C_P vs. ln T for HCl(*g*)	13.47
298.15	Correction from the real gas at 1 atm to the ideal gas standard state[b]	0.84
	$\bar{S}_{298.15} = 186.61 \pm 0.42 \text{ J} \cdot \text{K}^{-1} \cdot \text{mol}^{-1}$	

[a] Data of E. D. Eastman, W. C. McGavork, and E. D. West.
[b] This correction is discussed in Chapter 9.

and at 25°C

$$\Delta S_{\text{rxn}} = (2\text{ mol})(69.91\text{ J} \cdot \text{K}^{-1} \cdot \text{mol}^{-1}) - (2\text{ mol})(130.15\text{ J} \cdot \text{K}^{-1} \cdot \text{mol}^{-1}) - (1\text{ mol})(205.03\text{ J} \cdot \text{K}^{-1} \cdot \text{mol}^{-1})$$

or

$$\Delta S_{\text{rxn}} = -325.51\text{ J} \cdot \text{K}^{-1}$$

Calorimetrically determined ΔS values can be combined with calorimetrically determined ΔH values for a reaction to obtain the Gibbs energy change for the reaction

$$\Delta G_{\text{rxn}} = \Delta H_{\text{rxn}} - T\,\Delta S_{\text{rxn}} \tag{6-19}$$

Equation 6-19 shows that *the third law enables us to compute the Gibbs energy change for a reaction from thermal data alone.* Involving a substance in a stoichiometrically well-defined and reversible reaction for which the equilibrium distribution of species can be conveniently determined may be difficult or in certain cases impossible, whereas it is usually possible to load the pure substance in a calorimeter and lower its temperature. Likewise, the determination of ΔH_{rxn} for a reaction does not usually require a precise knowledge of the equilibrium distribution of species for the calorimetric reaction. The observed agreement between calorimetrically determined ΔG_{rxn} values and ΔG_{rxn} values obtained from equilibrium studies provides a strong confirmation of the Third Law of Thermodynamics. Such calculations will be illustrated in Chapter 7.

TABLE 6-2

Experimental Test of the Third Law for Allotropic Modifications of a Substance. If the Third Law Holds, Then We Expect $\bar{S}_{0,\beta} - \bar{S}_{0,\alpha} = 0$ within the Experimental Error.[a]

	Sulfur	Phosphine	Tin
α	Rhombic	—	gray
β	Monoclinic	—	white
T_{tr}/K	368.6	49.43	286.4
$\bar{S}(\alpha \to \beta)$at T_{tr}	1.09	15.73	7.82
$\int_0^{T_{tr}} \left(\frac{\bar{C}_{P,\alpha}}{T}\right) dT$	36.86	18.33	38.62
$\int_0^{T_{tr}} \left(\frac{\bar{C}_{P,\beta}}{T}\right) dT$	37.82	34.02	46.74
$\bar{S}_{0,\beta} - \bar{S}_{0,\alpha}$	0.13 ± 0.63	0.04 ± 0.21	-0.30 ± 0.84

[a] All entropies are in $\mathrm{J \cdot K^{-1} \cdot mol^{-1}}$.

Another experimental check on the third law is provided by the entropies of two allotropic modifications of a pure substance that can both be cooled to very low temperatures. Let us suppose that above the 1-atm transition temperature β is the stable modification, whereas below the transition temperature α is the stable modification. Suppose also that neither α nor β is internally metastable below the transition temperature. If the third-law entropies of both the α and β forms are determined from 0 K to the transition temperature, and $\Delta\bar{S}_{tr}$ is also determined in separate experiments, then we can check whether or not $\bar{S}_0(\alpha) - \bar{S}_0(\beta) = 0$, as required by the third law. At the 1-atm transition temperature we have

$$\bar{S}_T(\beta) - \bar{S}_T(\alpha) = \Delta\bar{S}_{tr,T}(\alpha \to \beta)$$

or

$$\bar{S}_T(\beta) = \Delta\bar{S}_{tr,T}(\alpha \to \beta) + \bar{S}_T(\alpha)$$

Using the third law for $\bar{S}_T(\beta)$ and $\bar{S}_T(\alpha)$, we have

$$\bar{S}_0(\beta) + \int_0^T \frac{\bar{C}_P(\beta)\,dT}{T} = \Delta\bar{S}_{tr,T}(\alpha \to \beta) + \bar{S}_0(\alpha) + \int_0^T \frac{\bar{C}_P(\alpha)\,dT}{T}$$

$$0 = \bar{S}_0(\beta) - \bar{S}_0(\alpha) = \Delta\bar{S}_{tr,T}(\alpha \to \beta) + \int_0^T \frac{\bar{C}_P(\alpha)\,dT}{T} - \int_0^T \frac{\bar{C}_P(\beta)\,dT}{T} \qquad (6\text{-}20)$$

The data in Table 6-2 show that Eq. 6-20 holds within the experimental error.

If we determine the third-law entropy of a substance calorimetrically, and for some reason a phase transition is passed over on cooling without the solid undergoing the phase change to the form more stable at low temperatures, then

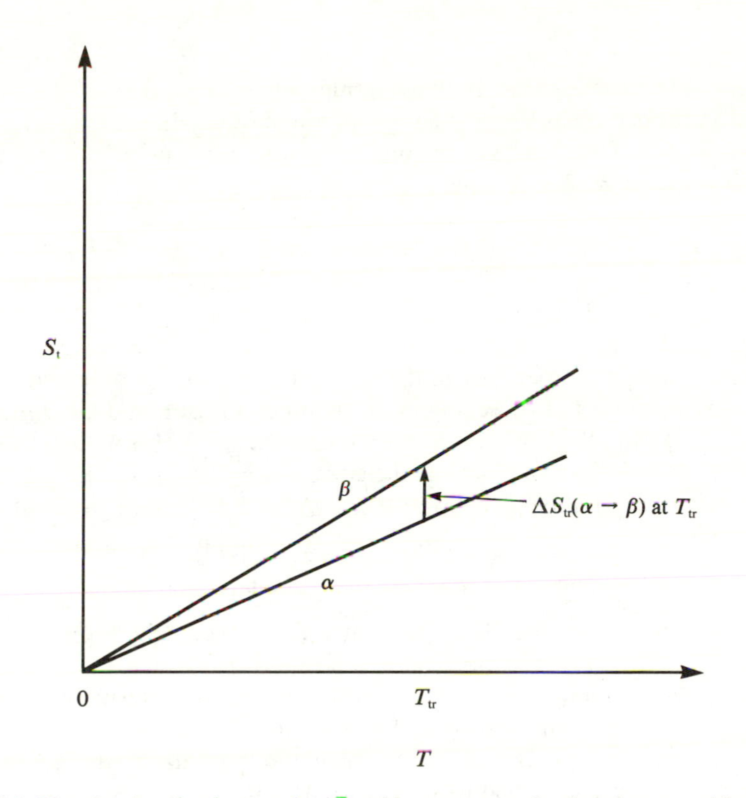

Figure 6-2. Plot showing the decrease in $\Delta\overline{S}_{tr}$ with temperature for the phase transition $\alpha \to \beta$. Note that $\lim_{T\to 0} \Delta\overline{S}_{tr} = 0$ and that at the transition temperature $\overline{S}_\beta = \overline{S}_\alpha + \Delta\overline{S}_{tr(\alpha\to\beta)}$. Note that the entropy at point *a* is the same whether the substance remains as the β phase as $T \to 0$ or the β phase converts to the α phase as $T \to 0$.

the correct third-law entropy of the form of the substance stable above T_{tr} will still be obtained because $\overline{S}_0(\alpha) - \overline{S}_0(\beta) = 0$ (see Figure 6-2).

6-5 *Internal Metastability Can Lead to Frozen-In Entropy as* $T \to 0$

The Third Law of Thermodynamics as originally stated by Nernst had a serious defect in that it did not specify that the phases must be in a state of internal equilibrium; this defect was removed by G. N. Lewis who qualified the Nernst statement of the third law by restricting it to substances that did not involve internal metastability (frozen-in entropy) in the limit $T \to 0$. Any state in which the arrangement of atoms or molecules is disordered will have a higher entropy than a perfectly ordered arrangement. If the disorder persists in the condensed phase down to the lowest temperatures of the calorimetric measurements, then the third-law entropy of the substance will be *lower* than that obtained for the substance as a liquid or gas from equilibrium measurements. In other words, if that amount of entropy due to the disorder remains in the crystal at low T, then it will not show up in the calorimetrically determined S_T. The first experimental work to show this effect was that of G. E. Gibson and W. F. Giauque on glycerin. It is possible to supercool glycerin to very low temperatures, and thus it is possible to obtain and compare third-law entropies for glycerin that (1) has been taken as a crystalline solid from a few degrees kelvin up through the normal melting point to the normal liquid at, say, 25°C and (2) has been taken as a supercooled (glassy) liquid from a few degrees Kelvin up to the normal liquid at 25°C. The entropy between 0 K and the lowest temperature of the measurements can be obtained in either case by extrapolation. Proceeding in this manner, Gibson and Giauque found that $\bar{S}_{298} - \bar{S}_0$ for the crystalline case *exceeded* $\bar{S}_{298} - \bar{S}_0$ for the glassy case by $23.4 \pm 0.4\ \mathrm{J \cdot K^{-1} \cdot mol^{-1}}$. Thus the entropy of glassy glycerin at 0 K must exceed that of crystalline glycerin at 0 K by $23.4\ \mathrm{J \cdot K^{-1} \cdot mol^{-1}}$. That is,

$$[\bar{S}_{298}(\text{liquid}) - \bar{S}_0(\text{crystal})] - [\bar{S}_{298}(\text{liquid}) - \bar{S}_0(\text{glass})]$$
$$= \bar{S}_0(\text{glass}) - \bar{S}_0(\text{crystal}) = 23.4\ \mathrm{J \cdot K^{-1} \cdot mol^{-1}}$$

The higher entropy in the glassy state at 0 K is due to the more disordered arrangement of the glycerin molecules in the glassy state relative to the crystalline state. We conclude that because a glassy material is internally metastable, its conventional absolute entropy at 0 K is greater than zero (see Figure 6-3).

Frozen-in entropy can arise for magnetic crystals, if the calorimetric measurements are not carried to sufficiently low temperatures (sometimes 1 K or less) to remove the magnetic entropy that arises from a random arrangement of the individual magnetic moments (spin disorder). If the measurements are taken down through the magnetic-ordering transition region in which the spins become aligned, then the third-law entropy should agree with that determined from equilibrium data; whereas if the measurements are not carried to a sufficiently low temperature such that the magnetic entropy is removed (spins aligned), then the third-law value of $\bar{S}_T$ will be lower than the "equilibrium" value by as much as $R \ln(2J + 1)$.

Another type of configurational randomness that has been found arises when a molecule can enter a position in the crystal lattice in more than one physically distinguishable, but approximately energetically equivalent, orientation. If N is the maximum number of such possible orientations per molecule, then the entropy per mole arising from this configurational randomness in the crystal is $R \ln N$. For example, J. O. Clayton and W. F. Giauque found that the calorimetrically determined third-law entropy of $CO(g)$ at 298.15 K was about $4.2\ \mathrm{J \cdot K^{-1} \cdot mol^{-1}}$ less than the entropy of $CO(g)$ at 298.15 K determined from equilibrium data. They ascribed the discrepancy to end-for-end disorder (CO versus OC) of the CO molecules in the crystal lattice. In other words, CO cannot tell its O end from its C end when it goes into the crystal lattice. The discrepancy predicted from configuration considerations is $R \ln 2 = 5.76\ \mathrm{J \cdot K^{-1} \cdot mol^{-1}}$. The actual residual entropy is about 73% of this value, which presumably indicates that the molecular orientation is not completely

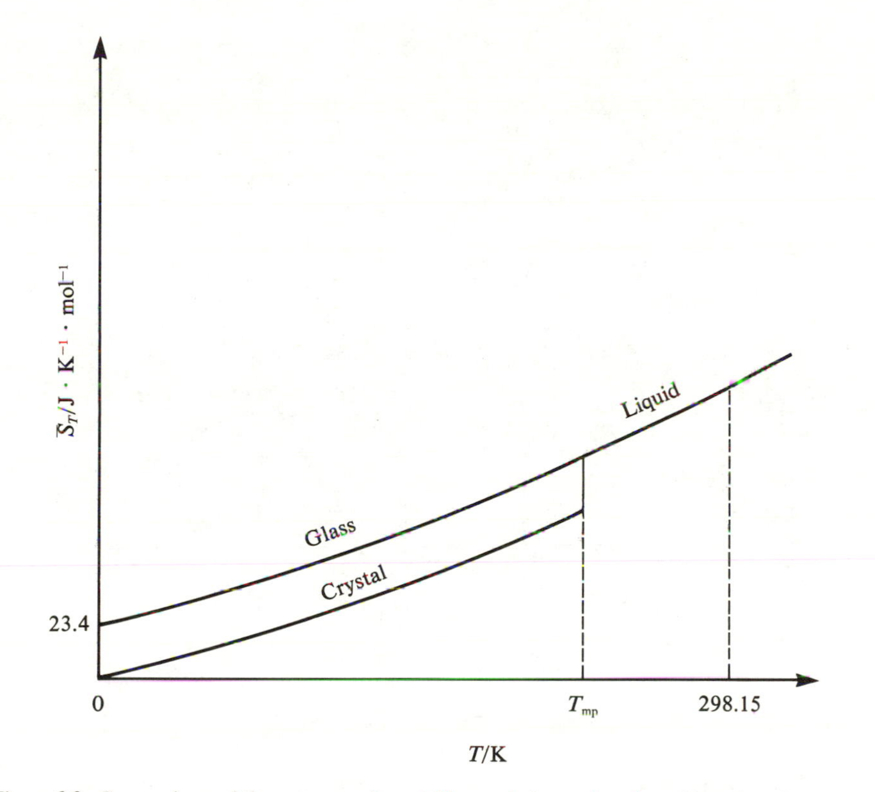

Figure 6-3. Comparison of the entropy of crystalline and glassy glycerine. Note that the entropy of glassy glycerin is 23.4 $J \cdot K^{-1} \cdot mol^{-1}$ higher than the entropy of crystalline glycerin at absolute zero. The difference in entropies at absolute zero arises from the frozen-in disorder (entropy) of glassy glycerin at absolute zero.

random, as required to give $\overline{S}_0 = 5.76\ J \cdot K^{-1} \cdot mol^{-1}$. The case of NO is different than CO because NO goes into the crystal lattice as a *dimer*. The dimers evidently are oriented randomly in the manner

$$\begin{matrix} |\bar{N}-\bar{O}| \\ |\quad\ | \\ |O=N| \end{matrix} \quad \text{or} \quad \begin{matrix} |\bar{O}-\bar{N}| \\ |\quad\ | \\ |N=O| \end{matrix}$$

including appropriate resonance forms. The observed residual entropy of NO(s) is $3.01 \pm 0.12\ J \cdot K^{-1} \cdot mol^{-1}$, and the predicted value is $\frac{1}{2}R \ln 2 = 2.89\ J \cdot K^{-1} \cdot mol^{-1}$, because 1 mol of NO yields $\frac{1}{2}$ mol of $(NO)_2$.

The case of frozen-in entropy in water is particularly interesting. In this case the observed residual entropy at 0 K is $3.43 \pm 0.21\ J \cdot K^{-1} \cdot mol^{-1}$. Ice (Figure 6-4) consists of tetrahedral units of O atoms, with a fifth O atom in the center of the tetrahedron; the whole structure is held together by hydrogen bonds. If we consider the H_2O molecule in the center of the tetrahedron, then each one of its protons can be directed at any one of three different apical O

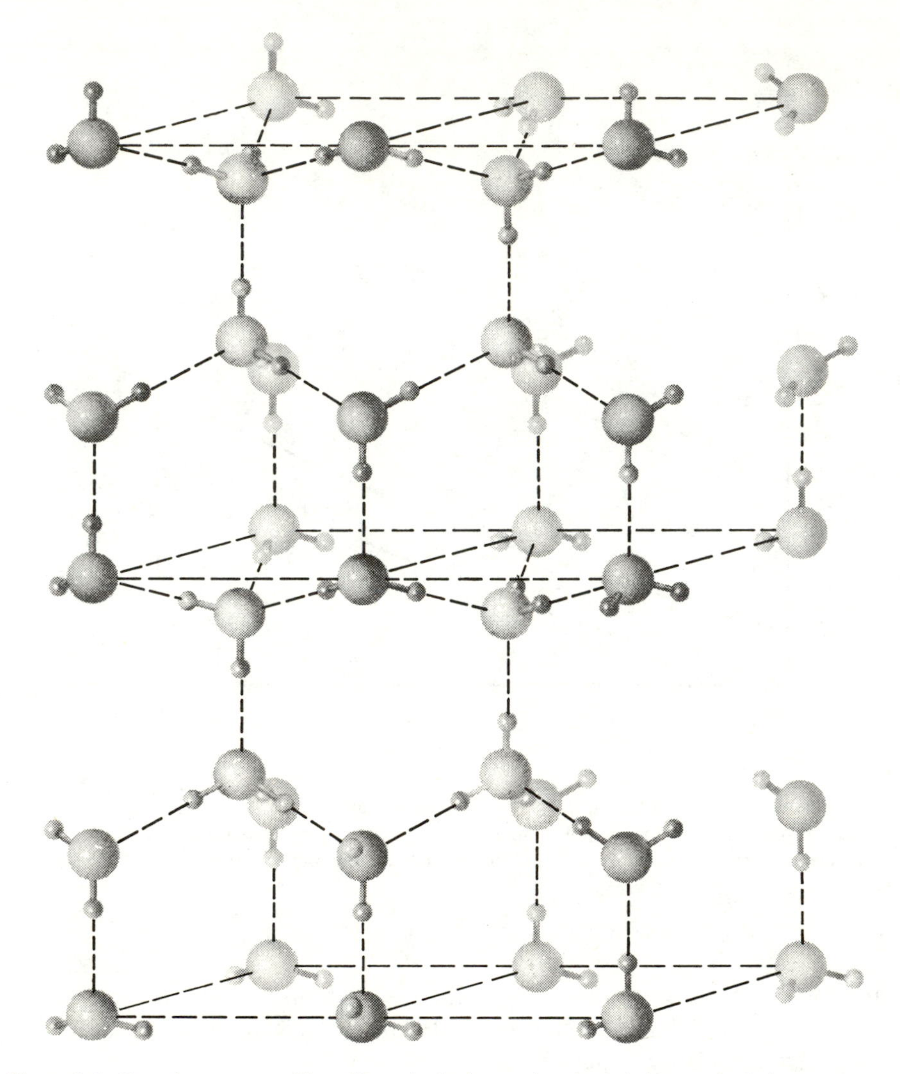

Figure 6-4. Crystal structure of ice. Note the hydrogen bonds and the tetrahedral arrangement of the oxygen atoms. Each oxygen atom is located in the center of a tetrahedron formed by four other oxygen atoms, and each oxygen atom is involved in four hydrogen bonds: two as H-atom donor and two as H-atom acceptor. The entire ice structure is held together by hydrogen bonds. (Reproduced by permission from L. Pauling, *General Chemistry*, 3rd Ed., 1970, W. H. Freeman & Company, Publishers, New York.)

atoms with the second proton held in a fixed orientation, for a total of six possibilities in all. However, each of the O atoms at the apices of the tetrahedron also has two protons, and thus the chance that one of the protons of an apical H_2O is occupying the position in question between two O atoms is $\frac{1}{2}$ or a factor of $\frac{1}{4}$ for the two protons. Thus the total number of possible configurations of the hydrogens is $6 \times (\frac{1}{4})$ and the predicted (Linus Pauling) frozen-in entropy in ice is $R \ln(\frac{6}{4}) = 3.37\ \mathrm{J \cdot K^{-1} \cdot mol^{-1}}$, which is in excellent agreement with the value calculated from experimental results [$\bar{S}$(equilibrium data) − $\bar{S}$(heat capacity data)].

EXAMPLE 6-3

Nitrous oxide, NNO, a linear molecule, has a frozen-in (residual) entropy at 0 K of $4.77\ J \cdot K^{-1} \cdot mol^{-1}$. That is, for $N_2O(g)$ at 25°C and 1 atm,

$$\bar{S}(\text{equilibrium data}) - \bar{S}(\text{heat capacity data}) = 4.77\ J \cdot K^{-1} \cdot mol^{-1}$$

Suggest an explanation for the above discrepancy based on configurational disorder in $N_2O(s)$ at 0 K.

Solution: If the linear NNO molecules are randomly oriented in the crystal lattice, that is, if each NNO molecule can enter the crystal in either of the two orientations

NNO or ONN

then the residual entropy per mole in the crystal at absolute zero would be

$$\bar{S} = R \ln N = R \ln 2 = 5.76\ J \cdot K^{-1} \cdot mol^{-1}$$

The fact that the observed residual entropy is less than $5.76\ J \cdot K^{-1} \cdot mol^{-1}$ suggests that the end-for-end disorder of N_2O in the crystal is not completely random.

6-6 *The Entropy of a Molecule Depends on the Type, Number, and Arrangement of the Atoms in the Molecule*

In Chapter 4 we described the dependence of the entropy of a substance on configurational and thermal randomness factors. We shall now describe briefly the dependence of the entropy of a molecule on the type, the number, and the arrangement of the atoms in a molecule. The entropy increases with increasing mass of the atoms. For example, at 1 atm and 25°C we have the following:

Substance	$\bar{S}/J \cdot K^{-1} \cdot mol^{-1}$	*Substance*	$\bar{S}/J \cdot K^{-1} \cdot mol^{-1}$	*Substance*	$\bar{S}/J \cdot K^{-1} \cdot mol^{-1}$
He(*g*)	126.0	$F_2(g)$	202.7	$CH_4(g)$	186.2
Ne(*g*)	146.2	$Cl_2(g)$	223.0	$CH_3Cl(g)$	234.5
Ar(*g*)	154.7	$Br_2(g)$	245.4	$CCl_4(g)$	309.7
Kr(*g*)	164.0	$I_2(g)$	260.6		

The entropy of a *diamagnetic* (no unpaired electrons) gaseous species arises from the absorption of energy into the translational, rotational, vibrational, and electronic motions of the molecule. At a given temperature and pressure, the contributions to the entropy of a gaseous species arising from the various motions in the molecule depend on the following factors: atomic masses, bond strengths, and molecular structure. In general, the more complex the molecule, the higher the entropy. Thus a molecule like $F_2(g)$ (molecular mass of 38) has a higher entropy than $Ar(g)$ (molecular mass of 40) because the $F_2(g)$ molecules rotate as well as translate. The vibrational contributions to the entropy are small, relative to the contributions from translation and rotation, around 25°C.

EXAMPLE 6-4

Make a prediction about the relative values of the molar entropies at 25°C and 1 atm of the following compounds:

$$NH_3(g) \qquad H_2O(g) \qquad CH_4(g)$$

Solution: The molecular masses of the three molecules are about the same, and all the molecules have three rotational degrees of freedom; thus we predict that the molar entropies of the three molecules will have roughly the same values. The actual $\bar{S}$ values at 25°C follow:

Species	$\bar{S}/J \cdot K^{-1} \cdot mol^{-1}$
$H_2O(g)$	188.7
$NH_3(g)$	192.3
$CH_4(g)$	186.2

The interpretation of entropy as a measure of disorder or randomness can be applied to chemical reactions. For a reaction involving *only phases of a given type*, be they solid, liquid, or gas, we expect $\Delta S_{rxn} > 0$, if the number of moles of products exceeds the number of moles of reactants, and we expect $\Delta S_{rxn} < 0$, if the reverse is true; whereas we expect $\Delta S \approx 0$ if the number of moles of reactants and products are the same. The foregoing predictions are generally, though not always, observed, because we are restricting our entropy considerations to configurational factors. If the reaction involves both gaseous and condensed phases, then we expect $\Delta S_{rxn} > 0$ if there are more moles of gas in the products than in the reactants and $\Delta S_{rxn} < 0$ if the reverse is true. Recall that the entropy of a gaseous species is generally much greater than the entropy of a condensed-phase species of similar mass.

EXAMPLE 6-5

Arrange the following reactions in order of increasing ΔS_{rxn} values (all species at 25°C, 1 atm).

(a) $C(s) + O_2(g) \rightleftharpoons CO_2(g)$
(b) $C(s) + H_2O(g) \rightleftharpoons H_2(g) + CO(g)$
(c) $C(s) + CO_2(g) \rightleftharpoons 2\,CO(g)$
(d) $2\,CO(g) + O_2(g) \rightleftharpoons 2\,CO_2(g)$

Solution: The change in the number of moles of gas (products minus reactants) for the four cases are

(a) 0 (b) +1 (c) +1 (d) −1

Thus we predict

$$\Delta S_d < \Delta S_a < \Delta S_b \approx \Delta S_c$$

which is the actual order of the ΔS_{rxn} values found experimentally.

The practical absolute entropy scale enables us to specify a number for the molar entropy of a substance at a particular temperature and pressure. Although practical absolute entropies are not true absolute entropies, in that they do not include the nuclear spin entropy or the entropy of isotope mixing, practical absolute entropies are unambiguously defined and they can be used to compute ΔS_{rxn} values for chemical reactions. The nuclear spin entropies and isotope-mixing entropies both cancel out for chemical reactions. The calorimetrically determined value of the practical absolute entropy of a substance may be less than the actual value, if there is any unaccounted for frozen-in entropy (disorder) at the lowest temperature of the calorimetric measurements.

6-7 *Very Low Temperatures Can Be Obtained by Adiabatic Demagnetization*

Liquid helium has the lowest normal boiling point (4.2 K at 1 atm) of any known substance. Liquid helium is produced from gaseous helium by a Joule-Thomson expansion at 14 K (the boiling point of liquid hydrogen). Temperatures as low as 0.7 K can be obtained by evaporative cooling of He(ℓ), using large, high-speed pumps to evaporate the He(ℓ). The temperature of the evaporating liquid helium drops as a result of the utilization of internal energy in the liquid to supply the energy necessary to vaporize the liquid ($\Delta \bar{H}_{vap}$). Temperatures below about 0.7 K cannot be attained by evaporative cooling of He(ℓ), because the equilibrium vapor pressure of He(ℓ) is less than 0.1 torr at 0.7 K and the rate of evaporation of He(ℓ) becomes too low to overcome the flow of heat into the system from the surroundings. Also, below 2.17 K, He(ℓ) becomes a *superfluid* (Chapter 9) and superfluid He(ℓ) flows up and out of the low-temperature containment vessel to higher-temperature regions where the He(ℓ) evaporates and then recondenses, thereby releasing heat ($-\Delta \bar{H}_{vap}$) in the low-temperature containment vessel.

Temperatures below 0.7 K are produced by *adiabatic demagnetization.* The possibility of using adiabatic demagnetization to achieve very low temperatures was proposed independently by W. F. Giauque (pronounced "joke") and P. Debye in 1926. The basic idea is as follows. Moving charges give rise to a magnetic field. The electrons in matter move about the nucleus (orbital motion) and also rotate about their own axes (electron spin). A substance with unpaired electrons is called *paramagnetic.* At ordinary temperatures the individual electron spin magnetic moments are disordered as a result of the atomic vibrations in the crystal (Figure 6-5). When a magnetic field is applied to a paramagnetic crystal, the magnetic lines of force tend to align the tiny electron spin magnets in the direction of the applied field. The stronger the applied magnetic field, the greater the extent of alignment of the electron spins; the lower the temperature of the crystal, the greater the extent of alignment for a given applied field, because of the decrease in the amplitude of the atomic vibrations with decreasing temperature. The extent to which a substance becomes magnetized by a given magnetic field is called the *magnetic susceptibility* χ:

$$\chi = \frac{M}{\mathscr{H}} \tag{6-21}$$

where M is the *magnetization* and $\mathscr{H}$ is the applied magnetic field strength.

In crystals like gadolinium sulfate, $Gd_2(SO_4)_3 \cdot 8\,H_2O(s)$, and ferric ammonium sulfate, $Fe_2(SO_4)_3 \cdot (NH_4)_2SO_4 \cdot 24\,H_2O(s)$, the electron spin magnetic moments are separated by 50 pm to 100 pm (1 pm $= 10^{-12}$ m) and *Curie's law*,

$$\bar{\chi} = \frac{c_m}{T} \qquad (c_m \text{ is a positive constant}) \tag{6-22}$$

which gives the dependence of the molar magnetic susceptibility $\bar{\chi}$ on the absolute temperature, holds down to very low temperatures. *As long as Curie's law holds, the spins are completely disordered in the absence of an applied magnetic*

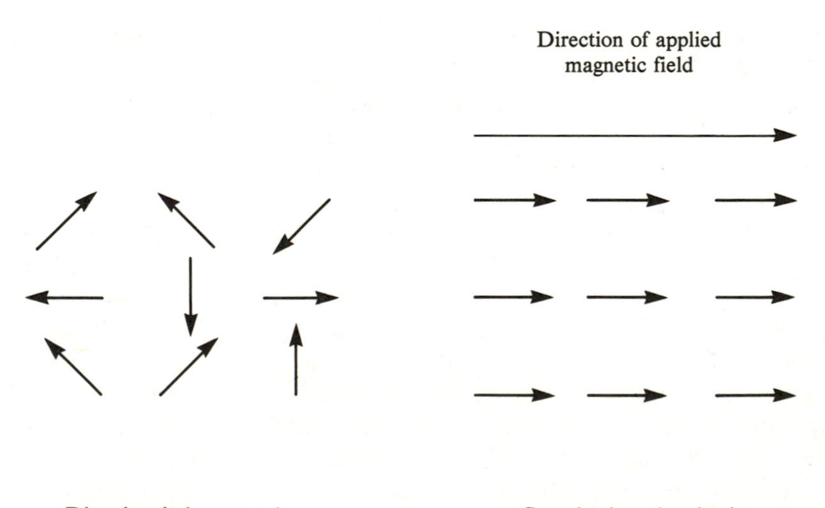

Figure 6-5. Effect of the application of a magnetic field to a system of electron spins.

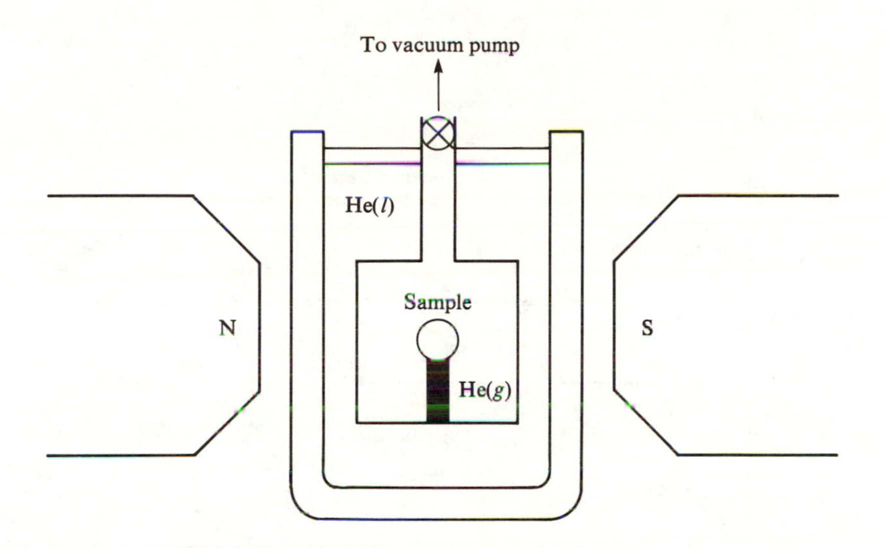

Figure 6-6. Schematic of the experimental setup used in an adiabatic demagnetization experiment. Following isothermal magnetization, in which heat is evolved and boils away some of the liquid helium, the helium gas in the sample chamber is evacuated, thereby achieving an adiabatic isolation of the magnetized sample. When the sample is adiabatically isolated, the magnetic field is turned off and the resulting spontaneous disordering of the spins produces a drop in the temperature.

field, and the spin entropy remains at its maximum value of

$$\bar{S}_{\text{spin}} = R \ln(n + 1) \tag{6-23}$$

where n is the number of unpaired electrons per formula unit. Magnetic cooling to a very low temperature requires a salt that obeys Curie's law to very low temperatures.

The basic idea involved in the adiabatic demagnetization method for obtaining very low temperatures is as follows. A salt that obeys Curie's law down to below 1 K is cooled to 1 K with liquid helium. The salt, in contact with He(g), which acts as a heat transfer medium, is *isothermally magnetized*, say, to 10 kilogauss (kG) at 1 K; the magnetized salt is then adiabatically isolated by evacuating the He(g) from the chamber (Figure 6-6) and the magnetic field is turned off (adiabatic demagnetization). The spins that were aligned in the isothermal demagnetization step become disordered once again on adiabatic demagnetization. The adiabatic disordering of the spins draws energy from the crystal lattice and thereby lowers the lattice temperature. The adiabatic process occurs without change in the entropy (second law), and thus the increase in entropy associated with the spin disordering is exactly compensated by the decrease in entropy arising from the decrease in the temperature of the crystal lattice.

In Figure 6-7(a) the molar entropy is plotted against the absolute temperature for several values of the applied magnetic field $\mathscr{H}_i$. The higher the applied magnetic field at a particular temperature, the lower the entropy, because of

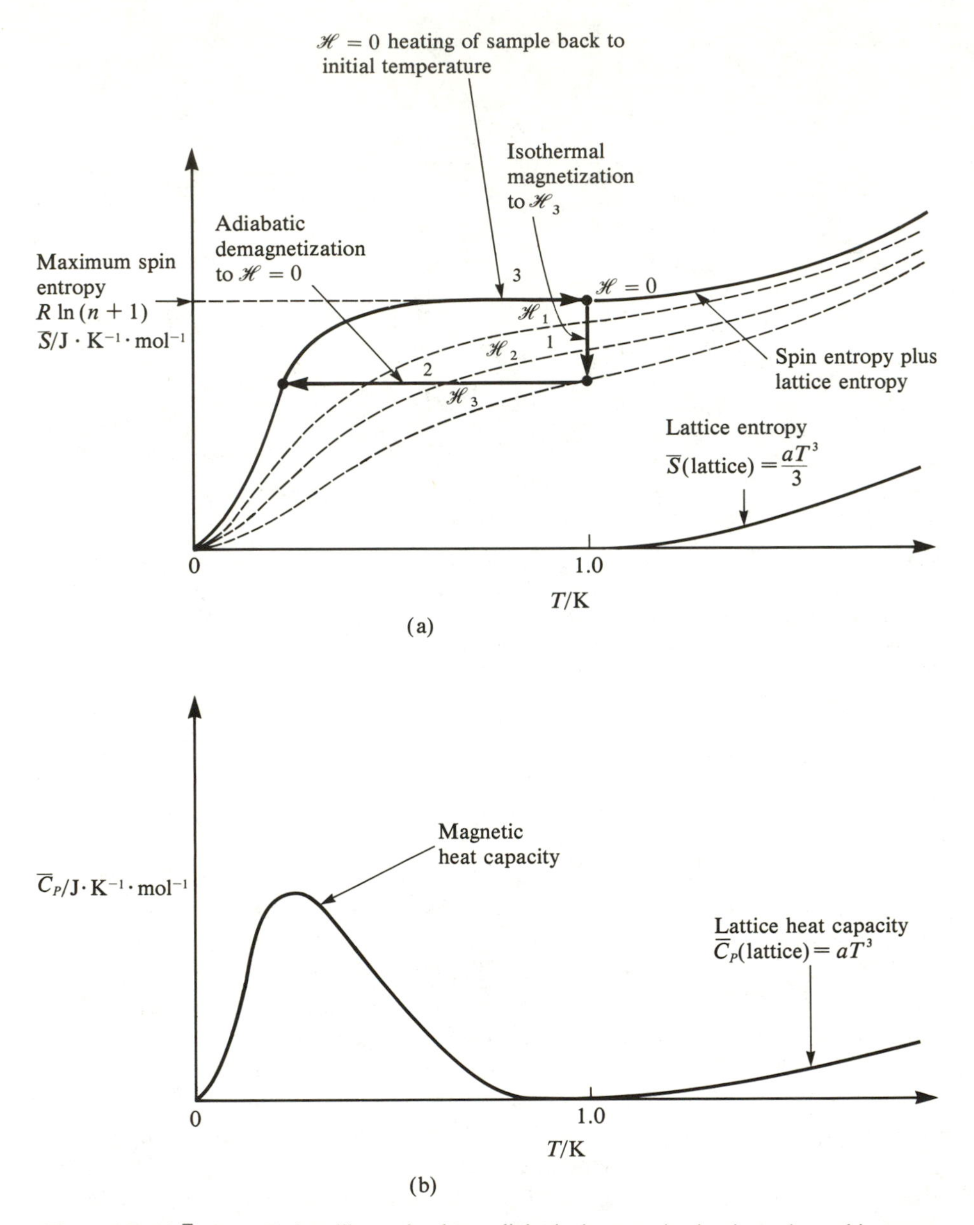

Figure 6-7. (a) $\overline{S}$ versus T plots illustrating how adiabatic demagnetization is used to achieve very low temperatures. (b) $\overline{C}_P$ versus T at zero magnetic field illustrating the magnetic heat capacity associated with the electron spin disordering in the crystal.

the increased extent of ordering of the spins at higher field strengths. Note that the spins order as $T \to 0$ even in the absence of an applied magnetic field (third law). As $T \to 0$, the lattice thermal energy becomes too small to overcome the spin-spin interactions and Curie's law fails as $T \to 0$. Note the drop in $\overline{S}$

on isothermal magnetization and the drop in temperature on adiabatic demagnetization. Temperatures as low as 2×10^{-3} K can be obtained by adiabatic demagnetization of a field-aligned electron spin system.

In Figure 6-7(b) the molar heat capacity at constant pressure and zero magnetic field is plotted against the absolute temperature. The hump in the $\bar{C}_P$ versus T curve below 1 K is associated with the magnetic spin disordering as T increases from 0 K; the disordering of the spins as T increases requires energy, and thus there is an associated magnetic heat capacity. Once the spins are completely disordered, the magnetic heat capacity drops to zero. Above 1 K the lattice heat capacity $\bar{C}_P(\text{lattice}) = aT^3$ becomes discernible on the scale of Figure 6-7(b).

6-8 *The Entropy of a Paramagnetic Substance Depends on the Applied Magnetic Field*

The magnetic work done on a substance in a magnetic field is given by

$$\delta w = \mathscr{H}\, dM \tag{6-24}$$

where $\mathscr{H}$ is the applied magnetic field and M is the magnetization. Application of the First and Second Laws of Thermodynamics to a sýstem involving magnetic work yields

$$dU = \delta q + \delta w = T\,dS - P\,dV + \mathscr{H}\,dM \tag{6-25}$$

The most convenient independent variables for magnetic systems are $\mathscr{H}$ and P rather than V and M as in Eq. 6-25, because $\mathscr{H}$ and P are much easier to control and to vary in the laboratory than V and M. We find the energy state functions especially convenient for the thermodynamic analysis of magnetic systems as follows. Equation 6-25 can be rearranged to

$$dU + P\,dV - \mathscr{H}\,dM = T\,dS \tag{6-26}$$

Adding $V\,dP - M\,d\mathscr{H}$ to both sides of Eq. 6-26 yields

$$dU + P\,dV + V\,dP - \mathscr{H}\,dM - M\,d\mathscr{H} = T\,dS + V\,dP - M\,d\mathscr{H} \tag{6-27}$$

or

$$d(U + PV - \mathscr{H}M) = T\,dS + V\,dP - M\,d\mathscr{H} \tag{6-28}$$

We now define the *magnetic enthalpy* H_m as

$$H_m \equiv U + PV - \mathscr{H}M \tag{6-29}$$

Substitution of Eq. 6-29 in Eq. 6-28 yields

$$dH_m = T\,dS + V\,dP - M\,d\mathscr{H} \tag{6-30}$$

Note from Eq. 6-30 that for a process occurring at constant pressure and constant magnetic field we have

$$dH_m = T\,dS = \delta q_{P,\mathscr{H}}$$

and thus

$$\Delta H_m = q_{P,\mathscr{H}} \qquad \text{(at constant } P \text{ and } \mathscr{H}\text{)}$$

We also note from Eq. 6-30 that because dH_m is an exact differential, we have for the absolute temperature of a magnetic system

$$T = \left(\frac{\partial H_m}{\partial S}\right)_{P,\mathscr{H}}$$

We can obtain from Eq. 6-30 a thermodynamic energy state function with the natural independent variables T, P, and $\mathscr{H}$, which are the set of intensive state variables that are most convenient from an experimental viewpoint. To do this we rewrite Eq. 6-30 as

$$dH_m - T\,dS = V\,dP - M\,d\mathscr{H} \tag{6-31}$$

and add $-S\,dT$ to both sides of Eq. 6-31:

$$dH_m - T\,dS - S\,dT = -S\,dT + V\,dP - M\,d\mathscr{H} \tag{6-32}$$

or

$$d(H_m - TS) = -S\,dT + V\,dP - M\,d\mathscr{H} \tag{6-33}$$

We define the *magnetic Gibbs energy* G_m as

$$G_m = H_m - TS \tag{6-34}$$

Substitution of Eq. 6-34 into Eq. 6-33 yields

$$dG_m = -S\,dT + V\,dP - M\,d\mathscr{H} \tag{6-35}$$

EXAMPLE 6-6

Derive the relationship

$$\left(\frac{\partial S}{\partial \mathscr{H}}\right)_{T,P} = \left(\frac{\partial M}{\partial T}\right)_{P,\mathscr{H}} \tag{6-36}$$

Solution: The total differential dG_m is an exact differential and therefore the second-cross-partial derivatives in Eq. 6-35 must be equal. From the first and third terms on the right-hand side of Eq. 6-35, we obtain

$$\left[\frac{\partial(-S)}{\partial \mathscr{H}}\right]_{T,P} = \left[\frac{\partial(-M)}{\partial T}\right]_{P,\mathscr{H}}$$

which is Eq. 6-36. Note that the partial derivative $(\partial S/\partial \mathscr{H})_{T,P}$ tells us that, in general, S is a function of $S(\mathscr{H}, T, P)$. If we faced the problem of deriving Eq. 6-36 without Eq. 6-35 at hand, then we would note that Eq. 6-36 is a Maxwell relation and proceed to develop an expression for dX (where X is a thermodynamic state function) for which $\mathscr{H}$, T, and P are the natural independent variables.

For the thermodynamic analysis of adiabatic demagnetization, we first want to know how the entropy varies with the applied magnetic field at fixed T and P, because the first step of the analysis is to determine the entropy change for the isothermal magnetization of the sample. With $S = S(T, P, \mathscr{H})$, we have

$$dS = \left(\frac{\partial S}{\partial T}\right)_{P,\mathscr{H}} dT + \left(\frac{\partial S}{\partial P}\right)_{\mathscr{H},T} dP + \left(\frac{\partial S}{\partial \mathscr{H}}\right)_{T,P} d\mathscr{H} \tag{6-37}$$

At fixed T and P, we obtain from Eq. 6-38

$$dS = \left(\frac{\partial S}{\partial \mathscr{H}}\right)_{T,P} d\mathscr{H} \qquad \text{(fixed } T \text{ and } P\text{)} \tag{6-38}$$

Substitution of Eq. 6-36 into Eq. 6.38 yields

$$dS = \left(\frac{\partial M}{\partial T}\right)_{P,\mathscr{H}} d\mathscr{H} \qquad \text{(fixed } T \text{ and } P\text{)} \tag{6-39}$$

The molar magnetization $\bar{M}$ is given by

$$\bar{M} = \bar{\chi}\mathscr{H} \tag{6-40}$$

and if Curie's law ($\bar{\chi} = c_m/T$) holds, then

$$\bar{M} = \frac{c_m \mathscr{H}}{T} \tag{6-41}$$

Conversion of Eq. 6-39 to molar quantities (divide both sides by n, the number of moles), followed by substitution of Eq. 6-41 into the molar version of Eq. 6-39, yields

$$d\bar{S} = -\frac{c_m \mathscr{H}}{T^2} d\mathscr{H} \qquad \text{(fixed } T \text{ and } P\text{)} \tag{6-42}$$

From Eq. 6-42 we can obtain an expression for the change in entropy per mole $\Delta\bar{S}$ when a Curie's-law paramagnetic substance, initially at zero magnetic field, is isothermally magnetized to a magnetic field of strength $\mathscr{H}$. Integration of Eq. 6-42 yields

$$\Delta\bar{S} = -\frac{c_m}{T^2}\int_0^{\mathscr{H}} \mathscr{H}\, d\mathscr{H} = -\frac{c_m}{2}\left(\frac{\mathscr{H}}{T}\right)^2 \tag{6-43}$$

Because c_m is a positive constant for a given paramagnetic substance, and because $\mathscr{H}$ and T are necessarily positive, we conclude from Eq. 6-43 that $\Delta\bar{S} < 0$ for the isothermal magnetization step.

EXAMPLE 6-7

Given a Curie's law paramagnetic solid for which $c_m = 2 \times 10^{-7}\ \text{J}\cdot\text{K}^{-1}\cdot\text{mol}^{-1}\cdot\text{Oe}^{-2}$ (Oe = oersted), compute $\Delta\bar{S}$ for this substance when it is isothermally magnetized at 1.0 K to a magnetic field of 1.0×10^4 Oe.

Solution: From Eq. 6-43, we have

$$\Delta\bar{S} = -\frac{c_m}{2}\left(\frac{\mathscr{H}}{T}\right)^2 = -\frac{(2 \times 10^{-7}\ \text{J}\cdot\text{K}^{-1}\cdot\text{mol}^{-1}\cdot\text{Oe}^{-2})}{2}\left(\frac{1.0 \times 10^4\ \text{Oe}}{1.0\ \text{K}}\right)^2$$

and

$$\Delta\bar{S} = -10\ \text{J}\cdot\text{K}^{-1}\cdot\text{mol}^{-1}$$

Reference to Figure 6-7(a) shows that for the closed cycle (Figure 6-8) we have per mole of substance

$$\Delta\bar{S}_1 + \Delta\bar{S}_2 + \Delta\bar{S}_3 = 0$$

because $\bar{S}$ is a state function. Because ② is an adiabatic process, $\Delta\bar{S}_2 = 0$; further, we know from Eq. 6-43 that for step ①

$$\Delta\bar{S}_1 = -\frac{c_m}{2}\left(\frac{\mathscr{H}}{T_1}\right)^2$$

Thus

$$\Delta\bar{S}_3 = \frac{c_m}{2}\left(\frac{\mathscr{H}}{T_1}\right)^2 \qquad (6\text{-}44)$$

If we can express $\Delta\bar{S}_3$ in terms of T_1 and T_2, then we can compute T_2, the final temperature in the adiabatic demagnetization. The heat capacity on the high-temperature side of the $\bar{C}_P$ maximum in Figure 6-7(b) is known to fall

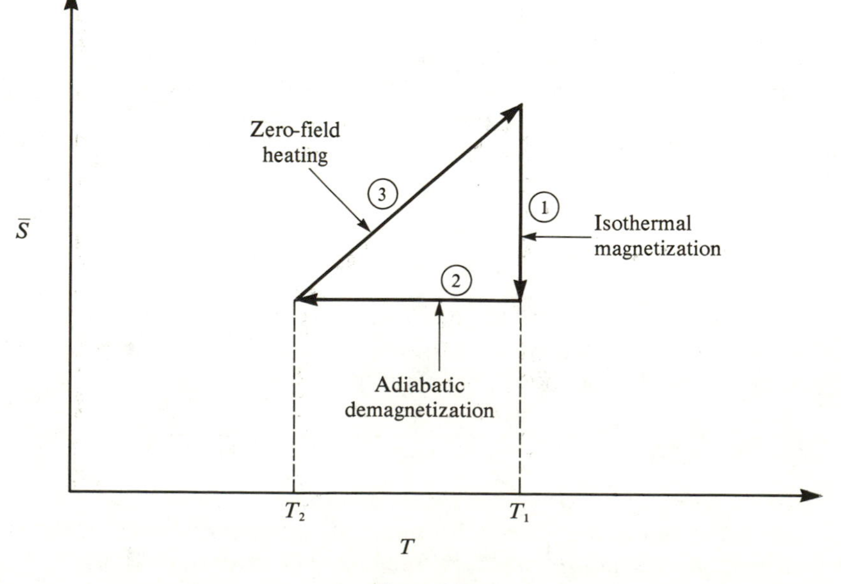

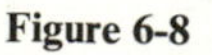
Figure 6-8

off with temperature as T^{-2}. Consider a paramagnetic solid for which

$$\bar{C}_P = \frac{R}{4}\left(\frac{\varepsilon}{kT}\right)^2 \tag{6-45}$$

where R is the gas constant, k is Boltzmann's constant, and ε is also a constant. At fixed P and $\mathscr{H}$, we have from Eq. 6-37

$$dS = \left(\frac{\partial S}{\partial T}\right)_{P,\mathscr{H}} dT = \frac{C_{P,\mathscr{H}}}{T}\, dT \qquad \text{(fixed } P \text{ and } \mathscr{H}) \tag{6-46}$$

Thus for process ③ we have from Eq. 6-46

$$\Delta\bar{S}_3 = \int_{T_2}^{T_1} \frac{C_{P,\mathscr{H}}}{T}\, dT = \int_{T_2}^{T_1} \frac{R}{4}\left(\frac{\varepsilon}{k}\right)^2 \frac{dT}{T^3}$$

and

$$\Delta\bar{S}_3 = -\frac{R}{8}\left(\frac{\varepsilon}{k}\right)^2\left(\frac{1}{T_1^2} - \frac{1}{T_2^2}\right) \tag{6-47}$$

Combination of Eq. 6-47 with Eq. 6-44 yields for the final temperature T_2 in the adiabatic demagnetization

$$T_2 = T_1\left[1 + \frac{4c_m}{R}\left(\frac{k}{\varepsilon}\right)^2 \mathscr{H}^2\right]^{-1/2} \tag{6-48}$$

The term in the brackets on the right-hand side of Eq. 6-48 is greater than unity, and thus $T_2 < T_1$.

Equation 6-43 is only an approximation because it depends on Curie's law. Equation 6-48 is also an approximation and, further, it is not general because it depends on Eq. 6-45, which is not general, and also on Eq. 6-43. A more general and accurate analysis of adiabatic demagnetization can be carried out using a more accurate equation of state for $\bar{M}$ (the *Brillouin equation*, see Problem 33) and a more accurate expression for $\bar{C}_P$, which gives $\bar{C}_P$ for the magnetic spin system over the entire range of temperature. These more general equations for $\bar{M}$ and $\bar{C}_P$ are obtained from statistical thermodynamics. Once these more accurate equations are available, the thermodynamic analysis proceeds exactly as outlined above, the only difference being the complexity of the resulting equations for $\Delta\bar{S}$ and T_2 and, of course, the increased reliability of the calculated values of $\Delta\bar{S}$ and T_2.

6-9 *Negative Absolute Spin Temperatures Can Be Obtained by Magnetic Field Reversals on a Set of Nuclear Spins*

The lowest bulk-lattice temperature obtained in an adiabatic demagnetization experiment involving a paramagnetic substance is about 10^{-3} K. By using two samples connected by a heat conductor, say, a piece of copper wire, and two magnetic fields (Figure 6-9), it is possible to decrease the

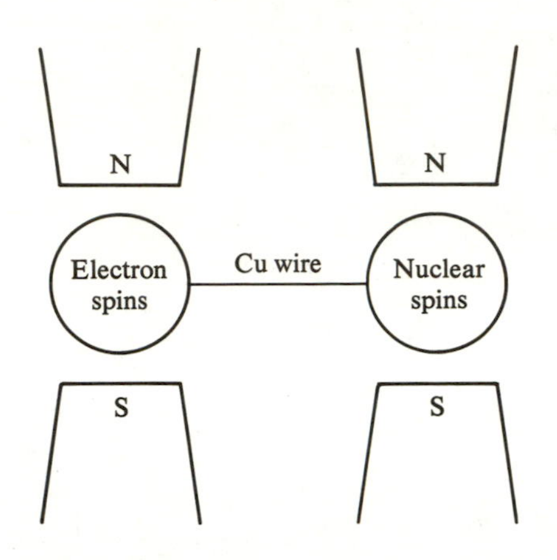

Figure 6-9. Schematic of experimental setup for successive adiabatic demagnetizations of an electron spin system (left) and a nuclear spin system (right). The temperature drop produced by the adiabatic demagnetization of the electron spin system is used to drop the temperature of the magnetized sample on the right to 10^{-3} K. The subsequent adiabatic demagnetization of the field-aligned nuclear spin system drops the nuclear spin temperature to about 10^{-6} K.

temperature of a set of nuclear spins (analogous to electron spins) in a *nuclear adiabatic demagnetization* experiment to about 10^{-6} K. The adiabatic demagnetization of the electron spins produces a temperature of about 1×10^{-3} K; this low temperature is used to drop the temperature of a second *magnetized* sample to about 2×10^{-3} K (Figure 6-9) at which temperature the nuclear spin magnetic moments become aligned with the applied magnetic field. An adiabatic demagnetization of the nuclear spin system leads to a drop in the temperature of the nuclear spin system to about 10^{-6} K. It is important to recognize, however, that the *bulk-lattice temperature* of the sample on the right in Figure 6-9 *remains at* 2×10^{-3} K *after the nuclear adiabatic demagnetization. It is only the temperature of the set of nuclear* spins, and *not* the crystal lattice, that drops to 10^{-6} K. In effect, for at least a short period of time, the demagnetized sample has two temperatures, a lattice temperature of 10^{-3} K and a nuclear spin temperature of 10^{-6} K. The system is not in thermal equilibrium. Energy begins to flow from the lattice vibrations to the nuclear spin system immediately following demagnetization and the nuclear spin temperature rises rapidly to a value equal to the bulk-lattice temperature of about 10^{-3} K.

Consider a set of nuclear spins confined to the set of two energy levels shown in Figure 6-10. The *equilibrium* ratio of the populations of the two spin states is given by statistical thermodynamics as

$$\frac{n_1}{n_0} = e^{-\varepsilon/kT} \tag{6-49}$$

where n_1 is the number of spins in the upper level and n_0 is the number of spins

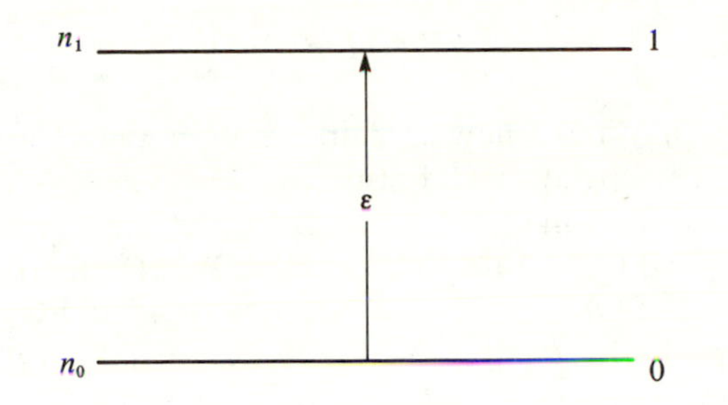

Figure 6-10. Nuclear spin system with two energy levels having populations n_0 and n_1.

in the lower level. Note that as $T \to 0$,

$$\frac{n_1}{n_0} \longrightarrow 0$$

and all the spins are in the ground (lowest) energy level. Further, note that as $T \to \infty$,

$$\frac{n_1}{n_0} \longrightarrow 1$$

that is, the levels become equally populated at high temperatures. Given that Eq. 6-49 applies, the only way to have

$$\frac{n_1}{n_0} > 1$$

a so-called *inverted population*, is for $T < 0$, that is, to have a negative absolute temperature.

Suppose a magnetic field is applied to a sample containing nuclei with nonzero nuclear spins. The nuclear spins aligned with the field have a lower energy than those nuclear spins not aligned with the field, and $n_1/n_0 < 1$. Now suppose that the direction of the applied magnetic field is reversed rapidly. The field reversal inverts the two energy levels while leaving the level populations unchanged; thus we have $n_1'/n_0' > 1$, following field reversal. The inverted spin population, when plugged into Eq. 6-49, leads to the conclusion that $T < 0$. Because of the slow energy transfer between the nuclear spin system and the crystal lattice, the inverted spin population persists for some time.

Nuclear spin population inversions have led to the espousal of the concept of *negative absolute temperatures*, in violation of the Third Law of Thermodynamics. However, it should be noted that an inverted nuclear spin population is not in thermal equilibrium with the crystal lattice in which the nuclear spins are located, and thus the macroscopic thermodynamic temperature of the total system is undefined. The temperature defined via Eq. 6-49 applied to a system not in internal thermal equilibrium is not a thermodynamic temperature and, as such, although of intrinsic interest, has no bearing on the Third Law of Thermodynamics, which rules out the attainability of $T \leqslant 0$.

Problems

1. Predict for each of the following pairs of compounds the species with the higher molar entropy at 25°C, 1 atm.
   (a) $H_2O(g)$ and $D_2O(g)$ (b) $CH_3OH(\ell)$ and $CH_3CH_2OH(\ell)$
   (c) $CH_3Cl(g)$ and $CH_2Cl_2(g)$ (d) $NH_3(g)$ and $NF_3(g)$
   (e) $ICl(g)$ and $BrF(g)$ (g) $SO_2(g)$ and $SO_3(g)$
   (h) $KCl(s)$ and $KClO_4(s)$ (i) $N_2(g)$ and $F_2(g)$
   (j) $F_2O(g)$ and $H_2O(g)$ (k) $^6Li(s)$ and $^7Li(s)$

2. Arrange the following reactions in order of increasing ΔS_{rxn} values:
   (a) $N_2(g) + 3\,H_2(g) \rightleftharpoons 2\,NH_3(g)$ (b) $2\,H_2(g) + O_2(g) \rightleftharpoons 2\,H_2O(g)$
   (c) $H_2(g) + I_2(g) \rightleftharpoons 2\,HI(g)$ (d) $CO(g) + 2\,H_2(g) \rightleftharpoons CH_3OH(g)$
   (e) $CO(g) + 2\,H_2(g) \rightleftharpoons CH_3OH(\ell)$

3. Show that for a chemical reaction

$$\left(\frac{\partial \Delta S_{rxn}}{\partial T}\right)_P = \frac{\Delta C_{P,rxn}}{T}$$

4. Given the following data, compute ΔS_{rxn} for the reaction

$$N_2(g) + 3\,H_2(g) \rightleftharpoons 2\,NH_3(g)$$

   at 25°C and 50°C. Assume that the $\bar{C}_P$ values are constant.

Species	$\bar{S}/J \cdot K^{-1} \cdot mol^{-1}$ *at 25°C*	$\bar{C}_P/J \cdot K^{-1} \cdot mol^{-1}$
$N_2(g)$	191.50	29.12
$H_2(g)$	130.57	28.82
$NH_3(g)$	192.34	35.06

5. Take $\bar{C}_P = 23.64 + 4.79 \times 10^{-2}T - 1.92 \times 10^5 T^{-2}$ and $\bar{S}_{298.15} = 186.15\ J \cdot K^{-1} \cdot mol^{-1}$ for $CH_4(g)$, and compute $\bar{S}$ at 500 K and 1 atm.

6. Given that

$$\bar{C}_P = \gamma T + aT^3$$

   for a metal at low temperatures, derive an expression for the entropy change $\bar{S}_T - \bar{S}_0$.

7. Calculate the molar nuclear spin entropies for the following compounds:

$$^{12}C^1H_4 \qquad ^{13}C^1H_4 \qquad ^{12}C^2H_4 \qquad ^{12}C^1H_3{}^2H$$

   Given that for ^1H, $I = \frac{1}{2}$; ^2H, $I = 1$; ^{12}C, $I = 0$, and ^{13}C, $I = \frac{1}{2}$.

8. Given that the crystal structure of D_2O (heavy water) is the same as that of normal water, estimate the frozen-in entropy of D_2O at 0 K.

9. Oxygen gas is paramagnetic with two unpaired electrons. The total angular momentum of an oxygen molecule is $J = 1$. Estimate the contribution of the unpaired electron spin disorder to the entropy of oxygen. Would you expect oxygen in the solid phase to remain paramagnetic as $T \to 0$? Explain.

10. The compound Cl_3PO has a tetrahedral structure. What do you predict for the entropy difference

$$\bar{S}_{298}(\text{equilibrium data}) - \bar{S}_{298}(\text{third-law data})$$

for gaseous Cl_3PO, if in the crystalline state Cl_3PO is completely disordered with respect to the positioning of the oxygen and chlorine atoms?

11. Suppose that we determine by calorimetric measurements the entropy at 25°C of a solid *solution* of AgBr(*s*) in AgCl(*s*) composed of 1.00 mol of AgBr(*s*) and 1.00 mol of AgCl(*s*).

(a) What do you predict for the discrepancy between the third law entropy of the solid solution as compared to the calorimetrically measured entropy of 1.00 mol of AgCl(*s*) plus 1.00 mol of AgBr(*s*)?

(b) Compute ΔS_{mix} for 1.00 mol of AgCl(*s*) + 1.00 mol of AgBr(*s*) to form a solid solution of AgCl in AgBr.

12. Compute the maximum value of $\bar{S}_{mag}$ arising solely from unpaired electron spins in substances with 0, 3, and 7 unpaired electrons.

13. The entropy at 298 K of the diamagnetic compound $MX_4Y_2(\ell)$ with the structure

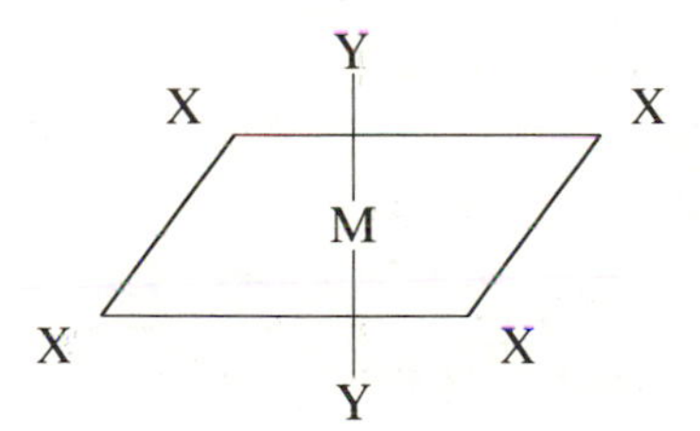

determined from heat capacity measurements was found to be 490.95 ± 0.34 $J \cdot K^{-1} \cdot mol^{-1}$, whereas equilibrium studies yield an entropy of 499.57 ± 0.50 $J \cdot K^{-1} \cdot mol^{-1}$. Assuming that the experimental data are reliable in both cases, suggest a possible explanation for the discrepancy.

14. Given that for an equilibrium phase transition

$$\frac{dP}{dT} = \frac{\Delta S}{\Delta V}$$

where P and T are the equilibrium pressure and temperature for the phase transition and ΔS and ΔV are the entropy change and the volume change, respectively, show that a plot of P versus T comes into the P axis with zero slope as $T \to 0$.

15. Prove that

$$\lim_{T \to 0} C_V = 0$$

16. What do you predict for the entropy difference between two optical isomers at the same T and P?

17. Show that for an electrochemical cell

$$\lim_{T \to 0} \left(\frac{\partial \mathscr{E}}{\partial T}\right)_{P,Z} = 0$$

(*Hint:* Write the ΔG expression for a cell.)

18. Discuss why it is impossible to have a superheated solid in thermal equilibrium, whereas supercooled liquids are well known.

19. At 4.23 K helium gas was adsorbed on the surface of a solid. The equilibrium adsorption pressure (at monolayer coverage) was found to be 25.0 torr. Calculate the heat of adsorption in joules per mole of the helium in equilibrium with the above pressure. Assume ideal gas behavior and also assume that the entropy of adsorbed helium is negligible. Take $\bar{C}_P = \frac{5}{2}R$ and $\bar{S}_{298.15} = 126.0\ \text{J}\cdot\text{K}^{-1}\cdot\text{mol}^{-1}$ for He(g). Also estimate the equilibrium adsorption pressure (at monolayer coverage) of helium on this solid at 15.0 K. State any assumptions that you make in addition to those mentioned above.

20. When a rubber band is stretched, the long macromolecules undergo a change from a kinky-coiled configuration to a more-or-less linear arrangement. X-ray diffraction patterns show that the stretched rubber band is more crystalline than the unstretched band.

(a) If a rubber band is stretched quickly, will the temperature of the band increase or decrease?

(b) Show that

$$\left(\frac{\partial S}{\partial L}\right)_{T,P} = -\left(\frac{\partial \mathscr{T}}{\partial T}\right)_{L,P}$$

(L is the length and $\mathscr{T}$ the tension of the rubber band).

(c) Show, using the results from parts (a) and (b), that a rubber band becomes stronger as its temperature is increased.

(d) Repeat parts (a) and (c) for a Hooke's law metallic spring. Explain, in microscopic terms, why a metallic spring becomes weaker as its temperature is increased.

21. The Debye theory for the heat capacity of solids gives the following expression for $\bar{C}_V$ at low temperatures:

$$\bar{C}_V = \frac{12\pi^4 R}{5}\left(\frac{T}{\theta}\right)^3 \qquad 0 < \frac{T}{\theta} < \frac{1}{12}$$

where θ is the characteristic Debye temperature. For silicon, $\theta = 625$ K. Evaluate ΔS for silicon over the range 0 K to 50 K.

22. Explain why ΔH does not equal q_P for a magnetic system. Does $\Delta G = w_{npv}$ for a magnetic system? Explain.

23. Show that if Curie's law holds, then the *relative* reduction in the entropy of a paramagnetic substance for a given isothermal increase in the magnetic field is greater the lower the temperature.

24. List the factors of importance in obtaining the lowest possible temperature in an adiabatic demagnetization.

25. Suggest an explanation for the following observation: Very high magnetic fields are not of much value relative to moderate magnetic fields in obtaining low temperatures in an adiabatic demagnetization experiment, but very high fields are particularly useful in making it possible to keep the system for a longer time at low temperatures. (*Hint:* See Figure 6-7.)

26. (a) Show that for a Curie's law paramagnetic solid

$$\left(\frac{\partial U}{\partial \mathscr{H}}\right)_{T,V} = \left[\mathscr{H} - T\left(\frac{\partial \mathscr{H}}{\partial T}\right)_{M,V}\right]\left(\frac{\partial M}{\partial \mathscr{H}}\right)_{T,V}$$

(b) Given $M = M(\mathscr{H}/T)$ and using the result in part (a), show that $U = U(T)$ for a Curie's law paramagnetic solid.

(c) Given the results in part (b), show that

$$C_{V,M} = C_{V,M}(T)$$

27. Consider a substance for which Curie's law holds and for which $\bar{C}_{M,V} = A/T^2$ (A = constant), show that

$$\bar{C}_{\mathscr{H},V} = \frac{A}{T^2} + \frac{c_m \mathscr{H}^2}{T^2}$$

28. Derive the following relationships:

(a) $$\left(\frac{\partial V}{\partial \mathscr{H}}\right)_{P,T} = -\left(\frac{\partial M}{\partial P}\right)_{\mathscr{H},T}$$

(b) $$\left(\frac{\partial M}{\partial S}\right)_{P,\mathscr{H}} = -\left(\frac{\partial T}{\partial \mathscr{H}}\right)_{P,S}$$

29. Prove that absolute zero is unattainable using internally metastable phases even if the process removes the internal metastability.

30. Explain why Eq. 6-43 fails when the field $\mathscr{H}$ is made very large.

31. Derive the relationships

(a) $$M = -\left(\frac{\partial H_m}{\partial \mathscr{H}}\right)_{S,P}$$

(b) $$-\left(\frac{\partial M}{\partial S}\right)_{P,\mathscr{H}} = \left(\frac{\partial T}{\partial \mathscr{H}}\right)_{P,S}$$

32. Given that

$$\frac{\bar{C}_{P,\mathscr{H}}}{R} = \left(\frac{\varepsilon}{kT}\right)^2 \frac{e^{-\varepsilon/kT}}{(1 + e^{-\varepsilon/kT})^2}$$

for a mole of electron spins confined to a pair of energy levels separated by the energy gap ε:
(a) Find the low- and high-temperature limiting values of $\bar{C}_{P,\mathscr{H}}$.
(b) Obtain an expression for the temperature at which $\bar{C}_P$ is a maximum.
(c) Show that $\bar{C}_{P,\mathscr{H}} \approx R(\varepsilon/kT)^2/4$ for $T \gg \varepsilon/k$.

33. The *Brillouin equation* gives $\bar{M}$ as a function of $\mathscr{H}$, T and molecular properties for a paramagnetic substance (see Figure 6-11)

$$\bar{M} = N_0 g\beta\left[\left(\frac{2J+1}{2}\right)\coth\left[\frac{(2J+1)X}{2}\right] - \frac{1}{2}\coth\frac{X}{2}\right]$$

where $X = g\beta\mu\mathscr{H}/kT$; N_0 is Avogadro's number; g, β, k, and μ are constants; and $J = L + S$ is the total angular-momentum quantum number.
(a) First show that

$$\lim_{y\to 0} \coth y = \frac{1}{y} + \frac{y}{3} \qquad \left(\textit{Note}: \coth y = \frac{e^y + e^{-y}}{e^y - e^{-y}}\right)$$

(b) Using the result in part (a), derive Curie's law and show that

$$c_m = N_0 \mu g^2 \beta^2 J(J+1)/3k$$

Identify the Curie's law regions in Figure 6-11.

34. Ideal gas thermometers do not work at 1 K and below. How would you *measure* (as opposed to calculate) $T_2 < 1$ K following an adiabatic demagnetization? [*Hint:* Show that

$$T = \left(\frac{\partial H_m}{\partial S}\right)_{P,\mathscr{H}} = \frac{\bar{C}^*_{P,\mathscr{H}=0}}{(\partial\bar{S}/\partial T^*)_{P,\mathscr{H}=0}}$$

where an asterisk denotes a quantity at a measured but nonkelvin temperature based on Curie's law.]

35. Given the following values of n_1/n_0 in the equation $n_1/n_0 = e^{-\varepsilon/kT}$, compute T for $\varepsilon = 1.38 \times 10^{-23}$ J: $n_1/n_0 = 0, 0.5, 1, 2$.

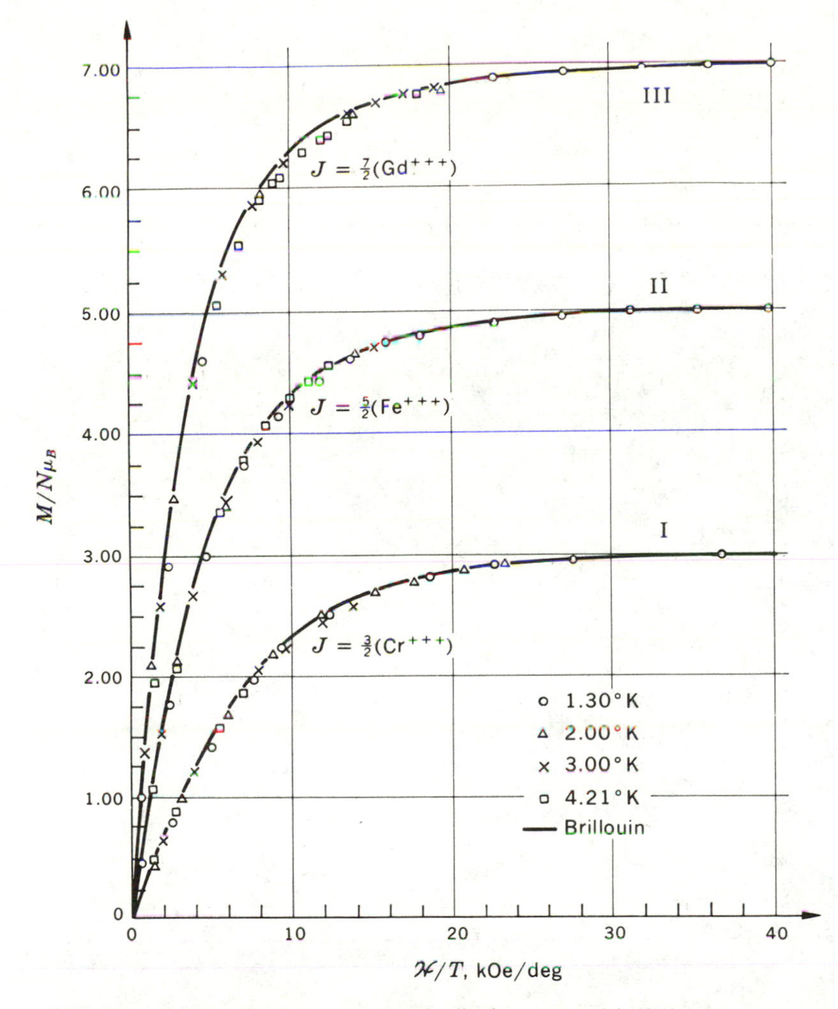

Figure 6-11. Plot of magnetization per magnetic dipole, expressed in Bohr magnetons, against $\mathscr{H}/T$ for (I) chromium potassium alum ($J = \frac{3}{2}$); (II) iron ammonium alum ($J = \frac{5}{2}$); and (III) gadolinium sulfate ($J = \frac{7}{2}$). The points are experimental results of W. E. Henry (1952), and the solid curves are graphs of the Brillouin equation. (From M. W. Zemansky, *Heat and Thermodynamics*, 5th Ed., New York: McGraw-Hill Book Co., 1968.) In our notation

$$\frac{M}{N\mu_B} = \frac{\bar{M}}{N_0 g\beta}$$

Liftoff of the reusable spaceship Columbia (*space shuttle*) *on its third space trip* (*22 March 1982*).

7

Thermodynamics of Chemical Reactions

It is highly convenient, especially for purposes of concise tabulation, to know the heats of reaction when various substances are formed from the elements. It is therefore desirable to choose some one standard reference state for each element. For this purpose, we shall, at all temperatures, take the element at a pressure of one atmosphere, and in that form which is most stable or most common at room temperature. Thus liquid mercury, gaseous oxygen, solid iodine, rhombic sulfur and graphitic carbon, all under a pressure of one atmosphere, will be considered to be in their standard reference state.

We speak, then, of the ΔH of formation of a given substance . . . meaning the increase in heat content [enthalpy] in the reaction by which one mole of that substance is prepared from the elements in their standard states.

G. N. Lewis and M. Randall, *Thermodynamics.* New York: McGraw-Hill, 1923.

THE ANALYSIS of the thermodynamics of chemical reactions is one of the primary objectives of chemical thermodynamics. The emphasis in this chapter is on the thermodynamic analysis of chemical reactions for which the reactants are converted quantitatively into the products. The thermodynamic analysis of chemical reactions involves the calculation of the enthalpy change for a reaction ΔH_{rxn}, the entropy change for a reaction ΔS_{rxn}, and the Gibbs energy change for a reaction ΔG_{rxn} from the enthalpies, the entropies, and the Gibbs energies of the products and the reactants, respectively. In the general case, the values of ΔH_{rxn}, ΔS_{rxn}, and ΔG_{rxn} for a given chemical reaction depend on the temperature and the pressure at which the reaction takes place.

The values of ΔH_{rxn}, ΔS_{rxn}, and ΔG_{rxn} for a particular chemical reaction are related. The Gibbs energy is defined by the relation

$$G = H - TS \tag{7-1}$$

For a chemical reaction that occurs at a fixed temperature T, we have from Eq. 7-1

$$\Delta G_{rxn} = \Delta H_{rxn} - T\Delta S_{rxn} \tag{7-2}$$

Thus, if we know the values of any two of the three quantities ΔG_{rxn}, ΔH_{rxn}, and ΔS_{rxn} for a chemical reaction, then Eq. 7-2 can be used to calculate the value of the other quantity.

7-1 *The Values of the Thermodynamic Energy Functions of a Substance Are Specified Relative to a Standard State of the Substance*

The enthalpy change for a chemical reaction ΔH_{rxn} ("delta H for the reaction") is equal to the value of the heat absorbed or evolved when the reaction occurs at constant pressure (see Eq. 3-36):

$$\Delta H_{rxn} = q_{P,rxn} \tag{7-3}$$

If heat is evolved ($q_{P,rxn} < 0$), then the reaction is said to be *exothermic* ("exo" means *out*); whereas if heat is absorbed, then the reaction is said to be *endothermic* ("endo" means *in*). The value of ΔH_{rxn} depends, in general, on both the temperature and on the pressure:

$$\Delta H_{rxn} = \Delta H_{rxn}(T, P) \tag{7-4}$$

Because the enthalpy of a substance also depends on T and P, we shall restrict our analysis, at least initially, to chemical reactions that occur at a fixed temperature and pressure; that is, the reactants are converted to products at constant temperature and at constant pressure.

The value of $\Delta H_{rxn}(T, P)$ for a chemical reaction can be determined in a calorimetric experiment. It is convenient for the purpose of economies in tabulation to refer all ΔH values to a particular pressure, to avoid the necessity of tabulating ΔH as a function of both pressure and temperature. The reference pressure chosen is one standard atmosphere (1.01325×10^5 Pa). The particular one-atmosphere pressure reference states chosen for substances are called *standard states*. The choice of a standard state for each substance at each temperature amounts to the arbitrary choice of a thermodynamic state relative to which the numerical values of all the thermodynamic energy state functions (U, H, A, and G) can be specified. Recall that we cannot specify an absolute value for the internal energy U of a substance in any thermodynamic state of the substance. Further, because

$$H = U + PV \qquad A = U - TS \qquad G = U + PV - TS$$

we also cannot specify absolute values for H, A, and G. Because we cannot specify absolute values for the energy state functions, we are free to choose arbitrarily a standard state for a substance and then determine the value of, say, the enthalpy relative to the standard state. There is nothing fundamental involved in the choice of standard states, and, as long as we maintain internal consistency in our definitions, it makes no difference what state is chosen as a standard state. The standard state need not even be an *actual* (i.e., experimentally attainable) state of the substance. The particular choices made for standard states are based solely on convenience.

Our choices of standard states for solids, liquids, and gases are as follows.

1. *Solids and Liquids: The standard state for a solid or a liquid is the pure solid or the pure liquid phase in a reproducible equilibrium state at one standard atmosphere pressure and the temperature of interest.*

Thus, the standard state for liquid water at 25°C is pure $H_2O(\ell)$ at 1 atm and 25°C; the standard state for crystalline sodium chloride at 25°C is pure NaCl(s) at 1 atm and 25°C. The standard states for solids and for liquids are experimentally attainable thermodynamic states.

2. *Gases: The standard state for a gas is the hypothetical ideal gas at one standard atmosphere pressure and the temperature of interest.*

The standard states for gases are not real states because a real gas at 1 atm is not ideal; that is, a real gas does not obey the ideal gas equation of state $PV = nRT$, except in the limit of zero pressure. A hypothetical ideal gas standard state is chosen for gases, because an ideal gas has no intermolecular interactions and the ideal gas molecules occupy a totally negligible fraction of the total volume occupied by the gas. Thus the values of the thermodynamic energy state functions of an ideal gas at a given T and P depend only on the *intramolecular* properties of the gas molecules and are independent of *intermolecular* effects. Corrections from the real gas state to the ideal gas standard state are made using the equation of state for the gas. The connection between the real gas in the actual laboratory state (T, P) and the ideal gas in the hypothetical standard state at (T, $P = 1$) is shown in Figure 7-1. All that is needed to compute the enthalpy change for the process labeled ① in Figure 7-1 is the equation of state for the gas.

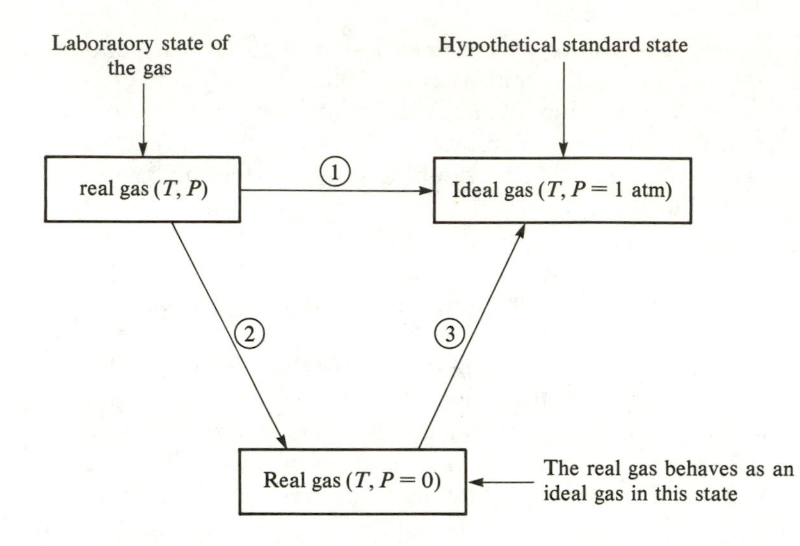

Figure 7-1. Diagram of method used to calculate the change in a thermodynamic state function on going from a real gas state to the ideal gas standard state at the same temperature. Note that for any thermodynamic state function X we have

$$\Delta X_1 = \Delta X_2 + \Delta X_3$$

because the value of ΔX is independent of path.

EXAMPLE 7-1

Given the equation of state

$$PV = nRT + nBP \qquad B = 30\ \text{cm}^3\cdot\text{mol}^{-1}$$

for a gas, compute the change in enthalpy when 1.00 mol of the gas is taken from the state 298.15 K, 2.00 atm to the hypothetical 1-atm ideal gas standard state at 298.15 K.

Solution: From Figure 7-1 we have

$$\Delta H_1 = \Delta H_2 + \Delta H_3$$

where ΔH_1 is the desired enthalpy change. The dependence of enthalpy on the pressure at fixed temperature is given by Eq. 5-38:

$$\left(\frac{\partial H}{\partial P}\right)_T = V - T\left(\frac{\partial V}{\partial T}\right)_P$$

For an ideal gas, $PV = nRT$, and thus $(\partial H/\partial P)_T = 0$. The enthalpy of an ideal gas does not depend on the pressure, and therefore $\Delta H_3 = 0$; thus

$$\Delta H_1 = \Delta H_2$$

The value of ΔH_2 is given by

$$\Delta H_2 = \int_P^0 \left(\frac{\partial H}{\partial P}\right)_T dP$$

From the equation of state for the gas we obtain

$$\left(\frac{\partial H}{\partial P}\right)_T = V - T\left(\frac{nR}{P}\right) = \frac{nRT}{P} + nB - \frac{nRT}{P} = nB$$

and thus

$$\Delta H_2 = \int_P^0 nB\,dP = -nBP$$

Hence

$$\Delta H_2 = -(1.00\text{ mol})(30\text{ cm}^3\cdot\text{mol}^{-1})(2.00\text{ atm})\left(\frac{8.314\text{ J}}{82.05\text{ cm}^3\cdot\text{atm}}\right) = -6.1\text{ J}$$

7-2 *The Molar Enthalpy of Formation of a Compound in Its Standard State from the Elements in Their Standard States Is Denoted by $\Delta\bar{H}_f^\circ$*

Consider the general balanced chemical reaction

$$a\text{A} + b\text{B} + \cdots \longrightarrow x\text{X} + y\text{Y} + \cdots \tag{7-5}$$

where a is the number of moles of substance A, and so forth. The enthalpy change for reaction 7-5 is given by

$$\Delta H_{\text{rxn}} = x\bar{H}_{\text{X}} + y\bar{H}_{\text{Y}} + \cdots - (a\bar{H}_{\text{A}} + b\bar{H}_{\text{B}} + \cdots) \tag{7-6}$$

where $\bar{H}_i$ is the molar enthalpy of i. We can write Eq. 7-6 more compactly by introducing the symbol ν_i; that is, the ν_i values are the reaction balancing coefficients that are defined as positive for products ($\nu_x = x, \nu_y = y, \ldots$) and negative for reactants ($\nu_a = -a, \nu_b = -b$). Thus Eq. 7-6 becomes

$$\Delta H_{\text{rxn}} = \sum_i \nu_i \bar{H}_i \tag{7-7}$$

where the sum extends over all the products and the reactants.

We designate the value of ΔH_{rxn} for a reaction as $\Delta H^\circ_{\text{rxn}}$ ("delta H super zero of reaction") when all the products and reactants involved in the reaction are in their respective standard states. Thus we have from Eq. 7-7

$$\Delta H^\circ_{\text{rxn}} = \sum_i \nu_i \bar{H}_i^\circ \tag{7-8}$$

where the superscript "zeros" serve to remind us that we are dealing with substances in their standard states. Note that, although ΔH_{rxn} is a function of T and P, the value of ΔH_{rxn} depends, for a given choice of standard states, only on the temperature:

$$\Delta H^\circ_{\text{rxn}} = \Delta H^\circ_{\text{rxn}}(T) \tag{7-9}$$

because the pressure is fixed by convention at a value of 1 atm.

Although the value of $\Delta H^\circ_{\text{rxn}}$ for a reaction can be obtained from appropriate experimental data, there is no way to determine the value of $\bar{H}_i^\circ$ for a substance, because absolute enthalpies are not measurable. It would nonetheless be convenient, in the interest of economies of tabulation, to be able to tabulate *enthalpies of substances* as a function of temperature relative to some arbitrarily chosen zero of enthalpy rather than to tabulate ΔH°_{rxn} values, because the number of chemical substances is far less than the number of possible reactions between these substances.

We can achieve our objective by defining the standard molar enthalpy of formation of a chemical substance. *The value of the standard molar enthalpy of formation of a substance* $\Delta \bar{H}^\circ_f$ (where the subscript f denotes *formation*) *is defined as the enthalpy change for the reaction in which one mole of the substance is formed in its standard state from the elements in their standard states at the temperature of interest.*

Consider the reaction

$$C(s) + O_2(g) \longrightarrow CO_2(g) \qquad \Delta H^\circ_{rxn,298.15\,K} = -393.5\ kJ\cdot mol^{-1}$$

in which 1 mol of $CO_2(g)$ is formed from the elements. The value of ΔH°_{rxn} for this reaction is defined to be the standard *molar enthalpy of formation* of $CO_2(g)$ from the elements

$$\Delta \bar{H}^\circ_f[CO_2(g)] = -393.5\ kJ\cdot mol^{-1}$$

where the bar over the H denotes 1 mol of the substance and the subscript f stands for $f$ormation from the elements. Note that the above reaction involves the formation of 1 mol of $CO_2(g)$ in its 1-atm ideal gas standard state at 25°C from its constituent elements $C(s)$ and $O_2(g)$, in their respective 1-atm standard states at 25°C. In other words, 1.00 mol of $CO_2(g)$ in its standard state at 25°C lies 393.5 kJ *downhill* on the enthalpy scale relative to the elements $C(s)$ and $O_2(g)$ in their standard states at 25°C (see Figure 7-2). The standard molar enthalpies of formation of $CH_4(g)$ and $H_2O(g)$ from the elements at 25°C are (see Figure 7-2)

$$C(s) + 2\,H_2(g) \longrightarrow CH_4(g) \qquad \Delta H^\circ_{rxn} = \Delta \bar{H}^\circ_f[CH_4(g)] = -74.8\ kJ\cdot mol^{-1}$$

$$H_2(g) + \tfrac{1}{2}\,O_2(g) \longrightarrow H_2O(g) \qquad \Delta H^\circ_{rxn} = \Delta \bar{H}^\circ_f[H_2O(g)] = -241.8\ kJ\cdot mol^{-1}$$

As suggested by Figure 7-2, we can set up a table of $\Delta \bar{H}^\circ_f$ values for *compounds* by setting the $\Delta \bar{H}^\circ_f$ values for the *elements* equal to zero. That is, for each element in its most stable form at the temperature of interest, we set $\Delta \bar{H}^\circ_f$ equal to zero. For example, at 25°C we take

$$\Delta \bar{H}^\circ_f[C(s), \text{graphite}] = 0 \qquad \Delta \bar{H}^\circ_f[S(s), \text{rhombic}] = 0$$

$$\Delta \bar{H}^\circ_f[H_2(g)] = 0 \qquad \Delta \bar{H}^\circ_f[O_2(g)] = 0$$

The foregoing assignments of $\Delta \bar{H}^\circ_f = 0$ for the elements follows directly from the definition of $\Delta \bar{H}^\circ_f$. For example, for the "reaction"

$$O_2(g) \longrightarrow O_2(g)$$

we obviously have $\Delta H^\circ_{rxn} = 0$, and thus $\Delta \bar{H}^\circ_f[O_2(g)] = 0$. Note, however, that $\Delta \bar{H}^\circ_f$ for gaseous oxygen atoms $O(g)$ is not equal to zero, but rather $\Delta \bar{H}^\circ_f[O(g)] = +247$ kJ. Oxygen normally occurs as diatomic molecules and energy must be supplied to produce $O(g)$ from $O_2(g)$:

$$O_2(g) \longrightarrow 2\,O(g) \qquad \Delta H^\circ_{rxn} = +494\ kJ$$

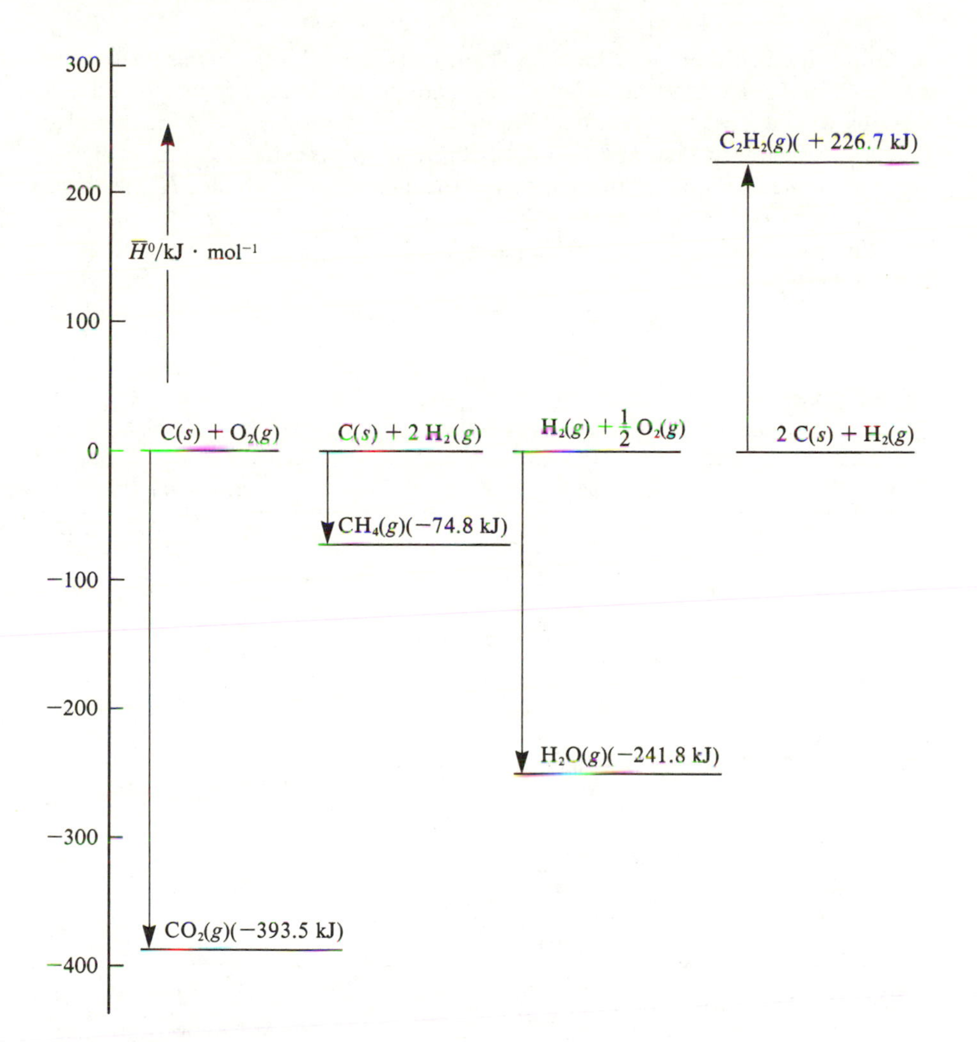

Figure 7-2. Enthalpy changes involved in the formation of the compounds $CO_2(g)$, $CH_4(g)$, $H_2O(g)$, and $C_2H_2(g)$ in their 1-atm standard states at 25°C from the respective elements in their 1-atm standard states at 25°C.

Thus the value of $\Delta H^\circ_{\text{rxn}}$ for the dissociation of $O_2(g)$ to 2 $O(g)$ is equal to

$$\Delta H^\circ_{\text{rxn}} = 2\Delta\bar{H}^\circ_f[O(g)] - \Delta\bar{H}^\circ_f[O_2(g)] = +494\ \text{kJ}$$

Solving for $\Delta\bar{H}^\circ_f[O(g)]$, we obtain

$$\Delta\bar{H}^\circ_f[O(g)] = \frac{494\ \text{kJ} + \Delta\bar{H}^\circ_f[O_2(g)]}{2\ \text{mol}} = \frac{494\ \text{kJ} + 0}{2\ \text{mol}} = 247\ \text{kJ}\cdot\text{mol}^{-1}$$

In assigning a value of $\Delta\bar{H}^\circ_f = 0$ at all temperatures to the elements, it is necessary to specify the particular form of the element, for example, gas, liquid, or solid (of which there may be several forms) is chosen. *As long as one is consistent, it makes no difference what form is chosen;* however, it is usually advantageous with regard to experimental convenience to set $\Delta \bar{H}_f^\circ = 0$ for the form that is most stable at 25°C and 1 atm pressure, and this is usually done. Table 7-1 contains some examples of those forms of the elements for which $\Delta \bar{H}_f^\circ$ has been taken as zero.

The relationship between $\Delta H_{\text{rxn}}^\circ$ and the $\Delta \bar{H}_{f,i}^\circ$ values of the products and the reactants is

$$\Delta H_{\text{rxn}}^\circ = \sum_i \nu_i \Delta \bar{H}_{f,i}^\circ \tag{7-10}$$

TABLE 7-1

Choices of $\Delta \bar{H}_f^\circ = 0$ for Some Representative Elements

Element	*Form*	$\Delta \bar{H}_f^\circ$ *(at all T)*[a]
Bromine	$Br_2(\ell)$	0
Carbon	C(*s*, graphite)	0
Sulfur	S(*s*, rhombic)	0
Selenium	Se(*s*, gray, hexagonal)	0
Phosphorus	P(*s*, white)	0
Hydrogen	$H_2(g)$	0
Deuterium[b]	$D_2(g)$	0
Silicon	Si(*s*, crystalline)	0
Tin	Sn(*s*, white)	0

These choices lead to the following $\Delta \bar{H}_f^\circ$ values at 25°C.

Element	*Form*	$\Delta \bar{H}_f^\circ/\text{kJ}\cdot\text{mol}^{-1}$ *(at* 298.15 K)
Hydrogen	H(*g*) (ground state)	217.94
Bromine	$Br_2(g)$	30.71
Carbon	C(*s*, diamond)	1.94
Sulfur	S(*s*, monoclinic)	0.30
Selenium	Se(*s*, red, monoclinic)	0.84
Phosphorus	P(*s*, red)	−18.41
Phosphorus	P(*s*, black)	−43.10
Silicon	Si(*s*, amorphous)	4.18
Tin	Sn(*s*, gray)	2.51

[a] At temperatures far removed from 25°C it may prove advantageous to change the form of an element for which we have set $\Delta \bar{H}_f^\circ = 0$ to some other form that has become more stable than the form that is most stable at 25°C. For example, at 500°C, $Br_2(g)$ is more stable than $Br_2(\ell)$, and thus between 25°C and 500°C we change from $\Delta H_f^\circ = 0$ for the latter to $\Delta \bar{H}_f^\circ = 0$ for the former. Considerable care must be exercised to make certain for which forms of the elements ΔH_f° has been set equal to zero at the temperature of interest.

[b] Except for D_2, by "element" is meant the element with its normal isotopic composition.

Equation 7-10 follows directly from the fact that the chemical reaction is balanced and thus the number of moles of the various elements needed to form the required numbers of moles of the various reaction products is exactly the same as the number of moles of the various elements needed to form the required numbers of moles of the various reactants (Figure 7-3).

In order to obtain $\Delta\bar{H}_f^\circ$ for a substance at a given temperature, we need only to obtain ΔH_{rxn}° for a reaction in which all the $\Delta\bar{H}_{f,i}^\circ$ values are known except the one of interest; then, using Eq. 7-10, we can obtain the desired $\Delta\bar{H}_{f,i}^\circ$ value. The procedure works because H is a thermodynamic coordinate and therefore ΔH is independent of path. Two examples of calculations involving the determination of $\Delta\bar{H}_{f,i}^\circ$ values follow.

1) $H_2O(\ell) \longrightarrow H_2O(g)$

Given that, at 25°C, $\Delta H_{rxn}^\circ = 44.011$ kJ for the above reaction and that $\Delta\bar{H}_f^\circ[H_2O(\ell)] = -285.838\ \text{kJ}\cdot\text{mol}^{-1}$, compute $\Delta\bar{H}_f^\circ[H_2O(g)]$ at 25°C. From Eq. 7-10 we have

$$\Delta H_{rxn}^\circ = (1\ \text{mol})\Delta\bar{H}_f^\circ[H_2O(g)] - (1\ \text{mol})\Delta\bar{H}_f^\circ[H_2O(\ell)]$$

and thus

$$44.011\ \text{kJ} = (1\ \text{mol})\,\Delta\bar{H}_f^\circ[H_2O(g)] - (1\ \text{mol})(-285.838\ \text{kJ}\cdot\text{mol}^{-1})$$

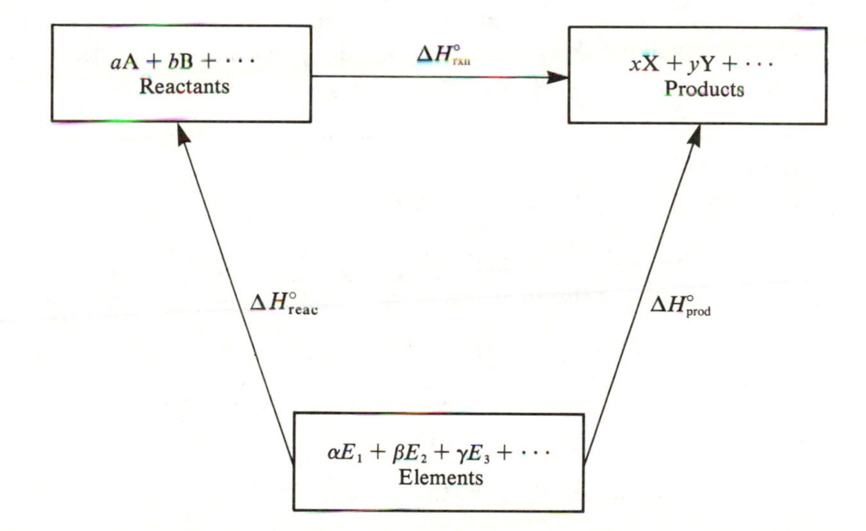

Figure 7-3. Schematic diagram illustrating the derivation of Eq. 7-10. Because ΔH_{rxn}° is independent of path, we have

$$\Delta H_{rxn}^\circ = \Delta H_{prod}^\circ - \Delta H_{reac}^\circ$$

Furthermore

$$\Delta H_{prod}^\circ = x\,\Delta\bar{H}_{f,X}^\circ + y\,\Delta\bar{H}_{f,Y}^\circ + \cdots$$

and

$$\Delta H_{reac}^\circ = a\,\Delta\bar{H}_{f,A}^\circ + b\,\Delta\bar{H}_{f,B}^\circ + \cdots$$

Thus

$$\Delta H_{rxn}^\circ = \sum_i \nu_i\,\Delta\bar{H}_{f,i}^\circ$$

or

$$\Delta\bar{H}_f^\circ[H_2O(g)] = \frac{44.011\ \text{kJ} - 285.838\ \text{kJ}}{1\ \text{mol}} = -241.827\ \text{kJ}\cdot\text{mol}^{-1}$$

2) $2\,H_2O_2(g) \longrightarrow 2\,H_2O(g) + O_2(g)$

Given that, at 25°C, $\Delta H_{rxn}^\circ = -211.46$ kJ and using the result from part 1, compute $\Delta\bar{H}_f^\circ[H_2O_2(g)]$ at 25°C. From Eq. 7-10 we have

$$\Delta H_{rxn}^\circ = 2\,\Delta\bar{H}_f^\circ[H_2O(g)] + \Delta\bar{H}_f^\circ[O_2(g)] - 2\,\Delta\bar{H}_f^\circ[H_2O_2(g)]$$

Thus with $\Delta\bar{H}_f^\circ[O_2(g)] = 0$, we have

$$-211.46\ \text{kJ} = (2\ \text{mol})(-241.83\ \text{kJ}\cdot\text{mol}^{-1}) - 0 - (2\ \text{mol})\,\Delta\bar{H}_f^\circ[H_2O_2(g)]$$

and

$$\Delta\bar{H}_f^\circ[H_2O_2(g)] = \frac{211.46\ \text{kJ} - 2(241.83\ \text{kJ})}{2\ \text{mol}} = -136.10\ \text{kJ}\cdot\text{mol}^{-1}$$

Proceeding as outlined in parts 1 and 2 above, we can build up a table of $\Delta\bar{H}_{f,i}^\circ$ values at 298.15 K (25°C). Of course, there is nothing special about 25°C and we could construct a table of $\Delta\bar{H}_{f,i}^\circ$ values at any temperature, if we so choose, using the same procedure as outlined for the 25°C table. From tabulated $\Delta\bar{H}_{f,i}^\circ$ values, we can compute ΔH_{rxn} for any reaction of interest for which we have available $\Delta\bar{H}_{f,i}^\circ$ values for all the products and reactants. A highly abridged compilation of $\Delta\bar{H}_{f,i}^\circ$ values is given in Appendix I.

EXAMPLE 7-2

Compute ΔH_{rxn}° at 25°C for the reaction

$$N_2H_4(\ell) + O_2(g) \longrightarrow N_2(g) + 2\,H_2O(\ell)$$

given the following $\Delta\bar{H}_{f,i}^\circ$ values at 25°C:

$$\Delta\bar{H}_f^\circ[N_2(g)] = 0 \qquad \Delta\bar{H}_f^\circ[H_2O(\ell)] = -285.838\ \text{kJ}\cdot\text{mol}^{-1}$$

$$\Delta\bar{H}_f^\circ[N_2H_4(\ell)] = 50.417\ \text{kJ}\cdot\text{mol}^{-1} \qquad \Delta\bar{H}_f^\circ[O_2(g)] = 0$$

Solution: From Eq. 7-10 we have

$$\Delta H_{rxn}^\circ = \Delta\bar{H}_f^\circ[N_2(g)] + 2\,\Delta\bar{H}_f^\circ[H_2O(\ell)] - \Delta\bar{H}_f^\circ[N_2H_4(\ell)] - \Delta\bar{H}_f^\circ[O_2(g)]$$

and thus

$$\begin{aligned}\Delta H_{rxn}^\circ &= 0 + (2\ \text{mol})(-285.838\ \text{kJ}\cdot\text{mol}^{-1}) - (1\ \text{mol})(50.417\ \text{kJ}\cdot\text{mol}^{-1}) - 0 \\ &= -622.09\ \text{kJ}\cdot\text{mol}^{-1} \qquad \text{at } 25°\text{C}\end{aligned}$$

A table of $\Delta\bar{H}_{f,i}^\circ$ values is particularly useful where it is not possible to carry out directly a reaction of interest. An example is the hydrocarbon isomerization reaction

$$\underset{n\text{-butane}}{CH_3CH_2CH_2CH_3(g)} \longrightarrow \underset{i\text{-butane}}{CH_3CH(CH_3)CH_3(g)}$$

This isomerization reaction cannot be achieved exclusively in the laboratory because of a number of competing reactions that also occur. Nonetheless, it is of importance to know ΔH°_{rxn} for reactions of the above type in connection with the production of gasolines from petroleum liquids. It is not at all difficult to carry out the combustion reactions

$$\underset{n\text{-butane}}{CH_3CH_2CH_2CH_3(g)} + \tfrac{13}{2}\,O_2(g) \longrightarrow 4\,CO_2(g) + 5\,H_2O(\ell) \qquad \Delta H^\circ_{rxn}(1)$$

$$\underset{i\text{-butane}}{CH_3CH(CH_3)CH_3(g)} + \tfrac{13}{2}\,O_2(g) \longrightarrow 4\,CO_2(g) + 5\,H_2O(\ell) \qquad \Delta H^\circ_{rxn}(2)$$

From the measured values of $\Delta H^\circ_{rxn}(1)$ and $\Delta H^\circ_{rxn}(2)$, together with $\Delta \bar{H}^\circ_f$ values for $CO_2(g)$ and $H_2O(\ell)$, we can calculate $\Delta \bar{H}^\circ_f$ values for *n*-butane and *i*-butane. The resulting $\Delta \bar{H}^\circ_f$ values can then be used to calculate ΔH°_{rxn} for the isomerization reaction. Alternatively, because ΔH is independent of path, we can write (Figure 7-4)

$$\Delta H^\circ(\text{isomerization}) = \Delta H^\circ_{rxn}(1) - \Delta H^\circ_{rxn}(2) = -6.862\text{ kJ}$$

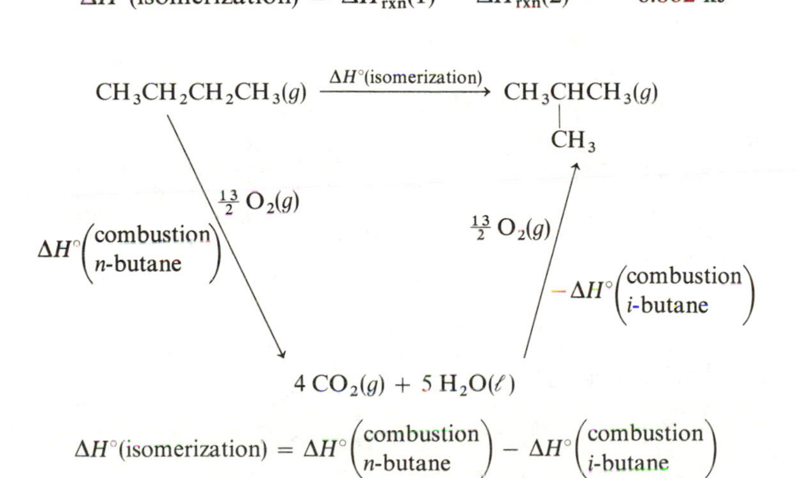

Figure 7-4. Illustration of the use of the path-independent property of the enthalpy function to obtain ΔH°_{rxn} for the isomerization reaction

$$CH_3CH_2CH_2CH_3(g) \longrightarrow CH_3CH(CH_3)CH_3(g)$$

from ΔH_{rxn} values for the combustion reactions of the two compounds.

The foregoing calculations are based on the additivity of ΔH_{rxn} values. If we add two chemical reactions together to obtain a third chemical reaction, then the ΔH_{rxn} value for the third reaction is equal to the sum of the ΔH_{rxn} values for the two reactions that were added (provided that the temperatures and pressures are the same for each ΔH_{rxn} value); thus

$$\Delta H_{rxn}(3) = \Delta H_{rxn}(1) + \Delta H_{rxn}(2)$$

The additivity of ΔH_{rxn} values is sometimes referred to as Hess's law. Note that Hess's law is a direct consequence of the fact that the enthalpy is a state function. Note also that if ΔH_{rxn} is the value of ΔH for a reaction as written, then the value of ΔH_{rxn} for the reverse reaction is $-\Delta H_{rxn}$; that is,

$$\Delta H_{rxn}(\text{forward reaction}) = -\Delta H_{rxn}(\text{reverse reaction})$$

We utilized the above equation in the analysis of the butane isomerization reaction.

7-3 *The Value of ΔH°_{rxn} Depends on the Temperature*

Consider the variation in the enthalpy with temperature and pressure. Taking $H_i = H_i(T, P)$ we have

$$dH_i = \left(\frac{\partial H_i}{\partial T}\right)_P dT + \left(\frac{\partial H_i}{\partial P}\right)_T dP \tag{7-11}$$

or

$$dH_i = C_{P,i}\,dT + \left(\frac{\partial H_i}{\partial P}\right)_T dP \tag{7-12}$$

For 1 mol of a substance held at 1 atm pressure $dP = 0$, $H_i = \bar{H}_i$ and $C_{P,i} = \bar{C}_{P,i}$; thus

$$d\bar{H}_i = \bar{C}_{P,i}\,dT \qquad \text{(constant pressure)} \tag{7-13}$$

If the change in temperature takes place with the substance in its 1-atm standard state at each T, then Eq. 7-13 becomes

$$d\bar{H}_i^\circ = \bar{C}_{P,i}^\circ\,dT \tag{7-14}$$

where $\bar{C}_{P,i}^\circ = \bar{C}_{P,i}^\circ(T)$ is the heat capacity of substance i in the 1-atm standard state. Integration of Eq. 7-14 from T_1 to T_2 yields

$$\bar{H}_{i,T_2}^\circ - \bar{H}_{i,T_1}^\circ = \int_{T_1}^{T_2} \bar{C}_{P,i}^\circ\,dT \tag{7-15}$$

Application of Eq. 7-14 to ν_i mol of substance i yields

$$d(\nu_i\bar{H}_i^\circ) = \nu_i\bar{C}_{P,i}^\circ\,dT \tag{7-16}$$

Application of Eq. 7-16 to a balanced chemical reaction yields

$$d\left(\sum_i \nu_i\bar{H}_i^\circ\right) = \left(\sum_i \nu_i\bar{C}_{P,i}^\circ\right)dT \tag{7-17}$$

or

$$d(\Delta H^\circ_{\text{rxn}}) = \Delta C^\circ_{P,\text{rxn}}\, dT \qquad \text{(constant pressure)} \tag{7-18}$$

where

$$\Delta C^\circ_{P,\text{rxn}} = \sum_i \nu_i \bar{C}^\circ_{P,i} \tag{7-19}$$

Integration of Eq. 7-18 from T_1 to T_2 yields

$$\Delta H^\circ_{\text{rxn},T_2} - \Delta H^\circ_{\text{rxn},T_1} = \int_{T_1}^{T_2} \Delta C^\circ_{P,\text{rxn}}\, dT \tag{7-20}$$

Equation 7-20 shows that the temperature dependence of $\Delta H^\circ_{\text{rxn}}$ is determined by $\Delta C^\circ_{P,\text{rxn}}$.

EXAMPLE 7-3

Derive Eq. 7-20 starting from $\Delta H_{\text{rxn}} = \Delta H_{\text{rxn}}(T, P)$.

Solution: ΔH_{rxn} is a state function because it is equal to an algebraic sum of state functions:

$$\Delta H_{\text{rxn}} = \sum_i \nu_i \bar{H}_i$$

Thus we can write

$$d(\Delta H_{\text{rxn}}) = \left[\frac{\partial(\Delta H_{\text{rxn}})}{\partial T}\right]_P dT + \left[\frac{\partial(\Delta H_{\text{rxn}})}{\partial P}\right]_T dP$$

with all reactants and products in their 1-atm standard states $\Delta H_{\text{rxn}} = \Delta H^\circ_{\text{rxn}}$ and $dP = 0$; thus

$$d(\Delta H^\circ_{\text{rxn}}) = \left[\frac{\partial(\Delta H^\circ_{\text{rxn}})}{\partial T}\right]_P dT = \Delta C^\circ_{P,\text{rxn}}\, dT$$

from which Eq. 7-20 follows directly. The method of derivation of Eq. 7-20 outlined in this example is more direct, but less explicit, than the method given in the preceding text. The more detailed derivation was given in the text because it is more explicit, and the derivation of Eq. 7-20 was our first encounter with a derivation involving a ΔX_{rxn} function, where X is a thermodynamic state function. In subsequent derivations involving ΔX_{rxn} functions we shall use the more direct approach employed in this example.

Equation 7-20 is used to compute a value of $\Delta \bar{H}^\circ_{\text{rxn}}$ at temperature T_2 from a value of $\Delta H^\circ_{\text{rxn}}$ at temperature T_1, and $\bar{C}^\circ_{P,i}(T)$ data (valid over the range $T_1 \rightarrow T_2$) for the product and the reactant species. Note that $\bar{C}^\circ_{P,i}$ depends only on the temperature, because the pressure in the standard state is fixed by definition at 1 atm.

EXAMPLE 7-4

Consider the reaction

$$N_2(g) + 3\,H_2(g) \longrightarrow 2\,NH_3(g)$$

for which at 298.15 K

$$\Delta H^\circ_{rxn} = 2\Delta \bar{H}^\circ_f[NH_3(g)] = (2\text{ mol})(-46.11\text{ kJ}\cdot\text{mol}^{-1}) = -92.22\text{ kJ}$$

Compute ΔH°_{rxn} for this reaction at 1000 K given the following heat capacity data (with T in kelvin, $\bar{C}^\circ_P$ has the units $J\cdot K^{-1}\cdot mol^{-1}$):

$$\begin{aligned}
\bar{C}^\circ_P[NH_3(g)] &= 29.75 + 25.10 \times 10^{-3}T - 1.55 \times 10^5 T^{-2} \\
\bar{C}^\circ_P[N_2(g)] &= 28.58 + 3.77 \times 10^{-3}T - 0.50 \times 10^5 T^{-2} \\
\bar{C}^\circ_P[H_2(g)] &= 27.28 + 3.26 \times 10^{-3}T + 0.50 \times 10^5 T^{-2}
\end{aligned}$$

The above $\bar{C}^\circ_P$ equations are valid over the range 298 K–2000 K.

Solution: The value of $\Delta C^\circ_{P,rxn}$ is computed using Eq. 7-19; thus

$$\begin{aligned}
\Delta C^\circ_{P,rxn} = {} & [2(29.75) - 3(27.28) - 28.58] + [2(25.10) - 3(3.26) - (3.77)] \\
& \times 10^{-3}T + [2(-1.55) - 3(0.50) - (-0.50)] \times 10^5 T^{-2}
\end{aligned}$$

or

$$\Delta C^\circ_P = -50.92 + 36.65 \times 10^{-3}T - 4.10 \times 10^5 T^{-2}$$

From Eq. 7-20 we have

$$\Delta H^\circ_{rxn,1000K} = \Delta H^\circ_{rxn,298.15K} + \int_{298.15K}^{1000K} \Delta C^\circ_{P,rxn}\, dT$$

thus

$$\begin{aligned}
\Delta H^\circ_{rxn,1000} = {} & -92{,}220 - 50.92(1000 - 298) + 18.33 \times 10^{-3}[(1000)^2 \\
& - (298)^2] + 4.10 \times 10^5\left(\frac{1}{1000} - \frac{1}{298}\right)
\end{aligned}$$

and

$$\begin{aligned}
\Delta H^\circ_{rxn,1000} &= -92{,}220 - 35{,}746 + 16{,}698 - 966 \\
&= -112{,}234\text{ J} = -112.23\text{ kJ}
\end{aligned}$$

The calculation of

$$\Delta H^\circ_{rxn,T_2} - \Delta H^\circ_{rxn,T_1}$$

also can be carried out using tabulated $\bar{H}^\circ_T - \bar{H}^\circ_{298.15}$ or $\bar{H}^\circ_T - \bar{H}^\circ_0$ data; these calculations are simpler and usually more accurate than those involving $\bar{C}^\circ_{P,i}$ data. From such tables one can compute for a reaction

$$\Delta H^\circ_{rxn,T} - \Delta H^\circ_{rxn,298} = \sum_i \nu_i(\bar{H}^\circ_T - \bar{H}^\circ_{298})_i \qquad (7\text{-}21)$$

(where 298 is shorthand for 298.15 K) or

$$\Delta H^\circ_{rxn,T} - \Delta H^\circ_{rxn,0} = \sum_i \nu_i(\bar{H}^\circ_T - \bar{H}^\circ_0)_i \tag{7-22}$$

provided, of course, that $(\bar{H}^\circ_T - \bar{H}^\circ_{298})$ or $(\bar{H}^\circ_T - \bar{H}^\circ_0)$ data are available for all the products and the reactants. To compute $\Delta H^\circ_{rxn,T}$ from Eq. 7-21 or 7-22, we also need to know $\Delta H^\circ_{rxn,298}$ or $\Delta H^\circ_{rxn,0}$, respectively.

Values of $\bar{H}_{T_2} - \bar{H}_{T_1}$ for a substance are obtained directly from calorimetric experiments on individual substances, and such data are available for a wide variety of compounds. The most direct method for measuring $\bar{H}^\circ_{T_2} - \bar{H}^\circ_{T_1}$ for a solid or liquid is simply to place a known mass of the substance into a calorimeter and determine the amount of electrical energy (Joule heat, I^2Rt) necessary to raise the temperature of the substance from T_1 to T_2. For high values of T_2, *drop calorimetry* is used to measure $\bar{H}^\circ_{T_2} - \bar{H}^\circ_{T_1}$. A solid sample of known mass is heated in an oven (equipped with a trapdoor) to T_2. The trapdoor is then opened and the sample drops into an adiabatic calorimeter at $T \ll T_2$. The energy is transferred as heat from the sample to the calorimeter at constant pressure; thus

$$(\bar{H}^\circ_{T_1} - \bar{H}^\circ_{T_2}) + C_P(T_1 - T) = 0$$

provided $T_1 - T$ is small. The value of C_P for the calorimeter and contents including the sample is determined in a separate experiment using resistance heating.

EXAMPLE 7-5

From tabulated thermodynamic data, we obtain

Species	$(\bar{H}^\circ_{1000} - \bar{H}^\circ_{298.15})/\text{kJ}\cdot\text{mol}^{-1}$
$H_2(g)$	20.686
$N_2(g)$	21.460
$NH_3(g)$	32.581

For the reaction

$$N_2(g) + 3\,H_2(g) \longrightarrow 2\,NH_3(g)$$

$\Delta H^\circ_{rxn,298.15\text{K}} = -92.22$ kJ. Compute $\Delta H^\circ_{rxn,1000}$.

Solution: From Eq. 7-21 we have

$$\Delta H^\circ_{rxn,1000} = \Delta H^\circ_{rxn,298.15} + \sum \nu_i(\bar{H}^\circ_{1000} - \bar{H}^\circ_{298.15})_i$$

and thus

$$\Delta H^\circ_{rxn,1000} = -92.22 + 2(32.581) - 3(20.686) - (21.460) = -110.90\text{ kJ}$$

The above calculation of $\Delta H^{\circ}_{rxn,1000}$ is more reliable than that carried out using $\bar{C}^{\circ}_P$ data in Example 7-4; these two calculations illustrate how the use of data from different sources can lead to discrepancies, which arise in part from differences in source data and in part from differences in the way in which the data were treated numerically.

7-4 *The Value of ΔH_{rxn} Is Determined Primarily by the Difference in Bond Enthalpies of the Reactant and the Product Molecules*

A key question in the analysis of ΔH_{rxn} values is What molecular property(ies) of the reactants and the products give(s) rise to the observed value of ΔH_{rxn}? What determines the sign and the magnitude of ΔH_{rxn}? The answer is that ΔH_{rxn} is determined primarily (but not solely) by the difference in bond enthalpies of the reactant and the product molecules. The bond enthalpy is approximately equal to the bond energy at 1 atm:

$$\bar{H} = \bar{U} + P\bar{V} = \bar{U} + (1\text{ atm})\bar{V} \approx \bar{U}$$

The enthalpy change for the reaction

$$H_2O(g) \longrightarrow O(g) + 2\,H(g)$$

at 25°C is $\Delta H^{\circ}_{rxn} = +925$ kJ. The input of 925 kJ of energy as heat is required to break the 2 mol of O—H bonds in 1 mol of gaseous water molecules:

$$925\text{ kJ} + \mathrm{H{-}O{-}H} \longrightarrow \mathrm{O} + 2\,\mathrm{H}$$

The average O—H bond enthalpy per mol, $\bar{H}$(O—H), for water is equal to one-half of the total energy input as heat that is required to break the 2 mol of O—H bonds; that is

$$\bar{H}(\mathrm{O{-}H}) = \frac{925\text{ kJ}}{2} = 463\text{ kJ}$$

Note that 463 kJ is *not* the enthalpy change at 25°C for the reaction

$$H_2O(g) \longrightarrow OH(g) + H(g) \qquad \Delta H^{\circ}_{rxn} = 502\text{ kJ}$$

The *bond dissociation enthalpy* for an O—H bond in water is $502\text{ kJ}\cdot\text{mol}^{-1}$. The bond dissociation enthalpy and the bond enthalpy are identical only for a diatomic molecule. A *bond dissociation enthalpy* is a rigorously defined thermodynamic quantity, being in fact the enthalpy change for the reaction in which the particular bond of interest is the only bond broken. Bond enthalpies are *average* quantities that do not have a rigorous thermodynamic significance but which are useful for estimating ΔH_{rxn} values when precise thermochemical data are not available.

Thermodynamic data can be used to construct a table of *average bond enthalpies*. For example, from $\Delta H^{\circ}_{f,298}$ data, we can compute ΔH°_{rxn} at 25°C for the reaction

$$H_2O_2(g) \longrightarrow 2\,H(g) + 2\,O(g)$$

The value of ΔH°_{rxn} can be expressed in terms of bond enthalpies as follows:

$$\Delta H^\circ_{rxn} = 2\,\bar{H}(O{-}H) + \bar{H}(O{-}O) = 1071\text{ kJ}$$

The value of $\bar{H}(O{-}O)$ can be calculated using the average $\bar{H}(O{-}H)$ value (Table 7-2) of 464 kJ; the computed value of $\bar{H}(O{-}O)$ is

$$\bar{H}(O{-}O) = 1071 - 2(464) = 143\text{ kJ}\cdot\text{mol}^{-1}$$

Proceeding in the above manner, several values of $\bar{H}(O{-}O)$ can be obtained

TABLE 7-2
Average Bond Enthalpies at 25°C

Bond	$\bar{H}(X{-}Y)/kJ\cdot mol^{-1}$	*Bond*	$\bar{H}(X{-}Y)/kJ\cdot mol^{-1}$
O—H	464	C—I	234
O—O	138	C—Si	285
O=O	494	C—N	259
O—F	188	C≡N	879
O—Cl	188	C—S	259
O—Br	188	C=S	477
S—F	285	H—H	435
S—Cl	255	F—F	155
S—Br	218	Cl—Cl	243
B—F	644	Br—Br	192
B—Cl	456	H—F	565
Si—F	565	H—Cl	431
Si—Cl	381	H—Br	368
Si—Br	310	H—I	297
Si—I	234	N—N	159
C—H	414	N=N	418
C—C	335	N≡N	946
C—O	339	N—H	389
C=O	732	N—F	272
C=C	607	N—Cl	192
C≡C	828	P—F	490
C—F	427	P—Cl	326
C—Cl	322	P—Br	264
C—Br	268		

from different molecules that contain O—O bonds. The average of the resulting bond enthalpy values is designated as $\bar{H}$(O—O). Table 7-2 lists $\bar{H}$(X—Y) values for a variety of chemical bonds.

The value of ΔH_{rxn} *for a reaction is given approximately by the difference in the energy input as heat required to break all the chemical bonds in the reactants and the energy released on the formation of all the chemical bonds in the products* (Figure 7-5):

$$\Delta H_{rxn} \approx \begin{bmatrix} \text{total energy input as} \\ \text{heat at constant pressure} \\ \text{required to break all} \\ \text{the bonds in the reactants} \end{bmatrix} - \begin{bmatrix} \text{total energy released as} \\ \text{heat at constant pressure} \\ \text{on formation of all} \\ \text{the bonds in the products} \end{bmatrix} \tag{7-23}$$

or

$$\Delta H_{rxn} \approx \underbrace{\sum_i \alpha_i \bar{H}(\text{X—Y})_i}_{\text{reactants}} - \underbrace{\sum_i \alpha_i \bar{H}(\text{X—Y})_i}_{\text{products}} \tag{7-24}$$

where α_i is the total number of bonds of the type X—Y. If more energy is released as heat on the formation of all the bonds in the products than is required to break all the bonds in the reactants, then the value of ΔH_{rxn} is negative (energy released as heat). If less energy is released on the formation of the bonds in the products than is required to break the bonds in the reactants, then the value of ΔH_{rxn} is positive (energy consumed). The value of ΔH_{rxn} is determined primarily by the difference in the bond energies in the reactants and products.

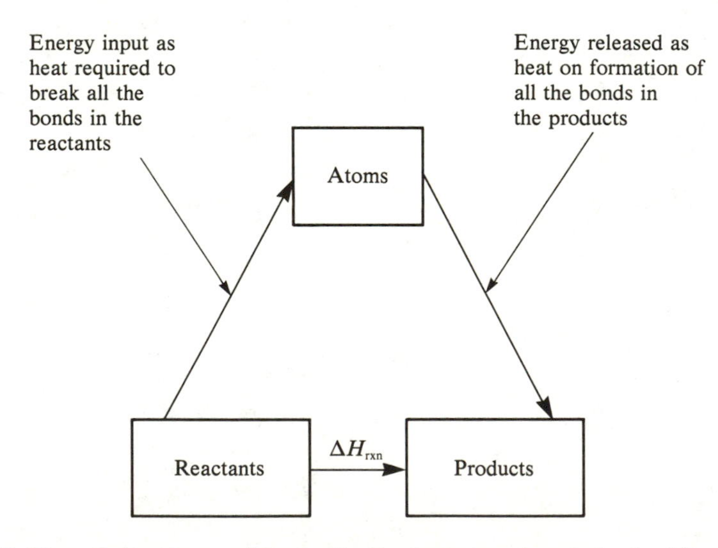

Figure 7-5. The enthalpy change of the reaction is given approximately by the difference in the energy input as heat at constant pressure required to break all the chemical bonds in the reactants and the energy released as heat at constant pressure on the formation of all the chemical bonds in the products:

$$\Delta H_{rxn} \approx \underbrace{\sum_i \alpha_i \bar{H}(\text{X—Y})_i}_{\text{reactants}} - \underbrace{\sum_i \alpha_i \bar{H}(\text{X—Y})_i}_{\text{products}}$$

EXAMPLE 7-6

Use the bond enthalpy data in Table 7-2 to estimate $\Delta H^\circ_{rxn,298}$ for the reaction

$$C_2H_2(g) + 2\,H_2(g) \longrightarrow C_2H_6(g)$$

and compare your result with that calculated using $\Delta \bar{H}^\circ_{f,i}$ data in Appendix I.

Solution: The above reaction is first rewritten to show all the bonds in the molecules:

$$\text{H—C}{\equiv}\text{C—H} + 2\,\text{H—H} \longrightarrow \text{H}_3\text{C—CH}_3 \;(\text{H—}\overset{\text{H}}{\underset{\text{H}}{\text{C}}}\text{—}\overset{\text{H}}{\underset{\text{H}}{\text{C}}}\text{—H})$$

From Eq. 7-24 we have

$$\begin{aligned}\Delta H^\circ_{rxn} \approx\ & 2\bar{H}(\text{C—H}) + \bar{H}(\text{C}{\equiv}\text{C}) + 2\bar{H}(\text{H—H}) \\ & - [\bar{H}(\text{C—C}) + 6\bar{H}(\text{C—H})]\end{aligned}$$

Using the $\bar{H}$(X—Y) data in Table 7-2 yields

$$\Delta H^\circ_{rxn} \simeq 2(414) + 828 + 2(435) - [335 + 6(414)] = -293\ \text{kJ}$$

From $\Delta \bar{H}^\circ_{f,298}$ data in Appendix I we compute

$$\Delta H^\circ_{rxn} = \Delta \bar{H}^\circ_f[C_2H_6(g)] - \Delta \bar{H}^\circ_f[C_2H_2(g)] - 2\Delta \bar{H}^\circ_f[H_2(g)]$$

$$\Delta H^\circ_{rxn} = -84.68 - 226.73 - 2(0) = -311.41\ \text{kJ}$$

Note that the bond enthalpy estimate for ΔH°_{rxn} is about 6% in error.

In general, ΔH°_{rxn} values that are estimated from average bond enthalpies for reactions involving all gaseous species are usually within 10% of the actual value. The major sources of the discrepancies arise from the fact that bond enthalpies are *averages* of values obtained from data on a variety of compounds. If $\Delta H^\circ_{f,i}$ data are available for all the reactants and products, then, of course, the use of bond enthalpies to estimate ΔH°_{rxn} is unnecessary and undesirable.

Although the value of ΔH_{rxn} is determined primarily by the difference in bond energies of the reactants and the products, interactions (attractive forces) *between* chemical species in liquid and in solid phases can also make significant contributions to the value of ΔH_{rxn}. As an example, consider the vaporization of liquid water:

$$H_2O(\ell) \longrightarrow H_2O(g) \qquad \Delta H^\circ_{rxn} = +44.0\ \text{kJ} \qquad (25^\circ\text{C})$$

In the vaporization of water it is the attractive forces *between* the water molecules that must be overcome by the input of energy that gives rise to the observed ΔH°_{rxn} value. The estimation of ΔH°_{rxn} values from bond energies is more accurate for gas-phase reactions than for reactions involving liquid or solid phases, because intermolecular forces are relatively minor for gas-phase species, but intermolecular forces are large for liquids and solids. Large errors in ΔH°_{rxn} values estimated from bond enthalpies can arise for reactions involving solids and/or liquids.

7-5 *Exothermic Chemical Reactions Can Be Used as Energy Sources*

An essential criterion for the quality of a chemical that is used as a fuel in a combustion reaction is the quantity of heat evolved per gram of fuel when the fuel is burned in oxygen. The more energy evolved as heat per unit mass of fuel, the greater the utility of the fuel as an energy source. Other important criteria for fuels are cost, ease of transport, and utilization hazards.

The heats of combustion of a variety of fuels are given in Table 7-3. Combustible fuels are used extensively as thermal energy sources. Note that in every case the fuel combustion reactions are highly exothermic (Table 7-3). A large negative value of ΔH_{rxn} for a combustion reaction is an essential prerequisite for an effective fuel.

Conventional liquid fuels, such as gasoline, jet fuel, diesel fuel, aviation gasoline, and heating oil, are mixtures of hydrocarbons, that is, compounds composed solely of carbon and hydrogen. Gasolines consist of mixtures of

TABLE 7-3
Heats of Combustion of Fuels

Reaction	$\Delta H_{rxn}/kJ \cdot mol^{-1}$	$\Delta H_{rxn}/kJ \cdot g^{-1}$
$H_2(g) + \frac{1}{2}O_2(g) \rightarrow H_2O(g)$	−242	−121
$CH_4(g) + 2\,O_2(g) \rightarrow CO_2(g) + 2\,H_2O(g)$	−802	−50
$C(s) + O_2(g) \rightarrow CO_2(g)$	−394	−33
$C_2H_2(g) + \frac{5}{2}O_2(g) \rightarrow 2\,CO_2(g) + H_2O(g)$	−1256	−48
$CH_3OH(g) + \frac{3}{2}O_2(g) \rightarrow CO_2(g) + 2\,H_2O(g)$	−676	−21
$C_2H_5OH(\ell) + 3\,O_2(g) \rightarrow 2\,CO_2(g) + 3\,H_2O(g)$	−1235	−27
$C_6H_6(\ell) + \frac{15}{2}O_2(g) \rightarrow 6\,CO_2(g) + 3\,H_2O(g)$	−3135	−40
$C_8H_{18}(\ell, i\text{-octane}) + \frac{25}{2}O_2(g) \rightarrow 8\,CO_2(g) + 9\,H_2O(g)$	−5456	−48
Oil (refined heating oil)	—	−44
Gasoline	—	−48
Kerosene (diesel fuel)	—	−48
Coal	—	−28
Wood (dry, seasoned)	—	−25

over 100 different hydrocarbon compounds in variable proportions. The hydrocarbons range from C_4 (four-carbon) hydrocarbons to C_{14} hydrocarbons. Various blends are produced depending on the environmental conditions of usage (hot versus cold weather and high versus low elevations) and the quality of the gasoline.

Methane, CH_4, propane, C_3H_8, and butane, C_4H_{10}, are also hydrocarbon fuels. Propane and butane are stored as liquids in tanks and the vapor over the liquids has a sufficiently high pressure to flow out of the tank as a gas to the combustion zone. Disposable cigarette lighters are charged with liquid butane that expands out the lighter valve and is ignited by a spark when the lighter wheel is rotated over the flint:

$$C_4H_{10}(g) + \tfrac{13}{2}\,O_2(g) \longrightarrow 4\,CO_2(g) + 5\,H_2O(g) \qquad \Delta H^\circ_{rxn} = -2658\text{ kJ}$$

Gasohol is a mixture of gasoline and ethyl alcohol. The most common blend is 90% conventional gasoline plus 10% ethyl alcohol, C_2H_5OH. Ethyl alcohol is produced by fermentation of sugars from plants using yeasts as the fermentation agent; for example,

$$\underset{\text{Glucose, a sugar}}{C_6H_{12}O_6(aq)} \xrightarrow{\text{Yeast}} \underset{\text{Ethyl alcohol}}{2\,C_2H_5OH(aq)} + 2\,CO_2(g)$$

There are several critical points to be made regarding the use of ethanol as a gasoline additive. The most important point is that the heat of combustion per gram of ethyl alcohol is significantly less than that for a hydrocarbon. The reason for this is that ethyl alcohol is already partially oxidized relative to a hydrocarbon:

$$\underset{\text{Ethane}}{C_2H_6(g)} + \tfrac{1}{2}\,O_2(g) \longrightarrow \underset{\text{Ethyl alcohol}}{C_2H_5OH(\ell)}$$

EXAMPLE 7-7

The ΔH°_{rxn} values at 25°C for the combustion of ethane gas, $C_2H_6(g)$, and liquid ethyl alcohol, $C_2H_5OH(\ell)$, are

$$C_2H_6(g) + \tfrac{7}{2}\,O_2(g) \longrightarrow 2\,CO_2(g) + 3\,H_2O(g) \qquad \Delta H^\circ_{rxn} = -1428\text{ kJ}$$

$$C_2H_5OH(\ell) + 3\,O_2(g) \longrightarrow 2\,CO_2(g) + 3\,H_2O(g) \qquad \Delta H^\circ_{rxn} = -1235\text{ kJ}$$

(a) Compute the heats of combustion per gram for C_2H_6 and C_2H_5OH.
(b) Compute ΔH°_{rxn} for the reaction

$$C_2H_6(g) + \tfrac{1}{2}\,O_2(g) \longrightarrow C_2H_5OH(\ell)$$

Solution:

(a) The ΔH°_{rxn} values given are for 1 mol of fuel in each case. The value of ΔH°_{rxn} per gram of fuel can be obtained by dividing by the formula mass in grams. For ethane we have

$$\Delta H^\circ_{rxn}\begin{pmatrix}\text{per gram}\\ \text{of ethane}\end{pmatrix} = \frac{-1428\text{ kJ}\cdot\text{mol}^{-4}}{30.07\text{ g}\cdot\text{mol}^{-1}} = -47.5\text{ kJ}\cdot\text{g}^{-1}$$

and for ethyl alcohol we have

$$\Delta H^\circ_{rxn}\begin{pmatrix}\text{per gram}\\ \text{of ethanol}\end{pmatrix} = \frac{-1235\ \text{kJ}\cdot\text{mol}^{-1}}{46.07\ \text{g}\cdot\text{mol}^{-1}} = -26.8\ \text{kJ}\cdot\text{g}^{-1}$$

The ratio of ΔH°_{rxn} (per gram) values is 47.5/26.8 = 1.77, and thus ethane has a heat of combustion per gram that is 77% greater than ethyl alcohol.

(b) The value of ΔH°_{rxn} for the oxidation of C_2H_6 to C_2H_5OH can be obtained from the ΔH°_{rxn} values for the two combustion reactions using Hess's law. Subtraction of the ethyl alcohol combustion reaction from the ethane combustion reaction yields the reaction

$$C_2H_6(g) + \tfrac{1}{2}O_2(g) \longrightarrow C_2H_5OH(\ell)$$

and thus for this reaction

$$\Delta H^\circ_{rxn} = -1428\ \text{kJ} - (-1235\ \text{kJ}) = -193\ \text{kJ}$$

An automobile running on pure C_2H_5OH fuel would use about 75% more ethyl alcohol per mile than gasoline. Another factor is that the production of ethyl alcohol requires, even under favorable conditions, a total energy input roughly equal to the energy value of the ethyl alcohol produced. Finally, the conversion of prime agricultural land to the production of, say, corn for use in fermentation to alcohol fuel might lead to a major increase in food prices. Nonetheless, in certain areas the supply of fermentable agricultural products may be great enough that, in combination with solar energy heating for distillation, alcohol can be a competitively priced fuel additive.

The food we eat constitutes the fuel necessary to maintain our body temperature and other physiological functions and, in addition, to provide the energy necessary to move about. Food is fuel. The energy value of the daily food intake necessary to maintain body weight for a normally active, healthy adult is about 75 *kJ per pound of body weight.* The total daily energy intake required to maintain various body weights is as follows:

Body weight	*Required daily energy input*	
110 lb (50 kg)	8250 kJ	1970 kcal
175 lb (80 kg)	13,125 kJ	3140 kcal
250 lb (114 kg)	18,750 kJ	4480 kcal

The above values are rough average values. Actual values can vary significantly from person to person depending on genetically determined physiological factors. Some individuals have much higher than average metabolic rates and readily burn off excess intake of foods. About 45 *kJ per pound per day* are required to keep the body functioning at the minimum possible activity level.

If we consume more food than we require for our normal activity level, then the excess that is not eliminated is stored in the body as fat. One pound of body fat contains about 18,000 kJ of stored energy. The chemical energy values per gram of fats, proteins, and carbohydrates are as follows:

Fats	39 kJ/g	9.3 kcal/g
Proteins	13 kJ/g	3.1 kcal/g
Carbohydrates	16 kJ/g	3.8 kcal/g

Note that proteins have only one-third the energy content of fats.

Exercise is great for improving muscle tone and thereby firming up sagging tissues, but exercise is not an effective way to lose weight. A 1-hour brisk walk over average terrain consumes only about 580 kJ of stored energy, which corresponds to about one-fifth of the energy content of a quarter-pound hamburger on a roll. The two most effective weight reduction exercises are (1) the isometric exercise in which the teeth are tightly clenched in the presence of fattening foods and (2) the two-handed pushaway from the dinner table before the second helping or the dessert arrives. The energy values of some common foods are given in Table 7-4. The message is that calories (or better, kilojoules) count.

7-6 *Rockets and Explosives Utilize Highly Exothermic Reactions with Gaseous Products*

A study of the ΔH°_{rxn} data in Table 7-3 leads to the conclusion that the most energy-rich fuel on a mass basis is hydrogen, which has an energy content per gram of well over twice that of the next best fuel. Because of its unusually high energy content per gram, hydrogen was used in the first stage of the Apollo series spaceships that traveled to the moon (the fuel must be lifted as part of the space vehicle). The main disadvantages of liquid H_2 as a fuel is that H_2 can be maintained as a liquid only at very low temperatures (about 20 K at 1 atm) and that hydrogen forms *explosive* mixtures with air over a wide composition range (2% to 98% H_2). The second and third stages of the Apollo spaceships were powered by the reaction between kerosene and liquid oxygen ("LOX"), both of which were stored on the spaceship.

The Apollo Lunar Lander spaceships were powered to and from the surface of the moon by the energy released on the reaction of the liquid fuel *N*,*N*-dimethylhydrazine, $H_2NN(CH_3)_2(\ell)$, with the liquid oxidizer nitrogen tetroxide, $N_2O_4(\ell)$:

$$H_2NN(CH_3)_2(\ell) + 2\,N_2O_4(\ell) \longrightarrow 3\,N_2(g) + 2\,CO_2(g) + 4\,H_2O(g)$$
$$\Delta H^\circ_{rxn} = -29\text{ kJ/g fuel}$$

The above reaction is especially suitable for a lunar escape vehicle because *the reaction starts spontaneously on mixing.* No battery or spark plugs and associated electrical circuitry are required.

TABLE 7-4

Average Energy Values of Some Common Foods[a] (1 kcal = 4.18 kJ; 1 oz = 28 g)

Substance	*kJ/100 g*	*kcal/4 oz*
Green vegetables	115	31
Beer	200	54
Fruits	250	67
Milk	300	80
Seafood (steamed)	400	107
Cottage cheese, low-fat yogurt	450	120
Chicken (broiled, meat only)	600	160
Steak (broiled, no fat)	900	240
Yogurt	1000	270
Liquor (80 proof)	1000	270
Ice cream (bulk)	1100	295
Bread, cheese	1200	320
Hamburger, hot dogs	1300	350
Popcorn, sugar	1600	430
Potato chips, nuts (roasted)	2400	640
Butter, cream, margarine, mayonnaise	3000	800
Fat	3900	1045

[a] The values given in the table assume complete combustion to $CO_2(g)$ and $H_2O(\ell)$; this is a good approximation for, say, vodka or mayonnaise but not for whole peanuts.

The nutritionist's calorie is actually a kilocalorie. The reason for this is presumably psychological—there are 2,560,000 calories in a pound of cashews.

The principal *solid* fuel used in the *Minuteman* *i*nter*c*ontinental *b*allistic *m*issiles, ICBM's; the *Polaris* missile that is fired from submarines (Figure 7-6); and the airplane-to-airplane *Sidewinder* missiles typically consists of a mixture of 70% ammonium perchlorate, 18% aluminum metal powder, and 12% organic polymer. Ammonium perchlorate is a self-contained solid fuel—the fuel NH_4^+ and the oxidizer ClO_4^- are together in the crystal. The overall reaction stoichiometry for the NH_4ClO_4 decomposition reaction is

$$2\,NH_4ClO_4(s) \longrightarrow N_2(g) + 2\,HCl(g) + 3\,H_2O(g) + \tfrac{5}{2}\,O_2(g)$$

Note that the NH_4ClO_4 is oxygen rich; that is, O_2 is a reaction product. To utilize this available oxygen and thereby to provide more rocket thrust, Al powder and an organic polymer, which also acts as a binder, are added. The Al is oxidized to aluminum oxide:

$$2\,Al(s) + \tfrac{3}{2}\,O_2(g) \longrightarrow Al_2O_3(s)$$

Figure 7-6. Underwater firing of a Polaris A-3 missile from a nuclear submarine. (U.S. Navy photograph)

and the organic polymer is oxidized to CO_2 and H_2O. The aluminum powder also promotes a more rapid and even decomposition of the NH_4ClO_4.

The powerful and dangerous explosive nitroglycerin is made by the reaction of a mixture of nitric acid, HNO_3, and sulfuric acid, H_2SO_4, with glycerin:

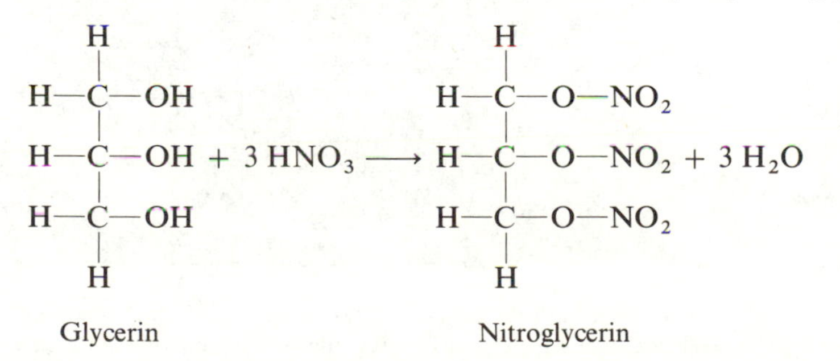

Nitroglycerin can detonate on even minor shock. In 1867, the Swedish chemist Alfred Nobel (1833–1896) tamed nitroglycerine by adding the solid absorbent material *diatomaceous earth* (a naturally occurring material consisting of the remains of one-celled plants called *diatoms*) to produce *dynamite*. The Nobel Prizes are funded by interest that accrues on a prize fund established by Nobel using part of the money that he made on the licensing of his numerous patents on explosives.

7-7 *Sunlight Is a Major Energy Source*

A major and largely untapped energy source, at least by humans, is sunlight. The sunlight that reaches the surface of the United States on a clear day has a power level of about 1 kilowatt per square meter ($kW \cdot m^{-2}$). The energy requirements of a typical U.S. home are about 30 kW hr/day. If 30% of the incident solar energy could be collected and utilized, then for 8 hr of sunlight in a day, all the household energy requirements could be satisfied by about 12 m^2 (say, 3 m × 4 m) of collector surface.

There are a wide variety of methods for the collection and utilization of solar energy. One particularly simple method is large south-facing windows, through which sunlight is allowed to pass in winter but not in summer. Another

Figure 7-7. A solar water heater basically consists of two components—the solar panels and the water heater. The two black solar panels (right) are mounted on the roof of a building at an angle to receive the most sunlight possible. The panels, known as collectors, have rows of tubing inside and are covered with glass. Water circulates through the tubes, is heated by the sun and returns to the water heater (left) where it is stored until needed. (Courtesy of Pacific Gas and Electric Company)

simple method involves using sunlight to heat water. The sunlight is allowed to pass through a transparent cover and is absorbed by a blackened underlying surface. The water that is placed between the two surface absorbs heat from the blackened surface and is thereby heated (Figure 7-7).

One of the major drawbacks of solar energy is the fact that the sunshine is inherently intermittent in nature. An energy storage device is necessary so that the energy can be used when the sun is not shining or the day is overcast. The energy collected can be stored in basically two ways: (1) by heating up a large mass of substance, such as water or rocks, and (2) by storing the energy in chemicals that, on reaction, will release the stored energy. In effect, a suitable chemical reaction is driven uphill energetically by the solar energy input.

Salt hydrates are the simplest types of chemicals that can be used to store energy in a chemical process. An example is Glauber's salt, $Na_2SO_4 \cdot 10\ H_2O(s)$, which at 32.3°C (90°F) dissolves in its own waters of hydration to form a solution:

$$Na_2SO_4 \cdot 10\ H_2O(s) \rightleftharpoons (Na_2SO_4 + 10\ H_2O)_{soln}$$

When the temperature of the salt solution drops below 32.3°C, the $Na_2SO_4 \cdot 10\ H_2O(s)$ crystallizes out of the solution with the evolution of about 0.21 kJ/g of salt formed, that is, 354 kJ/L of salt solution. As the salt crystallizes, the heat can be drawn off and used for space or water heating.

Another approach to solar energy conversion is to use mirrors to concentrate the sunlight on a reactor vessel and use the high temperatures produced to drive a chemical reaction uphill on the enthalpy scale. An example is the reaction

$$CO_2(g) + CH_4(g) \xrightarrow{750^\circ C} 2\ CO(g) + 2\ H_2(g) \qquad \Delta H_{rxn} = 247\ \text{kJ}$$

The 247 kJ of stored energy per mole of CO_2 or CH_4 reacted can be released by running the reaction in reverse. The energy evolved can be used to provide steam for power generation. Focused sunlight can also be used to convert liquid water to steam that is then used to drive a turbine that produces electricity.

7-8 *Entropy Changes for Chemical Reactions Can Be Computed from Third-Law Entropies*

The Third Law of Thermodynamics forms the basis of the *practical absolute entropy scale.* The entropy of a perfect crystalline substance on the practical absolute entropy scale is set equal to zero at the absolute zero of temperature. Nuclear entropies and entropies of isotope mixing are not included in the *practical absolute entropy scale*, and this is the reason for the designation *practical.* The neglect of nuclear entropies and isotope-mixing entropies are of no consequence in the thermodynamic analysis of balanced chemical reactions, because the nuclei are conserved and therefore there is no net nuclear entropy change for the reaction (Chapter 6).

The choice of standard states for entropies is the same as that discussed for enthalpies, except that, in contrast to the enthalpy, the entropy is not set equal to zero for the most stable form of the elements, owing to the existence of the absolute entropy scale. The superscript zero in the symbol S° denotes the practical absolute entropy in the 1-atm standard state at the temperature of interest. Corrections from a nonstandard state to the standard state are made using the equation of state for the substance.

EXAMPLE 7-8

Suppose that the equation of state for a gas is

$$PV = nRT + \frac{naP}{T}$$

where $a = 4.00 \text{ L}\cdot\text{K}\cdot\text{mol}^{-1}$. Obtain an expression for ΔS for the process

$$\text{real gas}(T, P) \longrightarrow \text{ideal gas}(T, 1 \text{ atm})$$

Solution: From Eq. 5-34 we have

$$\left(\frac{\partial S}{\partial P}\right)_T = -\left(\frac{\partial V}{\partial T}\right)_P$$

and thus, from the equation of state, we obtain

$$\left(\frac{\partial S}{\partial P}\right)_T = -\left(\frac{nR}{P} - \frac{na}{T^2}\right)$$

The value of ΔS for the change in state (Figure 7-1)

$$\text{real gas}(P, T) \longrightarrow \text{ideal gas}(P', T)$$

where P' is some low pressure where the gas behaves ideally, is

$$\Delta S_2 = -\int_P^{P'} \left(\frac{nR}{P} - \frac{na}{T^2}\right) dP$$

$$\Delta S_2 = -nR \ln \frac{P'}{P} + \frac{na}{T^2}(P' - P)$$

The change in the entropy of an ideal gas for the change in state

$$\text{ideal gas}(T, P') \longrightarrow \text{ideal gas}(T, 1 \text{ atm})$$

is ($PV = nRT$)

$$\left(\frac{\partial S}{\partial P}\right)_T = -\left(\frac{\partial V}{\partial T}\right)_P = -\frac{nR}{P}$$

and therefore

$$\Delta S_3 = -\int_{P'}^{1} \frac{nR}{P}\, dP = -nR \ln \frac{1}{P'}$$

Combination of ΔS_3 and ΔS_2 yields ΔS_1 for the process (see Figure 7-1):

$$\text{real gas}(T, P) \longrightarrow \text{ideal gas}(T, P = 1 \text{ atm})$$

$$\Delta S_1 = \Delta S_3 + \Delta S_2 = nR \ln P + \frac{na}{T^2}(P' - P)$$

All real gases behave ideally in the limit of zero pressure and thus for $P' = 0$ or, in other words, for the process

$$\text{real gas}(T, P) \longrightarrow \text{ideal gas } (T, 1 \text{ atm})$$

we obtain

$$\Delta S = nR \ln P - \frac{naP}{T^2}$$

Thus, for $P = 2.00$ atm, $T = 298$ K, $n = 1.00$ mol, we obtain

$$\Delta S = (1.00 \text{ mol})\left\{8.314 \text{ J}\cdot\text{K}^{-1}\cdot\text{mol}^{-1})\ln\left(\frac{2.00 \text{ atm}}{1.00 \text{ atm}}\right) - \frac{(4.00 \text{ L}\cdot\text{K}\cdot\text{mol}^{-1})(2.00 \text{ atm})}{(298 \text{ K})^2}\left(\frac{8.314 \text{ J}}{0.0821 \text{ L}\cdot\text{atm}}\right)\right\}$$

and

$$\Delta S = 5.75 \text{ J}\cdot\text{K}^{-1}$$

Thus the entropy of 1.00 mol of the gas in the ideal gas 1-atm standard state at 298 K is 5.75 $\text{J}\cdot\text{K}^{-1}$ higher than that of the real gas in the state (298 K, 2.00 atm).

The entropy change for an isothermal chemical reaction is given by

$$\Delta S_{\text{rxn},T} = \sum_i v_i \bar{S}_{T,i} \tag{7-25}$$

where, as usual, the v_i values are the reaction-balancing coefficients (positive for products and negative for reactants). For the conversion of reactants in their standard states to products in their standard states, Eq. 7-25 becomes

$$\Delta S^\circ_{\text{rxn},T} = \sum_i v_i \bar{S}^\circ_{T,i} \tag{7-26}$$

Values of $\bar{S}^\circ_{298,i}$ for various substances are tabulated in Appendix I.

An example of the use of tabulated $\bar{S}^\circ$ values for the calculation of $\Delta S^\circ_{\text{rxn}}$ values follows. For the water-vaporization reaction

$$H_2O(\ell) \longrightarrow H_2O(g)$$

we have

$$\Delta S^\circ_{\text{rxn}} = \bar{S}^\circ[H_2O(g)] - \bar{S}^\circ[H_2O(\ell)]$$

and at 25°C (Appendix I)

$$\Delta S^\circ_{\text{rxn}} = 188.72 \text{ J}\cdot\text{K}^{-1}\cdot\text{mol}^{-1} - 69.91 \text{ J}\cdot\text{K}^{-1}\cdot\text{mol}^{-1} = 118.81 \text{ J}\cdot\text{K}^{-1}\cdot\text{mol}^{-1}$$

The value of $\Delta C^\circ_{P,\text{rxn}}$ at 25°C for the water-vaporization reaction is (see Appendix I for $\bar{C}^\circ_P$ data)

$$\Delta C^\circ_{P,\text{rxn}} = \bar{C}^\circ_P[H_2O(g)] - \bar{C}^\circ_P[H_2O(\ell)] = 33.58 - 75.29$$
$$= -41.71 \text{ J}\cdot\text{K}^{-1}\cdot\text{mol}^{-1}$$

Assuming that $\Delta C^\circ_{P,\text{rxn}}$ for the vaporization reaction is independent of temperature over the range 25°C to 30°C, we compute the value of $\Delta S^\circ_{\text{rxn}}$ for the water-vaporization reaction at 30°C as follows:

$$\frac{d(\Delta S^\circ_{\text{rxn}})}{dT} = \frac{\Delta C^\circ_{P,\text{rxn}}}{T} \tag{7-27}$$

and thus

$$\Delta S^\circ_{\text{rxn},T_2} - \Delta S^\circ_{\text{rxn},T_1} = \Delta C^\circ_{P,\text{rxn}} \ln\left(\frac{T_2}{T_1}\right)$$

Hence

$$\Delta S^\circ_{\text{rxn},303.15} = 118.81 \text{ J}\cdot\text{K}^{-1}\cdot\text{mol}^{-1} - (41.71 \text{ J}\cdot\text{K}^{-1}\cdot\text{mol}^{-1}) \ln\left(\frac{303.15}{298.15}\right)$$

$$\Delta S^\circ_{\text{rxn},303.15} = 118.12 \text{ J}\cdot\text{K}^{-1}\cdot\text{mol}^{-1}$$

The positive value of $\Delta S^\circ_{\text{rxn}}$ is a consequence of the increased disorder of water on vaporization.

EXAMPLE 7-9

Use data in Appendix I to compute $\Delta S^\circ_{\text{rxn}}$ for the reaction

$$\text{C}(s, \text{graphite}) + \text{H}_2\text{O}(g) \longrightarrow \text{CO}(g) + \text{H}_2(g)$$

The value of $\Delta S^\circ_{\text{rxn}}$ is given by

$$\Delta S^\circ_{\text{rxn}} = \bar{S}^\circ[\text{CO}(g)] + \bar{S}^\circ[\text{H}_2(g)] - \bar{S}^\circ[\text{C}(s)] - \bar{S}^\circ[\text{H}_2\text{O}(g)]$$

and at 25°C we have from Appendix I

$$\Delta S^\circ_{\text{rxn}} = 197.56 + 130.57 - 5.74 - 188.72 = 133.67 \text{ J}\cdot\text{K}^{-1}\cdot\text{mol}^{-1}$$

The positive value of $\Delta S^\circ_{\text{rxn}}$ is a consequence of the fact that the reaction involves an increase in the number of moles of gas, $\Delta n_{\text{gas}} = +1$.

The value of $\Delta S^\circ_{\text{rxn}}$ obtained from calorimetric data should be equal to the value of $\Delta S^\circ_{\text{rxn}}$ for the same reaction obtained from equilibrium measurements, provided the third law holds. That is, from calorimetric data we obtain $(\Delta S^\circ_{\text{rxn},T} - \Delta S^\circ_{\text{rxn},0})_{\text{cal}}$, whereas from equilibrium data we obtain

$$(\Delta S^\circ_{\text{rxn},T})_{\text{equil}} = -\left(\frac{\partial \Delta G^\circ_{\text{rxn}}}{\partial T}\right)_P \tag{7-28}$$

If the third law holds, then we have

$$(\Delta S^\circ_{\text{rxn},T})_{\text{equil}} = (\Delta S^\circ_{\text{rxn},T} - \Delta S^\circ_{\text{rxn},0})_{\text{cal}} = (\Delta S^\circ_{\text{rxn},T})_{\text{cal}} \tag{7-29}$$

because the third law requires that $\Delta S^\circ_{\text{rxn},0} = 0$.

Table 7-5 gives a comparison of $\Delta S^\circ_{\text{rxn}}$ values obtained from equilibrium data with $\Delta S^\circ_{\text{rxn}}$ values obtained from calorimetric data. In the first three cases

TABLE 7-5
Comparison of ΔS°_{rxn} Values from Equilibrium Measurements with ΔS°_{rxn} Values from Calorimetric Data

Reaction	T/K	$\Delta S^\circ_{rxn}/J\cdot K^{-1}\cdot mol^{-1}$ (*equilibrium*)	$\Delta S^\circ_{rxn}/J\cdot K^{-1}\cdot mol^{-1}$ (*third law*)	$(\Delta S^\circ_{rxn})_{equil} - (\Delta S^\circ_{rxn})_{third\ law}$
$Ag(s) + \frac{1}{2}Br_2(\ell) \rightleftharpoons AgBr(s)$	265.9	-12.59 ± 1.67	-12.84 ± 0.42	0.25 ± 2.09
$Zn(s) + \frac{1}{2}O_2(g) \rightleftharpoons ZnO(s)$	298.15	-100.71 ± 1.05	-101.42 ± 0.21	0.29 ± 1.26
$CaCO_3(s) \rightleftharpoons CaO(s) + CO_2(g)$	298.15	160.67 ± 0.84	159.12 ± 0.84	1.55 ± 1.68
$Mg(OH)_2(s) \rightleftharpoons MgO(s) + H_2O(g)$	298.15	153.39 ± 0.42	150.00 ± 0.33	3.39 ± 0.75

Figure 7-8. Fog rolling eastward from the Pacific Ocean through (right to left) the Golden Gate. The City of San Francisco is under the fog beyond the south end of the Golden Gate Bridge in this typical summer pattern. (Bill Teas/Aero Photographers, Sausalito, Calif.)

in Table 7-5, as well as in numerous other cases, the agreement between equilibrium and calorimetric values is within the experimental error. The discrepancy for the reaction

$$Mg(OH)_2(s) \rightleftharpoons MgO(s) + H_2O(g)$$

has been traced to the formation of microcrystals of MgO(*s*), which are sufficiently small that the associated surface entropy is not negligible relative to the total entropy of the phase. Because $\bar{S}^\circ$ for the MgO(*s*) microcrystals is greater than $\bar{S}^\circ$ for MgO(*s*) macrocrystals, the equilibrium ΔS°_{rxn} value is greater than the calorimetric ΔS°_{rxn} value. In effect the production of the MgO(*s*) microcrystals makes ΔS°_{rxn} larger than would be the case if MgO(*s*) macrocrystals were produced. Microcrystals (radii less than about 10^{-3} mm) of a substance have a higher entropy than macrocrystals of the same substance at the same temperature and pressure because the surface entropy of microcrystals makes a significant contribution to the total entropy of the substance. A similar situation occurs in fog formation (Figure 7-8), where $\bar{S}^\circ_{fog} > \bar{S}^\circ_{H_2O(\ell)}$.

7-9 *Gibbs Energy Changes for Chemical Reactions Can Be Computed from $\Delta\bar{G}^\circ_{f,i}$ Values*

The value of ΔG_{rxn} for a chemical reaction at a particular temperature gives the difference in Gibbs energy between the products and reactants at that temperature; that is,

$$\Delta G_{rxn,T} = \sum_i \nu_i \bar{G}_{i,T} \tag{7-30}$$

where, as usual, the ν_i values are the reaction-balancing coefficients, taken as positive for products and negative for reactants. If the reactants in their respective standard states at a particular temperature are converted into products in their respective standard states at the same temperature, then the Gibbs energy change for the reaction is referred to as the *standard Gibbs energy change* for the reaction ΔG°_{rxn}, where

$$\Delta G^\circ_{rxn,T} = \sum_i \nu_i \bar{G}^\circ_{i,T} \tag{7-31}$$

Introduction of $\Delta\bar{G}^\circ_{f,i}$ values, that is, the standard Gibbs energy of formation of *i* from the elements, which are defined in a manner strictly analogous to $\Delta\bar{H}^\circ_{f,i}$ values, enables us to rewrite Eq. 7-31 as

$$\Delta G^\circ_{rxn,T} = \sum_i \nu_i \Delta\bar{G}^\circ_{f,i,T} \tag{7-32}$$

We establish a $\Delta\bar{G}^\circ_{f,i}$ scale by setting $\Delta\bar{G}^\circ_{f,i} = 0$ at all temperatures for some arbitrarily chosen state of each element. The form of each element for which we set $\Delta\bar{G}^\circ_{f,i} = 0$ at all T is usually (but not invariably) the form that is most stable, that is, the form that has the lowest molar Gibbs energy at 1 atm and 25°C. The form of the element for which we take $\Delta\bar{G}^\circ_f = 0$ *must* be the same form for which we take $\Delta\bar{H}^\circ_f = 0$, because G and H are related, $G = H - TS$. As long as one is consistent, it makes no difference from a computational viewpoint which form is chosen; however, the choice of the most stable form (lowest value of $\bar{G}$ at a given T) of the element is usually the most advantageous for experimental work.

As an example of the use of $\Delta\bar{G}^\circ_{f,i}$ values, consider the reaction

$$H_2(g) + \tfrac{1}{2}\,O_2(g) \longrightarrow H_2O(\ell) \qquad \Delta G^\circ_{rxn} = -237.183 \text{ kJ}$$

Using Eq. 7-32 we have

$$\Delta G^\circ_{rxn} = \Delta\bar{G}^\circ_f[H_2O(\ell)] - \tfrac{1}{2}\Delta\bar{G}^\circ_f[O_2(g)] - \Delta\bar{G}^\circ_f[H_2(g)] = -237.183 \text{ kJ}$$

If we set $\Delta\bar{G}^\circ_f[H_2(g)] = \Delta\bar{G}^\circ_f[O_2(g)] = 0$, then

$$\Delta G^\circ_{rxn} = \Delta\bar{G}^\circ_f[H_2O(\ell)] = -237.183 \text{ kJ}\cdot\text{mol}^{-1}$$

The procedure for building up a table of $\Delta\bar{G}^\circ_{f,i}$ values is strictly analogous to that discussed in Section 7-2 for $\Delta\bar{H}^\circ_{f,i}$ values.

EXAMPLE 7-10

Use data in Appendix I to calculate ΔG°_{rxn} at 298.15 K for the reaction

$$C_2H_5OH(\ell) + 3\,O_2(g) \longrightarrow 2\,CO_2(g) + 3\,H_2O(\ell)$$

Solution: From Eq. 7-32 we have

$$\Delta G^\circ_{rxn} = 2\,\Delta\bar{G}^\circ_f[CO_2(g)] + 3\,\Delta\bar{G}^\circ_f[H_2O(\ell)] - 3\,\Delta\bar{G}^\circ_f[O_2(g)] - \Delta\bar{G}^\circ_f[C_2H_5OH(\ell)]$$

From Appendix I we obtain

$$\Delta G^\circ_{rxn} = (2 \text{ mol})(-394.36 \text{ kJ}\cdot\text{mol}^{-1}) + (3 \text{ mol})(-237.18 \text{ kJ}\cdot\text{mol}^{-1}) - 3(0) - (1 \text{ mol})(-174.89 \text{ kJ}\cdot\text{mol}^{-1})$$

and thus

$$\Delta G^\circ_{rxn} = -1325.37 \text{ kJ} \qquad (\text{at } 298.15 \text{ K})$$

Values of ΔG°_{rxn} can also be computed from values of ΔH°_{rxn} and ΔS°_{rxn}. By definition,

$$G = H - TS$$

and thus for a chemical reaction occurring under isothermal conditions

$$\Delta G_{rxn} = \Delta H_{rxn} - T\Delta S_{rxn}$$

If the reactants and product are in their respective 1-atm standard states at temperature T, then

$$\Delta G^\circ_{rxn} = \Delta H^\circ_{rxn} - T\Delta S^\circ_{rxn} \qquad (7\text{-}33)$$

Further, if we determine ΔG°_{rxn} and $\Delta S^\circ_{rxn} = -d\Delta G^\circ_{rxn}/dT$ from equilibrium measurements, the value of ΔH°_{rxn} can be calculated using Eq. 7-33.

EXAMPLE 7-11

For the reaction

$$CH_4(g) + 2\,O_2(g) \longrightarrow CO_2(g) + 2\,H_2O(g)$$

at 25°C, $\Delta H^\circ_{rxn} = -802.3$ kJ and $\Delta S^\circ_{rxn} = -5.13\ \text{J}\cdot\text{K}^{-1}$. Compute ΔG°_{rxn} at 25°C.

Solution: From Eq. 7-33 we have

$$\Delta G^\circ_{rxn} = \Delta H^\circ_{rxn} - T\Delta S^\circ_{rxn}$$

Thus

$$\Delta G^\circ_{rxn} = -802.3 \times 10^3\ \text{J} - (298.15\ \text{K})(-5.13\ \text{J}\cdot\text{K}^{-1})$$

and

$$\Delta G^\circ_{rxn} = -800.8\ \text{kJ}$$

7-10 *The Free-Energy Functions $-(\bar{G}^\circ_T - \bar{H}^\circ_{298})/T$ and Their Use in Thermodynamic Calculations*

The quantities $\bar{H}^\circ_T - \bar{H}^\circ_{298}$ and $\bar{S}^\circ_T$ can be obtained directly from calorimetric data, and these quantities can, in turn, be combined with $\Delta\bar{H}^\circ_{f,298}$ data to calculate $\Delta G^\circ_{rxn,T}$ values for a reaction. Note that because $G = H - TS$, we have

$$\bar{G}^\circ_T = \bar{H}^\circ_T - T\bar{S}^\circ_T \tag{7-34}$$

Subtracting $\bar{H}^\circ_{298}$ from both sides of Eq. 7-34 and dividing by T, we obtain

$$\frac{\bar{G}^\circ_T - \bar{H}^\circ_{298}}{T} = \frac{\bar{H}^\circ_T - \bar{H}^\circ_{298}}{T} - \bar{S}^\circ_T$$

Occasionally $\bar{H}^\circ_T - \bar{H}^\circ_0$ is tabulated, and in such a case we have

$$\frac{\bar{G}^\circ_T - \bar{H}^\circ_0}{T} = \frac{\bar{H}^\circ_T - \bar{H}^\circ_0}{T} - \bar{S}^\circ_T$$

where the use of $\bar{H}^\circ_0$ requires the tabulation of $\Delta\bar{H}^\circ_{f,0}$ rather than $\Delta\bar{H}^\circ_{f,298}$ values [or alternatively the simultaneous tabulation of $(\bar{H}^\circ_{298} - \bar{H}^\circ_0)$] for substances. Note that

$$\frac{\bar{G}^\circ_T - \bar{H}^\circ_0}{T} - \frac{(\bar{H}^\circ_{298} - \bar{H}^\circ_0)}{T} = \frac{\bar{G}^\circ_T - \bar{H}^\circ_{298}}{T}$$

The functions

$$\frac{-(\bar{G}^\circ_T - \bar{H}^\circ_{298})}{T} \quad \text{and} \quad \frac{-(\bar{G}^\circ_T - \bar{H}^\circ_0)}{T}$$

are known as *free-energy functions.* The free-energy functions vary only slowly with temperature and are convenient for tabulational purposes. It is, of course, absolutely essential to convert all free-energy functions to ones based all on either $\bar{H}^\circ_{298}$ or $\bar{H}^\circ_0$ before combining them to obtain $\Delta\bar{G}^\circ_{rxn,T}$. In Appendix I we have assembled a highly abridged list of free-energy function data. The subscript 298 is shorthand for 298.15 K.

EXAMPLE 7-12

Use the free-energy function data in Appendix I to compute $\Delta G^\circ_{rxn,1000K}$ for the reaction

$$2\,Fe_3O_4(s) + CO_2(g) \rightleftharpoons 3\,Fe_2O_3(s) + CO(g)$$

Solution: From the data in Appendix I we obtain

Species	$[(\bar{G}^\circ_{1000} - \bar{H}^\circ_{298})/1000]/J\cdot K^{-1}\cdot mol^{-1}$	$\Delta\bar{H}^\circ_{f,298}/kJ\cdot mol^{-1}$
$Fe_3O_4(s)$	−240.484	−1118.4
$Fe_2O_3(s)$	−152.360	−824.2
$CO(g)$	−212.735	−110.5
$CO_2(g)$	−235.806	−393.5

Thus

$$\frac{\Delta G^\circ_{rxn,1000} - \Delta H^\circ_{rxn,298}}{1000} = +(-212.735) + 3(-152.360) - (-235.806) - 2(-240.484)$$

and

$$\Delta G^\circ_{rxn,1000} - \Delta H^\circ_{rxn,298} = 46{,}959\ \text{J} = 46.96\ \text{kJ}$$

The value of $\Delta H^\circ_{rxn,298}$ is given by

$$\Delta H^\circ_{rxn,298} = 3(-824.2) + (-110.5) - 2(-1118.4) - (-393.5) = 47.2\ \text{kJ}$$

and therefore

$$\Delta G^\circ_{rxn,1000} = 47.2\ \text{kJ} + 47.0\ \text{kJ} = 94.2\ \text{kJ}$$

7-11 *Thermodynamic Properties Depend on Molecular Structure*

The influence of molecular structure on thermodynamic properties is illustrated in Figure 7-9 for gaseous isomers with the formula C_3H_6O. Note that the total entropy is not strongly dependent on molecular structure, whereas the values of $\Delta\bar{H}^\circ_f$ and thus $\Delta\bar{G}^\circ_f$ are sensitive to the details of the arrangement

$\Delta\bar{H}^\circ_f/kJ \cdot mol^{-1}$ *minimum energy*	$\bar{S}^\circ/J \cdot K^{-1} \cdot mol^{-1}$ *maximum entropy*	$\Delta\bar{G}^\circ_f/kJ \cdot mol^{-1}$ *maximum stability*
$CH_3\underset{\underset{O}{\Vert}}{C}CH_3$ acetone −217.57	$H_2C{=}CHCH_2OH$ allyl alcohol 307.57	$CH_3\underset{\underset{O}{\Vert}}{C}CH_3$ acetone −153.06
$CH_3CH_2\underset{\underset{O}{\Vert}}{C}H$ propionaldehyde −192.05	$CH_3CH_2\underset{\underset{O}{\Vert}}{C}H$ propionaldehyde 304.72	$CH_3CH_2\underset{\underset{O}{\Vert}}{C}H$ propionaldehyde −130.46
$H_2C{=}CHCH_2OH$ allyl alcohol −132.01	$CH_3\underset{\underset{O}{\Vert}}{C}CH_3$ acetone 294.93	$H_2C{=}CHCH_2OH$ allyl alcohol −71.25
O / \\ H_2C—$CHCH_3$ propylene oxide −92.76	O / \\ H_2C—$CHCH_3$ propylene oxide 285.98	O / \\ H_2C—$CHCH_3$ propylene oxide −25.79
O—CH_2 | | H_2C—CH_2 trimethylene oxide −80.54	O—CH_2 | | H_2C—CH_2 trimethylene oxide 273.89	O—CH_2 | | H_2C—CH_2 trimethylene oxide −9.74

Figure 7-9. Thermodynamic properties at 25°C of isomeric gaseous molecules with the formula $C_3H_6O(g)$. Adapted from D. R. Stull, *American Scientist* 59: 734–743 (1971).

of atoms in the molecules. For the C_3H_6O isomers, energetic factors are predominant in the determination of relative stability. For the reaction

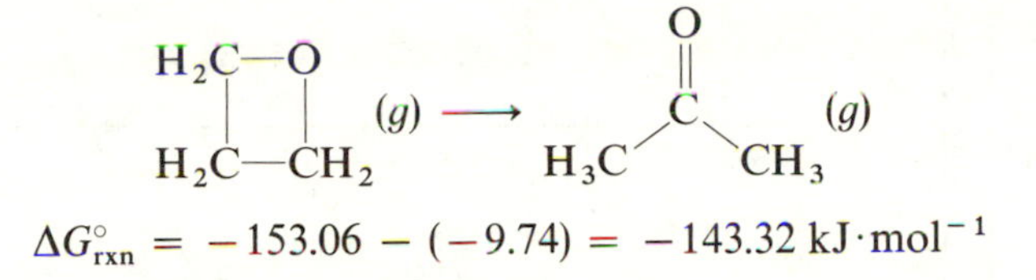

$$\Delta G^\circ_{\text{rxn}} = -153.06 - (-9.74) = -143.32 \text{ kJ}\cdot\text{mol}^{-1}$$

In other words, acetone is -143.32 kJ "downhill" on the Gibbs energy scale from trimethylene oxide. Acetone is 143.32 kJ·mol^{-1} more stable with respect to the elements than is trimethylene oxide. Note also the increase in the molar entropy with increasing freedom of internal motion. The "floppier" the molecule, the higher the entropy.

7-12 *Compilations of Thermodynamic Data*

The calculation of a value of $\Delta H^\circ_{\text{rxn}}$, $\Delta S^\circ_{\text{rxn}}$, or $\Delta G^\circ_{\text{rxn}}$ for a chemical reaction of interest necessitates the use of reference tables of thermodynamic data. There are several particularly useful compilations of thermodynamic data. Some of these compilations are described below.

1. D. D. Wagman, W. H. Evans, V. B. Parker, I. Halow, S. M. Bailey, and R. H. Schumm. *Selected Values of Chemical Thermodynamic Properties.* U.S. Department of Commerce, National Bureau of Standards; NBS Technical Notes 270-3, 270-4, 270-5, 270-6, 270-7, and 270-8.

A sample page from NBS Technical Note 270-3 is shown in Figure 7-10. The values given are all molar values. Thus for $H_2O(g)$ at 298.15 K we find from the figure (where 298 denotes 298.15 K)

$$\Delta\bar{H}^\circ_{f,0} = -57.102 \text{ kcal}\cdot\text{mol}^{-1} \qquad \Delta\bar{H}^\circ_{f,298} = -57.796 \text{ kcal}\cdot\text{mol}^{-1}$$

$$\Delta\bar{G}^\circ_{f,298} = -54.634 \text{ kcal}\cdot\text{mol}^{-1} \qquad \bar{H}^\circ_{298} - \bar{H}^\circ_0 = 2.3667 \text{ kcal}\cdot\text{mol}^{-1}$$

$$\bar{S}^\circ_{298} = 45.104 \text{ cal}\cdot\text{K}^{-1}\cdot\text{mol}^{-1} \qquad \bar{C}^\circ_{P,298} = 8.025 \text{ cal}\cdot\text{K}^{-1}\cdot\text{mol}^{-1}$$

The calorie units of the above values can be converted to joules by multiplication by 4.184 J·cal^{-1}; thus

$$\bar{S}^\circ_{298} = (45.104 \text{ cal}\cdot\text{K}^{-1}\cdot\text{mol}^{-1})(4.184 \text{ J}\cdot\text{cal}^{-1}) = 188.715 \text{ J}\cdot\text{K}^{-1}\cdot\text{mol}^{-1}$$

2. D. R. Stull and H. Prophet. *JANAF Thermochemical Tables, 2nd Edition.* NSRDS-NBS Publication 37. 1974 plus Supplements.

A sample page from the JANAF tables is shown in Figure 7-11. The JANAF Tables give $\bar{C}^\circ_{P,T}$, $\bar{S}^\circ_T$, $-(\bar{G}^\circ_T - \bar{H}^\circ_{298})/T$, $\bar{H}^\circ_T - \bar{H}^\circ_{298}$, $\Delta\bar{H}^\circ_{f,T}$, $\Delta\bar{G}^\circ_{f,T}$, and $\log_{10} K_P = -\Delta\bar{G}^\circ_{f,T}/4.5756T$. The JANAF tables contain data over a range of temperature as opposed to NBS Technical Notes 270-X that are confined primarily to 25°C with the exception of $\bar{H}^\circ_{298} - \bar{H}^\circ_0$ and $\Delta\bar{H}^\circ_{f,0}$ data.

SELECTED VALUES OF CHEMICAL THERMODYNAMIC PROPERTIES - SERIES I

National Bureau of Standards — Washington, D. C.

Table 2(2)

Enthalpy and Gibbs Energy of Formation; Entropy and Heat Capacity

HYDROGEN

Substance			ΔHf_0°	ΔHf°	ΔGf°	$H_{298}^\circ - H_0^\circ$	S°	C_p°
Formula and Description	State	Formula Weight	0 °K	298.15 °K (25 °C)				
			kcal/mol				cal/deg mol	
HO_2^+	g		272.	271.				
HO_2^- std. state, m = 1	aq			-38.32	-16.1		5.7	
H_2O	liq	18.0153		-68.315	-56.687		16.71	17.995
^2H_2O	liq	20.0276		-70.411	-58.195		18.15	20.16
$^1H^2HO$	liq	19.0213		-69.285	-57.817		18.95	
H_2O	g	18.0153	-57.102	-57.796	-54.634	2.3667	45.104	8.025
^1H_2O	g	18.0150	-57.102	-57.796	-54.634	2.3667	45.103	8.025
^2H_2O	g	20.0276	-58.855	-59.560	-56.059	2.3801	47.378	8.19
$^1H^2HO$	g	19.0213	-57.927	-58.628	-55.719	2.3721	47.658	8.08
H_2O^+	g	18.0153	233.5	234.3				
H_2O_2	liq	34.0147		-44.88	-28.78		26.2	21.3
	g		-31.08	-32.58	-25.24	2.594	55.6	10.3
undissoc.; std. state, m = 1	aq			-45.69	-32.05		34.4	
in 0.1 H_2O	aq			-44.965				
0.5 H_2O	aq			-45.198				
1 H_2O	aq			-45.365				
2 H_2O	aq			-45.520				
3 H_2O	aq			-45.585				
4 H_2O	aq			-45.620				

H_2O_2

Figure 7-10. Sample page from NBS Technical note 270-3. Edited by D. D. Wagman, et al.

Calcium Oxide (CaO)

(Crystal) GFW = 56.0794

T, °K	Cp° (gibbs/mol)	S° (gibbs/mol)	-(G°-H°298)/T (gibbs/mol)	H°-H°298 (kcal/mol)	ΔHf° (kcal/mol)	ΔGf° (kcal/mol)	Log Kp
0	.000	.000	INFINITE	-1.413	-150.995	-150.995	INFINITE
100	3.517	1.487	16.531	-1.504	-151.467	-149.165	325.998
200	8.049	5.507	9.997	-.898	-151.750	-146.726	160.334
298	10.067	9.133	9.133	.000	-151.790	-144.247	105.735
300	10.096	9.195	9.133	.019	-151.790	-144.200	105.049
400	11.144	12.261	9.544	1.087	-151.714	-141.680	77.410
500	11.707	14.814	10.350	2.232	-151.581	-139.184	60.837
600	12.045	16.982	11.279	3.421	-151.420	-136.720	49.800
700	12.322	18.862	12.231	4.641	-151.242	-134.284	41.925
800	12.524	20.521	13.166	5.884	-151.391	-131.845	36.018
900	12.694	22.006	14.067	7.145	-151.389	-129.402	31.423
1000	12.843	23.351	14.929	8.422	-151.431	-126.957	27.746
1100	12.978	24.581	15.751	9.713	-151.517	-124.507	24.737
1200	13.105	25.716	16.535	11.017	-153.429	-121.894	22.200
1300	13.224	26.770	17.282	12.334	-153.261	-119.273	20.051
1400	13.339	27.754	17.996	13.662	-153.085	-116.665	18.212
1500	13.450	28.678	18.677	15.001	-152.901	-114.070	16.620
1600	13.558	29.550	19.330	16.352	-152.709	-111.487	15.228
1700	13.665	30.375	19.955	17.713	-152.509	-108.917	14.002
1800	13.769	31.159	20.556	19.085	-189.195	-105.678	12.831
1900	13.873	31.906	21.134	20.467	-188.760	-101.050	11.623
2000	13.975	32.620	21.691	21.859	-188.317	-96.446	10.539
2100	14.076	33.305	22.228	23.262	-187.869	-91.863	9.560
2200	14.177	33.962	22.746	24.674	-187.417	-87.301	8.673
2300	14.277	34.594	23.248	26.097	-186.960	-82.761	7.864
2400	14.376	35.204	23.733	27.530	-186.501	-78.240	7.125
2500	14.475	35.793	24.204	28.972	-186.041	-73.739	6.446
2600	14.574	36.362	24.661	30.425	-185.580	-69.256	5.821
2700	14.672	36.914	25.104	31.887	-185.122	-64.791	5.244
2800	14.770	37.450	25.536	33.359	-184.666	-60.342	4.710
2900	14.868	37.970	25.955	34.841	-184.215	-55.909	4.213
3000	14.966	38.475	26.364	36.333	-183.772	-51.493	3.751
3100	15.063	38.968	26.763	37.834	-183.337	-47.090	3.320
3200	15.161	39.447	27.152	39.345	-182.913	-42.702	2.916
3300	15.258	39.915	27.532	40.866	-182.502	-38.328	2.538
3400	15.355	40.372	27.903	42.397	-182.105	-33.965	2.183
3500	15.452	40.819	28.265	43.937	-181.724	-29.614	1.849

June 30, 1971

CALCIUM OXIDE (CaO) (CRYSTAL) GFW = 56.0794 CaO

ΔHf°_0 = -150.99 ± 0.21 kcal/mol

$S^\circ_{298.15}$ = 9.133 ± 0.03 gibbs/mol $\Delta Hf^\circ_{298.15}$ = -151.79 ± 0.21 kcal/mol

Tm = 3223 ± 15 K ΔHm° = [19 ± 2] kcal/mol

Heat of Formation

Huber and Holley (1) determined the heat of combustion of calcium metal in a bomb calorimeter and derived the heat of formation of calcium oxide (c) as -151.79 ± 0.21 kcal/mol which is adopted in the tabulation. The adopted value is in good agreement with the value, -151.9 kcal/mol (2) derived from solution calorimetry.

Heat Capacity and Entropy

Gmelin (3) measured low temperature Cp data from 4 to 300 K in an adiabatic calorimeter. We use his smoothed Cp values to derive S°_{298} = 9.133 ± 0.03 eu based on S° = 0.0001 eu at 4 K. Lander (4) determined high temperature enthalpy data from 563.6 to 1176.4 K by drop calorimetry. The low temperature Cp and high temperature enthalpy data are smoothly joined at 298 K by a polynomial curve fitting method. The deviations of the observed enthalpies from the adopted values are about 0.2-1%, except the enthalpy value at 753 K (2.0%). Heat capacities above 1200 K are extrapolated from the adopted Cp functions. The extrapolated Cp at the melting point (2887 K), 14.8 gibbs/mol, is in reasonable agreement with the value 2 x 7.25 gibbs/mol suggested by Kubaschewski (5).

Combination of the earlier low temperature Cp measurements of Nernst and Schwers (28-90 K) (6) and Parks and Kelley (87-293 K) (7) yields S°_{298} = 9.5 ± 0.2 eu, based on S°_{28} = 0.04 eu (8). These Cp measurements are less accurate than those of Gmelin (3), and are not adopted in the tabulation.

Fischer and Ertmer (9) determined high temperature enthalpy data by drop calorimetry in the temperature range from 0° to 1716°C. The accuracy was claimed to be approximately ±4%. We have not adopted their enthalpy data in the tabulation since the heat capacities which we derive from their data are always less than those of MgO (10) when the temperature is above 1000 K. The deviations between their enthalpy data and the adopted values are approximately 1.8% at 693 K, 3.3% at 1283 K and 5.3% at 1989 K.

Melting Data

Schneider (11) reviewed literature data (12, 13, 14, 15) for the melting point of CaO and selected the value 2887 K based on Kanolt's observations (12) with proper corrections for the temperature scale change. However, Foex (16) determined recently the melting point as 3223 K in a solar furnace using a calibrated pyrometer. His method was relatively free of contamination between sample holder and sample at high temperatures. Foex also found the measurement of Kanolt would be falsified by the presence of tungsten supports in contact with calcium oxide. The latter will react with metallic tungsten to form $WO_3 \cdot 3CaO$ at high temperature. This may be the reason leading to a lower melting point in Kanolt's measurement. The value, 3223 K is tentatively adopted in the tabulation.

The heat of melting is assumed to be 19 ± 2 kcal/mol which is calculated from the estimated ΔSm° = 6 eu at the melting point. The latter is estimated to be the same as that of MgO (10).

References

1. E. J. Hubber, Jr. and C. E. Holley, Jr., J. Phys. Chem. 60, 498 (1956).
2. U. S. Natl. Bur. Std. Circ. 500, 1952.
3. E. Gmelin, Z. Naturforsch. 24a, 1794 (1969).
4. J. J. Lander, J. Amer. Chem. Soc. 73, 5794 (1951).
5. O. Kubaschewski, E. Ll. Evans and C. B. Alcock, "Metallurgical Thermochemistry," 4th Ed., Pergamon Press, London, 1967, p. 306.
6. W. Nernst and F. Schwers, 'Untersuchungen über die spezifischen Wärme bei tiefen Temperaturen,' Sitzb. Konig. Preuss. Akad. Wiss. 1914, p. 355.
7. G. S. Parks and K. K. Kelley, J. Phys. Chem. 30, 47 (1926).
8. U. S. Bur. Mines Bulletin 592, 1961.
9. W. A. Fischer and W. Ertmer, Archiv für das Eisenhuttenwesen, 37, 275 (1966).
10. JANAF MgO(c) table dated Dec. 31, 1965.
11. S. J. Schneider, 'Compilation of the Melting Points of the Metal Oxides," NBS Monograph 68, 1963.
12. C. W. Kanolt, J. Wash. Acad. Sci. 3, 315 (1913), Z. Anorg. Chem. 85, 1411 (1914).
13. Ya. I. Ol'shanskii, Dokl. Akad. Nauk. SSSR 59, 1105 (1948).
14. R. C. Doman, J. B. Barr, N. R. McNally, and A. M. Alper, Bull. Amer. Ceram. Soc. 41, 584 (1962).
15. E. E. Schumacher, J. Amer. Chem. Soc. 48, 396 (1926).
16. M. Foex, Elec. MHD, Proc. Symp. 5, 3139 (1968) (CA 70, 108848S, 1969); Solar Energy 9, 61 (1965).

CaO

Figure 7-11. Sample page from the JANAF Thermochemical Tables, 2nd Edition, D. R. Stull and H. Prophet.

Problems

1. Use data in Appendix I to compute ΔH°_{rxn}, ΔS°_{rxn}, and ΔG°_{rxn} at 25°C for the following reactions:
(a) $SO_2(g) + \frac{1}{2} O_2(g) \rightarrow SO_3(g)$
(b) $CO(g) + 2 H_2(g) \rightarrow CH_3OH(g)$
(c) $CH_3OH(g) \rightarrow HCHO(g) + H_2(g)$

2. Given that $\Delta H^\circ_{rxn} = -2815.8$ kJ at 25°C for the reaction

$$\underset{\text{Glucose}}{C_6H_{12}O_6(s)} + 6 O_2(g) \longrightarrow 6 CO_2(g) + 6 H_2O(\ell)$$

Use $\Delta \bar{H}^\circ_f$ data in Appendix I, together with the above ΔH°_{rxn} value, to compute $\Delta \bar{H}^\circ_f$ at 25°C for crystalline glucose.

3. Use data in Appendix I to compute ΔH°_{rxn}, ΔS°_{rxn}, and ΔG°_{rxn} at 25°C for the following reactions:
(a) $ZnS(s, \text{sphalerite}) + \frac{3}{2} O_2(g) \rightarrow ZnO(s) + SO_2(g)$
(b) $3 Fe_2O_3(s) \rightarrow 2 Fe_3O_4(s) + \frac{1}{2} O_2(g)$
(c) $0.947 Fe(s) + \frac{1}{2} O_2(g) \rightarrow Fe_{0.947}O(s)$

4. Use bond enthalpies to estimate $\Delta H^\circ_{rxn,298}$ for the reactions
(a) $CH_4(g) + Br_2(g) \rightarrow CH_3Br(g) + HBr(g)$
(b) $H_2C{=}CH_2(g) + H_2(g) \rightarrow CH_3CH_3(g)$
(c) $CH_3CH_2OH(\ell) + 3 O_2(g) \rightarrow 2 CO_2(g) + 3 H_2O(g)$
Compare your results with $\Delta H^\circ_{rxn,298}$ values calculated using data in Appendix I.

5. The value of ΔH°_{rxn} at 25°C for the reaction

$$CH_3Br(g) + H_2(g) \longrightarrow CH_4(g) + HBr(g)$$

is -76.11 kJ. Use this result together with data in Appendix I for $H_2(g)$, $CH_4(g)$ and $HBr(g)$, to compute $\Delta \bar{H}^\circ_f$ for $CH_3Br(g)$.

6. The enthalpies of combustion per mole of sucrose, glucose, and fructose at 25°C are as follows [combustion to $CO_2(g)$ and $H_2O(\ell)$]:

Compound	$\Delta H_{comb}/kJ \cdot mol^{-1}$
Sucrose(*s*)	−5646.7
Glucose(*s*)	−2815.8
Fructose(*s*)	−2826.7

Use these data to compute $\Delta \bar{H}^\circ_{rxn}$ for the reaction

$$\underset{\text{Sucrose}}{C_{12}H_{22}O_{11}(s)} + H_2O(\ell) \longrightarrow \underset{\text{Glucose}}{C_6H_{12}O_6(s)} + \underset{\text{Fructose}}{C_6H_{12}O_6(s)}$$

7. Take $\bar{C}_P^\circ = (29.75 + 25.10 \times 10^{-3}T - 1.55 \times 10^5 T^{-2})\ \mathrm{J \cdot K^{-1} \cdot mol^{-1}}$ for $NH_3(g)$. Derive an expression for $\bar{H}_T^\circ - \bar{H}_{298.15}^\circ$ for $NH_3(g)$ and use it to compute $\bar{H}_{1000}^\circ - \bar{H}_{298.15}^\circ$.

8. Take $\bar{C}_P^\circ = (23.64 + 47.87 \times 10^{-3}T - 1.93 \times 10^5 T^{-2})\ \mathrm{J \cdot K^{-1} \cdot mol^{-1}}$ for $CH_4(g)$, and also take $\bar{S}_{298.15}^\circ = 186.15\ \mathrm{J \cdot K^{-1} \cdot mol^{-1}}$ for $CH_4(g)$. Compute $\bar{S}_T^\circ$ for $CH_4(g)$ at 500 K and 1000 K.

9. Compute $\Delta G_{\mathrm{rxn}}^\circ$ for the reactions

$$C(s, \text{graphite}) + H_2O(g) \longrightarrow CO(g) + H_2(g)$$
$$C(s, \text{graphite}) + 2\,H_2(g) \longrightarrow CH_4(g)$$

at 298 K, 500 K, 700 K, and 1000 K, and plot your results as a function of T over the range 300 K to 700 K.

10. Suggest an explanation for the difference in the $\Delta H_{\mathrm{rxn}}^\circ$ values for the reactions

$$H_2O(g) \longrightarrow HO(g) + H(g) \qquad \Delta H_{\mathrm{rxn}}^\circ = 502\ \mathrm{kJ}$$
$$HO(g) \longrightarrow H(g) + O(g) \qquad \Delta H_{\mathrm{rxn}}^\circ = 423\ \mathrm{kJ}$$

Hint: Consider the bonding in H_2O and OH.

11. Without consulting thermodynamic tables, arrange the following reactions in order of increasing $\Delta S_{\mathrm{rxn}}^\circ$ values:
(a) $C(s) + H_2O(g) \rightarrow CO(g) + H_2(g)$
(b) $C(s) + 2\,H_2(g) \rightarrow CH_4(g)$
(c) $C(s) + O_2(g) \rightarrow CO_2(g)$
(d) $CH_3OH(\ell) + \frac{3}{2}\,O_2(g) \rightarrow CO_2(g) + 2\,H_2O(g)$
(e) $CO_2(s) \rightarrow CO_2(g)$
(f) $C(s) + \frac{1}{2}\,O_2(g) \rightarrow CO(g)$
(g) $CO(g) + 3\,H_2(g) \rightarrow CH_4(g) + H_2O(g)$
Compare your predictions with calculated values.

12. Given the following data:

Compound	$\Delta \bar{H}_{f,298}^\circ/kcal \cdot mol^{-1}$	$\bar{H}_{298}^\circ - \bar{H}_0^\circ/kcal \cdot mol^{-1}$
$N_2O_4(g)$	2.19	3.918
$N_2H_4(g)$	22.80	2.743
$H_2O(g)$	−57.80	2.367
$N_2(g)$	0	2.072

(a) Compute ΔH_0° for the reaction

$$2\ \mathrm{H_2N{-}NH_2}(g) + \mathrm{O_2N{-}NO_2}(g) \longrightarrow 3\,N_2(g) + 4\,H_2O(g)$$

(b) Given the following bond enthalpies (in kilocalories per mole at 0 K)

$$\bar{H}(N{\equiv}N) = 226 \qquad \bar{H}(N{-}H) = 93$$
$$\bar{H}(O{-}H) = 111 \qquad \bar{H}(N{-}N) = 38 \quad (\text{in } N_2H_4)$$
$$\bar{H}(N{=}O) = 142 \qquad \bar{H}(N{-}O) = 112$$

Estimate the N—N bond enthalpy in N_2O_4.

13. A 1.00-g sample of benzene, $C_6H_6(\ell)$, is burned in oxygen in a bomb calorimeter; the observed temperature increase was 1.681 K and the final temperature was 25°C. The heat capacity of the calorimeter assembly is $23.80\ \text{kJ}\cdot\text{K}^{-1}$. Compute the value of ΔU for the combustion reaction per gram and per mole of $C_6H_6(\ell)$. Also compute ΔH_{rxn} and, from the value of ΔH_{rxn} and data in Appendix I, compute $\Delta\bar{H}_f^\circ$ at 298 K for $C_6H_6(\ell)$.

14. Given $\bar{H}(O{-}H) = 464\ \text{kJ}\cdot\text{mol}^{-1}$ and $\Delta H_{rxn}^\circ = 502$ kJ for

$$H_2O(g) \longrightarrow OH(g) + H(g)$$

estimate ΔH_{rxn}° for

$$HO(g) \longrightarrow O(g) + H(g)$$

15. The standard enthalpies of combustion of ethanol(ℓ) and methyl ether(ℓ) at 25°C are $-1235\ \text{kJ}\cdot\text{mol}^{-1}$ and $-1247\ \text{kJ}\cdot\text{mol}^{-1}$, respectively. Compute ΔH_{rxn}° at 25°C for the reaction

$$CH_3OCH_3(\ell) \longrightarrow CH_3CH_2OH(\ell)$$

Suggest a qualitative explanation for the sign and magnitude of ΔH_{rxn}°.

16. Calculate the average bond enthalpy of the Si—H bonds in $SiH_4(g)$ from the following data. The molar enthalpy of combustion of $SiH_4(g)$ to $SiO_2(s)$ and $H_2O(\ell)$ is $-326.84\ \text{kcal}\cdot\text{mol}^{-1}$. The standard enthalpies of formation of $SiO_2(s)$ and $H_2O(\ell)$ are $-205.00\ \text{kcal}\cdot\text{mol}^{-1}$ and $-68.32\ \text{kcal}\cdot\text{mol}^{-1}$, respectively. Further, for

$$Si(s) \longrightarrow Si(g) \qquad \Delta H_{rxn,298}^\circ = 88.04\ \text{kcal}$$
$$\tfrac{1}{2}H_2(g) \longrightarrow H(g) \qquad \Delta H_{rxn,298}^\circ = 52.09\ \text{kcal}$$

17. Arrange the following compounds in order of decreasing enthalpies of combustion per mole of the compounds

$$CH_3CH_3, \quad CH_3CH_2OH, \quad HOCH_2CH_2OH$$
$$CH_3COOH, \quad HOCH_2COOH, \quad HOOC{-}COOH$$

Discuss the significance of your predictions in terms of the combustion process.

18. (a) Show that for a chemical reaction

$$\Delta H_{rxn} = \Delta U_{rxn} + \sum_i \nu_i P_i \bar{V}_i$$

where the ν_i values are the reaction-balancing coefficients.

(b) If all reactants and products are in their 1-atm standard states, then show that

$$\Delta H^\circ_{\text{rxn}} = \Delta U^\circ_{\text{rxn}} + (1\ \text{atm}) \sum_i \nu_i \bar{V}_i^\circ$$

(c) For a reaction involving only gaseous species, show that

$$\Delta H^\circ_{\text{rxn}} = \Delta U^\circ_{\text{rxn}} + RT\Delta n$$

where Δn is the number of moles of gaseous products minus the number of moles of gaseous reactants.

(d) For a reaction involving gases and condensed phases, show that

$$\Delta H^\circ_{\text{rxn}} \approx \Delta U^\circ_{\text{rxn}} + RT\Delta n$$

[*Hint:* Compare $\bar{V}$ for a gas at 25°C, 1 atm with $\bar{V}$ for, say, $H_2O(\ell)$ and NaCl(s) at 25°C, 1 atm.]

19. Show that for a chemical reaction

$$\left(\frac{\partial \Delta U_{\text{rxn}}}{\partial T}\right)_V = \Delta C_{V,\text{rxn}}$$

20. Derive the Gibbs-Helmholtz equation

$$\Delta H = \Delta G - T\left(\frac{\partial \Delta G}{\partial T}\right)_P$$

21. Show that for a chemical reaction

$$\Delta G^\circ_{\text{rxn},T_2} = \Delta G^\circ_{\text{rxn},T_1} - \Delta S^\circ_{\text{rxn},T_1}(T_2 - T_1) + \int_{T_1}^{T_2} \Delta C^\circ_{P,\text{rxn}}\, dT - T_2 \int_{T_1}^{T_2} \frac{\Delta C^\circ_{P,\text{rxn}}\, dT}{T}$$

22. It might appear in setting $\Delta \bar{H}_f^\circ = 0$ for the elements in their standard states *at all* T that we have violated a basic principle, namely, that the enthalpy is a function of temperature and, therefore, that $\Delta \bar{H}_f^\circ$ can be set equal to zero at only one T. Resolve the above paradox by first showing that the following tabulated thermodynamic data is internally consistent.

Species	$\bar{H}^\circ_{298.15} - \bar{H}^\circ_0$ *kJ·mol*$^{-1}$	$\Delta \bar{H}_f^\circ$(*at* 0 *K*) *kJ·mol*$^{-1}$	$\Delta \bar{H}_f^\circ$(*at* 298.15 *K*) *kJ·mol*$^{-1}$
$Br_2(\ell)$	24.51	0	0
$Br_2(g)$	9.72	45.70	30.91

Then analyze the method of your calculations to show in general that it is not a violation of thermodynamic principles to set $\Delta \bar{H}_f^\circ = 0$ at all T for the most stable form of an element.

23. Suppose 2.00 mol of $CH_4(g)$ is mixed with 5.00 mol of $O_2(g)$ in an adiabatic enclosure at 25°C. A spark is produced in the mixture and the CH_4 is completely burned in the oxygen to produce $CO_2(g)$ and $H_2O(g)$. Assume ideal gas behavior and estimate the final temperature of the gas mixture. Use the following $\bar{C}_P^\circ$ values.

Species	$\bar{C}_P^\circ/J \cdot K^{-1} \cdot mol^{-1}$
$O_2(g)$	$29.96 + 4.18 \times 10^{-3}T$
$CO_2(g)$	$44.23 + 8.79 \times 10^{-3}T$
$H_2O(g)$	$30.54 + 10.29 \times 10^{-3}T$
$CH_4(g)$	$23.64 + 47.87 \times 10^{-3}T$

24. Certain chemical reactions can be used to produce high-temperature flames. For example, a flame temperature of about 4850 K can be obtained from the reaction

$$C_2N_2(g) + O_2(g) = 2\,CO(g) + N_2(g)$$

An especially violent flame ($T > 6000$ K, which exceeds the surface temperature of the sun) can be produced from the reaction

$$H_2(g) + F_2(g) = 2\,HF(g)$$

Discuss the various thermodynamic properties of reactants and products that contribute to the production of high-temperature flames.

25. Discuss qualitatively why the entropy of acetone is greater than the entropy of trimethylene oxide.

These gem diamonds were created in the laboratory from graphite, the soft black substance used in "lead" pencils, by scientists at the General Electric Research and Development Center. The larger crystals are approximately one carat in weight. Although these diamonds have undergone slight polishing, they have not been cut and retain the shape in which they were "grown" in a special apparatus for subjecting graphite (the pile of black powder) to extreme pressures and temperatures.

8

The Chemical Potential and the Phase Rule

If a homogeneous body has c independently variable components, the phase of the body is evidently capable of c + 1 independent variations. A system of p coexistent phases, each of which has the same c independently variable components is capable of c − p + 2 variations of phase. For the temperature, the pressure, and the potentials for the actual components have the same values in the different phases, and the variations of these quantities are subject to as many conditions as there are different phases. Therefore, the number of independent variations in the values of these quantities, that is, the number of independent variations of phase of the system, will be c − p + 2.

The Scientific Papers of J. Willard Gibbs, Vol. I, *Thermodynamics*. New York: Dover Publications, 1961, p. 96.

THE THERMODYNAMIC methods for the characterization of equilibrium states of heterogeneous systems involving any number of substances were developed by J. Willard Gibbs. Prior to Gibbs's derivation of the phase rule, the basic problem confronting the experimentalist centered around the determination of the *minimum* number of intensive variables (including the variables necessary to describe the composition of the system) that had to be specified in order to determine unambiguously the thermodynamic state of the system. From our development of the principles of thermodynamics in preceding chapters, we know that a knowledge of the number of independent variables of a system is a necessary prerequisite to the application of thermodynamics. Throughout the development in those chapters we have restricted the application of thermodynamics to systems of fixed composition. We now turn to a consideration of systems of variable composition or, putting it another way, to chemical thermodynamics. In chemical thermodynamics the phase rule and chemical potentials play indispensable roles.

8-1 *Determination of the Number of Phases and Components of a System*

Each physically or chemically distinct, homogeneous (i.e., uniform throughout in chemical composition and physical state) *and mechanically separable part of a system is called a "phase."* A phase need not be continuous in a physical sense. For example, a mass of a crystalline solid constitutes only one phase, whether it is a single crystal or a large number of crystals; also, a collection of droplets of a particular liquid constitutes a single liquid phase. Provided the degree of subdivision of a liquid or solid phase is not too fine, we need not consider the surface area of a phase in our thermodynamic analysis. However, it should be remembered that if the degree of subdivision of a phase is very fine, as is the case with colloids and extremely small crystals and droplets, then surface area effects ($\sigma\, d\mathscr{A}$) must be included in the thermodynamic analysis of the system. We shall assume in the discussion to follow that surface effects are unimportant for the systems of interest.

It usually is not difficult to enumerate the phases of a system. Some examples of phase enumerations are given below.

1. Clean air (i.e., air free of particulate matter) constitutes a single phase because the component gases are miscible in all proportions. A homogeneous gas constitutes a single phase no matter how many chemical species are present in the gas. Note that gases are easy to separate mechanically from condensed (i.e., solid or liquid) phases.
2. The phases in a system composed of a saturated aqueous solution of NaCl in equilibrium with excess NaCl(*s*) and ice in a closed, air-free vessel (Figure 8-1) follow:
   (a) $H_2O(s)$
   (b) NaCl(*s*)
   (c) Aqueous solution of NaCl
   (d) Water vapor

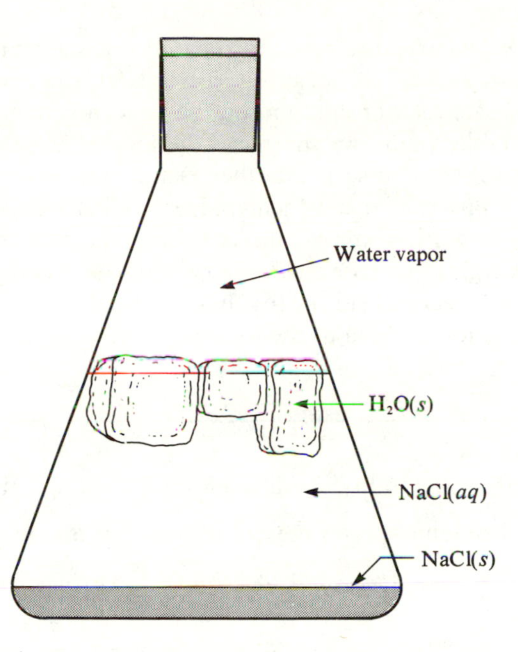

Figure 8-1. A saturated aqueous solution of NaCl in equilibrium with ice, NaCl(s), and water vapor.

EXAMPLE 8-1

The system formed in the high-temperature reduction of solid zinc oxide by solid carbon is found to contain the following constituents at equilibrium:

$$ZnO(s), \quad C(s), \quad Zn(g), \quad CO(g), \quad \text{and} \quad CO_2(g)$$

How many phases are present?

Solution: There are three phases present, namely, the solid phases $ZnO(s)$ and $C(s)$, and a gas phase containing $Zn(g)$, $CO(g)$, and $CO_2(g)$.

In addition to a specification of the number of phases, it is also necessary in chemical thermodynamics to specify the number of components of a system. *The number of components of a system is the minimum number of composition variables necessary to describe all possible variations in the composition of the system.* Algebraically, the number of components c of a system is given by

$$c = s - r \tag{8-1}$$

where s is the total number of chemically distinct constituents and *r is the number of restrictive conditions.* That is to say, r is equal to the number of algebraic relationships among the composition variables.

Each *independent* chemical reaction at equilibrium gives rise to a restrictive condition because, as we shall see more clearly later, the reaction equilibrium imposes an algebraic relationship among the concentrations of certain of the composition variables of the system (recall the Law of Mass Action from your introductory chemistry courses). Another example of a restrictive condition is the *electroneutrality condition*, which applies to all electrolyte solutions. The electroneutrality condition follows directly from the fact that an electrolyte solution does not give rise to an electric field in the vicinity of the solution. The absence of an electric field means that the total positive charge per unit volume of the solution is equal to the total negative charge per unit volume of the solution.

EXAMPLE 8-2

Write the electroneutrality condition for a $CaCl_2(aq)$ solution.

Solution: The ionic species present in the solution are

$$Ca^{2+}(aq), \quad Cl^{-}(aq), \quad CaCl^{+}(aq), \quad H^{+}(aq) \text{ and } OH^{-}(aq)$$

Thus

$$2[Ca^{2+}] + [CaCl^{+}] + [H^{+}] = [Cl^{-}] + [OH^{-}]$$

where the brackets denote concentrations. Note that $[Ca^{2+}]$ is multiplied by 2 because each mole of $Ca^{2+}(aq)$ carries 2 mol of positive charge. The $H^{+}(aq)$ and $OH^{-}(aq)$ species arise from the autoprotolysis of $H_2O(\ell)$; $H_2O(\ell) \rightleftharpoons H^{+}(aq) + OH^{-}(aq)$.

Some examples of the determination of the number of components in a system follow.

1. The number of components in a homogeneous liquid mixture (i.e., a solution) of $CCl_4(\ell)$ and $C_6H_6(\ell)$ is

$$c = 2 - 0 = 2$$

   The significance of the number 2 in this case is that we can vary at will the number of moles of CCl_4 in the solution while holding the number of moles of C_6H_6 constant and vice versa. There are no restrictive conditions in the $CCl_4 + C_6H_6$ solution, and thus $r = 0$.
2. The thermal decomposition of calcium carbonate leads to the formation of calcium oxide and gaseous carbon dioxide:

$$CaCO_3(s) \rightleftharpoons CaO(s) + CO_2(g)$$

   The equilibrium system contains the three constituents $CaCO_3(s)$, $CaO(s)$, and $CO_2(g)$ and involves one restrictive condition (the reaction equilibrium). The number of components is $c = 3 - 1 = 2$. Note that the system also contains three phases.

EXAMPLE 8-3

In the nitrogen-hydrogen-ammonia system, several distinct possibilities arise for the number of components. Determine the number of components in each of the following cases.

(a) A nonreacting mixture of $N_2(g)$, $H_2(g)$, and $NH_3(g)$ at 25°C. In this case $r = 0$, and thus

$$c = 3 - 0 = 3$$

(b) Any mixture of $N_2(g)$, $H_2(g)$, and $NH_3(g)$ at a temperature where the following equilibrium is established:

$$N_2(g) + 3\,H_2(g) \rightleftharpoons 2\,NH_3(g)$$

In this case the reaction equilibrium constitutes a restrictive condition, and thus

$$c = 3 - 1 = 2$$

Note that $r = 1$ because the reaction equilibrium is established, whereas in part (a) the reaction equilibrium was not established (the reaction rate is negligible at 25°C).

(c) Any system obtained by heating $NH_3(g)$ to a temperature where the reaction equilibrium is established:

$$N_2(g) + 3\,H_2(g) \rightleftharpoons 2\,NH_3(g)$$

In this case, we have, in addition to the restriction imposed by the chemical reaction, the condition

$$P_{H_2} = 3P_{N_2}$$

which arises from the reaction stoichiometry. Thus $r = 2$, and the number of components is

$$c = 3 - 2 = 1$$

Note that the number of components may be reduced if the system is not prepared in an entirely arbitrary manner.

3. Let us assume for the moment that the species involved in a solution of acetic acid in water are

$$\left\{\begin{matrix} CH_3COOH(aq), & CH_3COO^-(aq), & H_2O(\ell), \\ H^+(aq) \text{ and } OH^-(aq) & & \end{matrix}\right\} \quad \text{(1 phase)}$$

The restrictive conditions are

(1) $CH_3COOH(aq) \rightleftharpoons CH_3COO^-(aq) + H^+(aq)$

(2) $H_2O(\ell) \rightleftharpoons H^+(aq) + OH^-(aq)$

(3) $[H^+] = [CH_3COO^-] + [OH^-]$

Thus

$$c = 5 - 3 = 2$$

The last condition (in which the brackets denote species concentrations) is the *electroneutrality* condition. The restrictive condition arising from electroneutrality must always be included in r for any electrolyte solution.

It may be argued at this point that no such species as H^+ exists in aqueous solution but that the proton is associated with one or more water molecules, and thus we should consider along with H^+ the species H_3O^+, $H_5O_2^+$, $H_7O_3^+$, $H_9O_4^+$, and so forth. The answer to this objection is that for the purpose of application of the phase rule it is unnecessary to specify the *actual* nature of the molecular species involved, as long as such species remain in rapid equilibrium with one another. In other words, each of the species noted above, if included along with H^+ in the list of species, gives rise to an additional restrictive condition and thus does not affect the number of components. The relevant restrictive conditions are

$$H^+(aq) + H_2O(\ell) \rightleftharpoons H_3O^+(aq)$$
$$H_3O^+(aq) + H_2O(\ell) \rightleftharpoons H_5O_2^+(aq)$$
$$H_5O_2^+(aq) + H_2O(\ell) \rightleftharpoons H_7O_3^+(aq)$$
$$H_7O_3^+(aq) + H_2O(\ell) \rightleftharpoons H_9O_4^+(aq)$$

Thus, for a determination of the number of components in the system, it is sufficient to designate all such species as $H^+(aq)$. Similar comments apply to $CH_3COO^-(aq)$, $OH^-(aq)$, and $CH_3COOH(aq)$. For example, suppose the dimer $(CH_3COOH)_2(aq)$ is an actual species as well as $CH_3COOH(aq)$; dimerization will not change c if the equilibrium

$$2\,CH_3COOH(aq) \rightleftharpoons (CH_3COOH)_2(aq)$$

is rapidly established.

Arguments similar to those above also apply to the various isotopic species; for example, H_2O represents H_2O^{16}, H_2O^{17}, H_2O^{18}, D_2O^{16}, D_2O^{17}, D_2O^{18}, HDO^{16}, HDO^{17}, and HDO^{18}. Among the foregoing nine species one can write eight independent isotope-exchange equilibria, such as

$$H_2O^{16} + D_2O^{16} \rightleftharpoons 2\,HDO^{16}$$

and

$$H_2O^{16} + HDO^{17} \rightleftharpoons HDO^{16} + H_2O^{17}$$

Consequently, the value of c is unaffected by the presence of isotopic species, provided that we do not vary the isotopic composition of the system. Note that $c = 2$ for H_2O/D_2O mixtures of variable isotope ratio.

We now turn to a development of the concepts and equations necessary for the derivation of the phase rule and the application of thermodynamics to systems of variable composition.

8-2 *At Equilibrium the Chemical Potential of Each Component Has the Same Value Throughout the Entire System*

If the composition and/or total mass of a thermodynamic system is subject to variations, then a complete thermodynamic description of the system requires the introduction of variables that characterize *the amounts of the components.* Assuming the absence of external force fields and surface effects, we have from the First and Second Laws of Thermodynamics

$$U = U(S, V, n_1, \ldots, n_c)$$

where the n_i values represent the number of moles of the various components and c is the total number of components. Further, because dU is an exact differential, we have

$$dU = \left(\frac{\partial U}{\partial S}\right)_{V,n_i} dS + \left(\frac{\partial U}{\partial V}\right)_{S,n_i} dV + \sum_{i=1}^{c} \left(\frac{\partial U}{\partial n_i}\right)_{S,V,n_j(j\neq i)} dn_i \tag{8-2}$$

where in the first two partial derivatives n_i represents all the composition variables, and the symbol $n_j(j \neq i)$ represents all composition variables except the one with respect to which U is being differentiated in each of the terms in the sum. We shall henceforth abbreviate this symbol somewhat, writing it as simply n_j. Making the identifications (Eq. 8-2 must hold with all $dn_i = 0$)

$$T = \left(\frac{\partial U}{\partial S}\right)_{V,n_i} \quad \text{and} \quad -P = \left(\frac{\partial U}{\partial V}\right)_{S,n_i} \tag{8-3}$$

we have

$$dU = T\,dS - P\,dV + \sum_{i=1}^{c} \left(\frac{\partial U}{\partial n_i}\right)_{S,V,n_j} dn_i \tag{8-4}$$

The intensive quantities $(\partial U/\partial n_i)_{S,V,n_j}$ were designated *chemical potentials* by Gibbs. Chemical potentials are intensive thermodynamic state functions, because they have the dimensions of energy per mole, and thus their values are independent of the size of the system. The chemical potentials are partial molar quantities, and hence their values depend on the *composition* of the system. The chemical potential has an important function analogous to temperature and pressure. The existence of a temperature difference is necessary for the transfer of energy as heat, and the existence of a pressure difference is necessary for the transfer of energy as pressure-volume work. The existence of a difference in chemical potential is necessary for the transfer of matter. A gradient in the chemical potential can be regarded as the driving force for a chemical reaction or for the spreading of a substance throughout a phase or for the transfer of a substance from one phase to another. Thus we see that the terms

$$\sum_{i=1}^{c} \left(\frac{\partial U}{\partial n_i}\right)_{S,V,n_j} dn_i$$

in Eq. 8-4 represent the amounts by which the total internal energy of a phase is changed by the transfer of material into or out of the phase or by the conversion of one component into another component. The chemical potential is thus a kind of chemical pressure.

Changing the independent variables in Eq. 8-4 from (S, V, n_i), where the symbol n_i represents the c composition variables, to (T, P, n_i), because the latter set of independent variables is more convenient for chemical thermodynamics, we obtain

$$d(U + PV - TS) = -S\,dT + V\,dP + \sum_{i=1}^{c} \left(\frac{\partial U}{\partial n_i}\right)_{S,V,n_j} dn_i \tag{8-5}$$

Making the identification

$$G = U + PV - TS = G(T, P, n_1, \ldots, n_c) \tag{8-6}$$

and recognizing from its definition that G is a thermodynamic state function of the system, we can write

$$dG = \left(\frac{\partial G}{\partial T}\right)_{P,n_i} dT + \left(\frac{\partial G}{\partial P}\right)_{T,n_i} dP + \sum_{i=1}^{c} \left(\frac{\partial G}{\partial n_i}\right)_{T,P,n_j} dn_i \tag{8-7}$$

Comparison of Eq. 8-7 with Eq. 8-5 leads to the following identifications:

$$-S = \left(\frac{\partial G}{\partial T}\right)_{P,n_i} \tag{8-8}$$

$$V = \left(\frac{\partial G}{\partial P}\right)_{T,n_i} \tag{8-9}$$

and

$$\left(\frac{\partial U}{\partial n_i}\right)_{S,V,n_j} = \left(\frac{\partial G}{\partial n_i}\right)_{T,P,n_j} \tag{8-10}$$

We shall use the symbol μ_i, where

$$\mu_i = \left(\frac{\partial G}{\partial n_i}\right)_{T,P,n_j} \tag{8-11}$$

for the chemical potential of the ith species.

Substitution of Eqs. 8-8, 8-9, and 8-11 into Eq. 8-7 yields for a single phase

$$dG = -S\,dT + V\,dP + \sum_{i=1}^{c} \mu_i\,dn_i \tag{8-12}$$

where G, S, and V are the total Gibbs energy, the total entropy, and the total volume of the phase. At constant T and P we obtain from Eq. 8-12

$$dG = \sum_{i=1}^{c} \mu_i\,dn_i \tag{8-13}$$

If we are dealing with a *closed* system at constant T and P, for which only pressure-volume work is possible, then at equilibrium the composition is fixed and $dn_i = 0$ for all i; hence

$$dG = \sum_{i=1}^{c} \mu_i \, dn_i = 0 \tag{8-14}$$

In the interests of simplicity we shall restrict our development at this stage to a system of two components (labeled 1 and 2) and two phases (labeled α and β). For such a case we have

$$G = G(T, P, n_{1\alpha}, n_{2\alpha}, n_{1\beta}, n_{2\beta})$$

and thus

$$dG = -S\,dT + V\,dP + \mu_{1\alpha} dn_{1\alpha} + \mu_{2\alpha} dn_{2\alpha} + \mu_{1\beta} dn_{1\beta} + \mu_{2\beta} dn_{2\beta} \tag{8-15}$$

At constant T and P, Eq. 8-15 becomes

$$dG = \mu_{1\alpha} dn_{1\alpha} + \mu_{2\alpha} dn_{2\alpha} + \mu_{1\beta} dn_{1\beta} + \mu_{2\beta} dn_{2\beta} \tag{8-16}$$

If the system is closed, then we have the material balance conditions

$$\left.\begin{aligned} dn_{1\alpha} + dn_{1\beta} &= 0 \\ dn_{2\alpha} + dn_{2\beta} &= 0 \end{aligned}\right\} \quad \text{(closed system)} \tag{8-17}$$

Equations 8-17 describe, for example, the fact that a decrease in the number of moles of component 1 in the α phase leads to an equal increase in the number of moles of component 1 in the β phase:

$$-dn_{1\alpha} = dn_{1\beta}$$

Combination of Eqs. 8-17 and 8-16 yields

$$dG = (\mu_{1\alpha} - \mu_{1\beta})\, dn_{1\alpha} + (\mu_{2\alpha} - \mu_{2\beta})\, dn_{2\alpha} \qquad \text{(constant } T \text{ and } P) \tag{8-18}$$

At equilibrium $dG = 0$, and thus

$$(\mu_{1\alpha} - \mu_{1\beta})\, dn_{1\alpha} + (\mu_{2\alpha} - \mu_{2\beta})\, dn_{2\alpha} = 0 \tag{8-19}$$

Because $dn_{1\alpha}$ and $dn_{2\alpha}$ are arbitrary, we conclude that, in order for Eq. 8-19 to hold at equilibrium, the values of the chemical potentials of the two components must be the same in both phases at equilibrium; that is

$$\left.\begin{aligned} \mu_{1\alpha} &= \mu_{1\beta} \\ \mu_{2\alpha} &= \mu_{2\beta} \end{aligned}\right\} \quad \text{(at equilibrium)} \tag{8-20}$$

Equation 8-20 tells us that *at equilibrium the chemical potential of each component is constant throughout the entire system.*

It is not difficult to generalize the foregoing development to any number of phases and any number of components. In general, *at equilibrium in a closed system, the temperature, the pressure, and the chemical potentials of all the components are uniform throughout the entire system.*

8-3 *The Gibbs-Duhem Equation Places a Restriction on the Simultaneous Variations in the Intensive Thermodynamic Variables*

The laws of thermodynamics require that any reversible infinitesimal changes in the internal energy of a phase in which only pressure-volume work is possible be given by

$$dU = T\,dS - P\,dV + \sum_{i=1}^{c} \mu_i\,dn_i \tag{8-21}$$

where dU is an exact differential. Let us now increase the size of the system to k times its original size, while keeping all the intensive variables (T, P, $\mu_1, \ldots, \mu_c$) fixed; this is certainly possible because intensive variables are by definition independent of the size of the system. From Eq. 8-21 we have

$$\int_U^{kU} dU = T\int_S^{kS} dS - P\int_V^{kV} dV + \sum_{i=1}^{c} \mu_i \int_{n_i}^{kn_i} dn_i$$

and thus

$$(k-1)U = T(k-1)S - P(k-1)V + \sum_{i=1}^{c} \mu_i(k-1)n_i$$

Cancellation of the $(k - 1)$ factor and rearrangement yields

$$U + PV - TS = \sum_{i=1}^{c} n_i\mu_i = G \tag{8-22}$$

Thus, the total Gibbs energy of the phase is given by

$$G = \sum_{i=1}^{c} n_i\mu_i \tag{8-23}$$

Equation 8-23 does not mean that we can calculate an absolute value for G given values of μ_i and n_i, because it is not possible to specify absolute values for the chemical potentials. Note also from Eq. 8-23 that, for a one-component system, $\mu_1 = G/n_1 = (\partial G/\partial n_1)_{T,P}$, and therefore the chemical potential of a pure chemical compound is simply the Gibbs energy per mole. For a two-component system,

$$\mu_1 = \left(\frac{\partial G}{\partial n_1}\right)_{T,P,n_2} = \mu_1(T, P, n_1, n_2) \neq \frac{G}{n_1}$$

From Eq. 8-23 we obtain on differentiation

$$dG = \sum_{i=1}^{c} \mu_i\,dn_i + \sum_{i=1}^{c} n_i\,d\mu_i \tag{8-24}$$

Equating the expressions for dG in Eqs. 8-12 and 8-24 yields the *Gibbs-Duhem equation*

$$S\,dT - V\,dP + \sum_{i=1}^{c} n_i\,d\mu_i = 0 \tag{8-25}$$

The Gibbs-Duhem equation is one of the most useful equations in chemical thermodynamics, because it gives the necessary relationship between simultaneous changes in the intensive variables ($T, P, \mu_1, \ldots, \mu_c$) of a phase. Thus, if there are c components in a phase, then we have $c + 2$ intensive variables, and of these only $c + 1$ can vary independently. We note here also that at constant T and P the Gibbs-Duhem equation reduces to

$$\sum_{i=1}^{c} n_i \, d\mu_i = 0 \tag{8-26}$$

8-4 *The Phase Rule Gives the Number of Independent Intensive Variables of a System*

Consider a pure substance. By a *pure substance* we mean a substance of fixed composition. The thermodynamic state of each of the phases of a pure substance is completely determined (in the absence of external force fields and surface effects) by specifying T and P. The specification of T and P determines all the other intensive variables; in particular, the specification of T and P completely determines the chemical potential. If we have two phases of a pure substance [e.g., $H_2O(\ell)$ and $H_2O(g)$], then the state of the combined system is completely determined by specifying the four variables $T_\alpha, P_\alpha, T_\beta, P_\beta$. However, if the two phases are coexisting in equilibrium, then there are three equations relating these four variables:

$$T_\alpha = T_\beta, \qquad P_\alpha = P_\beta, \qquad \mu_\alpha(T_\alpha, P_\alpha) = \mu_\beta(T_\beta, P_\beta) \tag{8-27}$$

and as a consequence only one of the four variables may be chosen arbitrarily. In such a case the system is said to have one *degree of freedom*, that is, one independent intensive variable. In other words, if we require that the two phases, say liquid water and water vapor, be in equilibrium *at some particular temperature* (assuming that such equilibrium is possible at the chosen T), then the equilibrium vapor pressure of water will be fixed at a definite value by the nature of water, and we have nothing whatsoever to say about the magnitude of that equilibrium vapor pressure.

The locus of points for which the three conditions in Eq. 8-27 are satisfied simultaneously is called a *phase-equilibrium curve.* A phase-equilibrium curve gives the pairs of values of T and P for which the two phases can coexist. If the two phases are liquid and vapor, then the phase-equilibrium curve is the *equilibrium vapor pressure curve* of the pure liquid. If the two phases are solid and vapor, then the phase-equilibrium curve is the *equilibrium sublimation pressure curve* of the pure solid. If the two phases are solid and liquid, then the phase-equilibrium curve is the *melting point curve* of the pure solid.

Consider the case of three phases (e.g., solid, liquid, vapor) of a pure substance coexisting in equilibrium. For such a system at equilibrium, we have six intensive state variables, $T_\alpha, P_\alpha, T_\beta, P_\beta, T_\gamma, P_\gamma$, and six relationships,

$$T_\alpha = T_\beta = T_\gamma \qquad P_\alpha = P_\beta = P_\gamma$$
$$\mu_\alpha(T_\alpha, P_\alpha) = \mu_\beta(T_\beta, P_\beta) = \mu_\gamma(T_\gamma, P_\gamma)$$

These six equations completely determine the values of the six intensive state variables, and as a consequence none of them can be chosen arbitrarily; the system is said to have no degrees of freedom. Geometrically this corresponds to the intersection of the three chemical potential surfaces at a point (μ_i, T_i, P_i), called the *triple point*, which occurs at a *unique* T and P for any given three phases of a pure substance. The values of the temperature and the pressure at the triple point are completely determined by the nature of the substance.

What the phase rule does is tell us the number of degrees of freedom (i.e., the number of independent intensive variables) for any equilibrium system having any number of components and phases. The state of each phase of a system is determined by the temperature, the pressure, and the chemical potentials of each of the c components. Counting the c chemical potentials plus the temperature and the pressure, there are $c + 2$ intensive variables *for each phase;* thus for p phases we have $p(c + 2)$ intensive variables in all:

$$\left.\begin{matrix} T_\alpha, P_\alpha, \mu_{1\alpha}, \mu_{2\alpha}, \ldots, \mu_{c\alpha} \\ T_\beta, P_\beta, \mu_{1\beta}, \mu_{2\beta}, \ldots, \mu_{c\beta} \\ T_p, P_p, \mu_{1p}, \mu_{2p}, \ldots, \mu_{cp} \end{matrix}\right\} \quad \begin{matrix} p(c+2) \\ \text{intensive variables} \end{matrix}$$

At equilibrium we have the following relationships among the above variables:

$$\left.\begin{matrix} T_\alpha & = T_\beta & = \cdots = & T_p \\ P_\alpha & = P_\beta & = \cdots = & P_p \\ \mu_{1\alpha} & = \mu_{1\beta} & = \cdots = & \mu_{1p} \\ \mu_{2\alpha} & = \mu_{2\beta} & = \cdots = & \mu_{2p} \\ \vdots & \vdots & \vdots & \vdots \\ \mu_{c\alpha} & = \mu_{c\beta} & = \cdots = & \mu_{cp} \end{matrix}\right\} \quad \begin{matrix} (p-1)(c+2) \\ \text{relationships} \end{matrix}$$

Application of the Gibbs-Duhem equation to each of the p phases yields

$$\left.\begin{matrix} S_\alpha\, dT_\alpha - V_\alpha\, dP_\alpha + \sum_{i=1}^{c} n_{i\alpha}\, d\mu_{i\alpha} = 0 \\ S_\beta\, dT_\beta - V_\beta\, dP_\beta + \sum_{i=1}^{c} n_{i\beta}\, d\mu_{i\beta} = 0 \\ \vdots \qquad \vdots \qquad \vdots \\ S_p\, dT_p - V_p\, dP_p + \sum_{i=1}^{c} n_{ip}\, d\mu_{ip} = 0 \end{matrix}\right\} \quad p \text{ relationships}$$

The number of independent intensive variables (number of degrees of freedom) f is thus

$$f = \begin{pmatrix} \text{total number of} \\ \text{intensive} \\ \text{state variables} \end{pmatrix} - \begin{pmatrix} \text{total number of} \\ \text{relationships among} \\ \text{the intensive state variables} \end{pmatrix}$$

that is,

$$f = p(c + 2) - [(p - 1)(c + 2) + p]$$

and

$$f = c - p + 2 \tag{8-28}$$

Equation 8-28 is known as *the phase rule.*

From Eq. 8-28 we compute for a $H_2O(\ell)$ in equilibrium with $H_2O(g)$

$$f = 1 - 2 + 2 = 1$$

The significance of the value $f = 1$ is that we can choose, for example, the value of the equilibrium temperature *or* the equilibrium pressure but not both. In other words, pure liquid water has a unique equilibrium pressure at each temperature. At the triple point of water, $H_2O(\ell)$, $H_2O(g)$, and $H_2O(s)$ are all in equilibrium, and

$$f = 1 - 3 + 2 = 0$$

Thus the triple point is a unique point; we are not free to pick either P or T at the triple point.

EXAMPLE 8-4

Compute the number of degrees of freedom for a solution of acetic acid, $CH_3COOH(aq)$, in water.

Solution: From the text discussion in Section 8-1, we know that $c = 2$ for the $CH_3COOH(aq)$ system, which consists of a single phase; therefore,

$$f = 2 - 1 + 2 = 3$$

Thus we must specify, for example, the temperature, the pressure, and the concentration of acetic acid in order to determine an equilibrium state of the $CH_3COOH(aq)$ system.

If the application of the phase rule to a system yields a negative value for f, then such a system is not capable of having all the phases enumerated coexisting in equilibrium. Thus two solid phases and a liquid and vapor phase of a pure substance cannot coexist in equilibrium in systems of the type under discussion.

The application of the phase rule does not require a knowledge of the *actual* constituents of a phase and, conversely, the phase rule does not yield any information about the nature of the molecular species in a system. Deductions about the internal structure of a phase necessarily involve a mechanistic viewpoint and equilibrium thermodynamics gives no information regarding the mechanisms of processes. The phase rule applies only to equilibrium systems and, like all other equilibrium thermodynamic equations, provides no information whatsoever regarding the rate of attainment of an equilibrium state.

8-5 *A Phase Diagram Displays the Phase Equilibrium Lines of a Substance*

Equilibrium Vapor Pressure Curve for Water

In Figure 8-2 we have displayed the low-pressure part of the phase diagram of water. The vapor pressure curve of liquid water is the locus of points representing states for which $H_2O(\ell)$ and $H_2O(g)$ are in equilibrium. The vapor

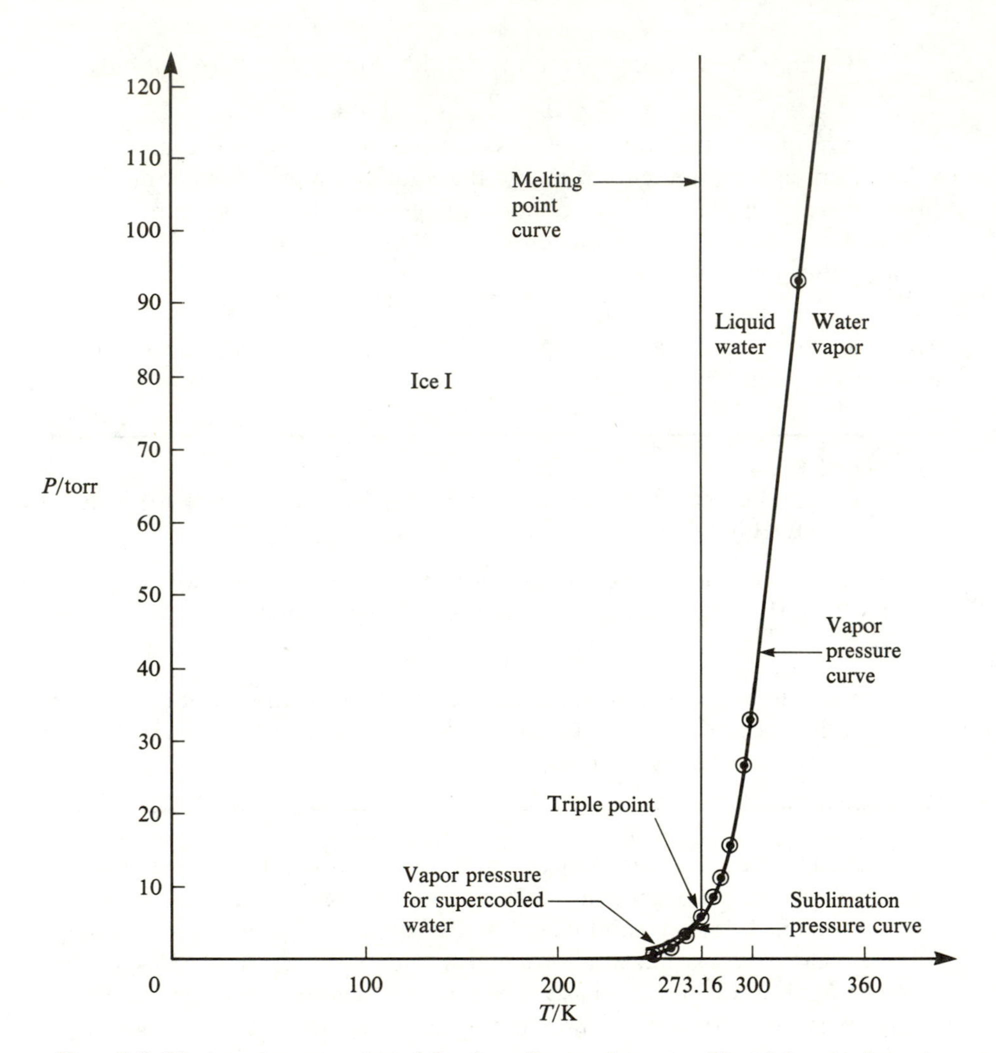

Figure 8-2. The low-pressure section of the phase diagram for water. The triple point for water occurs at 4.58 torr, 273.16 K.

pressure curve can be represented satisfactorily over a moderate range of temperature (say, 10°C–50°C, depending on the desired precision) by an equation of the form

$$P = be^{-a/T} \tag{8-29}$$

where a and b are constants characteristic of the liquid.

The *normal boiling point* of the liquid is defined as the temperature at which the vapor pressure of the liquid equals 1 standard atmosphere (101.325 kPa); thus the *normal* boiling point of a pure liquid can be placed unambiguously on the vapor pressure curve (Figure 8-3). Because a liquid open to the atmosphere boils when its equilibrium vapor pressure equals atmospheric pressure, the boiling point of a liquid is not a constant but depends on the atmospheric

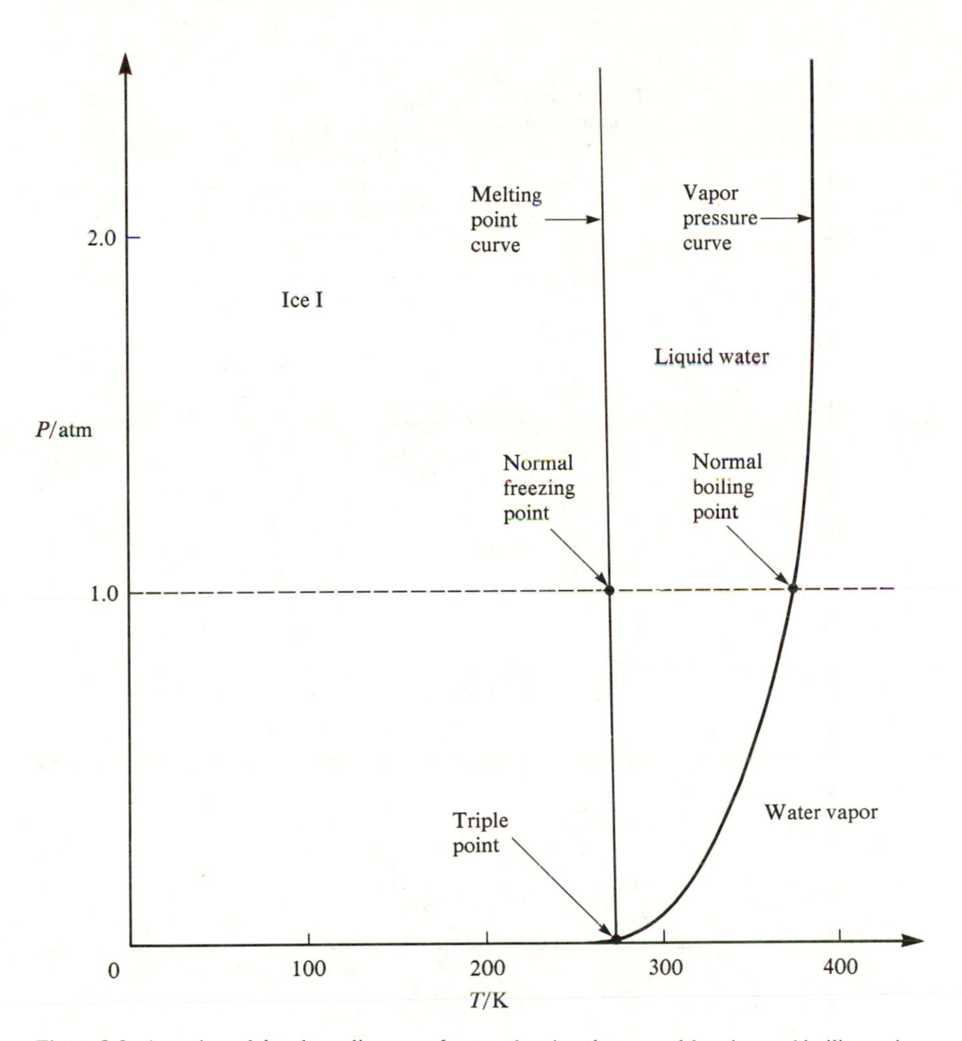

Figure 8-3. A section of the phase diagram of water showing the normal freezing and boiling points of water.

pressure. The boiling point of water at sea level is about 100°C and at 8000-ft elevation it is about 92°C. The vapor pressure curve ends abruptly at the critical point (see Figure 8-4).

Suppose a sample of gaseous water at 1.00-atm (760 torr) pressure and 120°C (463 K) is enclosed in a chamber equipped with a piston and immersed in a heat bath at 120°C. As the piston is pushed into the cylinder, the gas pressure increases until a pressure of 2.00 atm is reached (Figure 8-4). At 2.00-atm pressure the vapor begins to condense; if the piston is depressed further, then the gas pressure does not increase but rather vapor continues to condense until the pressure once again attains the equilibrium value. Continued depression of the piston produces no increase in the pressure of the

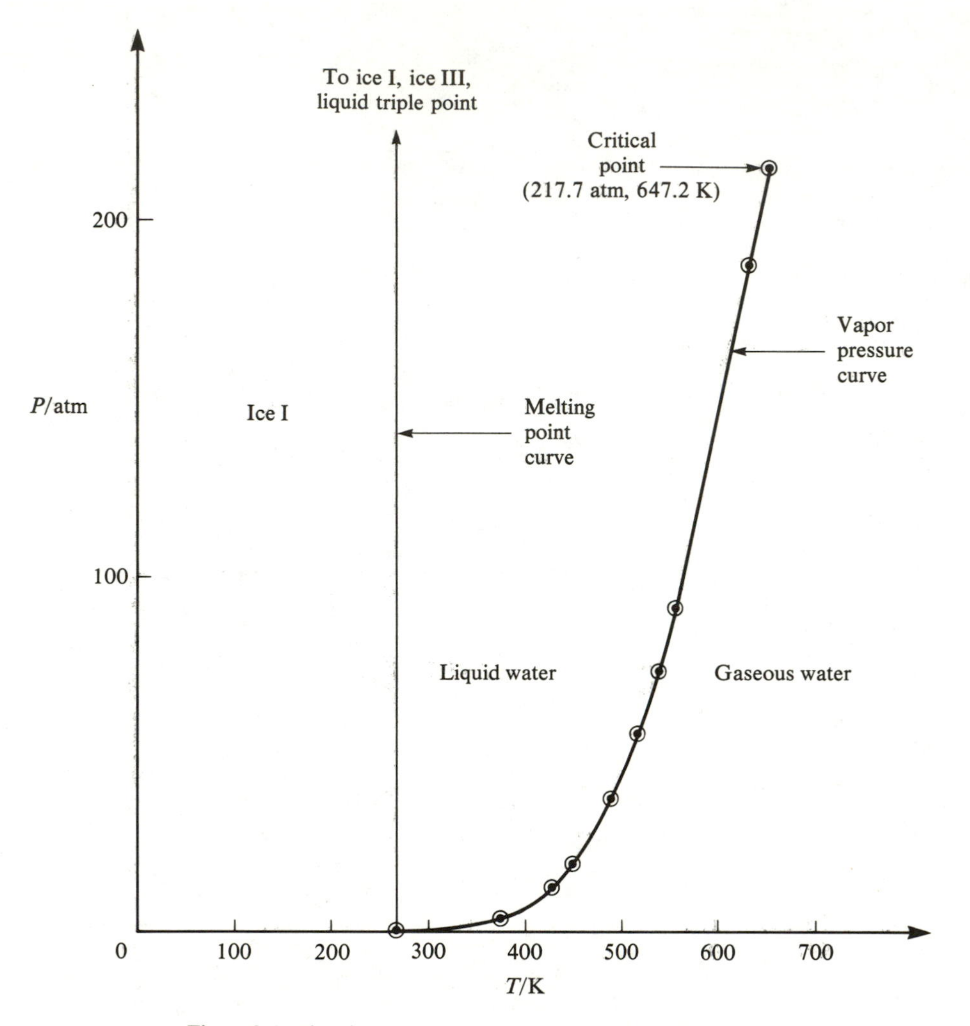

Figure 8-4. The phase diagram for water up to the critical point.

system until *all* the vapor has condensed to liquid, at which point the pressure begins to rise again as the piston is further depressed. The foregoing sequence of events occurs for any $H_2O(g)$ sample with a temperature between 273.16 K and 647.2 K. However, the lower the temperature of the gas, the lower the value of the pressure at which the liquid first appears. On the other hand, if a sample of $H_2O(\ell)$ is injected into an evacuated chamber maintained at a temperature lying between 273.16 K and 647.2 K, then the liquid will spontaneously evaporate until either the equilibrium value of the vapor pressure of water is attained or the liquid phase is exhausted.

The fact that $H_2O(\ell)$ has a definite equilibrium vapor pressure at all temperatures from 273.16 K to 647.2 K is the basis of the *percent relative humidity scale* used in meteorology. Thus, if the vapor pressure of water in

TABLE 8-1
Phase Equilibrium Data for Water[a]

$H_2O(s) \rightleftharpoons H_2O(g)$		$H_2O(\ell) \rightleftharpoons H_2O(g)$		$H_2O(\ell) \rightleftharpoons H_2O(g)$		$H_2O(s, I) \rightleftharpoons H_2O(\ell)$	
t/°C	P/torr	t/°C	P/torr	t/°C	P/atm	t/°C	P/atm
−98	0.000015	−16	1.34	100	1.00	0	0
−90	0.00007	−10	2.149	120.1	2.00	−5	590
−80	0.0004	0	4.578	152.4	5.00	−10	1093
−70	0.00194	0.01	4.580	180.5	10.00	−15	1539
−60	0.00808	10	9.209	213.4	20.00	−20	1907
−50	0.02955	20	17.535	251.1	40.00	−22	2047
−40	0.0966	25	23.756	276.5	60.00		
−30	0.2859	30	31.824	300.0	84.80		
−20	0.776	40	55.324	350.0	163.2		
−10	1.950	50	92.51	374.2	218.3		
−5	3.013	60	149.38				
0	4.579	70	233.7				
		80	355.1				
		90	525.76				
		100	760.00				

[a] Data on sublimation and vaporization from *CRC Handbook of Chemistry and Physics*, 48th ed., edited by Robert C. Weast and others. Cleveland; Chemical Rubber Co., 1967. Data on melting from *The Phase Rule*, 9th ed., by A. Findlay, A. N. Campbell, and N. O. Smith, Dover Publications, Inc. 760 torr = 101.3 kPa = 1.000 atm.

the atmosphere is 16.5 torr and the atmospheric temperature is 25.0°C, then the percent relative humidity (Table 8-1) is

$$\left(\frac{16.5 \text{ torr}}{23.76 \text{ torr}}\right) \times 100 = 69\%$$

If the atmospheric temperature in this case were to decrease to about 19°C, then the air would become saturated with water vapor and condensation would occur. Thus air containing 16.5 torr of water vapor is said to have a *dew point* of 19°C. A related phenomenon is the condensation of water vapor on objects at a lower temperature than the surroundings. For example, as in the case discussed above, if the object has sufficient heat capacity and thermal conductivity to lower the temperature of air immediately surrounding the object to at least 19°C, and the air contains 16.5 torr of water vapor, then condensation of water vapor will occur on the object's surface.

A sample of water at 25°C in contact with air at 25°C containing less than the equilibrium partial pressure of water (23.76 torr at 25°C) will evaporate spontaneously until either the air in contact with the liquid is saturated with water vapor or the liquid phase is exhausted by evaporation. The evaporation of water at 25°C requires 44.0 kJ·mol^{-1} of energy as heat. The evaporation of water from the surface of the body is one of the mechanisms used to cool the human body (evaporative cooling).

A gas above its critical temperature (647.2 K for water) (Figure 8-4) does not exhibit a phase change to liquid (cannot be liquefied) no matter how great the pressure is made. This situation is depicted graphically in Figure 8-5.

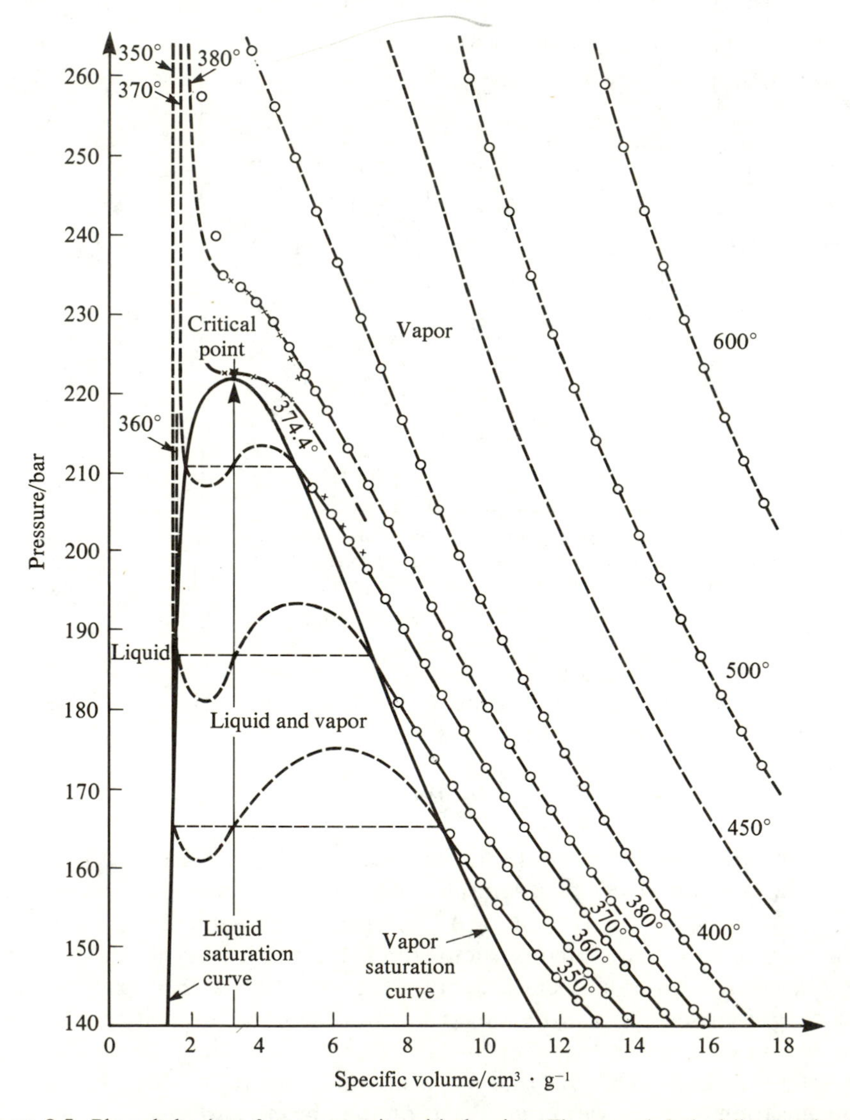

Figure 8-5. Phase behavior of water near its critical point. The curved dashed lines under the parabola represent the behavior of water treated as a van der Waals gas. Note that the plot consists of P versus $\bar{V}$ isotherms. The temperatures given are in degrees Celsius, and the pressures are in bar (1 bar = 100 kPa). (From Eisenberg and Kauzmann, 1969.)

In Figure 8-5, the area enclosed by the parabola is the two-phase liquid-gas region. Of particular interest is the P versus $\bar{V}$ curve that passes through the critical point (P_c, $\bar{V}_c$). The critical point on a P-$\bar{V}$ diagram is a horizontal inflection point, and, therefore, *at the critical point*

$$\left(\frac{\partial P}{\partial V}\right)_{T=T_c} = 0 \quad \text{and} \quad \left(\frac{\partial^2 P}{\partial V^2}\right)_{T=T_c} = 0 \tag{8-30}$$

The conditions expressed in Eq. 8-30 permit us to evaluate P_c, T_c, and V_c, given the equation of state of the gas; conversely, an equation of state that yields zero values for P_c, T_c, V_c is incapable of representing the behavior of a gas in the neighborhood of its critical point.

EXAMPLE 8-5

Obtain expressions for the critical constants of a van der Waals gas,

$$P = \frac{nRT}{V - nb} - \frac{n^2 a}{V^2}$$

where a, b, and R are positive constants.

Solution: From the equation of state, we obtain

$$\left(\frac{\partial P}{\partial V}\right)_T = \frac{-nRT}{(V - nb)^2} + \frac{2n^2 a}{V^3}$$

$$\left(\frac{\partial^2 P}{\partial V^2}\right)_T = \frac{2nRT}{(V - nb)^3} - \frac{6n^2 a}{V^4}$$

At the critical point we have from Eq. 8-30

$$\frac{-nRT_c}{(V_c - nb)^2} + \frac{2n^2 a}{V_c^3} = 0$$

and

$$\frac{2nRT_c}{(V_c - nb)^3} - \frac{6n^2 a}{V_c^4} = 0$$

Dividing the second equation into the first and solving for V_c yields

$$V_c = 3nb$$

Substitution of the result for V_c into the first equation yields for the critical temperature

$$T_c = \frac{8a}{27bR}$$

Substitution of the results for V_c and T_c into the equation of state yields for the critical pressure

$$P_c = \frac{4a}{27b^2} - \frac{a}{9b^2} = \frac{a}{27b^2}$$

The critical point constants (P_c, V_c, T_c) of a gas can be used to calculate a and b for a van der Waals gas.

Equilibrium Sublimation Pressure Curve

In Figure 8-2, the sublimation pressure curve, which extends from 0 K to the triple point (273.16 K), is the locus of all points representing states for which $H_2O(s)$ is in equilibrium with $H_2O(g)$. The sublimation curve can be represented analytically by an equation of the same form as the vapor pressure curve. At each temperature in the range 0 K to 273.15 K there is a unique value of the vapor pressure of ice for which equilibrium between the gas and solid exists. Thus, suppose we have a sample of $H_2O(g)$ at 2.0 torr and 268.15 K contained in a chamber fitted with a piston and immersed in a heat bath at 268.15 K. As the piston is steadily depressed, the $H_2O(g)$ pressure rises steadily until the pressure of the gas reaches 3.013 torr (Table 8-1) at which value of the pressure further depression of the piston leads to the formation of $H_2O(s)$ without further increase in pressure. The phase transformation of $H_2O(g)$ to $H_2O(s)$ continues until all the vapor phase is converted to solid, at which point further depression of the piston again leads to increased pressure. Conversely, if $H_2O(s)$ at 268.15 K is placed in a sealed evacuated space and maintained at 268.15 K, then the solid will spontaneously begin to sublime and will continue to do so until either the vapor pressure reaches the value 3.013 torr or the solid phase is exhausted. On the other hand, if the chamber is continuously swept out with carrier gas, thereby preventing the establishment of equilibrium with the $H_2O(s)$, then $H_2O(s)$ will continue to sublime until no ice remains.

The extension of the vapor pressure curve below the triple point of water lies *above* the sublimation pressure curve. Provided that care is taken to remove dust and other solid particles from the water and also to avoid agitation, then liquid water can be cooled to temperatures below the triple point of water. Liquid water below the triple point is said to be *supercooled*. Liquid water can be supercooled to $-45°C$. Supercooled liquid water has a higher vapor pressure than ice at the same temperature and thus supercooled water is thermodynamically unstable with respect to ice at the same temperature. If a small ice crystal is injected into a sample of supercooled water, or if the water is supercooled even further, then the water will turn into ice. Dust particles, or rough surfaces, act as nuclei where ice crystals can begin to form. Some liquids are particularly easy to supercool, and no special precautions, other than the absence of seed crystals, are necessary. An example is glycerin, which can be supercooled to the lowest attainable temperatures. However, once glycerin has been crystallized in a particular laboratory, it is extremely difficult to ever produce supercooled glycerin in that laboratory again, presumably because the presence of miniscule seed crystals in the laboratory.

The supercooled liquid-vapor equilibrium is *metastable*. For example, suppose we have $H_2O(\ell)$ at $-5°C$ in equilibrium with $H_2O(g)$, $P_{eq} = 3.163$ torr. The equilibrium vapor pressure of $H_2O(s)$ at $-5°C$ is 3.013 torr. Thus, if we place separate samples of $H_2O(\ell)$ and $H_2O(s)$ in a closed container at $-5°C$, then the $H_2O(\ell)$ phase will eventually be exhausted, because $H_2O(g)$ at any $P > 3.013$ torr (for $t = -5°C$) will deposit spontaneously on the $H_2O(s)$.

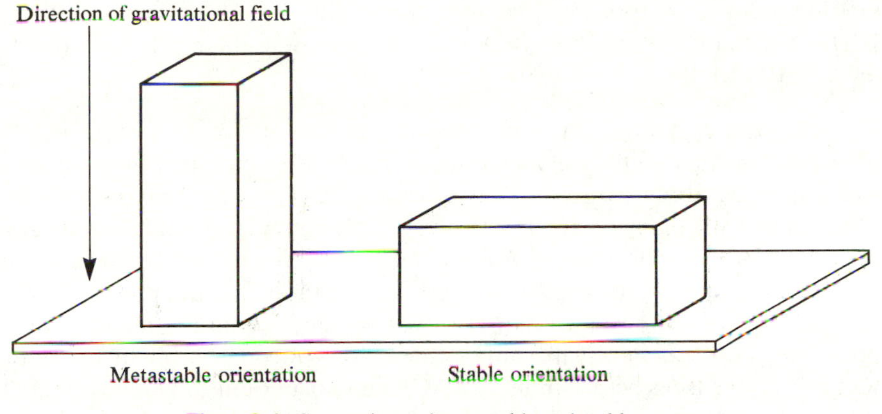

Figure 8-6. Comparison of metastable and stable states.

A useful mechanical analogy can be made between the equilibrium states of a rectangular parallelepiped and the terms *metastable* and *stable* as used in thermodynamics (see Figure 8-6). Such an object in state A can be subjected to small displacements from its equilibrium position without falling over; however, with sufficiently large displacements the object will not return to state A but rather will go over to the more stable state B. Comparatively large displacements from equilibrium to state B will not return the object to state A.

Liquid water can be *superheated*, that is, heated above the boiling point without boiling, provided care is taken to remove all dust particles and rough surfaces that act as nuclei for bubble formation. Water at 1 atm can be superheated to over 200°C without boiling. The bubbles that form in water as it is heated well below the boiling point arise from the evolution of dissolved gases (mostly N_2 and O_2).

The Equilibrium Melting Point Curve

The *melting point curve* is the locus of points representing states in which $H_2O(s)$ and $H_2O(\ell)$ coexist in equilibrium (Figures 8-2, 8-3, and 8-4). On the melting point curve, we have $f = 1 - 2 + 2 = 1$, and thus there is a unique value of the pressure for a given value of the temperature at which equilibrium obtains. It is worth noting that the pressure in the case of solid-liquid equilibrium, as opposed to solid-gas and liquid-gas equilibria, does not refer to the $H_2O(g)$ pressure. The pressure in this case acting on the condensed phases can be produced by a piston. An important feature of the melting point curve for water is its negative slope, $\bar{V}(\text{solid}) > \bar{V}(\text{liquid})$. Because the melting point curve for water has a negative slope, the equilibrium value of T decreases with increasing P, or, in other words, the melting point of Ice I is lower the higher the pressure. The melting point of ice decreases by about 0.01 K/atm. Thus if we have an ice cube at a temperature between 0°C and about -15°C and squeeze it hard enough (>1600 atm at -15°C), it will melt. A decrease in melting point with increasing pressure is very rare [other examples are Ga, Sb, Bi, Fe, Ge, C (diamond), and certain alloys containing these substances and a few compounds].

The overwhelming majority of substances have $\bar{V}_s < \bar{V}_l$ and thus exhibit increasing melting point with increasing pressure. The melting points for most solids increase about 0.01 K to 0.03 K/atm. In fact, six solid phases of water other than Ice I (namely, Ice II, III, IV, V, VI, and VII, Figure 8-7) all have $\bar{V}_s < \bar{V}_l$. For example, Ice V at 0°C and 6000 atm pressure sinks in $H_2O(\ell)$ at the same conditions. Below about −22°C ice cannot be brought into stable equilibrium with $H_2O(\ell)$ no matter how high the pressure is made. This is presumably the reason why it is impossible to ice-skate in very low-temperature weather. If the ice is below about −22°C, then no amount of blade pressure is sufficient to liquefy the ice, and the skate blades tend to stick, owing to the fact that a thin layer of water is not produced to aid in lubrication. The sensation is somewhat like trying to skate on glass.

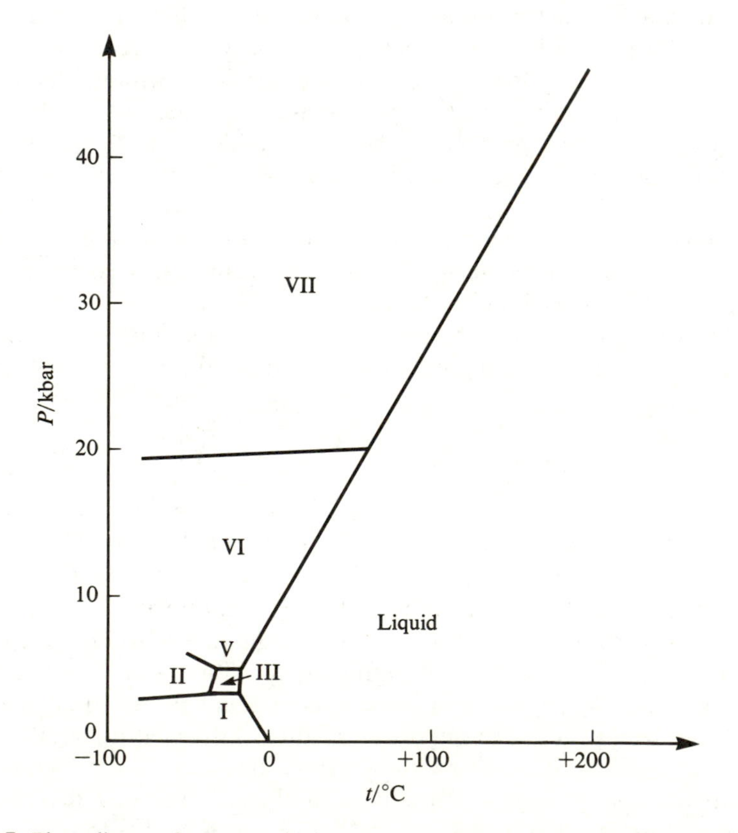

Figure 8-7. Phase diagram for water at high pressures, showing the regions of stability of the solid phases of water. (Data from J. W. Stewart, *The World of High Pressure*, Princeton, N.J.: D. van Nostrand, 1967.)

The point common to the vapor pressure, sublimation pressure, and melting point curves (point T, Figure 8-2) is the *triple point*. At the triple point, and only at the triple point ($f = 1 - 3 + 2 = 0$), $H_2O(s, \text{I})$, $H_2O(\ell)$, and $H_2O(g)$ can coexist in stable equilibrium with one another. For water the triple point occurs at 4.58 torr pressure of water vapor and 273.16 K. The triple point of water is the single most important fixed point in thermometry (Chapter 2). Water at its triple point can be produced by placing $H_2O(\ell)$ in an evacuated chamber that is then sealed. A part of the liquid is then frozen and the solid and liquid phases are allowed to come to equilibrium with the vapor phase. The temperature of the resulting three-phase system is exactly 273.16 K by definition. The water system has other triple points at higher pressures involving other phases (see Figure 8-7).

8-6 *Phase Diagram for Carbon Dioxide*

The carbon dioxide phase diagram is given in Figure 8-8. There are several noteworthy differences between this phase diagram and that for the water system. The first of these is that the melting point curve TX has a positive slope ($\bar{V}_\ell > \bar{V}_s$), and thus the melting point increases with increasing pressure. The second is that the solid-liquid-vapor triple point occurs above 1 atm (at 5.11 atm), and therefore CO_2 does not possess a normal melting or normal boiling point. Below 5.11 atm, CO_2 does not melt as we raise the temperature, but rather it sublimes—thus the designation "dry ice." However, as evident from the phase diagram, CO_2 does possess a normal sublimation point (which water does not) at 194.7 K. The normal sublimation point is the approximate temperature of $CO_2(s)$ stored in closed but not tightly sealed containers. If sufficient $CO_2(g)$ is pumped into a tank held at a temperature in the range 216.6 K to 304.2 K, then $CO_2(\ell)$ will be formed when the pressure of $CO_2(g)$ reaches the equilibrium value. Thus at 23.0°C (296.2 K) the pressure of CO_2 in a tank cannot be made to exceed 60.6 atm, providing there is a vapor space. As long as some $CO_2(\ell)$ remains in the tank, the tank pressure will remain at 60.6 atm, if the temperature is fixed at 23.0°C. (This is to be contrasted with, for example, tank N_2 or Ar, both of which are well above their critical points at room temperature and are sold as compressed gases.) As $CO_2(g)$ is withdrawn from the tank, more of the liquid evaporates in the process of restoring equilibrium.

Solid CO_2 can be produced from tank CO_2 (liquid plus gas) by allowing the gas to escape rapidly into an insulated partially enclosed space open to the atmosphere. Under such conditions the expansion of the gas is more or less adiabatic. The Joule-Thomson coefficient of CO_2 around 300 K is about $1\ \text{K}\cdot\text{atm}^{-1}$; a 60-atm pressure drop produces about a 60 K drop in temperature. As the continuously escaping gas reaches still lower T, owing to evaporative cooling in the tank, the temperature of the exiting gas eventually attains the value 194.9 K, and solid CO_2 begins to form.

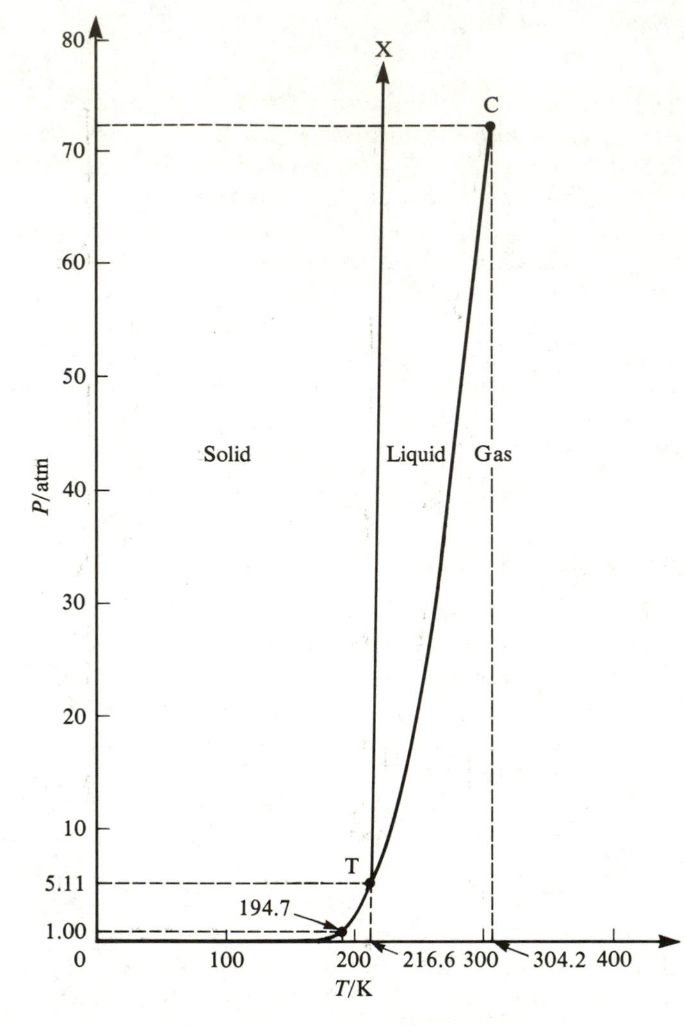

Figure 8-8. The phase diagram for CO_2.

Phase Equilibrium Data for Carbon Dioxide[a]

$CO_2(s) \rightleftharpoons CO_2(g)$		$CO_2(\ell) \rightleftharpoons CO_2(g)$		$CO_2(s) \rightleftharpoons CO_2(\ell)$	
t/°C	P/torr	t/°C	P/torr	t/°C	P/torr
−180	0.000013	−50	5128	−56.6	5.11
−160	0.0059	−40	7548		
−140	0.431	−30	10718		
−120	9.81	−20	14781		
−100	104.81	−10	19872		
−80	672.2	0	26142		
−78.2	760.0	10	33763		
−70	1486.1	20	42959		
−60	3073.1	30	54086		

[a] Data from *CRC Handbook of Chemistry and Physics*, 48th ed., edited by Robert C. Weast and others. Cleveland: Chemical Rubber Co., 1967.

An interesting use of dry ice is in the preparation of crushed dry-ice–acetone cold baths. Such a system has two components (CO_2 and CH_3COCH_3) and three phases [$CO_2(s)$, acetone saturated with carbon dioxide, and a gas-phase CO_2 plus CH_3COCH_3); therefore, $f = 2 - 3 + 2 = 1$. If we now fix the pressure at 1 atm, then such a system can exist in equilibrium at only one temperature. The gas phase is essentially pure CO_2, and the $CO_2(s)$ phase is pure; thus the temperature of the system can be read from the pure carbon dioxide phase diagram as 194.9 K (−78.2°C). Because the variation with temperature of the $CO_2(g)$ pressure in equilibrium with solid at this temperature is large (792.7 torr at −78.0°C and 730.3 torr at −79.0°C), fluctuations in atmospheric pressure, say, within 30 torr, either side of 760 torr will produce at maximum a 0.5°C variation in the cold-bath temperature. The only requirements for the substance used in preparing the slush (acetone in the case above) are that its normal freezing point be at least a few degrees below −78°C, that it not dissolve in $CO_2(s)$ to any appreciable extent, and that its equilibrium vapor pressure be less than about 1 torr at −78°C. The first and third requirements ensure that a liquid phase will persist at −78°C and that the vapor will be essentially pure CO_2. The second requirement is of consequence only if it is desired to read the temperature of the bath off the CO_2 phase diagram. Acetone and isopropyl alcohol meet these conditions nicely with normal freezing points of −95°C and −89°C and temperatures at which P_{eq} is 1 torr of −59°C and −26°C, respectively.

The reproducibility of the temperature of a dry-ice–acetone cold bath close to 1 atm pressure exceeds the accuracy with which its temperature can be measured using uncalibrated liquid-in-glass capillary or thermocouple thermometers. In principle, "cold" baths at other fixed temperatures can be prepared in the same manner using for the solid phase a substance whose solid-liquid-vapor triple point lies above 1 atm. Other cold baths (or "hot" baths for that matter) can be prepared by allowing any pure substance to reach its normal melting or boiling point and maintaining the presence of the two phases.

Below the critical point the continued compression of a gas leads to condensation (formation of liquid) when the equilibrium vapor pressure is attained. At a higher temperature the gas must be compressed to a higher pressure before condensation occurs, and the liquid that condenses has a lower density the higher the temperature. Thus the densities of the coexisting liquid and vapor phases become closer as the temperature increases. The critical point occurs when the densities of the coexisting phases become equal. For CO_2 this occurs at 31.8°C, 72.9 atm. Because the densities of the liquid and gas are equal (the *critical density*), the two phases are no longer distinct and the separating meniscus disappears. Thus the liquid-vapor equilibrium curve terminates abruptly at the critical point. The enthalpy of vaporization of a liquid also decreases rapidly as the critical point is approached, and $\Delta H_{vap} = 0$ at the critical point.

As the critical point is approached, the density becomes very sensitive to small variations in pressure and temperature because of fluctuations of regions

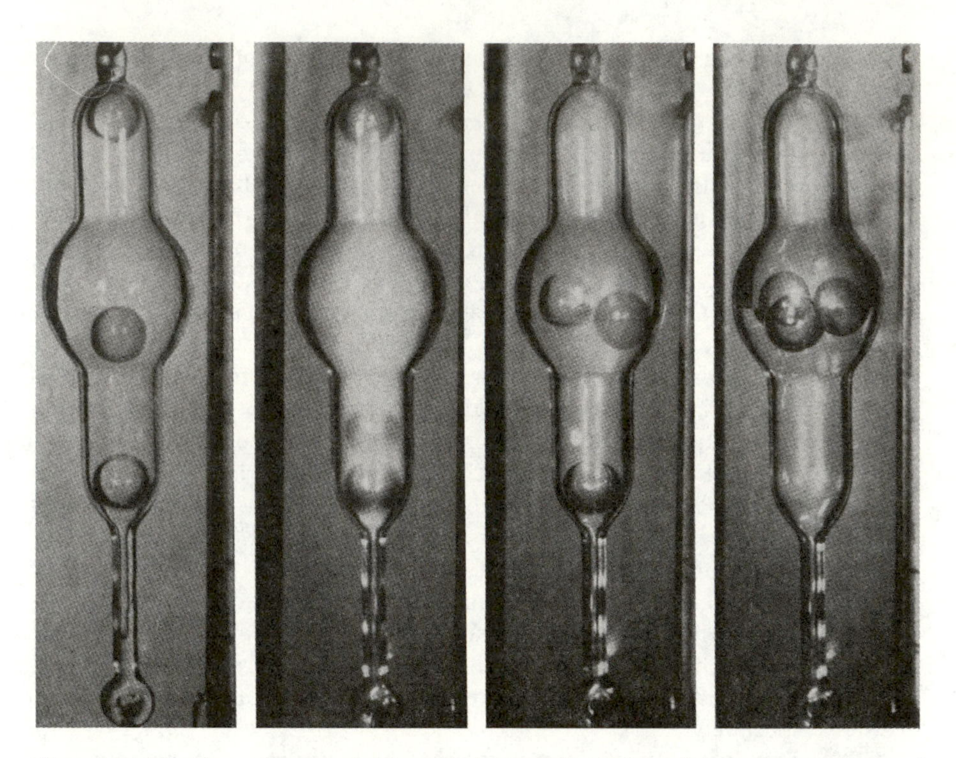

Figure 8-9. A fluid near its critical point exhibits anomalously large light scattering because of large fluctuations in the refractive index. The illustration shows a glass bulb constructed by E. F. Mueller and his co-workers at the National Bureau of Standards; it has been used for more than 25 years to demonstrate critical opalescence.

The glass bulb is filled with carbon dioxide with an average density close to the critical density ρ_c (pressure > 68 atm). The cell contains three balls with densities smaller than, approximately equal to, and larger than ρ_c. These balls indicate the variation of density with position and temperature. The four successive stages correspond to states at temperatures from slightly above to well below the critical point. In the first stage, at far left, carbon dioxide is in the gaseous phase at a temperature somewhat higher than the critical. The scattered light is sufficiently intense to be visible, the beginning of critical opalescence. The density is still rather uniform; thus the three balls are well separated, and the ball with density ρ_c floats in the center.

When the temperature is lowered to just above the critical, as in the second stage, opalescence becomes very intense. Since the expansion coefficient is large, the density distribution is very sensitive to temperature gradients; thus the middle ball is not in the center of the cell. As soon as the temperature passes through the critical temperature; a meniscus develops separating the gaseous and the liquid states. This meniscus is apparent in the third stage at the center of the middle ball. Since the density in the upper half of the cell corresponding to the gaseous state has decreased considerably, the lighter ball originally at the top of the cell has fallen to the meniscus. The local fluctuations of the density in both the gaseous and liquid phases are still very large, and critical opalescence persists below the critical point.

In the last stage, shown below at the far right, the temperature is well below the critical temperature. The density of the liquid state has increased, and all three balls are floating on the surface of the liquid. The fluctuations are now small, and scattering is no longer visible.

(Figure and caption from J. V. Sengers and A. L. Sengers, *Chemical and Engineering News*, June 10, 1968, p. 105, by permission.)

of the sample between the liquid and gaseous states. The isothermal compressibility

$$\beta = -\frac{1}{V}\left(\frac{\partial V}{\partial P}\right)_T$$

is infinite at the critical point, because at the critical point $(\partial P/\partial V)_T = 0$. The large compressibilities near the critical point give rise to large values of the heat capacity $\bar{C}_P$, to anomalous light scattering (*critical opalescence*, Figure 8-9), and to density gradients in the gravitational field. For CO_2 at its critical point (31.8°C, 72.9 atm), a pressure increase of only 1 torr or 1 part in 50,000 causes a density increase of more than 10%. The density of CO_2 at the critical point is about 0.5 $g \cdot cm^{-3}$; thus 1 torr of CO_2 corresponds to 1 in. In other words, a 1-in.-high column of CO_2 at its critical point will exhibit a 10% greater density at the bottom than the top, simply as a result of the weight of the gas.

8-7 *High Pressure Favors the Crystal Structure with the Smaller Molar Volume*

Numerous solids can exist in more than one crystalline form. The different forms have different physical properties, including different densities. Because high pressure favors the solid modification of higher density (smaller molar volume), the high density form is the more stable form at high pressure. For example, solid carbon is capable of existing as graphite ($\rho = 2.25\ g \cdot cm^{-3}$) or diamond ($\rho = 3.51\ g \cdot cm^{-3}$). The crystal structures of graphite and diamond are shown in Figure 8-10, and the graphite-diamond phase equilibrium curve is given in Figure 8-11.

Graphite is the stable form of carbon at ordinary temperatures and pressures. Graphite exists as a solid at a higher temperature than any other material known (4100 K). Thomas Edison discovered that when very pure graphite is heated above 1500°C and compressed, a material called *pyrolytic graphite* results. Pyrolytic graphite has a high thermal conductivity along its surface and a low thermal conductivity perpendicular to its surface. Because of the unusual thermal properties and exceptional stability to elevated temperatures, graphite is used to fabricate rocket nose cones, rocket nozzles, and pipe bowls. Ordinary graphite is used as the "lead" in lead pencils. The layered structure of graphite, together with the high thermal stability, makes graphite an excellent solid lubricant. The relatively weak interaction between adjacent layers in the graphite structure enables the layers to slip past one another with a minimum of applied force.

The graphite-diamond phase equilibrium curve in Figure 8-11 defines the minimum conditions necessary for the productions of diamonds from graphite. Diamond is the hardest substance known, and consequently diamond is extensively used as an abrasive and cutting material where a very high resistance to wear is required. Diamond is also the least compressible substance known with $\beta = 1.8 \times 10^{-7}\ atm^{-1}$. The industrial market for diamonds is much greater than the market for diamonds as jewels.

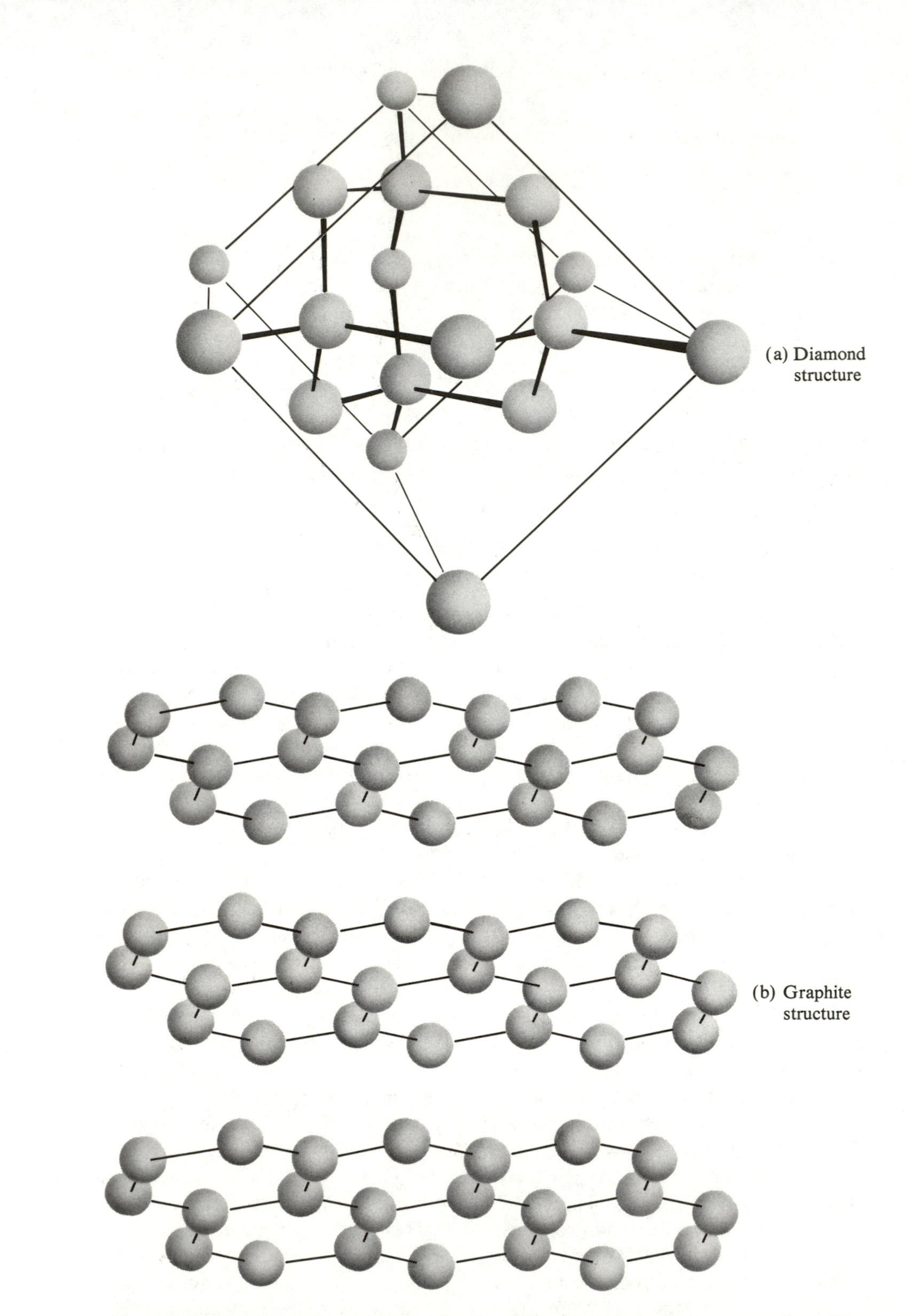

Figure 8-10. Crystal structures of diamond and graphite. Note the layered structure of graphite. The heavy lines in (a) represent bonds.

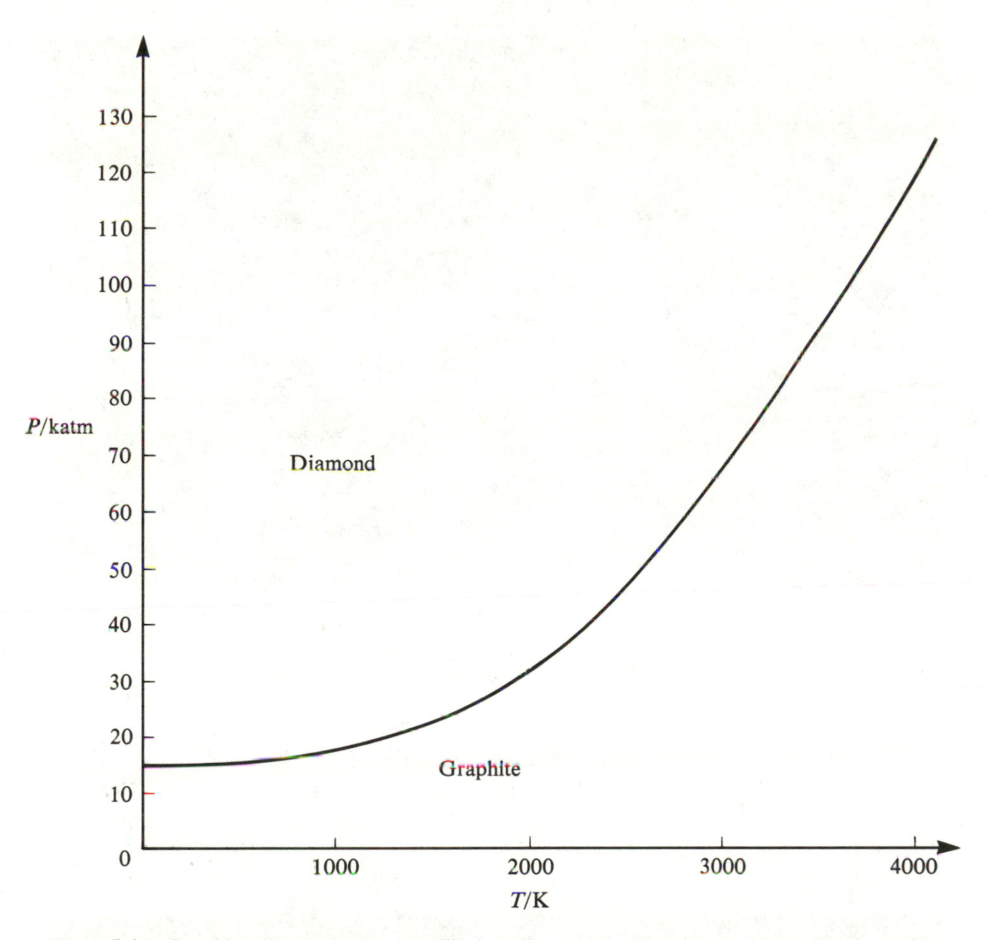

Figure 8-11. Graphite-diamond phase equilibrium line. Above the curve, diamond is the stable form of solid carbon; below the curve, graphite is the stable form of solid carbon. At 25°C, the equilibrium transition pressure for the conversion of graphite to diamond is 15,000 atm. Diamond has a higher density than graphite, and hence diamond is the stable form of C(s) at high pressures. Diamonds at 1 atm are unstable with respect to graphite, but the rate of conversion of diamonds to graphite at 25°C is completely negligible.

To produce diamonds from graphite it is necessary to subject graphite to a pressure and temperature that lies above the graphite-diamond equilibrium line in Figure 8-11. For example, at 300 K a pressure greater than 15,000 atm is required. Note that high pressure favors the solid form with the higher density (smaller molar volume). However, at 300 K and 15,000 atm, the rate of conversion of graphite to diamond is extremely slow. Higher pressures and temperatures are required to achieve the conversion in a reasonable time. The first successful synthesis of diamonds from graphite was carried out at General Electric Laboratories in 1955. At 2500 K and 150,000 atm, essentially complete conversion of graphite to diamond occurs in a few minutes. A rapid decrease in the pressure and temperature traps the carbon in the diamond

Figure 8-12. General Electric high-pressure apparatus used to make diamonds from graphite. This 1000-ton press achieves pressures up to 1.6 million psi (100,000 atm). (Courtesy of General Electric)

form. The rate of conversion of diamond to graphite at ordinary temperatures is completely negligible. However, diamonds are not forever. If diamonds are heated to 1500°C, then the conversion to graphite occurs in minutes. Figure 8-12 shows the apparatus used at General Electric Labs to generate the

Figure 8-13. A laboratory-made diamond shown just after being broken out of a superpressure chamber, where it was grown over a period of several days by subjecting Man-Made diamond powder and a metal catalyst to temperatures of 2500°F and pressures of nearly 1 million psi. (Courtesy of General Electric)

high pressures necessary for the production of diamonds from graphite. A photograph of a laboratory-made diamond is shown in Figure 8-13. Synthetic diamonds are not "artificial" diamonds; they are *real* diamonds in every scientific sense.

The production of diamonds from graphite illustrates some important points in chemical thermodynamics. The first point is that thermodynamics can be used to specify the minimum value of the pressure, at a given temperature, that is necessary to produce diamonds from graphite. However, thermodynamics gives no guarantee that if graphite is subjected to a pressure at which diamonds is the stable form, then diamond will form. The *YES* of thermodynamics is actually a *maybe*. On the other hand, the *NO* of thermodynamics is *emphatic*. There is no incantation or catalyst that can cause a sample of graphite to convert to diamond at a pressure and temperature below the graphite-to-diamond phase equilibrium line in Figure 8-11.

8-8 *Helium Becomes a Superfluid at about* 2 K

Helium is the only known substance that exhibits two different isotropic liquid phases, a normal liquid phase and a *superfluid* phase. The superfluid $He(\ell)$ phase has many interesting properties, the most startling of which is its ability to flow up the sides of containers. Thus if we partly immerse an empty container closed at the bottom into superfluid helium, then the container spontaneously fills up with liquid until the levels within and without the container are equal. If we now raise the level of the container, then the helium will flow back out over the rim of the container until the levels are again equal (assuming the container is not raised completely out of the liquid). Measurements indicate that the creeping superfluid helium film is about 5×10^{-6} cm thick and flows at a rate of about $1.2 \times 10^{-3}\ cm^3 \cdot sec^{-1} \cdot cm^{-1}$.

The phase diagram of the helium system is given in Figure 8-14. Besides the existence of the two different liquid phases, there are some other unusual features in this phase diagram. Among these are (1) the absence of a solid-liquid-vapor triple point; (2) the fact that liquid helium will not solidify at 1 atm pressure no matter how low its temperature, although liquid helium can be made to solidify at sufficiently high pressure (e.g., at 1.8 K, 30 atm pressure will suffice); (3) the presence of a *second-order phase transition line*, T_1T_2 (see Section 8-9); and (4) the existence of zero slope for the melting point curve as T approaches zero (as required by the third law, Chapter 6).

8-9 *First- and Second-Order Phase Transitions*

The line T_1T_2, called the *lambda line* (λ-line), in Figure 8-14 deserves special consideration because it is not susceptible to the same type of interpretation as the other equilibrium lines in Figure 8-14 or any of the other coexistence curves that we have considered. That is to say, the lambda line is not the locus of all points representing states for which the two phases of $He(\ell)$

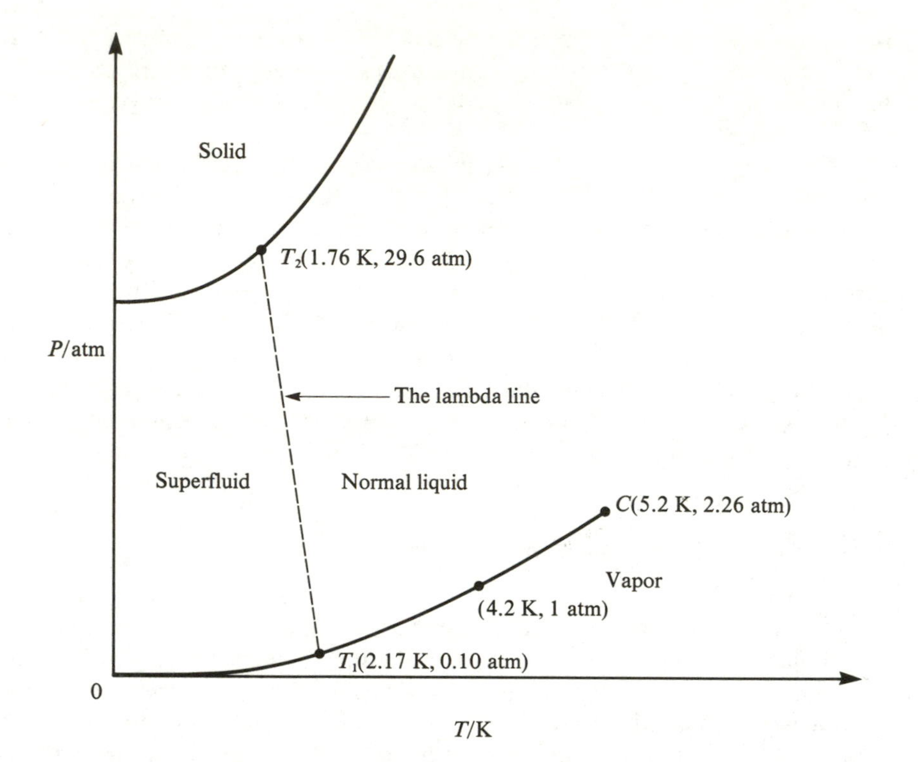

Figure 8-14. The phase diagram for helium (not to scale). The dashed line from T_1 to T_2 is the second-order phase transition line (λ-line) between normal and superfluid liquid helium.

are in equilibrium with one another; in addition, it cannot be concluded from the slope of the lambda line that $\rho_{\text{superfluid}} < \rho_{\text{liquid}}$.

The transition from normal He(ℓ) to superfluid He(ℓ) is a *second-order* phase transition as opposed to all the other phase transitions that we have considered up to this point in this chapter, which are *first-order* phase transitions. First-order phase transitions occur for a given pressure at a definite temperature and involve *phase separation* (i.e., *two* detectable distinct phases at the transition point), together with a definite volume, enthalpy, and entropy changes at the transition point. Second-order phase transitions, on the other hand, occur (for a given *P*) over a *range of temperature*; they do not involve phase separation (however, certain of the properties of the phase undergo a significant and apparently discontinuous change in the transition region); and further there are apparently no volume, enthalpy, or entropy changes at the transition point. The temperatures used to construct second-order phase transition lines are obtained from *maxima* in C_P versus T (or some other convenient property versus T) curves taken over a range of pressure (or magnetic or electric field). A comparison between the behavior of various thermodynamic quantities in first- and second-order phase transitions is presented in Figure 8-15. The origin of the designation *first-* and *second-order phase transitions* lies

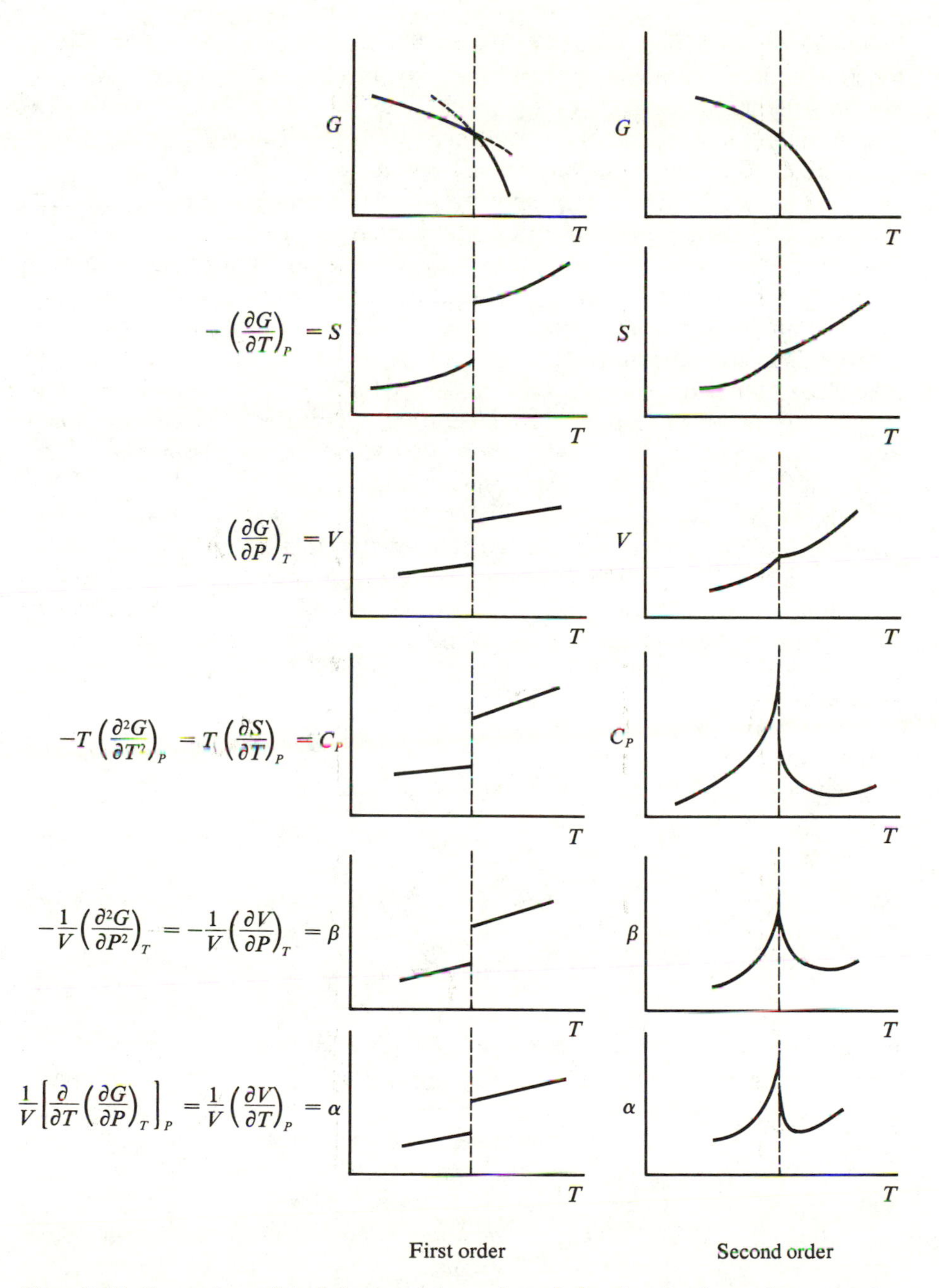

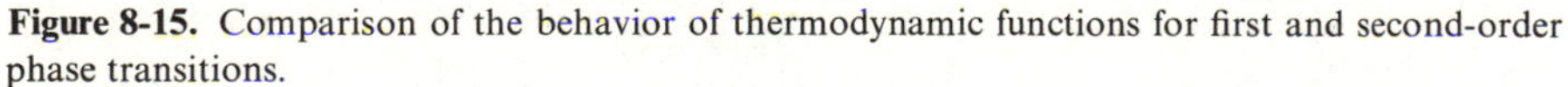

Figure 8-15. Comparison of the behavior of thermodynamic functions for first and second-order phase transitions.

in the recognition that both the first and second derivatives of the Gibbs function undergo a finite jump (step continuity) in a first-order phase transition, but only the second derivatives (C_P, β, and α) of the Gibbs function show such discontinuities in a second-order phase transition.

It remains a matter of contention, however, as to whether or not C_P, α, and β actually exhibit a step as opposed to an infinite discontinuity during a second-order phase transition or for that matter how precisely the transition temperature can be determined. Nonetheless it is clear from Figure 8-15 that the two types of phase transition can be distinguished experimentally.

The question that now arises is Why do first-phase transitions occur at a sharply defined temperature for a given P, whereas second-order transitions take place over a range of temperature? Consider a solid as its temperature is raised to the melting point. As the temperature is increased, molecules (or ions) vibrate more and more violently about their equilibrium positions, and, in addition, the intermolecular separations increase due to the expansion of the crystal. At the melting point the cohesive forces in the solid are just sufficient to hold the molecules in an ordered three-dimensional array; however, the solid on further absorption of energy can now go over to liquid without a concomitant change in the Gibbs energy of the substance owing to the higher entropy (i.e., higher configurational and thermal randomness) of the liquid. At the transition point (fixed T and P) the chemical potentials of the two phases are equal; thus

$$\mu_\alpha = \mu_\beta$$

Further, from the relation $\mu = \bar{U} + P\bar{V} - T\bar{S}$, we have

$$\bar{H}_\beta - \bar{H}_\alpha = \bar{U}_\beta - \bar{U}_\alpha + P(\bar{V}_\beta - \bar{V}_\alpha) = T(\bar{S}_\beta - \bar{S}_\alpha)$$

Because $\bar{H}_i$ and $\bar{S}_i$ are functions of T (and P), the temperature at which the condition

$$\bar{H}_\beta - T\bar{S}_\beta = \bar{H}_\alpha - T\bar{S}_\alpha$$

is met occurs only at the intersection (sharply defined) of two distinct Gibbs energy surfaces. First-order solid-solid phase transitions are basically of two types. The first of these is a change in crystal structure (like that of graphite to diamond) or molecular structure and is fairly common. The second type involves a change in electronic structure. Some examples of this type of transition are the nonmetallic $\rightarrow$ metallic transition in TlI at 25°C, and 160 kbar, and a $4f \rightarrow 5d$ electronic transition in cerium at 7.5 kbar and 25°C.

Consider $CH_4(s)$, which exhibits a second-order phase change. The second-order phase change in $CH_4(s)$ has, through the use of spectroscopic techniques, been identified with the onset of rotation of the CH_4 molecules in the solid. As the temperature is raised to the vicinity of the λ-point, some of the methane molecules begin to rotate in the solid, and it becomes easier for other CH_4 molecules to begin rotation once their neighbors are rotating. Thus, as the temperature is raised still further, we have a cascadelike (or *cooperative*) effect that reaches a plateau at the λ-point. The λ-point corresponds to the temperature where the transition is essentially complete (i.e., essentially all the CH_4 molecules are rotating), and thus the heat capacity associated with the second-order phase transition declines to zero as T increases. Because a second-order phase transition does not involve phase separation, we are not dealing with the intersection of two Gibbs energy surfaces but rather with what could probably be characterized as a flat region in the chemical potential surface (see Figure 8-14). Such a flat section in the chemical potential surface would not give rise to a ΔS, ΔV, or ΔH for the transition.

Several other types of second-order transitions are known. Among these are the ferromagnetic to paramagnetic transitions occurring at the Curie (λ) point, the transition of a superconductor to an ordinary conductor in the absence of a magnetic field, the transition from superfluid He(ℓ) to normal He(ℓ), and so-called *order-disorder transitions* in alloys (like β to α brass) that involve changes in packing arrangements.

Problems

1. Determine the number of components and the number of degrees of freedom for each of the following systems:
(a) $H_3PO_4(aq)$.
(b) A saturated aqueous solution in contact with $AgBrO_3(s)$.
(c) A saturated aqueous solution in contact with $Ag_2SO_3(s)$.
(d) An immiscible mixture of $CCl_4(\ell)$ and $H_2O(\ell)$.

2. Consider a mixture of silica (SiO_2), sawdust, iron filings, salt, and gold dust. Devise a method for the mechanical separation of the various phases of the system. Assume that sawdust is homogeneous.

3. Determine the number of degrees of freedom in each of the following cases (assume in each case that the system is placed in a closed air-free flask with a vapor space):
(a) A solution of LiBr in ether.
(b) A solution of LiBr in water.
(c) A saturated solution of LiBr in water.
(d) 50 mL of benzene, 50 mL of water, 1.0 g of KI, and 1.0 g of $I_2(s)$, which is shaken and allowed to come to equilibrium.
(e) A 50% by weight solution of ethylene glycol (antifreeze) in water that is cooled until some solid appears.

4. Calcite and aragonite are two different crystal modifications of calcium carbonate, $CaCO_3(s)$. The densities of the two forms are

$$\text{calcite } \rho = 2.71\ \text{g}\cdot\text{cm}^{-3} \qquad \text{aragonite } \rho = 2.94\ \text{g}\cdot\text{cm}^{-3}$$

Which form is more stable at high pressures? The calcite-to-aragonite phase transition has been proposed as a geobarometer, that is, as an indicator of the minimum pressure to which the $CaCO_3(s)$ was subjected.

5. The system formed in the high-temperature reduction of solid zinc oxide with solid carbon involves the following reaction equilibria:

$$ZnO(s) + C(s) \rightleftharpoons Zn(g) + CO(g)$$
$$2\,CO(g) \rightleftharpoons C(s) + CO_2(g)$$

Determine the number of components and the number of independent intensive thermodynamic variables for the general case.

Given the above two reaction equilibria, explain why we do not also include the reaction equilibrium

$$ZnO(s) + CO(g) \rightleftharpoons Zn(g) + CO_2(g)$$

as an additional restrictive condition.

6. The number of components of a system *may* be decreased if the system is not prepared in an entirely arbitrary manner.

(a) Suppose that the system described in Problem 5 is prepared solely from $ZnO(s)$ and $C(s)$. Show that $c = 2$, and write the additional restrictive condition.

(b) Suppose that the equilibrium system

$$CaCO_3(s) \rightleftharpoons CaO(s) + CO_2(g)$$

is prepared from $CaCO_3(s)$ alone. What is the number of components? Explain the difference between this case and that in part (a).

7. The temperature of the human body is about 37°C. Use data in the text to estimate the vapor pressure of water in exhaled air. Assume that air in the lungs is saturated with water vapor.

8. Compute the number of moles of water at 0°C that can be frozen by 1.00 mol of dry ice. ($\bar{C}_P = 46.3\ \mathrm{J\cdot K^{-1}\cdot mol^{-1}}$, for $CO_2(s)$) $\Delta\bar{H}_{sub} = 25.2\ \mathrm{kJ\cdot mol^{-1}}$

9. Moisture often forms spontaneously on the outside of a glass containing a mixture of ice and water or various aqueous solutions. Use the principles developed in this chapter to explain this phenomenon.

10. Commercial refrigeration units in the United States are rated in "tons." A 1-ton refrigeration unit is capable during 24 hr of operation of removing heat equal to that released when 1.00 ton of water at 0°C is converted to ice. Calculate the number of kilojoules of heat per hour that can be removed by a 4-ton home air conditioner.

11. Explain why the release of some of the contents of an aerosol can leads to a transitory decrease in the temperature of the liquid that remains in the can.

12. Suggest an explanation for the observation that the solid phase of a substance invariably has a higher surface tension than the liquid phase of the same substance.

13. The heat of vaporization of the refrigerant Freon-12 (CCl_2F_2) is about 155 $J \cdot g^{-1}$. Estimate the number of grams of Freon-12 that must be evaporated to freeze a tray of sixteen 1-oz (1 oz = 28 g) ice cubes with the water initially at 18°C.

14. Suppose that the relative humidity of the air is 50% at 25°C.
(a) Estimate the dew point of the air.
(b) Estimate the relative humidity if the air temperature is increased from 25°C to 30°C without gain or loss of water vapor.

15. Given the following data for iodine, I_2, sketch the phase diagram for I_2.

triple point	113°C	0.12 atm
critical point	512°C	116 atm
normal melting point	114°C	
normal boiling point	184°C	

$\rho_{solid} > \rho_{liquid}$

16. One of the methods used by the human body to maintain normal body temperature during hot weather is the evaporation of water from the skin surface. Compute the number of kilojoules of energy removed as heat from the body by the evaporation of 1.00 pt of water from the skin surface.

17. Suggest a simple explanation based on hydrogen bonding for the unusually high surface tension of liquid water relative to other nonmetallic liquids such as acetone, CH_3COCH_3, or ethyl alcohol.

18. The density of liquid water is greater than that of ice by about 11%, and thus ice floats on water. Liquid water has a maximum density at about 4°C. If a thermometer is submerged in an unstirred ice-water mixture below the ice level, then the thermometer reads 4°C. Explain.

19. Suppose an aqueous solution containing a dissolved protein is observed to begin freezing at +0.8°C at 1 atm. What can you say about the composition of the solid phase?

20. It is possible to supercool numerous liquids and maintain them in a supercooled state for prolonged periods. Is it possible to superheat a solid and to maintain it in the superheated condition? Explain your answer.

21. Find a substance not mentioned in the text whose triple point is above 1 atm and determine the temperature of a constant temperature bath open to the atmosphere obtained by mixing the solid form of your substance with a liquid whose vapor pressure is negligible at the bath temperature; also find a liquid suitable for this purpose.

22. Consider 1.00 mol of pure supercooled water at 263 K that is isolated from its surroundings. Suppose now that the water spontaneously begins to crystallize and that the crystallization process continues until equilibrium is attained.
(a) What is the final temperature of the system? Describe your reasoning.
(b) What phases are present at equilibrium?
(c) Estimate quantitatively the amounts of the phases present at equilibrium.
(d) Compute ΔS for the process given that
$\bar{C}_P^\circ = 75.31\ \text{J}\cdot\text{K}^{-1}\cdot\text{mol}^{-1}$ for $H_2O(\ell)$
$\bar{C}_P^\circ = 37.66\ \text{J}\cdot\text{K}^{-1}\cdot\text{mol}^{-1}$ for $H_2O(s)$
$\Delta\bar{H}_{\text{fusion}}^\circ$ of $H_2O(s) = 6.02\ \text{kJ}\cdot\text{mol}^{-1}$

23. Is it possible for a two-component gas-phase mixture to separate spontaneously into two gas phases?

24. Write all the relationships among the intensive state variables for a two-component, two-phase system.

25. Consider a two-phase, one-component system in which surface effects are important. Show that $f = 2$ and not $f = 1$. *Hint:* Begin by deriving the phase rule for systems in which surface effects are not negligible.

26. If diamonds are brought in contact with carbon vapor at 2000 K, it is observed that the mass of the diamonds increases. Does this observation contradict the discussion of diamond production in the text? Explain.

27. Show that for a system involving a magnetic field *or* an electric field that the phase rule becomes

$$f = c - p + 3$$

Is a *quadruple point* possible for such systems?

28. Derive the analog of Eq. 8-20 for a system of c components and p phases.

29. Show that for a system composed of p phases and c components that

$$dG = -S\,dT + V\,dP + \sum_{k=1}^{p} \sum_{i=1}^{c} \mu_{ik}\,dn_{ik}$$

30. (a) Show that the isothermal compressibility and isobaric thermal expansivity are given by

$$\beta = \frac{1}{\rho}\left(\frac{\partial \rho}{\partial P}\right)_T \qquad \alpha = -\frac{1}{\rho}\left(\frac{\partial \rho}{\partial T}\right)_P$$

where ρ is the density.
(b) Show that β and α become infinite at the critical point.

31. (a) Show that the maximum possible number of triple points for a system of p phases is given by

$$\frac{p!}{(p-3)!3!} \qquad p \geqslant 3$$

(b) Show that the maximum possible number of quadruple points for a system of p phases and two or more components is

$$\frac{p!}{(p-4)!4!} \qquad (p \geqslant 4, c \geqslant 2)$$

32. Show that for a first-order phase transition

$$\frac{d\,\Delta \bar{S}_{\mathrm{tr}}}{dT} = \frac{\Delta \bar{C}_{P,\mathrm{tr}}}{T} - \frac{\Delta \bar{S}_{\mathrm{tr}}}{T}$$

33. Show that $\Delta \bar{H}_{\mathrm{vap}} \to 0$ as the critical point is approached.

34. Apply van der Waal's equation to $CO_2(g)$ and plot P versus V at T values of $T_c + 50$, $T_c + 10$, $T_c - 10$, $T_c - 50$. Can van der Waal's equation be used to understand the vapor-liquid region on a PV diagram? (See Problems 39 and 40 for data on CO_2.)

35. Assuming $\bar{V}_\ell > \bar{V}_s$, what two points are required to make a rough sketch of the phase diagram of a pure substance with only one solid phase? Suppose there are two solid phases; what additional items of data are required?

36. Given the following data for the carbon system, sketch the carbon phase diagram (use log P versus T coordinates).
(a) The *normal* sublimation point of graphite occurs at about 4200 K.
(b) The graphite-liquid-vapor triple point occurs at 100 bar, 4300 K.
(c) The maximum melting point of graphite occurs at about 55 kbar, 4900 K.
(d) The diamond-graphite-liquid triple point occurs at 126 kbar, 4100 K.
(e) There is a third solid form of carbon (solid III) and the diamond-solid III-liquid triple point occurs at 800 kbar, 1100 K.
(f) At 300 K the graphite-to-diamond transition takes place at about 15 kbar.

37. It has been predicted that solid hydrogen will become metallic at about 5 megabars. What possible transition do you think was used as the basis of the calculation?

38. At 22 kbar and 25°C, cesium undergoes a first-order phase transition involving a change from a body-centered cubic structure to a more compact face-centered cubic structure. At 45 kbar and 25°C, cesium undergoes another first-order phase change involving a 10% decrease in volume. Can you suggest a possible explanation for the nature of this latter phase transition?

39. Show by using the critical point conditions that
(a) For a van der Waals gas,

$$b = \frac{\bar{V}_c}{3}, \qquad a = 3P_c\bar{V}_c^2, \quad \text{and} \quad \frac{P_c\bar{V}_c}{RT_c} = \frac{3}{8}$$

(b) For a Berthelot gas,

$$B = \frac{\bar{V}_c}{3}, \qquad A = 3P_c\bar{V}_c^2T_c, \quad \text{and} \quad \frac{P_c\bar{V}_c}{RT_c} = \frac{3}{8}$$

(c) For a Dieterici gas,

$$b' = \frac{\bar{V}_c}{2}, \qquad a' = e^2P_c\bar{V}_c^2 = 2RT_c\bar{V}_c, \quad \text{and} \quad \frac{P_c\bar{V}_c}{RT_c} = \frac{2}{e^2}$$

40. Given the following data, and using the results in Problem 39, compute (b, a), (A, B), (a', b'), and the compressibility factor $Z = P\bar{V}/RT$ (which measures the deviation of the gas from ideal behavior) at the critical point. Compare with the Z values predicted by your results in Problem 39.

Gas	T_c/K	P_c/atm	$\bar{V}_c/cm^3 \cdot mol^{-1}$
He	5.3	2.26	57.7
N_2	126.0	33.5	90.0
O_2	154.3	49.7	74.4
CO_2	304.2	72.9	95.6
H_2O	647.2	217.7	45.0
CH_4	190.6	45.8	98.8

41. Show by multiplying terms and rearranging that the compressibility factors (see Problem 40) of a van der Waals, Berthelot, and Dieterici gas are given by

(a) $$Z = 1 + \left(b - \frac{a}{RT}\right)\left(\frac{1}{\bar{V}}\right) + b^2\left(\frac{1}{\bar{V}^2}\right) + \cdots$$

(b) $$Z = 1 + \left(B - \frac{A}{RT^2}\right)\left(\frac{1}{\bar{V}}\right) + B^2\left(\frac{1}{\bar{V}^2}\right) + \cdots$$

(c) $$Z = 1 + \left(b' - \frac{a'}{RT}\right)\left(\frac{1}{\bar{V}}\right) + \left(b'^2 - \frac{a'b'}{RT} + \frac{a'^2}{2R^2T^2}\right)\left(\frac{1}{\bar{V}^2}\right) + \cdots$$

respectively.

42. Compare your results in Problem 41 with the virial equations of state (wherein B is not the same as the Berthelot equation B).

(a) $$\frac{P\bar{V}}{RT} = 1 + \frac{B}{\bar{V}} + \frac{C}{\bar{V}^2} + \frac{D}{\bar{V}^3} + \cdots$$

(b) $\dfrac{P\bar{V}}{RT} = 1 + B'P + C'P^2 + D'P^3 + \cdots$

which expresses the compressibility factor, $Z = P\bar{V}/RT$, as a power series in $1/\bar{V}$ and P, respectively. In these expressions the constants $B, C, D, \ldots$ and $B', C', D', \ldots$ are known as the second, third, fourth, ... virial coefficients, respectively. Show that if $B', C', D', \ldots$, and so forth, in part (b) are taken as temperature independent, then part (b) is incapable of representing gas behavior in the neighborhood of the critical point. Also show that the relationships between the second and third virial coefficients in parts (a) and (b) are

$$B' = \frac{B}{RT} \qquad C' = \frac{C - B^2}{R^2T^2}$$

Use the above equations, together with your results in Problems 39 and 41, to obtain the second and third virial coefficients B, C and B', C' for the gases listed in Problem 40, as predicted by the van der Waals, Berthelot, and Dieterici equations of state.

43. Using the results in Problem 39, show that if one defines the reduced state variables

$$P_r = \frac{P}{P_c}, \qquad T_r = \frac{T}{T_c} \quad \text{and} \quad V_r = \frac{V}{V_c}$$

then the van der Waals, Berthelot, and Dieterici equations of state expressed in terms of P_r, T_r, and V_r do not explicitly contain any constants characteristic of the particular gas. The foregoing is an illustration of the *principle of corresponding states*, which says that if we express the equation of state in terms of reduced variables, then the equation of state will be the same for all substances or, in other words, at equal reduced temperature and pressure all substances have the same reduced volume. Can the principle of corresponding states be rigorously valid?

44. The Boyle temperature of a gas is defined as that temperature for which the compressibility factor equals unity. Obtain expressions for the Boyle temperatures of a van der Waals, Berthelot, and Dieterici gas, and use them to estimate T_{Boyle} for He, N_2, O_2, and CO_2. The observed values are 35 K, 323 K, 423 K, and 650 K.

45. In the critical region, van der Waals's equation predicts the following behavior near the critical region ($v = V/V_c$, $p = P/P_c$, $t = T/T_c$):

(a) Liquid-vapor coexistence curve,

$$v_{\text{gas}} - v_{\text{liquid}} \approx 4(1 - t)^{1/2} \qquad (t \lesssim 1)$$

(b) Critical isotherm,

$$(p - 1) \approx \tfrac{3}{2}(1 - v)^3 \qquad (t = 1)$$

(c) Specific heat along the critical isochor (monatomic gas),

$$C_v - \frac{3R}{2} \approx \frac{9}{2} R \qquad (t \lesssim 1)$$

$$C_v = \frac{3R}{2} \qquad (t > 1)$$

Use the critical point conditions to derive these equations.

46. Indicate by circling the correct response whether each of the following statements is true (T), false (F), or cannot be answered as true *or* false (N), based on the information given.

(a) The equilibrium vapor pressure of pure water at a given temperature always decreases when a substance is dissolved in the water. T F N

(b) Any gas at a particular temperature can be liquefied by the application of a sufficiently high pressure. T F N

(c) Nitrogen, N_2, gas cannot be liquefied. T F N

(d) Pure ethyl alcohol cannot be frozen. T F N

(e) The equilibrium vapor pressure of a pure, stable liquid always increases with increasing temperature. T F N

(f) The percent relative humidity cannot exceed 100. T F N

(g) If ice is observed to form from an aqueous solution at 1 atm at $+1.0°C$, then the ice cannot be pure. T F N

(h) The melting point of a solid always decreases with increasing pressure. T F N

(i) Water cannot boil unless the temperature of the water is at least 100°C. T F N

(j) If a heavy-walled metal container is filled with dry ice, $CO_2(s)$, at $-78°C$ and the container is sealed and allowed to warm to room temperature, then liquid CO_2 will form in the container. T F N

47. Determine which of the following statements is not true in general. If the statement is not true, then explain why.

(a) Any pure substance has only one triple point.

(b) All fixed-composition gases undergo a temperature decrease when allowed to undergo an adiabatic free expansion (Joule-Thomson expansion).

(c) All adjacent phase transition equilibrium lines that meet at a triple point must intersect one another at an angle less than 180°.

(d) A second-order phase transition has no heat of transition, nor does it involve phase separation.

(e) It is possible to maintain certain supercooled liquids in thermal equilibrium with their surroundings, and under certain special conditions the same is true for a superheated solid.

(f) The heat capacity at constant pressure C_P becomes *infinite* at a first-order phase transition.
(g) The possibility of any actual (i.e., real) process whatsoever occurring reversibly is ruled out by the Second Law of Thermodynamics.
(h) The heat capacity of a pure substance must go to zero in the limit as T goes to zero.
(i) Helium of normal isotopic composition cannot be solidified under any conditions.
(j) The entropy of a paramagnetic substance in the limit of 0 K in the presence of a magnetic field exceeds the entropy of the same substance in the absence of a field by an amount that is a function of the field strength.
(k) Any chemical reaction involving only pure condensed phases can be brought into equilibrium at any values of P and T for which the various phases are capable of independent existence.
(l) A perfectly general thermodynamic equilibrium criterion for any process occurring at constant T and P is $\Delta G = 0$ (where G is the Gibbs energy function).
(m) Any phase equilibrium line on a P, T diagram that intercepts the pressure axis at $T = 0$ must do so with zero slope.
(n) The Second Law of Thermodynamics excludes all isothermal processes for which

$$\Delta S_{\text{sys}} < \frac{q_{\text{sys}}}{T}$$

(o) If for any process $\Delta H = \Delta H^\circ$, then $\Delta S = \Delta S^\circ$.
(p) For the purpose of establishing thermodynamic tables of standard Gibbs energies of formation of substances from the elements, one always sets $\Delta \bar{G}_f^\circ = 0$ for the thermodynamically most stable form of the element at the temperature of interest.
(q) The Third Law of Thermodynamics is applicable only to perfectly crystalline substances.
(r) The enthalpy (or heat content) ΔH is a direct measure of the heat in a body.
(s) Classical equilibrium thermodynamics is applicable only to reversible processes.
(t) Two soap bubbles of different radius are blown on the two ends of a tube with a stopcock in the middle. When the stopcock is opened, the two bubbles become the same size.
(u) The Gibbs energy of a very finely divided solid always exceeds that of the same solid when it is not finely divided or internally strained.

Pacific Gas and Electric Company's 17th geothermal unit at The Geysers in California has increased the utility's total generating capability from power plants using geothermal steam to 1,129,000 kilowatts. The facilities at The Geysers, about 90 miles north of San Francisco, comprise the world's largest power plant complex fueled by geothermal steam. At the unit shown here, electricity is produced in the turbine-generator building on the left. Spent steam is condensed to hot water, which goes to the cooling tower (right) for additional cooling before being reinjected into the earth.

9

Phase Equilibria: The Activity Function

It would evidently avoid confusion if once and for all we should choose for a given substance its standard state. This consideration, however, is outweighed by the practical advantage of being able at any time to choose the standard state or states best adapted for a particular problem. . . . we shall choose as standard now one state and now another, as convenience dictates, although to avoid confusion this choice must in each case be clearly stated. For as we proceed, we shall find it desirable to choose different standard states for a substance, not only in different problems, but even in a single problem. However, we may in advance lessen the arbitrariness of our procedure by setting down certain rules for choosing the standard state to which we shall invariably adhere.

G. N. Lewis and M. Randall,
Thermodynamics. New York:
McGraw-Hill, 1923.

9-1 *The Clapeyron Equation Gives the Slope of a Phase-Equilibrium Line*

Although the Gibbs phase rule predicts the existence of various phase equilibrium lines on a phase diagram, the phase rule does not tell us anything about the slopes of these lines. The equation governing the slopes of first-order phase transition lines on P, T phase diagrams is known as the *Clapeyron equation*, and we now proceed to its derivation.

Consider the equilibrium between two pure phases (α and β) of a single substance. The equilibrium condition for the phase transformation

$$\alpha \text{ phase} \rightleftharpoons \beta \text{ phase} \tag{9-1}$$

is

$$\mu_\alpha(T, P) = \mu_\beta(T, P) \tag{9-2}$$

where μ_i is the chemical potential of i. If the two-phase equilibrium is disturbed by a change in T or P, then the condition for the attainment of a new equilibrium state, that is, the condition that must be satisfied if the final state is to lie on the same equilibrium line as the initial state, is obtained from Eq. 9-2 as

$$d\mu_\alpha(T, P) = d\mu_\beta(T, P) \tag{9-3}$$

The chemical potentials are state functions, and thus their total differentials are exact. Consequently, from Eq. 9-3, we can write

$$\left(\frac{\partial \mu_\alpha}{\partial P}\right)_T dP + \left(\frac{\partial \mu_\alpha}{\partial T}\right)_P dT = \left(\frac{\partial \mu_\beta}{\partial P}\right)_T dP + \left(\frac{\partial \mu_\beta}{\partial T}\right)_P dT \tag{9-4}$$

or

$$\bar{V}_\alpha dP - \bar{S}_\alpha dT = \bar{V}_\beta dP - \bar{S}_\beta dT \tag{9-5}$$

Rearranging Eq. 9-5, we obtain

$$\frac{dP}{dT} = \frac{\bar{S}_\beta - \bar{S}_\alpha}{\bar{V}_\beta - \bar{V}_\alpha} = \frac{\Delta \bar{S}_{tr}}{\Delta \bar{V}_{tr}} \tag{9-6}$$

For a transition from one phase to another under equilibrium conditions at constant T and P, we have

$$\Delta \bar{G}_{tr} = \mu_\beta - \mu_\alpha = 0 = \bar{H}_\beta - \bar{H}_\alpha - T(\bar{S}_\beta - \bar{S}_\alpha) \tag{9-7}$$

or

$$\Delta \bar{H}_{tr} = T \Delta S_{tr} \tag{9-8}$$

Substitution of Eq. 9-8 into Eq. 9-6 yields the *Clapeyron equation*

$$\frac{dP}{dT} = \frac{\Delta \bar{H}_{tr}}{T \Delta \bar{V}_{tr}} \tag{9-9}$$

The Clapeyron equation is limited to (1) equilibria involving phases of fixed composition (because we have taken μ as independent of composition) and (2) first-order phase transitions (e.g., solid $\alpha \rightleftharpoons$ solid β, solid $\rightleftharpoons$ liquid,

liquid $\rightleftharpoons$ gas, and solid $\rightleftharpoons$ gas). Second-order phase transitions do not involve phase separation and the above derivation is, therefore, meaningless in such a case.

As an example of the use of the Clapeyron equation, consider the phase transformation

$$H_2O(s) \rightleftharpoons H_2O(\ell)$$

that is, the equilibrium between ice and liquid water. The molar enthalpy of fusion of ice *at* 1 *atm total pressure and* 0°C is $\Delta\bar{H}_{\text{fus}} = 6008\ J\cdot mol^{-1}$. The densities of $H_2O(s)$ and $H_2O(\ell)$ *at* 1 *atm pressure and* 0°C are $0.915\ \text{g}\cdot\text{cm}^{-3}$ and $1.000\ \text{g}\cdot\text{cm}^{-3}$, respectively. For the molar volume change on fusion we have

$$\Delta\bar{V}_{\text{fus}} = M\left(\frac{1}{\rho_\ell} - \frac{1}{\rho_s}\right)$$

where M is the molecular mass. Thus for water

$$\Delta\bar{V}_{\text{fus}} = 18.0\ \text{g}\cdot\text{mol}^{-1}\left(\frac{1}{1.000\ \text{g}\cdot\text{cm}^{-3}} - \frac{1}{0.915\ \text{g}\cdot\text{cm}^{-3}}\right) = -1.67\ \text{cm}^3\cdot\text{mol}^{-1}$$

The Clapeyron equation (Eq. 9-9) can be rewritten as

$$\frac{dT}{dP} = \frac{T\Delta\bar{V}_{\text{tr}}}{\Delta\bar{H}_{\text{tr}}}$$

and thus for the fusion of ice at 1 atm and 273.15 K we have

$$\frac{dT}{dP} = \frac{(273.15\ \text{K})(-1.67\ \text{cm}^3\cdot\text{mol}^{-1})}{(6008\ \text{J}\cdot\text{mol}^{-1})(82.06\ \text{cm}^3\cdot\text{atm}/8.314\ \text{J})} = -0.0077\ \text{K}\cdot\text{atm}^{-1}$$

where in the denominator we have utilized the ratio of two values of the gas constant to put dT/dP in the desired units. Note that an increase in pressure of 1.00 atm ($\Delta P = 1.00$ atm) *depresses* the freezing point of water by 7.7×10^{-3} K, around 0°C and about 1 atm pressure. The result $dT/dP < 0$ is rare for a solid → liquid transformation. For most solids the melting point increases with increasing pressure.

For a simple $\alpha \to \beta$ phase change where β lies on the high-temperature side of the equilibrium line, $\Delta\bar{S}(\alpha \to \beta)$ for the transition is necessarily greater than zero (the system absorbs energy as heat when the temperature of the surroundings is increased). Further, for the $\alpha \to \beta$ phase transition, we have from Eq. 9-8

$$\Delta\bar{H}(\alpha \to \beta) = T\Delta\bar{S}(\alpha \to \beta)$$

and thus $\Delta\bar{H}_{\text{tr}}$ for the transition is also necessarily greater than zero. Therefore the value of dT/dP for the fusion of a solid can be less than zero if, and only if, $\Delta\bar{V}_{\text{tr}} < 0$ (see Eq. 9-9). For sublimation and vaporization processes, dT/dP is always greater than zero, because both $\Delta\bar{H}_{\text{tr}}$ and $\Delta\bar{V}_{\text{tr}}$ are always positive in such cases. For solid-solid transformations, $\Delta\bar{H}_{\text{tr}} > 0$ (going to the high-temperature form), but $\Delta\bar{V}_{\text{tr}}$ can have either sign and, therefore, so can dT/dP.

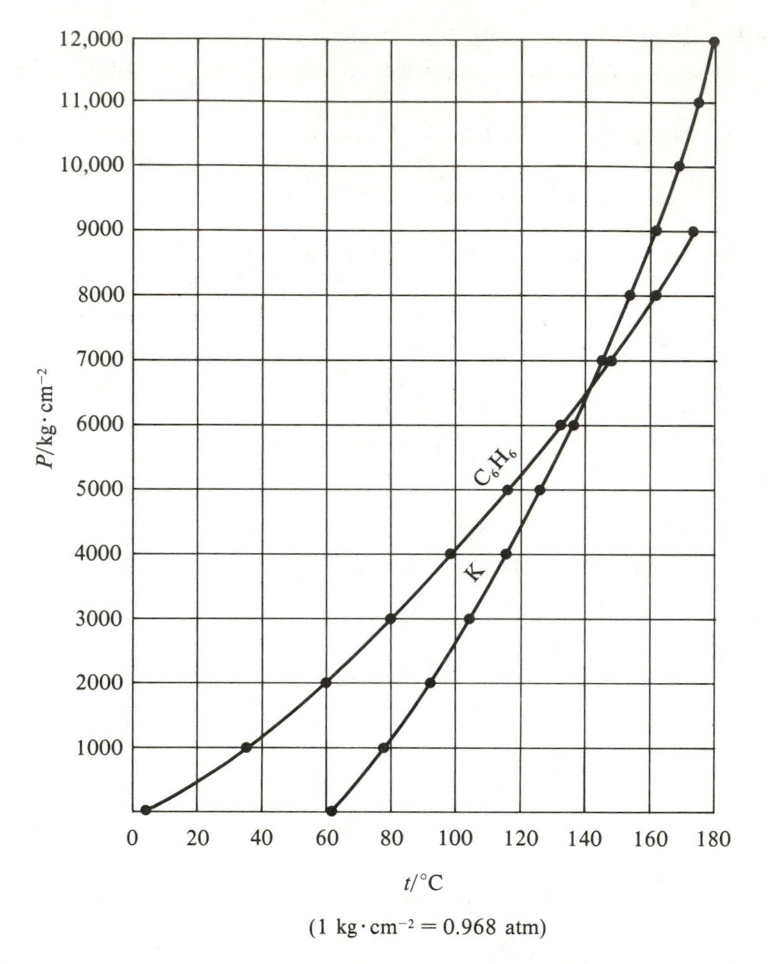

(1 kg·cm^{-2} = 0.968 atm)

Figure 9-1. Melting point curves for benzene and potassium. Note that the melting points increase with increasing pressure, which is the usual result. For both benzene and potassium, $\rho_s > \rho_l$, and thus an increase in pressure stabilizes the solid phase relative to the liquid phase ($\bar{V}_s < \bar{V}_l$).

EXAMPLE 9-1

Estimate the melting point of ice at 2000 atm.

Solution: From the above result for dT/dP for water we have

$$\Delta T \approx -(7.7 \times 10^{-3}\ \mathrm{K \cdot atm^{-1}})\Delta P$$

and thus

$$\Delta T \approx -(7.7 \times 10^{-3}\ \mathrm{K \cdot atm^{-1}})(2 \times 10^3\ \mathrm{atm}) = -15\ \mathrm{K}$$

Consequently we have ($\Delta T = T_{\mathrm{mp}} - 273.15$ K)

$$T_{\mathrm{mp}} = 273\ \mathrm{K} - 15\ \mathrm{K} = 258\ \mathrm{K\ or\ } -15^\circ\mathrm{C}$$

The actual melting point of ice I at 2×10^3 atm is -22°C. The reason for the discrepancy is that we have assumed that dT/dP is independent of

P and T; that is, we have assumed that the P, T solid-liquid equilibrium line has no curvature. This is not correct; both $\Delta\bar{H}_{fus}$ and $\Delta\bar{V}_{fus}$ are functions of P and T. For the $H_2O(s, I) \rightleftharpoons H_2O(\ell)$ transformation, the slope of the P, T equilibrium line becomes progressively more negative at higher pressure.

The assumption that dT/dP is constant over a few hundred atmospheres is a good approximation for solid-liquid and solid-solid transformations, but it fails miserably when one of the two phases is a gas. Even if there is good experimental evidence that dT/dP is a constant, the Clapeyron equation should not be applied to large pressure changes without first considering the possibility of additional phase changes. Thus we might crudely estimate the melting point of ice at 10,000 atm as $-75°C$. In actuality, however, the melting point of ice at 10,000 atm is $+30°C$. The reason for the discrepancy is that above about 2200 atm ice I is no longer a stable phase of water, and, in fact, the solid phase in equilibrium with $H_2O(\ell)$ at 10,000 atm is not ice I but ice IV. Ice IV has a molar volume less than that of $H_2O(\ell)$, and thus dT/dP is positive for ice IV at 10,000 atm.

As a rule of thumb, dT/dP for the melting point curves of most substances lie between $+0.01\ \text{K}\cdot\text{atm}^{-1}$ and $+0.03\ \text{K}\cdot\text{atm}^{-1}$, and thus a pressure increase of over 30 atm is required to raise the melting point by 1.0°C. As a consequence, atmospheric pressure fluctuations can be ignored in the determination of melting points or solid-solid transition points under atmospheric pressure. The dependence of melting point on pressure for benzene and potassium is shown in Figure 9-1.

9-2 *The Clapeyron-Clausius Equation Gives the Slope of the Equilibrium Vapor Pressure Curve*

Consider the application of the Clapeyron equation to phase equilibria of the types

(1) liquid $\rightleftharpoons$ gas (vapor pressure curve)

(2) solid $\rightleftharpoons$ gas (sublimation pressure curve)

For case 1, the Clapeyron equation (Eq. 9-9) can be written as

$$\frac{dP}{dT} = \frac{\Delta\bar{H}_{vap}}{T(\bar{V}_g - \bar{V}_\ell)} \tag{9-10}$$

Further, because $\bar{V}_g \gg \bar{V}_\ell$, then, provided that $T \ll T_c$, where T_c is the critical temperature, we obtain from Eq. 9-10

$$\frac{dP}{dT} \simeq \frac{\Delta\bar{H}_{vap}}{T\bar{V}_g} \tag{9-11}$$

The pressure P in Eq. 9-11 is the equilibrium vapor pressure of the substance. Suppose that the gas behaves ideally; that is, $\bar{V}_g = RT/P$. In this case, Eq. 9-11 can be written as

$$\frac{d \ln P}{dT} \approx \frac{\Delta \bar{H}_{\text{vap}}}{RT^2} \tag{9-12}$$

Equation 9-12 is known as the *Clapeyron-Clausius equation.* Analogous considerations for the sublimation pressure curve (case 2) lead to the result

$$\frac{d \ln P}{dT} \approx \frac{\Delta \bar{H}_{\text{sub}}}{RT^2} \tag{9-13}$$

Integration of Eq. 9-12, assuming $\Delta \bar{H}_{\text{vap}}$ is constant, yields

$$\int_{P_1}^{P_2} d \ln P = \int_{T_1}^{T_2} \frac{\Delta \bar{H}_{\text{vap}}}{RT^2} \, dT$$

$$\ln\left(\frac{P_2}{P_1}\right) = -\frac{\Delta \bar{H}_{\text{vap}}}{R}\left(\frac{1}{T_2} - \frac{1}{T_1}\right) = \frac{\Delta \bar{H}_{\text{vap}}}{R}\left(\frac{T_2 - T_1}{T_1 T_2}\right) \tag{9-14}$$

An analogous equation is obtained for sublimation with $\Delta \bar{H}_{\text{sub}}$ replacing $\Delta \bar{H}_{\text{vap}}$.

A knowledge of the equilibrium vapor pressure of a solid or a liquid at two temperatures (one of which, of course, could be the normal boiling point or normal sublimation point for which $P = 1.00$ atm) suffices for the estimation of $\Delta \bar{H}_{\text{vap}}$ or $\Delta \bar{H}_{\text{sub}}$. Conversely, a value of $\Delta \bar{H}_{\text{vap}}$ or $\Delta \bar{H}_{\text{sub}}$ together with the equilibrium vapor pressure at one temperature suffices for the estimation of the equilibrium vapor pressure at some other temperature, provided that the assumption that $\Delta \bar{H}_{\text{vap}}$ (or $\Delta \bar{H}_{\text{sub}}$) is a constant, is valid.

EXAMPLE 9-2

The vapor pressure of liquid mercury at 20°C is 1.20×10^{-3} torr; estimate the vapor pressure of mercury at 100°C. The enthalpy of vaporization of $Hg(\ell)$ is $61.5 \text{ kJ} \cdot \text{mol}^{-1}$.

Solution: From Eq. 9-14 we obtain

$$\ln\left(\frac{P}{1.20 \times 10^{-3} \text{ torr}}\right) = \frac{-61.5 \times 10^3 \text{ J} \cdot \text{mol}^{-1}}{8.314 \text{ J} \cdot \text{K}^{-1} \cdot \text{mol}^{-1}}\left(\frac{1}{373.15} - \frac{1}{293.15}\right)$$

and thus

$$P = 1.20 \times 10^{-3} \text{ torr} \times 2.23 \times 10^2 = 0.268 \text{ torr}$$

The measured value of the vapor pressure of mercury at 100°C is 0.273 torr (1 torr = 133.32 Pa).

Integration of Eq. 9-14 without definite limits for a vapor-liquid equilibrium assuming that $\Delta \bar{H}_{\text{vap}}$ is a constant yields

$$\ln P = -\frac{\Delta \bar{H}_{\text{vap}}}{RT} + b = -\frac{a}{T} + b \tag{9-15}$$

and thus, if $\Delta\bar{H}_{vap}$ is constant, then a plot of ln P or log P (where P is the equilibrium vapor pressure) versus $1/T$ will yield a straight line whose slope at any point is equal to $-\Delta\bar{H}_{vap}/R$ or $-\Delta\bar{H}_{vap}/2.303R$, respectively. If such a plot is linear, then $\Delta\bar{H}_{vap}$ can be regarded as constant within the precision of the data. Such a plot for the $Hg(\ell) \rightleftharpoons Hg(g)$ equilibrium is presented in Figure 9-2.

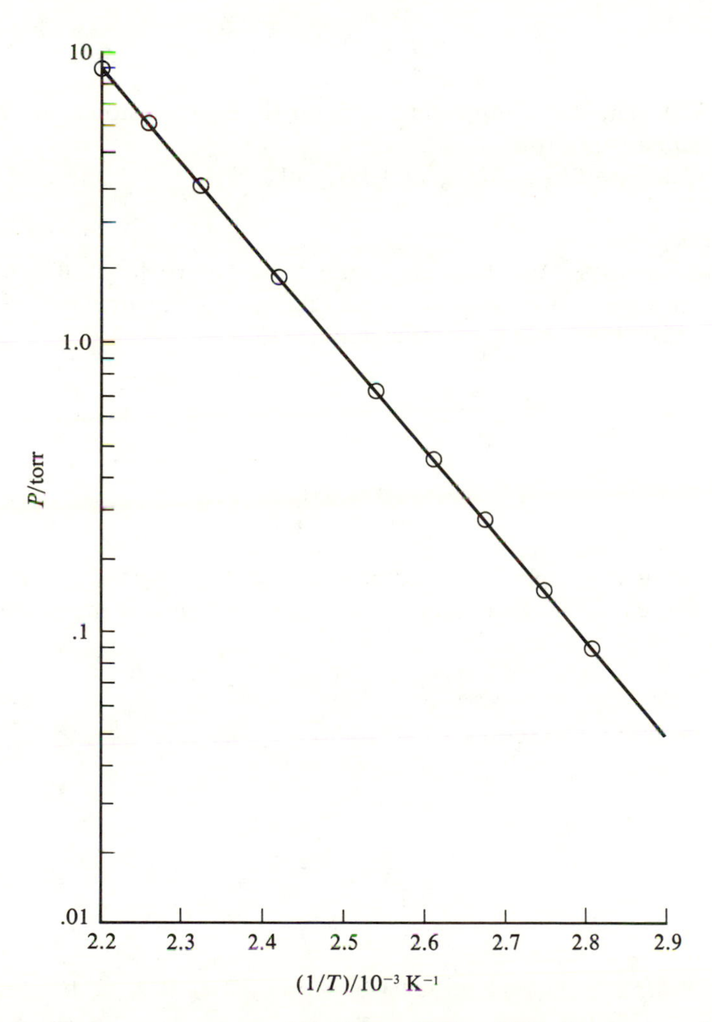

Figure 9-2. Equilibrium vapor pressure of $Hg(\ell)$ from 80°C to 180°C. For the slope of the curve we obtain

$$\text{slope} = -\frac{\Delta\bar{H}_{vap}}{2.303R} = \frac{\log 9.2_0 - \log 0.054_0}{(2.200 - 2.900) \times 10^{-3}}$$

and thus

$$\Delta\bar{H}_{vap} = \frac{2.23 \times 2.303 \times 8.314 \times 10^3}{0.700} = 61\ \text{kJ}\cdot\text{mol}^{-1}$$

EXAMPLE 9-3

The equilibrium vapor pressure (in torr) of $NH_3(s)$ is given by

$$\ln P = 23.03 - \frac{3754}{T}$$

whereas the equilibrium vapor pressure (in torr) of $NH_3(\ell)$ is given by

$$\ln P = 19.49 - \frac{3063}{T}$$

(a) Calculate the temperature and pressure of ammonia at the solid-liquid-vapor triple point.
(b) Calculate $\Delta\bar{H}_{vap}$, $\Delta\bar{H}_{sub}$, and $\Delta\bar{H}_{fus}$ of NH_3.

Solution:

(a) At the triple point (and only at the triple point), the equilibrium vapor pressures of the solid and liquid phases are equal; therefore we can equate the vapor pressure equations

$$23.03 - \frac{3754}{T_{tpt}} = 19.49 - \frac{3063}{T_{tpt}}$$

Thus

$$T_{tpt} = \frac{691}{3.54} = 195 \text{ K}$$

Using the triple point temperature of $T_{tpt} = 195$ K, T, and either $\ln P$ equation, we obtain for the vapor pressure at the triple point

$$\ln P_{tpt} = 23.03 - \frac{3754}{195} = 3.78$$

or

$$P_{tpt} = 43.8 \text{ torr} = 5.76 \times 10^{-2} \text{ atm}$$

(b) For $\Delta\bar{H}_{sub}$ we have

$$\frac{d \ln P}{dT} = \frac{\Delta\bar{H}_{sub}}{RT^2} = \frac{3754}{T^2}$$

Thus

$$\Delta\bar{H}_{sub} = (3754 \text{ K})(8.314 \text{ J}\cdot\text{K}^{-1}\cdot\text{mol}^{-1}) = 31.21 \times 10^3 \text{ J}\cdot\text{mol}^{-1}$$

For $\Delta\bar{H}_{vap}$ we have

$$\frac{d \ln P}{dT} = \frac{\Delta\bar{H}_{vap}}{RT^2} = \frac{3063}{T^2}$$

Thus

$$\Delta\bar{H}_{vap} = (3063 \text{ K})(8.314 \text{ J}\cdot\text{K}^{-1}\cdot\text{mol}^{-1}) = 25.47 \times 10^3 \text{ J}\cdot\text{mol}^{-1}$$

To evaluate $\Delta\bar{H}_{\text{fus}}$ we use the following cycle, which applies at any given T (ΔH is independent of path):

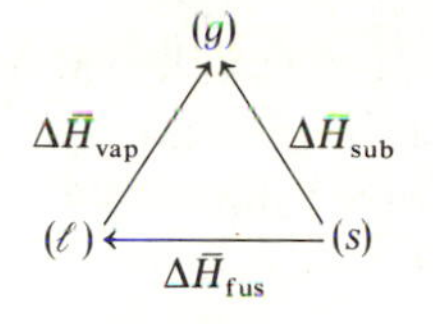

and from which we obtain the condition

$$\Delta\bar{H}_{\text{fus}} + \Delta\bar{H}_{\text{vap}} = \Delta\bar{H}_{\text{sub}}$$

Thus

$$\Delta\bar{H}_{\text{fus}} = (31.21 - 25.47)10^3\ \text{J}\cdot\text{mol}^{-1} = 5.74 \times 10^3\ \text{J}\cdot\text{mol}^{-1}$$

Equation 9-14 should not be used indiscriminately over a large range in T, because the assumptions that (1) $\Delta\bar{H}_{\text{vap}}$ is a constant independent of T and P and (2) the vapor is ideal are not necessarily reliable, especially near the critical point where $\Delta\bar{H}_{\text{vap}}$ approaches zero (Figure 9-3). Recall from our prior discussion of the critical point that $\Delta\bar{H}_{\text{vap}}$ and $\Delta\bar{V}_{\text{vap}}$ must go to zero at the critical

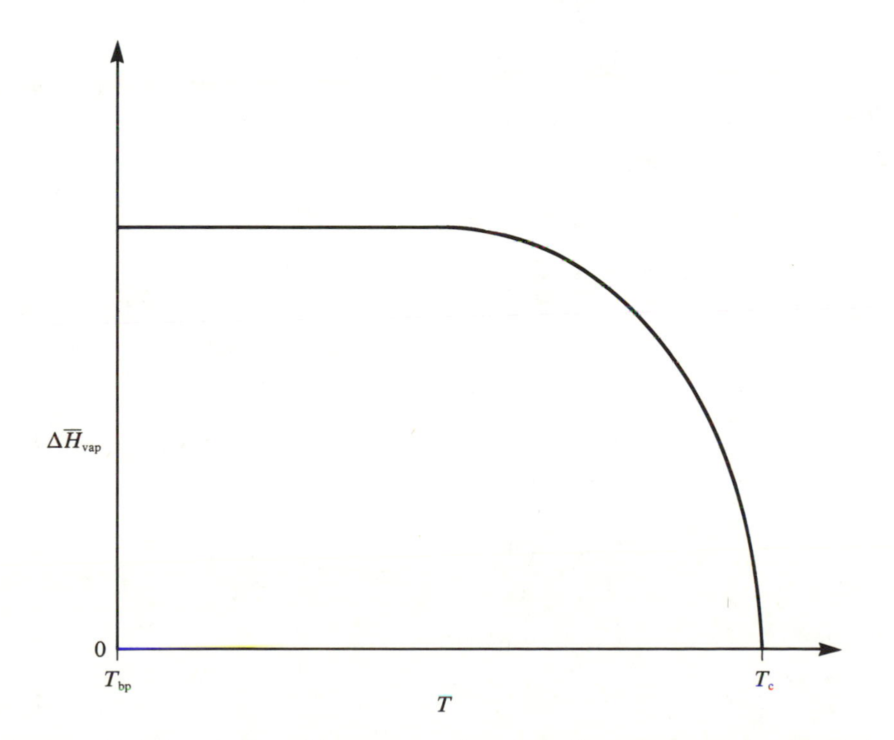

Figure 9-3. Variation of $\Delta\bar{H}_{\text{vap}}$ with temperature from the boiling point T_{bp} to the critical point T_c. Note that $\Delta\bar{H}_{\text{vap}}$ goes to zero as T goes to T_c, the critical temperature; this result is a consequence of the fact that if $T > T_c$, then the gas cannot be liquefied by application of pressure.

point, because this is the point at which the distinction between gas and liquid disappears. In these respects a critical point resembles a second-order phase transition.

It is often found that a simple vapor pressure equation like Eq. 9-15 cannot reproduce $P = P(T)$ within the precision of the data, especially as the applicable temperature range is increased. For example, the equilibrium vapor pressure of mustard gas in torr is given by

$$\log P = 38.525 - \frac{4500}{T} - 9.86 \log T$$

whereas the equilibrium pressure of $CO_2(g)$ over $CaCO_3(s) + CaO(s)$ is given in centimeters of mercury, cm Hg (torr/10), by the still more complicated expression

$$\log P = 9.3171 - 0.668 \times 10^{-3}\, T - \frac{9140}{T} + 0.382 \log T$$

Provided that the gas pressure is less than 1 atm and that $T \ll T_c$, the least tenable of the approximations used to obtain Eq. 9-15 is the assumption that $\Delta\bar{H}_{\text{vap}}$ is independent of temperature ($\Delta\bar{H}_{\text{vap}}$ is only weakly dependent on pressure). The temperature dependence of $\Delta\bar{H}_{\text{vap}}$ is given by

$$\frac{d\,\Delta\bar{H}_{\text{vap}}}{dT} = \Delta\bar{C}_{P,\text{vap}} = \bar{C}_{P(g)} - \bar{C}_{P(\ell)}$$

If we assume that $\Delta\bar{C}_{P,\text{vap}}$ is given by an expression of the type

$$\Delta\bar{C}_{P,\text{vap}} = k_1 + k_2 T \qquad (k_1, k_2 \text{ constants}) \tag{9-16}$$

then we obtain for $\Delta\bar{H}_{\text{vap}}$ on integration

$$\frac{d\,\Delta\bar{H}_{\text{vap}}}{dT} = k_1 + k_2 T$$

$$\Delta\bar{H}_{\text{vap}} = k_1 T + \frac{k_2}{2} T^2 + k_3 \tag{9-17}$$

The integration constant k_3 can be obtained from a knowledge of $\Delta\bar{H}_{\text{vap}}$ at a particular temperature. Substitution of the $\Delta\bar{H}_{\text{vap}}$ expression into the Clapeyron-Clausius equation yields

$$\frac{d \ln P}{dT} \simeq \frac{\Delta\bar{H}_{\text{vap}}}{RT^2} = \frac{k_3}{RT^2} + \frac{k_1}{RT} + \frac{k_2}{2R}$$

Integration of the above expression yields

$$\ln P = -\frac{k_3}{RT} + \frac{k_1}{R} \ln T + \frac{k_2 T}{2R} + k_4 \tag{9-18}$$

where k_4 is an integration constant. Equation 9-18 is of the form found for the equilibrium vapor pressure of $CO_2(g)$ over $CaCO_3(s) + CaCO(s)$, whereas with $k_2 = 0$ it reduces to the form found for mustard gas. Although Eq. 9-18 represents a considerable improvement over Eq. 9-15, as far as its capabilities for representing $P = P(T)$, Eq. 9-18 still has certain approximations built into it, namely, the assumptions that $\bar{V}_g \approx \bar{V}_g - \bar{V}_\ell$ and $\bar{V}_g = RT/P$. The first of these approximations is not difficult to eliminate; the second approximation, however, is a more difficult matter to deal with, and its elimination requires a fundamental rethinking of the problem, which leads to the introduction of the *activity* concept.

9-3 *The Activity Function Takes into Account Nonideal Behavior*

The use of chemical potentials in the thermodynamic analysis of gases and solutions is inconvenient in an important aspect; namely, the chemical potential decreases without bound as the gas pressure or solution concentration approaches zero. For example, suppose that we have a gas of fixed composition at constant temperature; then

$$d\bar{G} = d\mu = \bar{V}\,dP$$

If the gas pressure is low, then $\bar{V} = RT/P$ and

$$d\mu = RT\,d\ln P$$

Integration at constant temperature yields

$$\mu_B - \mu_A = RT\ln\frac{P_B}{P_A}$$

Setting P_A = a constant, we have

$$\lim_{P_B\to 0}\mu_B = \lim_{P_B\to 0}(\mu_A - RT\ln P_A + RT\ln P_B)$$

$$\lim_{P_B\to 0}\mu_B = \mu_A - RT\ln P_A + \lim_{P_B\to 0}(RT\ln P_B) \to -\infty$$

Or, in other words, in the limit as $P_B \to 0$, the chemical potential decreases without limit. To circumvent the problem of an unbounded chemical potential, G. N. Lewis invented a convenience function called the *activity*. The activity function not only facilitates the mathematical analysis of numerous types of problems in chemical thermodynamics but also enables us to cast several important equations of chemical thermodynamics in the same general form as found for systems that behave ideally.

From Eq. 5-89 we have for the change in the chemical potential at fixed composition

$$d\mu_i = -\bar{S}_i\,dT + \bar{V}_i\,dP \tag{9-19}$$

where $\mu_i = \mu_i(T, P, n_1, n_2, \ldots, n_c)$. An equation of the form of Eq. 9-19 holds for each particular composition $(n_1, n_2, \ldots, n_c)$ of the system for all c components of the system. From Eq. 9-19, we obtain for an isothermal process

$$d\mu_i = \bar{V}_i\,dP \tag{9-20}$$

If the substance is an ideal gas, then $\bar{V}_i = RT/P$ and

$$d\mu_i = RT\,d\ln P_i \tag{9-21}$$

Integration of Eq. 9-21 from state A to state B yields

$$\mu_{iB} - \mu_{iA} = RT\ln\frac{P_{iB}}{P_{iA}} \tag{9-22}$$

If the substance i is a solution-phase species in equilibrium with the same species in the gas phase, then for very dilute solutions experiments show that P_i is proportional to c_i, where c_i is the concentration of species i in solution. Substituting $P_i \propto c_i$ into Eq. 9-21 yields

$$d\mu_i = RT\,d\ln c_i \tag{9-23}$$

Integration of Eq. 9-23 from state A to state B yields

$$\mu_{iB} - \mu_{iA} = RT\ln\frac{c_{iB}}{c_{iA}} \tag{9-24}$$

The activity function is introduced to preserve the simple analytical form of Eqs. 9-22 and 9-24 for cases involving nonideal gases or nondilute solutions. The equation

$$d\mu_i = \bar{V}_i\,dP = RT\,d\ln a_i \qquad \text{(isothermal process)} \tag{9-25}$$

serves to define, to the extent of an as yet unspecified integration constant, the activity function a_i. Integration of Eq. 9-25 at constant temperature yields

$$\int_A^B d\mu_i = RT\int_A^B d\ln a_i$$

$$\mu_{iB} - \mu_{iA} = RT\ln\frac{a_{iB}}{a_{iA}} \tag{9-26}$$

If we now choose arbitrarily some convenient state (say, state A in Eq. 9-26) as a *standard state*, denoted by a superscript zero, and set $a_i^\circ \equiv 1$ for the standard state, then Eq. 9-26 becomes

$$\mu_i = \mu_i^\circ + RT\ln a_i \qquad \text{(isothermal process)} \tag{9-27}$$

In Eq. 9-27, the quantity μ_i° denotes the chemical potential of i in its standard state. Note that in Eq. 9-27 a_i is actually $a_i/a_i^\circ = a_i/1$, and thus we are free to pick the units of a_i because ($a_i/1$ units) is dimensionless.

For a pure (fixed-composition) condensed phase, we arbitrarily choose the standard state to be the pure substance in a reproducible defined state at one standard atmosphere pressure at the temperature of interest. The chemical potential μ is the molar Gibbs energy at that temperature in any defined state for which $P \neq 1$ atm. Note that at $P = 1$ atm we have $\mu = \mu^\circ$. From Eq. 9-25 we obtain for a pure condensed phase (subscript i unnecessary)

$$\int_{a=1}^{a} d\ln a = \int_1^P \frac{\bar{V}}{RT}\,dP$$

and

$$\ln a = \frac{1}{RT}\int_1^P \bar{V}\, dP \qquad \text{(constant } T\text{)} \tag{9-28}$$

In general, the molar volume $\bar{V}$ is a function of pressure, and thus the integral in Eq. 9-28 requires a knowledge $\bar{V} = \bar{V}(P)$ before it can be evaluated. If the pressure dependence of $\bar{V}$ can be neglected, then we obtain

$$\ln a \approx \frac{\bar{V}}{RT}(P - 1) \tag{9-29}$$

where $\bar{V}$ is taken as equal to $\bar{V}^\circ$, that is, the molar volume in the standard state.

EXAMPLE 9-4

Assume that $H_2O(\ell)$ is incompressible and compute the activity of $H_2O(\ell)$ at 10.0 atm, 25.0°C.

Solution: From Eq. 9-29 we have

$$\ln a_{H_2O} = \frac{\bar{V}_{H_2O}}{RT}(P - 1)$$

For water at 1 atm, $\bar{V}_{H_2O} = 18.0\ \text{cm}^3\cdot\text{mol}^{-1}$; thus

$$\ln a_{H_2O} = \frac{18.0\ \text{cm}^3\cdot\text{mol}^{-1}(10.0 - 1.00)\,\text{atm}}{82.1\ \text{cm}^3\cdot\text{atm}\cdot\text{K}^{-1}\cdot\text{mol}^{-1} \times 298.15\ \text{K}}$$

or

$$a_{H_2O} = 1.01$$

Note that the activity of $H_2O(\ell)$ is only weakly dependent on the pressure.

We choose the standard state of a gas as *the hypothetical* (i.e., nonattainable experimentally) *ideal gas at one atmosphere pressure and the temperature of interest*. We set $a \equiv 1$ in the standard state. We note from Eq. 9-25 that if $\bar{V} = RT/P$, then, and only then,

$$d\ln P = d\ln a$$

Integration of this equation from $P = 1.00$ atm (where $a = 1$ for an ideal gas) to pressure P

$$\int_{P=1}^{P} d\ln P = \int_{a=1}^{a} d\ln a$$

yields $a = P$ *for an ideal gas*.

We now proceed to the development of an expression relating a to P for a gas that is not ideal, and to this end we return to Eq. 9-25:

$$d\ln a = \frac{\bar{V}}{RT}\,dP \tag{9-30}$$

Subtracting $d \ln P$ from both sides of Eq. 9-30 yields

$$d \ln a - d \ln P = \frac{\bar{V}}{RT} dP - d \ln P \tag{9-31}$$

or

$$d \ln \frac{a}{P} = \left(\frac{\bar{V}}{RT} - \frac{1}{P} \right) dP \tag{9-32}$$

Integration of Eq. 9-32 from $P = 0$ to P yields

$$\int_0^P d \ln \frac{a}{P} = \int_0^P \left(\frac{\bar{V}}{RT} - \frac{1}{P} \right) dP \tag{9-33}$$

At the lower limit of the integral on the left side of Eq. 9-33 we have

$$\lim_{P \to 0} \left(\frac{a}{P} \right) = \frac{P}{P} = 1$$

because $a \to P$ as $P \to 0$; that is, a real gas behaves ideally ($a = P$) in the limit

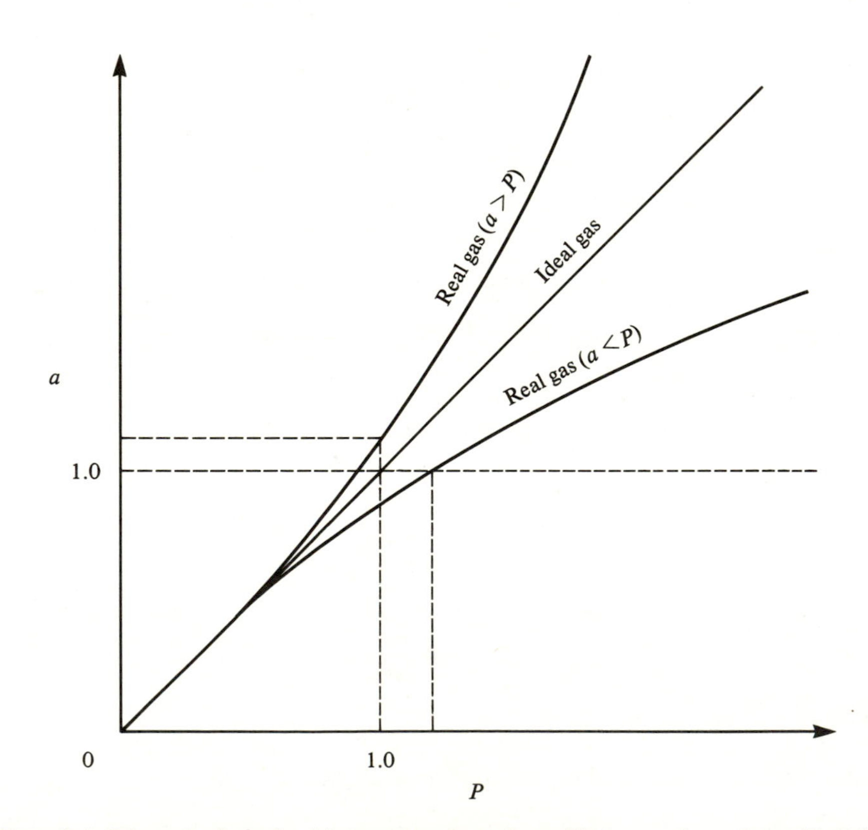

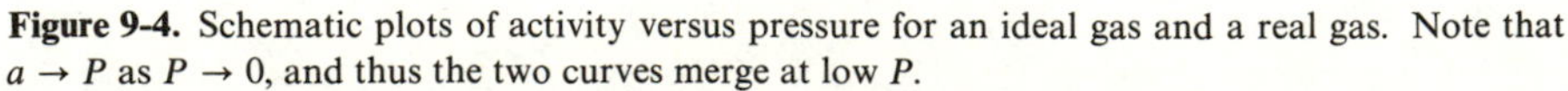
Figure 9-4. Schematic plots of activity versus pressure for an ideal gas and a real gas. Note that $a \to P$ as $P \to 0$, and thus the two curves merge at low P.

of zero pressure. Thus Eq. 9-33 becomes

$$\ln \frac{a}{P} = \int_0^P \left(\frac{\bar{V}}{RT} - \frac{1}{P} \right) dP \tag{9-34}$$

Note from Eq. 9-34 that $a \neq P$ in the case of a nonideal gas, because for a nonideal gas

$$\frac{\bar{V}}{RT} - \frac{1}{P} \neq 0$$

and thus the integral in Eq. 9-34 is nonzero. The distinction between the activity of a real gas and an ideal gas is depicted graphically in Figure 9-4.

We can evaluate the activity of a gas graphically using Eq. 9-34, if we have experimental P, $\bar{V}$, T data, by plotting $(\bar{V}/RT) - (1/P)$ versus P from 0 to P and computing the area under the curve. On the other hand, if the equation of state is available and is expressed in the form

$$P\bar{V} = RT + \sum_k C_k P^k$$

then Eq. 9-34 can be used to derive an analytical expression for the activity a.

EXAMPLE 9-5

Given that the equation of state for a real gas is

$$P\bar{V} = RT + BP$$

derive an analytical expression for the activity of the gas. Take $B = 60.0\ \text{cm}^3 \cdot \text{mol}^{-1}$ and evaluate a for the gas at 10.0 atm.

Solution: From the equation of state we have

$$\frac{\bar{V}}{RT} - \frac{1}{P} = \frac{B}{RT}$$

Using Eq. 9-34 we obtain

$$\ln \frac{a}{P} = \int_0^P \frac{B}{RT} dP$$

or

$$\ln \frac{a}{P} = \frac{BP}{RT}$$

Thus

$$a = Pe^{BP/RT}$$

with $B = 60.0\ \text{cm}^3 \cdot \text{mol}^{-1}$. We have, for the activity of the gas at $P =$ 10.0 atm, 25°C,

$$a = (10.0\ \text{atm}) \exp\left(\frac{60.0\ \text{cm}^3 \cdot \text{mol}^{-1} \times 10.0\ \text{atm}}{82.1\ \text{cm}^3 \cdot \text{atm} \cdot \text{K}^{-1} \cdot \text{mol}^{-1} \times 298.2\ \text{K}} \right) = 10.3\ \text{atm}$$

In this case the gas deviates from ideal gas behavior by about 3% at 10.0 atm.

At equilibrium the chemical potential of each substance must be the same in all phases in which it is a component. Thus we have

$$\mu_{iA} = \mu_{iB} \tag{9-35}$$

Using Eq. 9-27 we obtain

$$\mu_{i\alpha}^{\circ} + RT \ln a_{i\alpha} = \mu_{i\beta}^{\circ} + RT \ln a_{i\beta} \tag{9-36}$$

Rearrangement of Eq. 9-36 yields

$$\ln \frac{a_{i\beta}}{a_{i\alpha}} = \left[-\frac{(\mu_{i\beta}^{\circ} - \mu_{i\alpha}^{\circ})}{RT} \right]$$

or

$$a_{i\beta} = a_{i\alpha} \exp\left[-\frac{(\mu_{i\beta}^{\circ} - \mu_{i\alpha}^{\circ})}{RT} \right] \tag{9-37}$$

If the standard states are the same in the two phases (i.e., if $\mu_{i\beta}^{\circ} = \mu_{i\alpha}^{\circ}$), then at equilibrium

$$a_{i\beta} = a_{i\alpha} \tag{9-38}$$

whereas if the standard states are different in the two phases (e.g., gas and solid or gas and liquid), then because $\mu_{i\beta}^{\circ} - \mu_{i\alpha}^{\circ}$ is a constant at a given T, we have

$$a_{i\beta} = K a_{i\alpha}$$

where K,

$$K = \exp\left[-\frac{(\mu_{i\beta}^{\circ} - \mu_{i\alpha}^{\circ})}{RT} \right] = \frac{a_{i\beta}}{a_{i\alpha}} \tag{9-39}$$

is a constant at a given T. Thus, equality or proportionality of activities can be used as an equilibrium criterion in place of the equality of chemical potentials.

9-4 *The Thermodynamic Properties of a Gas Can Be Calculated from the Equation of State*

The hypothetical one-atmosphere ideal gas standard state of a real gas does not correspond to an actual state of the gas. However, given the equation of state for the gas, we can obtain expressions for the thermodynamic functions a (activity), $\mu - \mu^{\circ}$, $\bar{S} - \bar{S}^{\circ}$, $\bar{H} - \bar{H}^{\circ}$, and $\bar{C}_P - \bar{C}_P^{\circ}$ at any temperature directly from the equation of state by straightforward methods. For example, suppose that the equation of state is of the form

$$P\bar{V} = RT + BP + CP^2 + DP^3 + \cdots \tag{9-40}$$

where $B, C, D, \ldots$ are functions of T alone for a particular gas. The expression for the activity of the gas is obtained from Eq. 9-34:

$$\ln a = \ln P + \int_0^P \left(\frac{\bar{V}}{RT} - \frac{1}{P} \right) dP \tag{9-41}$$

Substitution of Eq. 9-40 into Eq. 9-41 yields

$$\ln a = \ln P + \frac{1}{RT} \int_0^P (B + CP + DP^2 + \cdots) dP$$

and thus

$$\ln a = \ln P + \frac{1}{RT}\left(BP + \frac{CP^2}{2} + \frac{DP^3}{3} + \cdots\right) \tag{9-42}$$

The chemical potential of the gas is given by Eq. 9-27 as

$$\mu = \mu^\circ + RT \ln a \tag{9-27}$$

Substitution of Eq. 9-42 into 9-27 yields

$$\mu - \mu^\circ = RT \ln P + BP + \frac{CP^2}{2} + \frac{DP^3}{3} + \cdots \tag{9-43}$$

We can obtain an expression for $\bar{S} - \bar{S}^\circ$ for the gas from Eq. 9-43. Because

$$\bar{S} - \bar{S}^\circ = -\left[\frac{\partial}{\partial T}(\mu - \mu^\circ)\right]_P \tag{9-44}$$

we have

$$\bar{S} - \bar{S}^\circ = -R \ln P - \left(P\frac{dB}{dT} + \frac{P^2}{2}\frac{dC}{dT} + \frac{P^3}{3}\frac{dD}{dT} + \cdots\right) \tag{9-45}$$

An expression for $\bar{H} - \bar{H}^\circ$ can be obtained from Eqs. 9-43 and 9-45 using the relation

$$\bar{H} - \bar{H}^\circ = \mu - \mu^\circ + T(\bar{S} - \bar{S}^\circ) \tag{9-46}$$

Thus

$$\bar{H} - \bar{H}^\circ = \left(B - T\frac{dB}{dT}\right)P + \left(C - T\frac{dC}{dT}\right)\frac{P^2}{2} + \left(D - T\frac{dD}{dT}\right)\frac{P^3}{3} + \cdots \tag{9-47}$$

Further, we have, for $\bar{C}_P - \bar{C}_P^\circ$,

$$\bar{C}_P - \bar{C}_P^\circ = \left[\frac{\partial(\bar{H} - \bar{H}^\circ)}{\partial T}\right]_P \tag{9-48}$$

and thus we obtain from Eq. 9-47

$$\bar{C}_P - \bar{C}_P^\circ = -T\left(P\frac{d^2B}{dT^2} + \frac{P^2}{2}\frac{d^2C}{dT^2} + \frac{P^3}{3}\frac{d^2D}{dT^2} + \cdots\right) \tag{9-49}$$

EXAMPLE 9-6

Given that the P, V, T behavior of a gas is described by Berthelot's equation of state

$$P\bar{V} = RT + \frac{9RT_cP}{128P_c}\left(1 - 6\frac{T_c^2}{T^2}\right)$$

where P_c and T_c are the critical pressure and critical temperature, respectively, derive expressions for the activity a, $\mu - \mu^\circ$, $\bar{S} - \bar{S}^\circ$, $\bar{H} - \bar{H}^\circ$, and $\bar{C}_P - \bar{C}_P^\circ$ for the gas.

Solution: The Berthelot equation of state is of the form

$$P\bar{V} = RT + BP$$

where

$$B = \frac{9RT_c}{128P_c}\left(1 - 6\frac{T_c^2}{T^2}\right)$$

and thus

$$\frac{dB}{dT} = \left(\frac{9RT_c}{128P_c}\right)\left(\frac{12T_c^2}{T^3}\right) = \frac{27RT_c^3}{32P_cT^3}$$

From Eqs. 9-42, 9-43, 9-45, 9-47, and 9-49 we obtain, respectively,

$$\ln a = \ln P + \frac{9T_c}{128P_cT}\left(1 - \frac{6T_c^2}{T^2}\right)P$$

$$\mu - \mu^\circ = RT\ln P + \frac{9RT_c}{128P_c}\left(1 - \frac{6T_c^2}{T^2}\right)P$$

$$\bar{S} - \bar{S}^\circ = -R\ln P - \frac{27RT_c^3P}{32P_cT^3}$$

$$\bar{H} - \bar{H}^\circ = \frac{9RT_c}{128P_c}\left(1 - \frac{18T_c^2}{T^2}\right)P$$

$$\bar{C}_P - \bar{C}_P^\circ = \frac{81RT_c^3P}{32P_cT^3}$$

Note that the above equations all reduce to ideal gas results when $B = 0$ or, equivalently, when $P_c \to \infty$. Thus for an ideal gas we have from the above equations ($P_c \to \infty$)

$$a = P \qquad \mu - \mu^\circ = RT\ln P \qquad \bar{S} - \bar{S}^\circ = -R\ln P$$

$$\bar{H} - \bar{H}^\circ = 0 \quad \text{and} \quad \bar{C}_P - \bar{C}_P^\circ = 0$$

Application of the expressions obtained in Example 9-6 to $CO_2(g)$ (P_c = 73.0 atm, T_c = 304.2 K) in the state 100.0 atm, 500.0 K yields the following results for $CO_2(g)$ assuming (1) ideal gas behavior and (2) Berthelot gas behavior.

CO_2(g) at 100.0 atm, 500K

Quantity	*Ideal gas*	*Berthelot gas*
a/atm	100.0	93.1
$(\mu - \mu^\circ)/\text{kJ}\cdot\text{mol}^{-1}$	19.15	18.85
$(\bar{S} - \bar{S}^\circ)/\text{J}\cdot\text{K}^{-1}\cdot\text{mol}^{-1}$	−38.28	−40.46
$(\bar{H} - \bar{H}^\circ)/\text{kJ}\cdot\text{mol}^{-1}$	0	−1.380
$(\bar{C}_P - \bar{C}_P^\circ)/\text{J}\cdot\text{K}^{-1}$	0	6.49

The above results show that possible deviations from ideal gas behavior must be considered carefully in thermodynamic calculations.

Problems

1. The enthalpy of vaporizarion of water at 25°C is 44.0 $\text{kJ}\cdot\text{mol}^{-1}$. Estimate the equilibrium vapor pressure of water at 120°C.

2. The following table gives the vapor pressure of liquid Ar from its melting point (−189.2°C) to its boiling point (−185.7°C).

$t/°C$	$P/torr$
−189.0	414
−188.0	466
−187.0	523
−186.0	585

Calculate the molar enthalpy of vaporization of argon.

3. The following table gives the vapor pressure of solid argon from −208°C to −189.2°C (its melting point).

$t/°C$	$P/torr$
−208	20.4
−206	31.4
−203	57.1
−200	98.9
−196	192.5
−190	463.5

Calculate the heat of sublimation of argon.

4. The vapor pressure (in torr) of solid zinc is given by the formula

$$\log P = -\frac{6946}{T} + 9.200$$

from 250°C to 419.4°C (the melting point of zinc). Calculate the heat of sublimation of zinc over this temperature range.

5. The vapor pressures (in torr) of solid and liquid chlorine are given by

$$\log P = -\frac{1640}{T} + 10.560 \qquad \text{(solid)}$$

$$\log P = -\frac{1159}{T} + 7.769 \qquad \text{(liquid)}$$

Calculate the temperature and the pressure at the triple point of chlorine.

6. The vapor pressures (in torr) of solid and liquid argon are given by the formulas

$$\log P_s = -\frac{408.2}{T} + 7.5741 \qquad (-208^\circ\text{C to } -189^\circ\text{C})$$

$$\log P_\ell = -\frac{356.5}{T} + 6.9605 \qquad (-189^\circ\text{C to } -186^\circ\text{C})$$

Calculate $\Delta\bar{H}_{sub}$, $\Delta\bar{H}_{vap}$, and $\Delta\bar{H}_{fus}$ for argon.

7. The vapor pressure of potassium from 260°C to 760°C (its boiling point) can be represented by the empirical formula

$$\log P = -\frac{4434}{T} + 7.183$$

where P is the vapor pressure in torr and T is the Kelvin temperature. Use this formula to determine the heat of vaporization of potassium.

8. The surface tension of water is 72 $\text{mJ}\cdot\text{m}^{-2}$. Calculate the energy that is required to disperse one spherical drop of radius 3.0 mm into spherical drops of radius 3.0×10^{-3} mm.

9. The vapor pressures (in torr) of solid and liquid uranium hexafluoride are given by

$$\log P_s = 10.646 - \frac{2559.1}{T} \qquad \text{(solid)}$$

$$\log P_\ell = 7.538 - \frac{1511}{T} \qquad \text{(liquid)}$$

Calculate the temperature and the pressure at the triple point of UF_6. Also calculate the equilibrium vapor pressure of UF_6 at 25°C.

10. The vapor pressures of solid chlorine are 3.52 torr at −110°C and 0.776 torr at −120°C. The vapor pressures of liquid chlorine are 27.6 torr at −90°C and 58.7 torr at −80°C. Calculate (a) $\Delta\bar{H}_{sub}$, $\Delta\bar{H}_{vap}$, $\Delta\bar{H}_{fus}$, and (b) the temperature and the pressure at the triple point of chlorine.

11. The phase diagram of sulfur is shown in Figure 9-5. How many triple points are there? Describe what happens if sulfur is heated from 40°C at 1 atm to 200°C at 1 atm. Below what pressure will sublimation occur?

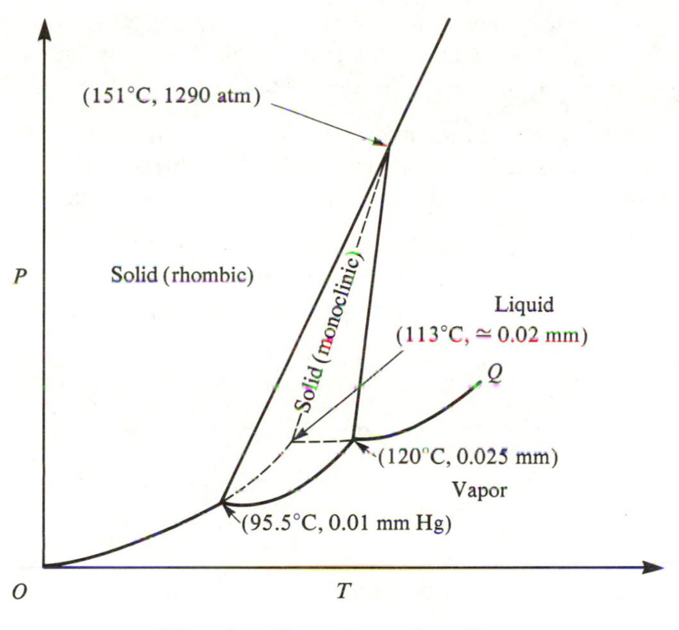

Figure 9-5. Phase diagram for sulfur.

12. Mercury vapors are very toxic. Suppose that some Hg(ℓ) is present in a sample of water that is heated to boiling. Given that the boiling point of Hg(ℓ) is 629 K and $\Delta \bar{H}_{vap} = 59.2\ \mathrm{kJ \cdot mol^{-1}}$ for Hg(ℓ), compute the vapor pressure of mercury at 100°C.

13. Trouton's rule states that the molar enthalpy of vaporization of a liquid that does not involve strong specific intermolecular interactions (such as hydrogen bonding or ion-ion attractions) is given by

$$\Delta \bar{H}_{vap} = (88\ \mathrm{J \cdot K^{-1} \cdot mol^{-1}}) T_{bp}$$

where T_{bp} is the normal boiling point of the liquid in degrees kelvin. Apply Trouton's rule to water and suggest a molecular-level explanation for any discrepancy that is found.

14. Trouton's rule can be combined with the Clapeyron-Clausius equation to yield the following equation

$$\log P = 4.60\left(1 - \frac{T_{bp}}{T}\right)$$

which can be used to estimate the vapor pressure P (in atmospheres) of a liquid at temperature T, given only the boiling point T_{bp} (in degrees kelvin) of the liquid. Given that the boiling point of the carcinogen carbon tetrachloride, CCl_4, is 77°C, (a) estimate the vapor pressure of CCl_4 at 20°C. (b) Derive the above equation.

15. Propylene carbonate has a normal boiling point of 160°C. Use Trouton's rule ($\Delta \bar{S}_{vap} = 88\ J \cdot K^{-1} \cdot mol^{-1}$ at the normal boiling point) to estimate the equilibrium vapor pressure of propylene carbonate liquid at 25°C.

16. Given that about 85% of the hydrogen atoms in liquid water are hydrogen bonded at 25°C, use the molar enthalpy of vaporization of liquid water ($44\ kJ \cdot mol^{-1}$) to estimate the hydrogen bond energy (kilojoules per mole) in liquid water.

17. Atmospheric pressure decreases with altitude. For example, on a clear day, when the atmospheric pressure at sea level is 760 torr, we have

Altitude/ft	*Atmospheric pressure/atm*
8000	0.75
10,000	0.67
15,000	0.50
20,000	0.35

Construct a plot of $\log P$ versus h and use your plot to estimate the boiling point of water at 12,000 ft. Explain how the boiling point of water can be used to determine altitude.

18. It is observed that a tighly capped bottle of a carbonated liquid soft drink at a temperature of -10°C begins to form ice when the cap is removed. Explain this observation.

19. The equilibrium vapor pressure of $CO_2(s)$ ("dry ice") is 439 torr at -85°C and 1009 torr at -75°C. Estimate $\Delta \bar{H}_{sub}$ for $CO_2(s)$.

20. For butane, which is used as a fuel in disposable cigarette lighters, the value of $\Delta \bar{H}_{vap}$ is $24.3\ kJ \cdot mol^{-1}$. The normal boiling point of butane is -0.5°C. Estimate the vapor pressure of liquid butane at 37°C. The result is the butane gas pressure in a cigarette lighter containing liquid butane that is in contact with the human body.

21. The equilibrium vapor pressure of benzene at various temperatures is as follows:

t/°C	20.6	39.1	49.1	60.8	74.0	80.9
P/torr	77.3	175.9	261.7	402.4	627.9	779.3

Construct a plot of $\log P$ versus $1/T$ and determine $\Delta \bar{H}_{vap}$ in kilojoules per mole for $C_6H_6(\ell)$. Also estimate the normal boiling point of $C_6H_6(\ell)$ and $\Delta \bar{S}_{vap}$ (1-atm gas) for $C_6H_6(\ell)$.

22. The equilibrium vapor pressure of solid carbon dioxide is 1.00 atm at $-78.3°C$. The enthalpy change for the reaction

$$CO_2(s) \rightleftharpoons CO_2(g)$$

is 25.2 kJ at $-78.3°C$. The polar caps on Mars are thought to be composed primarily of $CO_2(s)$. Given that the pressure of CO_2 in the Martian atmosphere is about 4.0 torr, estimate the temperature of the polar caps on Mars.

23. Derive an expression for the activity of a gas whose equation of state is

$$P\bar{V} = RT + BP + CP^2$$

24. Show that if $\Delta\bar{H}_{sub}$ is a constant, then

$$\ln P = -\frac{\Delta\bar{H}_{sub}}{RT} + b$$

where b is a constant.

25. The compressibility of $H_2O(\ell)$ at 25°C is $\beta = 4.67 \times 10^{-5}\ \text{atm}^{-1}$.
(a) Compute the activity of $H_2O(\ell)$ at 100 atm and at 1000 atm.
(b) Compare your results for a_{H_2O} in part a with those calculated assuming that $\beta = 0$ for $H_2O(\ell)$.

26. Show that the Clapeyron-Clausius equation is not limited to two-phase equilibria and that the equation can be applied to a case of the type

$$CaCO_3(s) \rightleftharpoons CaO(s) + CO_2(g)$$

with the qualification that in this case

$$\Delta\bar{S}_{tr} = \Delta S_{rxn} = \bar{S}_{CO_2(g)} + \bar{S}_{CaO(s)} - \bar{S}_{CaCO_3(s)}$$
$$\Delta\bar{H}_{tr} = \Delta H_{rxn} = \bar{H}_{CO_2(g)} + \bar{H}_{CaO(s)} - \bar{H}_{CaCO_3(s)}$$

Hint: Start the derivation using the equilibrium condition

$$\mu_{CO_2(g)} + \mu_{CaO(s)} = \mu_{CaCO_3(s)}$$

for the reaction.

27. Using the Clapeyron equation, we found that the normal freezing point of water lies about 0.0075 K below the triple point (4.58 torr, 273.16 K) of water. The *measured* value of the freezing point of water under an atmosphere of air is $-0.010°C$. Suggest an explanation for the small discrepancy.

28. Show that the integration constant in Eq. 9-15 is

$$b = \frac{\Delta\bar{S}'_{vap}}{R}$$

where $\Delta\bar{S}'_{vap}$ is the entropy change on vaporization of the liquid to the 1-atm gas. Use the above result to show that

$$P_{vap} = e^{\Delta S'_{vap}/R} e^{-\Delta H_{vap}/RT}$$

29. Use the vapor pressure data for water in Table 8-1 to construct a log P versus $1/T$ plot for the $H_2O(s) \rightleftharpoons H_2O(g)$ and $H_2O(\ell) \rightleftharpoons H_2O(g)$ equilibria, and compute $\Delta\bar{H}_{sub}$ and $\Delta\bar{H}_{vap}$ from your plot at several temperatures. Can you detect curvature in the $H_2O(\ell) \rightleftharpoons H_2O(g)$ plot?

30. Show that for a reaction of the type

$$\mathrm{A}\cdot n\mathrm{X}(s) \rightleftharpoons \mathrm{A}(s) + n\mathrm{X}(g)$$

that the Clapeyron-Clausius equation takes the form

$$\frac{d \ln P}{dT} \simeq \frac{\Delta\bar{H}}{nRT^2}$$

31. Given the following data for a certain compound, derive an expression for the equilibrium vapor pressure as a function of temperature.

$$\begin{aligned} \bar{C}_{P(\mathrm{gas})} &= (28.28 + 4.39 \times 10^{-3}T)\ \mathrm{J \cdot K^{-1} \cdot mol^{-1}} \\ \bar{C}_{P(\mathrm{liq})} &= (24.10 - 3.97 \times 10^{-3}T)\ \mathrm{J \cdot K^{-1} \cdot mol^{-1}} \\ \Delta\bar{H}_{\mathrm{vap}}(100^\circ\mathrm{C}) &= 41.84\ \mathrm{kJ \cdot mol^{-1}} \\ \text{normal boiling point} &= 120.0^\circ\mathrm{C} \end{aligned}$$

32. Derive an expression for the activity of a condensed phase whose compressibility at a particular T is given by

$$\beta = k_1 + k_2 P$$

where k_1 and k_2 are constants.

33. Take $\Delta\bar{H} = \Delta\bar{H}(T, P)$ and derive the equation

$$\frac{d\Delta\bar{H}}{dT} = \Delta\bar{C}_P + \frac{\Delta\bar{H}}{T} - \Delta\bar{H}\left(\frac{\partial \ln \Delta\bar{V}}{\partial T}\right)_P$$

34. The transition from the normal to the superconducting state (a superconductor has a zero resistance and is perfectly diamagnetic; i.e., it cannot be penetrated by a magnetic field) is a second-order phase change in the absence of a magnetic field, but the transition is first order in the presence of a magnetic field. The applied field necessary to cause the transition back to the normal state is given by

$$\mathscr{H} = \mathscr{H}_0\left(1 - \frac{T^2}{T_c^2}\right)$$

where $\mathscr{H}_0$ is the value of the field at $T = 0$ and T_c is the transition temperature at $\mathscr{H} = 0$. Derive the expression (analogous to the Clapeyron-Clausius equation)

$$\frac{d\mathscr{H}}{dT} = -\frac{(H_n - H_s)}{\mu_n V_n T \mathscr{H}}$$

which is valid under conditions of constant pressure and involves the approximation that $V_n = V_s$, where the subscripts n and s refer to the normal and superconducting phases, respectively. Note that $H = U + PV - \mu_0 \mathscr{H} M$ and $dw = \mathscr{H}\, dM$. Also, $M = \chi \mathscr{H} = [(\mu/\mu_0) - 1]V\mathscr{H}$, and $\mu = 0$ for a superconductor, where μ is the magnetic permeability.

35. Derive the expression $\bar{C}_P - \bar{C}_V = BT\bar{C}_V^2$ applicable to condensed phases where B is a constant. If we know $\bar{C}_P - \bar{C}_V$ at one T, then we can compute B and then use this equation to compute $\bar{C}_V$ from $\bar{C}_P$ data.

36. The Gibbs equation

$$\left(\frac{\partial \ln P_{\text{gas}}}{\partial P_{\text{tot}}}\right)_T = \frac{\bar{V}_\ell}{RT}$$

describes the variation of the equilibrium vapor pressure of a liquid (P_{gas}) as the function of the total pressure over the liquid phase, assuming that the inert gas ($P_{\text{tot}} = P_{\text{gas}} + P_{\text{inert}}$) does not dissolve in the condensed phase.
(a) Derive the Gibbs equation.
(b) Calculate the equilibrium vapor pressure of $H_2O(\ell)$ at 25°C subjected to a total pressure of 100 atm using an inert gas. Use your result to compute the percent increase in the equilibrium vapor pressure of water.

37. The diamond-to-graphite equilibrium on the carbon phase diagram has $dP/dT > 0$, with diamond being the high-pressure form; yet $\Delta H^\circ_{298} < 0$ and $\Delta V^\circ_{298} > 0$ for the transition diamond → graphite. How do you reconcile these data with the Clapeyron equation?

38. Show that for a gas as $P \to 0$, $\bar{H} \to$ constant, $\bar{C}_P \to$ constant, $\mu \to -\infty$ (i.e., undefined), $\bar{S} \to +\infty$ (i.e., undefined).

39. Show that if the principle of corresponding states (see Problem 43 in Chapter 8) holds, then all gases have the same value of the ratio of a/P, when compared at the same values of P_R and T_R.

40. Derive the expression

$$\bar{C}_P - \bar{C}_{\text{sat}} \approx \left(\frac{\alpha_\ell \bar{V}_\ell \Delta \bar{H}_{\text{vap}}}{RT}\right) P$$

where $\bar{C}_P$ is the heat capacity of a liquid, $\bar{C}_{\text{sat}}$ is the heat capacity of the liquid in equilibrium with its vapor, and α_ℓ is the thermal coefficient of expansion of the liquid. *Hints:* $\delta q_{\text{sat}} = dH - V\,dP$ and $\bar{C}_{\text{sat}} = \lim_{\Delta T \to 0} (q_{\text{sat}}/\Delta T)$.

41. Derive the Ehrenfest equations

$$\frac{dP}{dT} = \frac{\alpha_2 - \alpha_1}{\beta_2 - \beta_1} \quad \text{and} \quad \frac{dP}{dT} = \frac{C_{P_2} - C_{P_1}}{TV(\alpha_2 - \alpha_1)}$$

which are the second-order phase transition analogs of the Clapeyron-Clausius equation. Note that

$$\alpha_i = \frac{1}{V_i}\left(\frac{\partial V_i}{\partial T}\right)_P \quad \text{and} \quad \beta_i = -\frac{1}{V_i}\left(\frac{\partial V_i}{\partial P}\right)_T$$

42. Consider a gas whose equation of state is

$$P\bar{V} = RT + BP + CP^2$$

where B and C depend only on T. Obtain general expressions for $\Delta\bar{G}$, $\Delta\bar{H}$, and $\Delta\bar{S}$ for this gas that are applicable to the change in state

$$\text{gas}(T, a = P = 1) \longrightarrow \text{gas}(T, P)$$

43. The enthalpy difference $\bar{H} - \bar{H}^\circ$ at a particular temperature can be obtained from Joule-Thomson free-expansion ($P_2 \to 0$) data. Show that

$$\bar{H}(P \ll 1, T_2) = \bar{H}(P = 0, T_2) = \bar{H}^\circ(T_2)$$

and use this result to show that for a Joule-Thomson free expansion from (P_1, T_1) to $(P_2 \to 0, T_2)$

$$(\bar{H} - \bar{H}^\circ)_{\text{at } T_1} = \bar{H}(P_1, T_1) - \bar{H}^\circ(T_1) = \bar{H}^\circ(T_2) - \bar{H}^\circ(T_1)$$
$$= \int_{T_1}^{T_2} \bar{C}_P^\circ \, dT$$

where T_1 and T_2 are obtained from the Joule-Thomson free-expansion experiments. Suppose we have a gas for which $\bar{C}_P = 20.79\ \text{J}\cdot\text{K}^{-1}\cdot\text{mol}^{-1}$ and that this gas undergoes a Joule-Thomson free expansion from 500 atm, 350 K to 275 K. Compute $\bar{H}(500\ \text{atm}) - \bar{H}^\circ$ at 350 K.

44. Suppose we have an equation of state for a gas that can be solved explicitly for P but not for V, for example, a van der Waals gas

$$P = \frac{RT}{\bar{V} - b} - \frac{a}{\bar{V}^2}$$

In such a case it is more convenient to work with the Helmholtz energy function than the Gibbs energy function. At any particular temperature, $d\bar{A} = -P\,d\bar{V}$, and thus

$$\bar{A} - \bar{A}^\circ = -\int_{\bar{V}=RT, P=1}^{\bar{V}} P\,d\bar{V}$$

where the lower limit refers to the hypothetical ideal gas 1-atm standard state. We cannot simply insert $P = P(V, T)$ into the above expression and integrate at constant T, because the state $\bar{V} = RT$, $P = 1$ is not an actual state of the real gas. In the limit $P \to 0$ ($V \to \infty$), the real gas will behave ideally; therefore, we can rewrite the above expression for $\bar{A} - \bar{A}^\circ$ as

$$\bar{A} - \bar{A}^\circ = -\int_{\infty}^{\bar{V}} P\,d\bar{V} - \int_{\bar{V}}^{\infty} \frac{RT}{\bar{V}}\,d\bar{V}$$

Further

$$\bar{A} - \bar{A}^\circ = -\int_{\infty}^{\bar{V}} P\,d\bar{V} - \int_{RT}^{\bar{V}} \frac{RT}{\bar{V}}\,d\bar{V} - \int_{\bar{V}}^{\infty} \frac{RT}{\bar{V}}\,d\bar{V}$$

and thus

$$\bar{A} - \bar{A}^\circ = RT\ln\left(\frac{RT}{\bar{V}}\right) + \int_{\infty}^{\bar{V}} \left(\frac{RT}{\bar{V}} - P\right) d\bar{V}$$

Use the above expression for $\bar{A} - \bar{A}^\circ$ to obtain the following expression for the activity of a van der Waals gas:

$$\ln a = \ln\left(\frac{RT}{\bar{V} - b}\right) - \frac{2a}{RT\bar{V}} + \frac{b}{\bar{V} - b}$$

(*Hint:* Note that $\mu - \mu^\circ = \bar{A} - \bar{A}^\circ + P\bar{V} - P\bar{V}^\circ$.) Further, show that the above expression can be approximated by

$$\ln a = \ln\left[P\left(1 + \frac{aP}{R^2T^2}\right)\right] - \frac{2aP}{R^2T^2} + \frac{bP}{RT}\left(1 + \frac{aP}{R^2T^2}\right)$$

45. Consider the equations of state
(a) $P(\bar{V} - b) = RT$

(b) $\left(P + \frac{a'}{\bar{V}^2}\right)(\bar{V} - b) = RT$ (van der Waals gas)

Take $b = 42.7\ \mathrm{cm^3 \cdot mol^{-1}}$ and $a' = 3.59\ \mathrm{L^2 \cdot atm}$ for $CO_2(g)$, and compute a (activity), $\mu - \mu^\circ$, $\bar{S} - \bar{S}^\circ$, $\bar{H} - \bar{H}^\circ$, and $\bar{C}_P - \bar{C}_P^\circ$ for $CO_2(g)$ at 100.0 atm, 500 K. Compare your results for cases (a) and (b) with the analogous results for $CO_2(g)$ in Example 9-6. See Problem 44.

46. Derive a relationship between the compressibility factor $Z = P\bar{V}/RT$ and the activity of (a) a Berthelot gas, (b) a Dieterici gas, and (c) a van der Waals gas. (See Chapter 8 Problems 40–42.)

The Brea Plant of The Union Oil Company in Brea, California, produces liquid ammonia, dry ice, urea, nitric acid, ammonium nitrate, and aqueous solutions of ammonia, urea, and ammonium nitrate plus urea. The 750-ton-per-day ammonia production unit is located to the right of the two liquid ammonia storage spheres near the center of the photo. The hydrogen is obtained by the high-temperature reaction of methane and steam, with CO_2 as a by-product, which is used to make $CO_2(s)$ and H_2NCONH_2, the latter by the high-pressure reaction of CO_2 with NH_3. The nitrogen used is obtained from air. The nitrogen-hydrogen mixture is compressed to about 135 atm at about 480°C to form $NH_3(g)$. Nitric acid is made by oxidation of $NH_3(g)$ with $O_2(g)$ to produce an $NO(g) + NO_2(g)$ mixture that is dissolved in water to yield $HNO_3(aq)$. The largest tank shown is a $NH_3(l)$ storage tank, to the left of which are the nitric acid and ammonium nitrate plants. The urea plant is to the far left in the photo.

10

Equilibrium Constants

Every system in stable chemical equilibrium subjected to the influence of an external cause that tends to vary either its temperature or its condensation (pressure, concentration, number of molecules per unit volume) as a whole or merely in some of its parts, cannot experience other than those internal modifications which, if they were to be produced alone, would lead to a change of temperature or condensation of opposite sign to that resulting from the external cause.

Henry Le Châtelier, 1884

A narrow professional interest in the preparation of ammonia from the elements was based on the achievement of a simple result by means of special equipment. A more widespread interest was due to the fact that the synthesis of ammonia from its elements, if carried out on a large scale, would be a useful way of satisfying important economic needs. Such practical uses were not the principal purpose of my investigations. On the other hand, I would hardly have concentrated so hard on this problem had I not been convinced of the economic necessity of chemical progress in this field, and had I not shared to the full Fichte's conviction that while the immediate object of science lies in its own development, its ultimate aim must be bound up in the moulding influence which it exerts at the right time upon life in general and the whole human arrangement of things around us.

Fritz Haber, 1920,
Nobel Prize Lecture

10-1 *The Lewis Equation Gives the Dependence of ΔG_{rxn} on the Activities of the Products and Reactants*

What we wish to obtain in this section is a general relationship between ΔG_{rxn} and ΔG°_{rxn} for a chemical reaction or phase transformation. We know from earlier discussions (Section 5-7) that for a closed system (which may contain several phases) at constant T and P, for which only pressure-volume work is possible, we have as our equilibrium criterion $\Delta G = 0$ for any process; whereas $\Delta G < 0$ for any *spontaneous* process. Recall that the quantity ΔG° represents the standard Gibbs energy change for the reaction; that is, ΔG°_{rxn} is the Gibbs energy change when the reactants in their 1-atm standard states are converted into products in their 1-atm standard states at the temperature of interest. With fixed-pressure standard states, the quantity ΔG°_{rxn} is a function only of T. These distinctions are depicted schematically in Figure 10-1.

Figure 10-1 will now be used in the analysis of the general chemical reaction

$$m\text{M} + n\text{N} + \cdots \rightleftharpoons x\text{X} + y\text{Y} + \cdots \tag{10-1}$$

where $(m, n, \ldots)$ and $(x, y, \ldots)$ are the reaction-balancing coefficients and $(\text{M}, \text{N}, \ldots)$ and $(\text{X}, \text{Y}, \ldots)$ are the reactants and products, respectively. Because G is a thermodynamic coordinate, and ΔG_{rxn} values are just differences in G values, the sum of the ΔG values around any closed cycle must be zero. From Figure 10-1 we obtain the following condition:

$$\Delta G_{rxn} = \Delta G_1 + \Delta G^\circ_{rxn} + \Delta G_2 \tag{10-2}$$

For ΔG_1 we have (see Figure 10-1 and Eq. 10-1)

$$\Delta G_1 = m(\mu^\circ_{\text{M}} - \mu_{\text{M}}) + n(\mu^\circ_{\text{N}} - \mu_{\text{N}}) + \cdots \tag{10-3}$$

From Eq. 9-27 we have

$$\mu_i = \mu^\circ_i + RT\ln a_i \tag{10-4}$$

Substitution of Eq. 10-4 for each reactant species into Eq. 10-3 yields

$$\Delta G_1 = mRT\ln\frac{1}{a_{\text{M}}} + nRT\ln\frac{1}{a_{\text{N}}} + \cdots \tag{10-5}$$

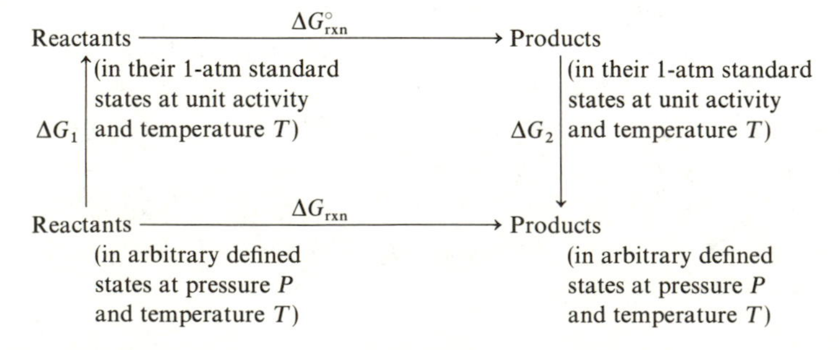

Figure 10-1. The distinction between ΔG_{rxn} and ΔG°_{rxn} for a reaction. Note that because ΔG is independent of path (G is a state function), we have

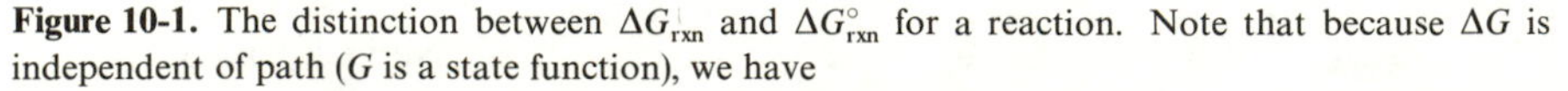

$$\Delta G_{rxn} = \Delta G_1 + \Delta G^\circ_{rxn} + \Delta G_2$$

Collecting terms in Eq. 10-5 yields

$$\Delta G_1 = RT\ln\left(\frac{1}{a_M^m a_N^n \cdots}\right) \tag{10-6}$$

Similarly, for ΔG_2 we have (see Figure 10-1 and Eq. 10-1)

$$\Delta G_2 = x(\mu_X - \mu_X^\circ) + y(\mu_Y - \mu_Y^\circ) + \cdots$$

Using Eq. 10-4 we obtain

$$\Delta G_2 = xRT\ln a_X + yRT\ln a_Y + \cdots \tag{10-7}$$

Collecting terms in Eq. 10-7 yields

$$\Delta G_2 = RT\ln(a_X^x a_Y^y \cdots) \tag{10-8}$$

Combination of Eqs. 10-8 and 10-5 with Eq. 10-2 yields the *Lewis equation**

$$\Delta G_{rxn} = \Delta G_{rxn}^\circ + RT\ln\left(\frac{a_X^x a_Y^y \cdots}{a_M^m a_N^n \cdots}\right) \tag{10-9}$$

It is convenient to define the *activity quotient* Q where

$$Q = \left(\frac{a_X^x a_Y^y \cdots}{a_M^m a_N^n \cdots}\right) \tag{10-10}$$

Note that the activity of each product species raised to a power equal to its stoichiometric coefficient appears in the numerator of Q, whereas the activity of each reactant species raised to a power equal to its stoichiometric coefficient appears in the denominator of Q.

Combination of Eqs. 10-10 and 10-9 yields the Lewis equation in a more compact form:

$$\Delta G_{rxn} = \Delta G_{rxn}^\circ + RT\ln Q \tag{10-11}$$

If the chemical reaction takes place with all the reactants and products in their standard states, then all the a_i values in Eq. 10-11 are unity; therefore, $Q = 1$, and $\Delta G_{rxn} = \Delta G_{rxn}^\circ$. If all the reactants and products are not in their respective standard states, then $Q \neq 1$ and $\Delta G_{rxn} \neq \Delta G_{rxn}^\circ$.

If a chemical reaction occurring at fixed T and P is at equilibrium, then $\Delta G_{rxn} = 0$; from the Lewis equation we obtain

$$0 = \Delta G_{rxn}^\circ + RT\ln Q_{eq}$$

or

$$\Delta G_{rxn}^\circ = -RT\ln Q_{eq} \tag{10-12}$$

where Q_{eq} denotes the value of Q at equilibrium. The standard Gibbs energy change for the reaction ΔG_{rxn}° is a constant at a given T, independent of P;

* Named by the author in honor of the great chemical thermodynamicist G. N. Lewis, who invented the activity function, pioneered the rigorous thermodynamic analysis of chemical equilibria, and was the first to write the equivalent of Eq. 10-9.

therefore, Q_{eq} is a constant at a given T independent of P. To emphasize this constancy, we denote Q_{eq} by K and call K the *equilibrium constant for the reaction*. Thus we can rewrite Eq. 10-12 as

$$\Delta G^{\circ}_{rxn} = -RT \ln K \quad (10\text{-}13)$$

The equilibrium constant expression for a general chemical reaction (see Eqs. 10-1 and 10-10) is

$$K = \left(\frac{a_X^x a_Y^y \cdots}{a_M^m a_N^n \cdots} \right)_{eq} \quad (10\text{-}14)$$

The algebraic form of K is exactly the same as that for Q, Eq. 10-10. There is a critical distinction, however, between Q and K: The a_i values that appear in the K expression are *equilibrium values*, whereas the a_i values that appear in Q are arbitrary; in particular, the a_i values in Q *need not be equilibrium values*. The experimental determination of K amounts to the determination of an *equilibrium set of a_i values*.

An equilibrium constant K (Eq. 10-14) is necessarily a nonzero, positive quantity because the activity of any substance present at equilibrium is necessarily nonzero and positive. *The equilibrium constant for a reaction is a dimensionless quantity* because each of the activities that appears in K is actually of the form $(a_X/1 \text{ units})^x$, $(a_M/1 \text{ units})^m$, and so forth, where the units in the denominator in each case are the same as the units of a_i. The fact that K does not have units does not mean that we do not have to be concerned with units in the formulation of K. The units of the a_i values must be the same as those chosen for the standard state of species i. For example, if we choose a 1-atm ideal gas standard state for a gas, then the activity of the gas that is used in K must be expressed in atmospheres. Similar considerations apply to solution-phase species.

Combination of the Lewis equation (Eq. 10-11) with Eq. 10-13 yields

$$\Delta G_{rxn} = -RT \ln K + RT \ln Q$$

or

$$\Delta G_{rxn} = RT \ln \left(\frac{Q}{K} \right) \quad (10\text{-}15)$$

The following possibilities can arise for the value of the ratio Q/K:

Q/K	ΔG_{rxn}	*Direction in which the reaction proceeds spontaneously toward equilibrium*
1	0	No net change—reaction at equilibrium
<1	<0	(left to right) $\longrightarrow$
>1	>0	(right to left) $\longleftarrow$

The value of the ratio Q/K is thus a measure of the degree of spontaneity of a chemical reaction. If $Q/K < 1$, then the chemical reaction is spontaneous from left to right as written. The reaction proceeds toward equilibrium with Q increasing (production of products, depletion of reactants) until either $Q = K$, that is, until equilibrium is attained, or until a reactant is completely exhausted, in which case the reaction does not attain equilibrium. If $Q/K > 1$, then the chemical reaction is spontaneous from right to left as written. The reaction proceeds toward equilibrium with Q decreasing (production of reactants, depletion of products) until either $Q = K$, that is, until equilibrium is attained, or until a product is completely exhausted, in which case the reaction does not attain equilibrium.

EXAMPLE 10-1

For the reaction

$$S(s) + O_2(g) \rightleftharpoons SO_2(g)$$

at 25°C, $\Delta H^\circ_{rxn} = -297$ kJ and $\Delta S^\circ_{rxn} = 11.3\ J\cdot K^{-1}$. Compute ΔG°_{rxn} at 25°C, and determine the direction for which the reaction is spontaneous with $O_2(g)$ and $SO_2(g)$ both at 1.0 atm.

Solution: The value of ΔG°_{rxn} at temperature T is given by

$$\Delta G^\circ_{rxn} = \Delta H^\circ_{rxn} - T\Delta S^\circ_{rxn}$$

Thus

$$\Delta G^\circ_{rxn} = -297\ \text{kJ} - \frac{(298\ \text{K})(11.3\ \text{J}\cdot\text{K}^{-1})}{(1000\ \text{J}\cdot\text{kJ}^{-1})} = -300\ \text{kJ}$$

The value of Q is

$$Q = \frac{a_{SO_2(g)}}{a_{S(s)} \cdot a_{O_2(g)}} \approx \frac{(1.00)}{(1.00)(1.00)} = 1.00$$

and, therefore,

$$\Delta G_{rxn} = \Delta G^\circ_{rxn} + RT\ln Q = \Delta G^\circ_{rxn} + RT\ln(1)$$

Thus

$$\Delta G_{rxn} = \Delta G^\circ_{rxn} = -300\ \text{kJ} < 0$$

The value of ΔG_{rxn} is negative, and therefore the reaction is spontaneous from left to right. Sulfur on ignition burns readily in air; however, in the absence of ignition (or sunlight), S(*s*) is metastable in air. The negative value of ΔG_{rxn} is no guarantee that the reaction will actually occur. Thermodynamics has nothing to say about the *rate* of a chemical reaction.

EXAMPLE 10-2

The reaction system

$$2\,SO_2(g) + O_2(g) \rightleftharpoons 2\,SO_3(g)$$

is prepared at 225°C such that $Q/K = 0.010$. Compute ΔG_{rxn} under these conditions.

Solution: From Eq. 10-15 we compute

$$\Delta G_{rxn} = RT\ln\left(\frac{Q}{K}\right) = (8.31\ \mathrm{J\cdot K^{-1}})(498\ \mathrm{K})\ln(0.010) = -19.1\ \mathrm{kJ}$$

The negative value of ΔG_{rxn} shows that the reaction is spontaneous from left to right at the stated conditions. A maximum of 1.91 kJ of work can be obtained from the reaction at 225°C with $Q/K = 0.010$, per 2 mol of $SO_2(g)$ that reacts.

10-2 *The Effect of Temperature on an Equilibrium Constant Can Be Computed from the van't Hoff Equation*

The dependence of an equilibrium constant on temperature is described by the *van't Hoff equation.* The derivation of the van't Hoff equation follows. The temperature dependence of the standard Gibbs energy change for a chemical reaction is given by the expression (Eq.7-28)

$$\left(\frac{\partial\,\Delta G^\circ_{rxn}}{\partial T}\right)_P = -\Delta S^\circ_{rxn} \tag{10-16}$$

Substitution of Eq. 10-13 into Eq. 10-16 yields

$$\left(\frac{\partial\,RT\ln K}{\partial T}\right)_P = \Delta S^\circ_{rxn}$$

from which we obtain

$$RT\left(\frac{\partial \ln K}{\partial T}\right)_P + R\ln K = \Delta S^\circ_{rxn} \tag{10-17}$$

Rearrangement of Eq. 10-17 gives

$$\left(\frac{\partial \ln K}{\partial T}\right)_P = \frac{\Delta S^\circ_{rxn}}{RT} - \frac{\ln K}{T} \tag{10-18}$$

We can eliminate $\ln K$ from the right side of Eq. 10-18 using the relationship

$$\ln K = -\frac{\Delta H^\circ_{rxn}}{RT} + \frac{\Delta S^\circ_{rxn}}{R} \tag{10-19}$$

which is obtained by equating the $\Delta G^\circ_{\text{rxn}}$ expressions in Eqs. 10-13 and 7-33:

$$\Delta G^\circ_{\text{rxn}} = -RT \ln K = \Delta H^\circ_{\text{rxn}} - T\Delta S^\circ_{\text{rxn}} \tag{10-20}$$

Thus we obtain from Eqs. 10-18 and 10-19

$$\left(\frac{\partial \ln K}{\partial T}\right)_P = \frac{\Delta S^\circ_{\text{rxn}}}{RT} - \frac{1}{T}\left(-\frac{\Delta H^\circ_{\text{rxn}}}{RT} + \frac{\Delta S^\circ_{\text{rxn}}}{R}\right) = \frac{\Delta H^\circ_{\text{rxn}}}{RT^2}$$

We have chosen fixed-pressure (1-atm) standard states; therefore, the equilibrium constant for a given reaction is independent of the total pressure. Consequently, the subscript P on $(\partial \ln K/\partial T)_P$ can be dropped. The result is the van't Hoff equation

$$\frac{d \ln K}{dT} = \frac{\Delta H^\circ_{\text{rxn}}}{RT^2} \tag{10-21}$$

The van't Hoff equation (Eq. 10-21) tells us that if $\Delta H^\circ_{\text{rxn}} > 0$, then K increases with increasing T; whereas, if $\Delta H^\circ_{\text{rxn}} < 0$, then K decreases with increasing T. Both $\Delta H^\circ_{\text{rxn}}$ and $\Delta S^\circ_{\text{rxn}}$ determine the magnitude of K at any particular temperature, as is clear from Eq. 10-19

$$\ln K = \frac{\Delta S^\circ_{\text{rxn}}}{R} - \frac{\Delta H^\circ_{\text{rxn}}}{RT}$$

with the term containing $\Delta H^\circ_{\text{rxn}}$ predominating at low T, because of the inverse T dependence, and the term containing $\Delta S^\circ_{\text{rxn}}$ predominating at high T. However, *the temperature dependence of K is determined solely by the sign and magnitude of* $\Delta H^\circ_{\text{rxn}}$ (Eq. 10-21).

Because the choice of standard states is arbitrary, there is nothing sacrosanct about the magnitude of K for a particular reaction. Different choices of standard states lead to different values of K, $\Delta G^\circ_{\text{rxn}}$, $\Delta H^\circ_{\text{rxn}}$, and $\Delta S^\circ_{\text{rxn}}$ for a reaction. It is essential, however, to be consistent in the choice of standard states. Erroneous results will be obtained if thermodynamic data for a substance based on different standard states are combined without adjustment to the same standard state.

If $\Delta H^\circ_{\text{rxn}}$ for the reaction is independent of temperature, then integration of the van't Hoff equation yields

$$\ln\left(\frac{K_2}{K_1}\right) = \frac{\Delta H^\circ_{\text{rxn}}}{R}\left(\frac{T_2 - T_1}{T_1 T_2}\right) \tag{10-22}$$

where K_1 is the reaction equilibrium constant at the Kelvin temperature T_1 and K_2 is the equilibrium constant at the Kelvin temperature T_2. Note from Eq. 10-22 that if $T_2 > T_1$, then the right-hand side of Eq. 10-22 will be positive if $\Delta H^\circ_{\text{rxn}} > 0$ (endothermic reaction), whereas the right-hand side will be negative if $\Delta H^\circ_{\text{rxn}} < 0$ (exothermic reaction). If the right-hand side of Eq. 10-22 is positive, then so also must be the left-hand side; that is, $K_2 > K_1$; whereas the reverse is true if $\Delta H^\circ_{\text{rxn}}$ is negative ($K_2 < K_1$ for $T_2 > T_1$, if $\Delta H^\circ_{\text{rxn}} < 0$). In other words, *the equilibrium constant for a reaction increases with increasing temperature, if* $\Delta H^\circ_{rxn} > 0$ (endothermic reaction), and *the equilibrium constant decreases with increasing temperature, if* $\Delta H^\circ_{rxn} < 0$ (exothermic reaction).

Note the similarity of Eq. 10-22 to the integrated form of the Clausius-Clapeyron equation:

$$\ln\left(\frac{P_2}{P_1}\right) = \frac{\Delta \bar{H}_{vap}}{R}\left(\frac{T_2 - T_1}{T_2 T_1}\right)$$

The similarity is a consequence of the fact that, for a vaporization reaction (assuming ideal gas behavior), $K \approx P$; thus $(K_2/K_1) = (P_2/P_1)$ and $\Delta H^\circ_{rxn} \approx \Delta \bar{H}_{vap}$.

EXAMPLE 10-3

The equilibrium constant for the ammonia synthesis reaction

$$N_2(g) + 3\,H_2(g) \rightleftharpoons 2\,NH_3(g)$$

based on 1-atm ideal gas standard states is 6.0×10^5 at 25°C and $\Delta H^\circ_{rxn} = -92.0$ kJ. Estimate K for the reaction at 50°C, assuming that ΔH°_{rxn} is independent of temperature.

Solution: Using the van't Hoff equation (Eq. 10-22), we have

$$\ln\left(\frac{K_2}{6.0 \times 10^5}\right) = \frac{-92.0 \times 10^3\ \text{J}}{8.314\ \text{J}\cdot\text{K}^{-1}}\left(\frac{323 - 298}{323 \times 298\ \text{K}}\right)$$

thus

$$\frac{K_2}{6.0 \times 10^5} = 5.6 \times 10^{-2}$$

and

$$K_2 = 6.0 \times 10^5 \times 5.6 \times 10^{-2} = 3.4 \times 10^4$$

Note that because the reaction is exothermic ($\Delta H^\circ_{rxn} < 0$), the value of K decreases with an increase in the temperature.

The temperature dependence of an equilibrium constant can be displayed graphically by plotting $\log K$ versus $1/T$. If ΔH°_{rxn} is constant, then a plot of $\log K$ versus $1/T$ will be linear (just like the $\log P$ versus $1/T$ plot for vapor pressure). Such a plot is shown for the reaction

$$N_2(g) + 3\,H_2(g) \rightleftharpoons 2\,NH_3(g)$$

in Figure 10-2.

A knowledge of K at several temperatures (at least three) can be used to obtain ΔH°_{rxn} from a plot of $\log K$ versus $1/T$. And, conversely, a knowledge of ΔH°_{rxn} and K at one temperature enables us to compute K at some other temperature (provided ΔH°_{rxn} is constant), as demonstrated in Example 10-3 for the ammonia synthesis reaction.

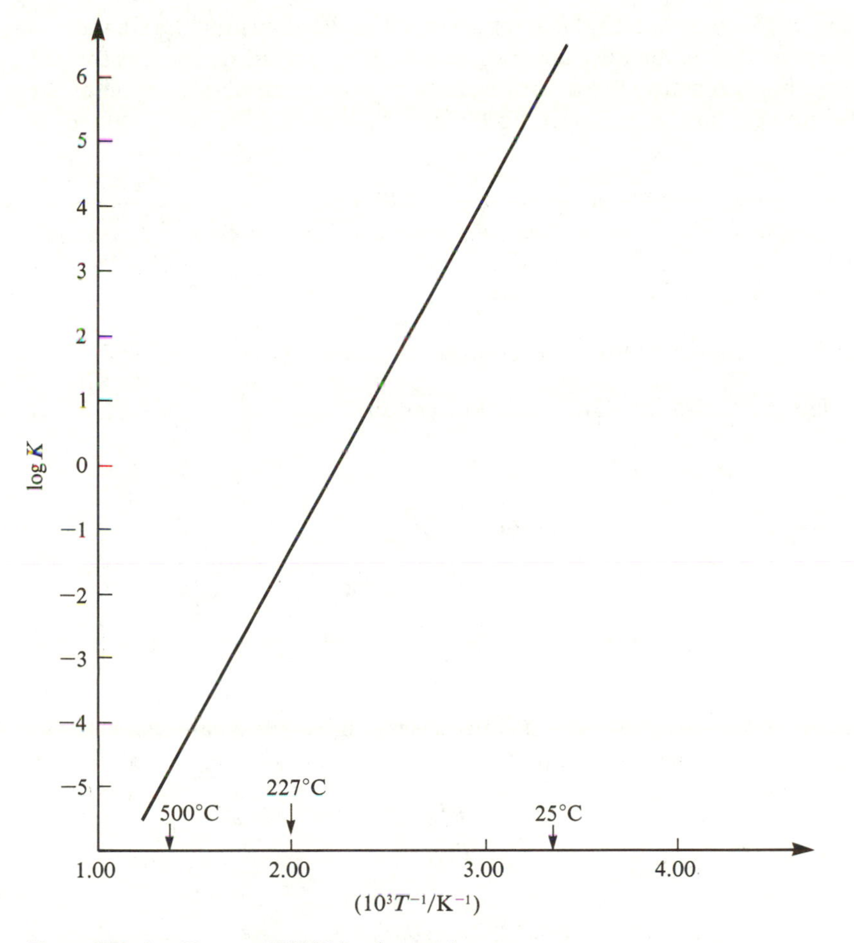

Figure 10-2. log K versus $1000/T$ for the reaction

$$N_2(g) + 3\ H_2(g) \rightleftharpoons 2\ NH_3(g)$$

Note that because we have plotted log K versus $1/T$, the temperature is decreasing from left to right on the x axis (lower T means higher $1/T$). The negative value of ΔH°_{rxn} makes the slope $(= -\Delta H^\circ_{rxn}/2.303R)$ of the line positive (increasing log K left to right).

10-3 *A Chemical Reaction That Is Displaced from Equilibrium Proceeds Toward a New Equilibrium State in the Direction That at Least Partially Offsets the Change in Conditions*

The direction in which an equilibrium chemical reaction shifts (left to right or right to left) in response to a change in the conditions that affect the reaction equilibrium can be predicted by using *Le Châtelier's principle.* In the application of Le Châtelier's principle, we consider the chemical reaction to be initially at equilibrium, and the equilibrium reaction is then subjected to a change in conditions that displaces the system from equilibrium. Le Châtelier's principle is used to predict the direction in which the reaction proceeds toward a new equilibrium state. We do not need to know the numerical value of the equilibrium constant K to apply Le Châtelier's principle, which is stated as follows:

> *If a chemical reaction at equilibrium is subjected to a change in conditions that displaces the reaction from equilibrium, then the direction in which the reaction proceeds toward a new equilibrium state will be such as to at least partially offset the change in conditions.*

The conditions that can affect a reaction equilibrium follow:

1. The concentration of a reactant or product.
2. The pressure.
3. The temperature.

Consider the reaction equilibrium

$$C(s) + CO_2(g) \rightleftharpoons 2\,CO(g)$$

for which

$$K = \frac{a^2_{CO(g)}}{a_{CO_2(g)}a_{C(s)}}$$

If the gas pressures are not high, then the gases can be treated as ideal gases to fairly good approximation. Further, $a_{C(s)}$ is only weakly dependent on the pressure; thus

$$K \approx \frac{P^2_{CO}}{P_{CO_2}}$$

For an ideal gas

$$P = \left(\frac{n}{V}\right)RT = [\text{gas}]\,RT$$

where the brackets denote concentration (moles per liter); thus

$$K = \frac{[CO]^2(RT)^2}{[CO_2]RT}$$

and

$$K_c = \frac{K}{RT} = \frac{[CO]^2}{[CO_2]}$$

Note that $K_c \neq K$; this is because K is based on 1-atm standard states, whereas K_c is based on 1-molar standard states for the gases CO and CO_2.

1. If we disturb the reaction equilibrium by injecting some additional $CO_2(g)$ into the reaction vessel, then the concentration of $CO_2(g)$ is increased. In response to the change in conditions, the reaction equilibrium shifts from left to right because this is the direction in which $CO_2(g)$ is consumed. Thus a shift from left to right leads to a partial offset of

the increase in the CO_2 concentration. In the new equilibrium state, the concentration of CO_2 and the concentration of CO are both greater than in the original equilibrium state, but the concentration of CO_2 in the new equilibrium state is less than it was immediately after the additional CO_2 was injected into the reaction vessel.

2. If we disturb the reaction equilibrium by injecting some additional $CO(g)$ into the reaction vessel, then the concentration of CO is increased and the reaction equilibrium shifts from right to left in response to the change in conditions.
3. If we inject some additional solid carbon $C(s)$ into the reaction vessel, while keeping $[CO]$ and $[CO_2]$ constant, then there is *no* shift in the reaction equilibrium, because the concentration of a solid is independent of the amount of the solid phase present. In other words, the injection (or removal) of some $C(s)$ does not affect the activity of $C(s)$ and thus does not displace the reaction from equilibrium.
4. If we remove some $CO(g)$ from the equilibrium reaction mixture, then the reaction equilibrium shifts from left to right to produce some more $CO(g)$, because this is the direction that partially offsets the change in conditions. The concentration (and, therefore, the partial pressure) of $CO(g)$ in the new equilibrium state is less than that in the original equilibrium state but greater than the concentration of $CO(g)$ in the reaction vessel immediately after the removal of $CO(g)$.

EXAMPLE 10-4

Consider the reaction equilibrium

$$N_2(g) + 3\,H_2(g) \rightleftharpoons 2\,NH_3(g)$$

Use Le Châtelier's principle to predict the effect on the equilibrium pressures of $NH_3(g)$ and of $N_2(g)$ produced by an increase in the pressure of $H_2(g)$.

Solution: An increase in the pressure of $H_2(g)$, which is a reactant, shifts the equilibrium from left to right and thus produces an *increase* in the equilibrium pressure of $NH_3(g)$. The equilibrium pressure of $N_2(g)$ decreases, because N_2 is consumed in the production of $NH_3(g)$.

If the total pressure in the reaction system is changed by the introduction of an inert gas, for example, by injecting $N_2(g)$ into a reaction vessel where the equilibrium

$$C(s) + CO_2(g) \rightleftharpoons 2\,CO(g)$$

exists, then, to a first approximation, the reaction equilibrium does not shift in either direction, because Q remains constant. Thus, because N_2 is neither a reactant nor a product, the addition of N_2 has no effect on the concentrations; thus the reaction equilibrium is unaffected.

EXAMPLE 10-5

Should the total pressure of the equilibrium reaction mixture

$$N_2(g) + 3\,H_2(g) \rightleftharpoons 2\,NH_3(g)$$

be increased or decreased to increase the extent of conversion of nitrogen and hydrogen to ammonia?

Solution: An increase in total pressure favors the side of the reaction with the smaller number of moles of gas. There are 4 mol of gas on the left and only 2 mol of gas on the right; thus $\Delta n_{gas} = 2 - 4 = -2$. An increase in pressure shifts the reaction equilibrium from left to right and thus increases the extent of conversion of nitrogen and hydrogen to ammonia.

Consider the effect of a change in the total pressure on the reaction equilibrium

$$C(s) + CO_2(g) \rightleftharpoons 2\,CO(g) \tag{10-23}$$

A change in the total pressure at fixed composition produces a change in the activity of species i, which is given by Eq. 9-25,

$$\left(\frac{\partial \ln a_i}{\partial P}\right)_{T,\text{comp}} = \frac{\bar{V}_i}{RT} \tag{10-24}$$

where the subscript "comp" denotes constant composition. For a chemical reaction we have

$$\left(\frac{\partial \ln Q}{\partial P}\right)_{T,\text{comp}} = \frac{\Delta V_{rxn}}{RT} \tag{10-25}$$

From Eq. 10-25 we conclude that if $\Delta V_{rxn} > 0$, then an increase in the pressure P produces an increase in the activity quotient Q; whereas if $\Delta V_{rxn} < 0$, then an increase in the pressure produces a decrease in Q. If $\Delta V_{rxn} > 0$, then an increase in the pressure produces a nonequilibrium system for which $Q/K > 1$; thus the reaction equilibrium will shift right to left toward a new equilibrium state. Conversely, if $\Delta V_{rxn} < 0$, then for $\Delta P > 0$, the value of Q decreases, $Q/K < 1$, and the reaction will shift left to right toward a new equilibrium state.

The volume change for a reaction is given by

$$\Delta V_{rxn} = \sum_i \nu_i \bar{V}_i \tag{10-26}$$

The molar volume of a liquid or a solid is small relative to the molar volume of a gas ($\bar{V}_\ell$ or $\bar{V}_s < 0.02\bar{V}_g$). Therefore, the major terms in Eq. 10-26 are those for gaseous species. The total pressure in the reaction mixture is proportional to the total number of moles of gas; thus, for the reaction in Eq. 10-23 we have

$$P_{tot} = P_{CO} + P_{CO_2} = n_{CO}\left(\frac{RT}{V}\right) + n_{CO_2}\left(\frac{RT}{V}\right)$$

and

$$P_{tot} = (n_{CO} + n_{CO_2})\left(\frac{RT}{V}\right) = n_{tot}\left(\frac{RT}{V}\right)$$

Because the total pressure is proportional to the total number of moles of gas, the value of P_{tot} decreases when n_{tot} decreases. The value of n_{tot} decreases when the reaction shifts toward the side of the reaction with the smaller number of moles of gas. Thus if

$$\Delta n_{gas} = (\text{moles of product gases}) - (\text{moles of reactant gases}) > 0$$

then an increase in P_{tot} will shift the reaction equilibrium to the left; whereas if $\Delta n_{gas} < 0$, then an increase in P_{tot} will shift the reaction equilibrium to the right.

If the number of moles of gas is the same on both sides, as in the reaction

$$H_2(g) + I_2(g) \rightleftharpoons 2\,HI(g)$$

then a change in the total pressure has no effect on the equilibrium, because the volume change for this reaction is zero, $\Delta \bar{V}_{rxn} = 0$. In this case a change in the total pressure does not disturb the equilibrium (assuming ideal gas behavior).

The value of ΔH°_{rxn} for the reaction

$$CaCO_3(s) \rightleftharpoons CaO(s) + CO_2(g)$$

at 500°C is $\Delta H^\circ_{rxn} = +158$ kJ. In other words, the conversion of 1.00 mol of $CaCO_3(s)$ to 1.00 mol of $CaO(s)$ plus 1.00 mol of $CO_2(g)$ requires an input of energy as heat of 158 kJ. Because an increase in temperature increases the availability of energy as heat to the reaction, an increase in the temperature of this reaction system leads to a shift in the equilibrium from left to right and thus to an increase in the equilibrium pressure of $CO_2(g)$ in the system (as predicted by the van't Hoff equation). If the temperature increase is maintained by continued energy input as heat, then the reaction continues to proceed from left to right until a new equilibrium state is achieved at an increased value of P_{CO_2}.

A reaction equilibrium with $\Delta H^\circ_{rxn} > 0$ (endothermic reaction) shifts to the right when the temperature is increased and shifts to the left when the temperature is decreased. If $\Delta H^\circ_{rxn} < 0$ (exothermic reaction), then the equilibrium shifts to the left when the temperature is increased and shifts to the right when the temperature is decreased.

EXAMPLE 10-6

For the reaction

$$N_2(g) + 3\,H_2(g) \rightleftharpoons 2\,NH_3(g)$$

$\Delta H^\circ_{rxn} = -92$ kJ (exothermic reaction). Will an increase in the temperature increase or decrease the extent of conversion of N_2 and H_2 to NH_3?

Solution: The value of ΔH°_{rxn} is negative, and thus the reaction evolves energy as heat:

$$N_2(g) + 3\,H_2(g) \rightleftharpoons 2\,NH_3(g) + 92\text{ kJ}$$

An increase in temperature favors the absorption of energy as heat, and thus an increase in temperature shifts the above reaction equilibrium to the left. A shift of the equilibrium to the left decreases the yield (extent of conversion of reactants to products) of ammonia.

The various possible changes in reaction conditions and their effects on reaction equilibria, as deduced from Le Châtelier's principle, are summarized in Table 10-1.

TABLE 10-1

Summary of Effects on Reaction Equilibria Arising from Various Possible Changes in Conditions as Predicted by Le Châtelier's Principle

Change in conditions	*Direction of shift in reaction equilibrium*
Concentration	
Increase concentration of a reactant	$\longrightarrow$
Increase concentration of a product	$\longleftarrow$
Decrease concentration of a reactant	$\longleftarrow$
Decrease concentration of a product	$\longrightarrow$
Temperature	
1. $\Delta H^\circ_{rxn} > 0$ (*endothermic reaction*)	
Increase in temperature	$\longrightarrow$
Decrease in temperature	$\longleftarrow$
2. $\Delta H^\circ_{rxn} < 0$ (*exothermic reaction*)	
Increase in temperature	$\longleftarrow$
Decrease in temperature	$\longrightarrow$
Total pressure of reaction gases	
1. $\Delta V_{rxn} > 0$ (*more moles of gas on right*, $\Delta n_{gas} > 0$)	
Increase total pressure	$\longleftarrow$
Decrease total pressure	$\longrightarrow$
2. $\Delta V_{rxn} < 0$ (*more moles of gas on left*, $\Delta n_{gas} < 0$)	
Increase total pressure	$\longrightarrow$
Decrease total pressure	$\longleftarrow$

10-4 *Le Châtelier's Principle Can Be Used to Select the Conditions That Maximize the Equilibrium Yield of a Reaction Product*

Over 16 million tons of ammonia are produced each year in the United States via the reaction

$$N_2(g) + 3\,H_2(g) \rightleftharpoons 2\,NH_3(g)$$

Ammonia production in the United States ranks second only to sulfuric acid in annual production. The ammonia synthesis reaction is carried out at about 500°C and a total pressure of about 300 atm to 500 atm in the presence of an iron-molybdenum *catalyst* (Haber process). A catalyst is a substance that increases the rate of a reaction without affecting the reaction stoichiometry. A catalyst acts as a reaction rate facilitator, but a catalyst does not affect the equilibrium constant of the reaction.

The variations in the mole percentages of N_2, H_2, and NH_3 with time for (1) the case in which N_2 and H_2 are mixed in a 1:3 ratio and reacted at 520°C at a total pressure of 500 atm and (2) the case in which $NH_3(g)$ is heated at 520°C at a total pressure of 500 atm are shown in Figure 10-3. Note that the

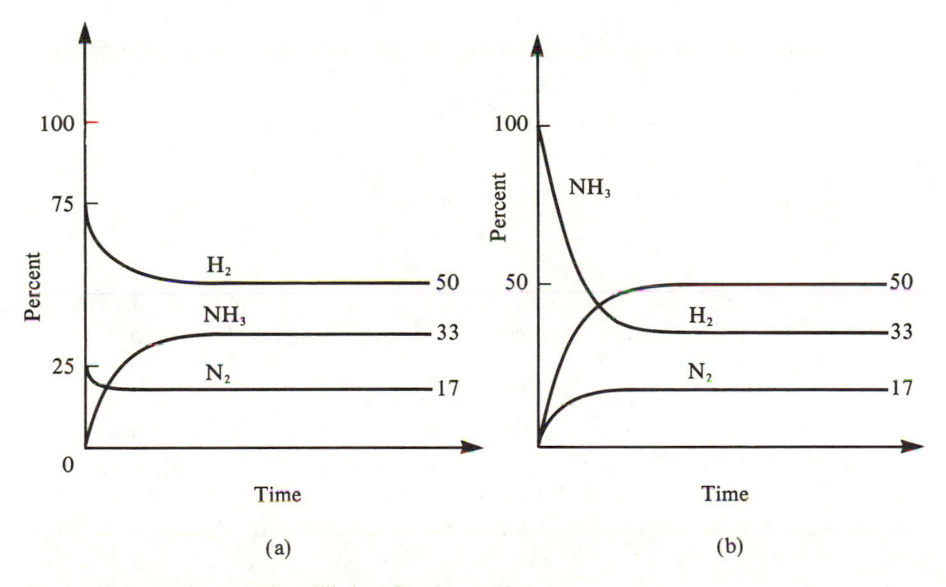

Figure 10-3. Attainment of equilibrium for the reaction

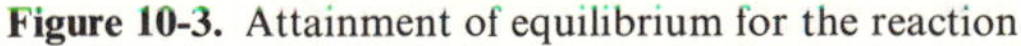

$$N_2(g) + 3\,H_2(g) \rightleftharpoons 2\,NH_3(g)$$

at 520°C, 500 atm total pressure. In case a, the N_2 and H_2 are mixed in a 1:3 ratio (1 mol of N_2 + 3 mol of H_2, or 25 mol % N_2 + 75 mol % H_2) and allowed to reach equilibrium. The equilibrium mole percentages are 17% N_2, 50% H_2, and 33% NH_3. In case b, 2 mol of NH_3 (100 mol % NH_3) is allowed to reach equilibrium. The equilibrium mole percentages of N_2, H_2, and NH_3 are exactly the same as in case a.

equilibrium state is independent of the direction from which the equilibrium is approached [either 1 mol of N_2 + 3 mol of H_2 at the start or 2 mol of NH_3 at the start yield the same equilibrium (final) distribution of species].

The economic importance of ammonia as a fertilizer and the tremendous scale of ammonia production facilities makes the choice of the economically most favorable reaction conditions for the ammonia synthesis reaction a matter of considerable importance. For the reaction

$$N_2(g) + 3\,H_2(g) \rightleftharpoons 2\,NH_3(g)$$
$$\Delta H^\circ_{rxn} = -92\text{ kJ} \quad \text{and} \quad \Delta n_{gas} = 2 - 4 = -2$$

The value of Δn_{gas} for the reaction is negative (4 mol gaseous reactants → 2 mol gaseous products), and thus an increase in the total pressure favors the conversion of reactants to products; that is, the percent conversion of N_2 and H_2 to ammonia is greater the higher the total pressure. The value of ΔH°_{rxn} for the reaction is negative (92 kJ of energy is liberated as heat when 1 mol of N_2 reacts with 3 mol of H_2 to produce 2 mol of NH_3). Thus an increase in temperature favors the conversion of products back to reactants; that is, the equilibrium shifts right to left when the temperature is increased. In other words, the equilibrium constant K for the reaction decreases with increasing temperature, as shown in the following table:

$t/^\circ C$	$K = a^2_{NH_3(g)}/a_{N_2(g)}a^3_{H_2(g)}$
25	5.94×10^5
227	0.100
500	6.93×10^{-5}

At a given total pressure the extent of conversion of nitrogen plus hydrogen to ammonia is *less* the higher the temperature, because K decreases with increasing temperature.

The application of Le Châtelier's principle to the ammonia synthesis reaction leads to the prediction that *at equilibrium* the yield of ammonia (percent conversion of N_2 or H_2, whichever is limiting, to NH_3) is greater the higher the total pressure and the lower the temperature. However, the *rate* of the ammonia synthesis reaction around 25°C is negligibly slow. A high yield of product is of no commercial value if it takes forever to achieve the conversion. The rate of most reactions increases with an increase in the temperature. The ammonia synthesis reaction is run at an elevated temperature, even though the equilibrium yield is not as favorable as at lower temperatures, in order to make the reaction proceed at an economically feasible rate. The Haber process is thus based on a compromise between equilibrium (yield) and rate (speed of reaction) considerations. The low value of K at the process temperature (500°C) is offset by using a very high pressure (500 atm), as shown in the following table:

$$N_2(g) + 3\,H_2(g) \rightleftharpoons 2\,NH_3(g) \qquad \text{at } 500°C$$

P_{tot}/atm	P_{NH_3}/atm	*Percent conversion to* NH_3
1.00	1.26×10^{-3}	0.25
500.	152	47.

Even at 500°C a catalyst is necessary to make the reaction proceed at an acceptable rate. At 700°C the reaction rate is fast enough without the catalyst, but K is so small that P_{tot} cannot be made large enough to make the yield economically acceptable.

The development of the industrial nitrogen-fixation process by the German chemist Fritz Haber had a profound influence of several segments of the world economy and, therefore, ultimately on world politics. One major effect was a severe disruption of the economy of Chile. Prior to the development of the Haber process, the major world supply of fixed nitrogen was in the form $NaNO_3$, "Chile saltpeter," which occurs naturally in large deposits in Chile. Because ammonia is easily converted to nitrates by oxidation, and because of the large mining and shipping costs involved in obtaining Chile saltpeter, the demand for naturally occurring $NaNO_3$ decreased tremendously following the development of the Haber process. One of the major uses of $NaNO_3$ in the early twentieth century was for the production of explosives. With the Haber process, Germany no longer needed to import $NaNO_3$ over highly vulnerable long-distance shipping lanes. The Haber process was thus a vital factor in Germany's ability to wage global war without control of the high seas. In the contemporary world, the Haber process is used primarily for benign purposes, namely, for the production of ammonia for use in fertilizers and for the production of nitric acid by the Ostwald process, in which ammonia is air oxidized to nitric acid.

An ambient-temperature process that permitted the synthesis of ammonia from the elements would rapidly make the Haber process obsolete. Such a process would eliminate (1) the need for fuel to maintain the high–reaction-vessel temperature, (2) the need to use compressors and heavy-walled apparatus to maintain the high pressures, and (3) the need to purify the H_2 and N_2 to avoid contamination of the expensive catalyst. Such a process is possible ($K = 6 \times 10^5$ at 25°C) and, in fact, exists in nature. Nature has developed a far more effective nitrogen-fixation catalyst than chemists have yet devised. These catalysts are called *nitrogenases* and are found in the bacteria of root nodules that attach to the roots of legumes like beans and peas. These bacteria can convert N_2 to NH_3 at ambient temperatures in a process involving several successive steps. The nitrogenase enzyme has a molecular weight of 270,000 and is composed of an iron-molybdenum protein and two different iron proteins. The details of the biological nitrogen-fixation reaction are not understood at the present time.

10-5 *Chemical Reactions Can Be Classified According to the Number of Degrees of Freedom*

For a chemical reaction involving only all pure phases, there is one and only one value of P at each T for which $\Delta G_{rxn} = 0$; that is, there is only one value of the pressure at each temperature for which the reaction can attain equilibrium. For a reaction with α reactants and β products, with all reactants and products having fixed composition, we have from the phase rule (one restrictive condition, the reaction equilibrium)

$$f = c - p + 2 = (\alpha + \beta - 1) - (\alpha + \beta) + 2 = 1$$

We shall denote such reactions as *Class I reactions* because $f = 1$. For a Class I reaction there is one, and only one, value of P at each T for which equilibrium can be attained. On the other hand, if one or more species involved in the reaction is a solution-phase species (be it a solid, liquid, or gaseous solution), then equilibrium can be attained at an infinite number of values of P for a given T. For example, suppose that the reaction of interest is a gas-phase reaction; then assuming only one restrictive condition, namely, the reaction equilibrium, we have with α reactants and β products

$$f = c - p + 2 = (\alpha + \beta - 1) - 1 + 2 = \alpha + \beta \geqslant 2$$

All chemical reactions that involve at least one solution phase have $f \geqslant 2$; we shall designate such reactions as *Class II reactions.* The thermodynamic behavior of Class I and Class II reactions is distinctly different, as we shall see in the examples that follow.

The Calcite-to-Aragonite Phase Transition

The calcite-aragonite phase transition is of special interest because of its potential use as a geobarometer. Calcite, $CaCO_3(s)$, is hexagonal with a density of 2.710 $g\cdot cm^{-3}$, whereas aragonite is orthorhombic with a density of 2.944 $g\cdot cm^{-3}$. Aragonite has the higher density and is thus the form of $CaCO_3(s)$ that is stable at high pressure. The values of ΔG°_{rxn} and ΔS°_{rxn} at 25°C for the transition

$$CaCO_3(s, \text{calcite}) \rightleftharpoons CaCO_3(s, \text{aragonite})$$

are

$$\Delta G^\circ_{rxn} = +1393\ J\cdot mol^{-1} \qquad \Delta S^\circ_{rxn} = -3.72\ J\cdot K^{-1}\cdot mol^{-1}$$

Note that the higher-density form has the lower entropy, as expected. The enthalpy change for the calcite-aragonite phase transition can be computed from the relation

$$\Delta G^\circ_{rxn} = \Delta H^\circ_{rxn} - T\Delta S^\circ_{rxn}$$

At 25°C we have

$$\Delta H^\circ_{rxn} = 1393\ J\cdot mol^{-1} + 298.15\ K(-3.72\ J\cdot K^{-1}\cdot mol^{-1}) = 284\ J\cdot mol^{-1}$$

Application of the Lewis equation to the calcite-aragonite phase transition yields

$$\Delta G_{rxn} = \Delta G^\circ_{rxn} + RT\ln Q = \Delta G_{rxn} + RT\ln\left(\frac{a_A}{a_C}\right)$$

At 1 atm pressure, $a_A = a_C = 1.00$, and assuming that ΔH°_{rxn} (and thus ΔS°_{rxn}) is independent of temperature, we obtain

$$\Delta G_{rxn} = \Delta G^\circ_{rxn} = \Delta H^\circ_{rxn} - T\Delta S^\circ_{rxn} = (284 + 3.72T)\ \mathrm{J\cdot mol^{-1}}$$

The number of degrees of freedom for the calcite-aragonite equilibrium system (1 component, 2 phases) is

$$f = c - p + 2 = 1 - 2 + 2 = 1$$

and thus there is at most one value of the temperature at each pressure for which the two phases can coexist in equilibrium. The equilibrium condition at fixed T and P is $\Delta G_{rxn} = 0$. For equilibrium at 1 atm pressure,

$$\Delta G_{rxn} = 0 = 284 + 3.72T$$

and thus

$$T = -\frac{284}{3.72} = -76\ \mathrm{K}$$

Negative absolute temperatures are impossible, and thus calcite and aragonite cannot coexist in equilibrium at 1 atm.

Can aragonite and calcite coexist in equilibrium at 25°C? From the Lewis equation applied to the equilibrium reaction we have

$$\Delta G_{rxn} = 0 = \Delta G^\circ_{rxn} + RT\ln\left(\frac{a_A}{a_C}\right)$$

Assuming that the solid phases are incompressible, we obtain from Eq. 9-29

$$\ln\left(\frac{a_A}{a_C}\right) = \left(\frac{\bar{V}^\circ_A - \bar{V}^\circ_C}{RT}\right)(P - 1) = \left(\frac{M_{CaCO_3}}{RT}\right)\left(\frac{1}{\rho_A} - \frac{1}{\rho_C}\right)(P - 1)$$

where M_{CaCO_3} is the molecular weight of $CaCO_3$ and the ρ's are the densities. Thus at equilibrium

$$0 = \Delta G^\circ_{rxn} + M_{CaCO_3}\left(\frac{\rho_C - \rho_A}{\rho_C\rho_A}\right)(P - 1)$$

or

$$P = 1 - \frac{\Delta G^\circ_{rxn}\rho_C\rho_A}{M_{CaCO_3}(\rho_C - \rho_A)}$$

from which we compute at 25°C

$$P = 1 - \frac{(1393\ \mathrm{J\cdot mol^{-1}})(2.710\ \mathrm{g\cdot cm^{-3}})(2.994\ \mathrm{g\cdot cm^{-3}})}{(100.09\ \mathrm{g\cdot mol^{-1}})(2.710 - 2.944)\ \mathrm{g\cdot cm^{-3}}}\left(\frac{82.1\ \mathrm{cm^3\cdot atm}}{8.341\ \mathrm{J}}\right)$$

$$P = 4.77 \times 10^3\ \mathrm{atm}$$

In other words, at 25°C, equilibrium between the two forms is attained at 4.77×10^3 atm. Above this pressure, aragonite becomes the stable form of $CaCO_3(s)$ at 25°C.

The negative value of ΔS°_{rxn} for the calcite-to-aragonite transition means that ΔG°_{rxn} increases with increasing temperature:

$$\left(\frac{\partial \Delta G^\circ_{rxn}}{\partial T}\right)_P = -\Delta S^\circ_{rxn} = +3.72 \text{ J}\cdot\text{K}^{-1}\cdot\text{mol}^{-1}$$

and thus the equilibrium transition pressure increases with increasing temperature. Assuming that ΔS°_{rxn} is independent of T (i.e., $\Delta C^\circ_{P,rxn} = 0$), the value of ΔG°_{rxn} at 125°C is

$$\Delta G^\circ_{rxn,T_2} = \Delta G^\circ_{rxn,T_1} - \Delta S^\circ_{rxn}(T_2 - T_1)$$

and

$$\Delta G^\circ_{rxn}(\text{at } 125°\text{C}) = 1393 \text{ J}\cdot\text{mol}^{-1} + (3.72 \text{ J}\cdot\text{K}^{-1}\cdot\text{mol}^{-1})(100 \text{ K})$$

Thus

$$\Delta G^\circ_{rxn} = 1765 \text{ J}\cdot\text{mol}^{-1}$$

The equilibrium pressure at 125°C is 6.04×10^3 atm.

The P, T equilibrium line for the calcite-to-aragonite phase transition can be used as a geobarometer in the sense that the presence of aragonite in rocks determines a minimum pressure to which the rock was subjected. The fact that aragonite is thermodynamically unstable with respect to calcite at low pressures does not mean that aragonite cannot exist or be formed at low pressure. The rate of conversion of aragonite to calcite around 25°C is negligibly slow, and thus aragonite can persist for eons at 1 atm and 25°C. The shells of certain marine organisms contain aragonite, which was formed at low pressure. Thermodynamics says that the direct conversion of calcite to aragonite at low

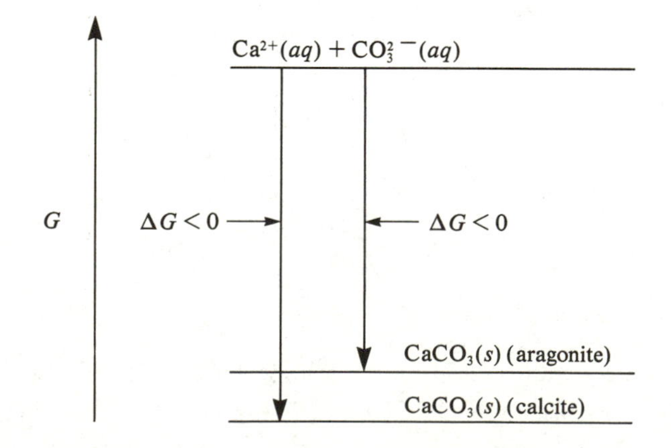

Figure 10-4. Either aragonite or calcite can form from the ions $Ca^{2+}(aq)$ and $CO_3^{2-}(aq)$. However, once formed, calcite cannot be converted directly to aragonite around 1 atm pressure.

pressure is impossible. Thermodynamics does not, however, rule out the formation of aragonite, say, from $Ca^{2+}(aq)$ and $CO_3^{2-}(aq)$ at low pressure, as illustrated in Figure 10-4.

The equilibrium constant at 25°C for the reaction

$$CaCO_3(\text{calcite}) \rightleftharpoons CaCO_3(\text{aragonite})$$

is

$$K = e^{(-\Delta G^\circ_{rxn}/RT)} = \exp\left(\frac{-1393\ \text{J}\cdot\text{mol}^{-1}}{8.314\ \text{J}\cdot\text{K}^{-1}\cdot\text{mol}^{-1} \times 298.15\ \text{K}}\right) = 0.570$$

In other words, at 25°C, equilibrium is attained when the activity ratio a_A/a_C is equal to 0.570; that is,

$$\frac{a_A}{a_C} = 0.570 = K \qquad (\text{at } 25^\circ\text{C})$$

For the calcite-to-aragonite reaction at 1 atm, 25°C,

$$Q = \frac{a_A}{a_C} = 1.00$$

and thus

$$\Delta G_{rxn} = \Delta G^\circ_{rxn} + RT\ln\left(\frac{a_A}{a_C}\right) = 1393\ \text{J} + R\ln(1) = 1393\ \text{J} > 0$$

The reaction cannot occur left to right (calcite → aragonite) at 25°C, 1 atm. If we start with calcite at 25°C, 1 atm, then $\Delta G_{rxn} > 0$, and thus *no aragonite whatsoever* can form from calcite under these conditions.

The Decomposition of Calcium Carbonate

Solid calcium carbonate can be brought into equilibrium with solid calcium oxide and carbon dioxide gas. The stoichiometry of the reaction is

$$CaCO_3(s) \rightleftharpoons CaO(s) + CO_2(g)$$

The above system has only one degree of freedom:

$$f = (3 - 1) - 3 + 2 = 1$$

Thus there is a unique value for the equilibrium pressure of $CO_2(g)$ at each temperature, *provided that the gas phase is pure* CO_2. The following equation gives the equilibrium partial pressure of $CO_2(g)$ (in atmospheres) over $CaCO_3(s)$:

$$\log(P/\text{atm}) = -\frac{9140}{T} + 0.382\log T - 0.668 \times 10^{-3}\,T + 7.4363 \qquad (10\text{-}27)$$

The equilibrium constant expression for the calcium carbonate decomposition reaction is

$$K = \frac{a_{CO_2(g)}a_{CaO(s)}}{a_{CaCO_3(s)}}$$

Taking the equation of state for $CO_2(g)$ at high temperature and low pressure as

$$P\bar{V} = RT + (42.7\ \text{cm}^3\cdot\text{mol}^{-1})P$$

we obtain for $a_{CO_2(g)}$ (see Example 9-5)

$$a_{CO_2(g)} = Pe^{42.7P/RT} = Pe^{0.520P/T}$$

where P is given by Eq. 10-27. Assuming that $\bar{V}$ for the solids is independent of P over moderate pressure ranges, we have (see Example 9-4)

$$\frac{a_{CaO(s)}}{a_{CaCO_3(s)}} \approx \exp\left\{\frac{[\bar{V}_{CaO(s)} - \bar{V}_{CaCO_3(s)}](P - 1)}{RT}\right\} \qquad (10\text{-}28)$$

Given that the densities and formula weights of $CaCO_3(s)$ and $CaO(s)$ are 2.93 $\text{g}\cdot\text{cm}^{-3}$, 100.09 $\text{g}\cdot\text{mol}^{-1}$ and 3.346 $\text{g}\cdot\text{cm}^{-3}$, 56.08 $\text{g}\cdot\text{mol}^{-1}$, respectively, we compute

$$\bar{V}_{CaCO_3} = \frac{100.09}{2.93} = 34.16\ \text{cm}^3\cdot\text{mol}^{-1}$$

$$\bar{V}_{CaO} = \frac{56.08}{3.346} = 16.76\ \text{cm}^3\cdot\text{mol}^{-1}$$

Thus, from Eq. 10-28 we have

$$\frac{a_{CaO(s)}}{a_{CaCO(s)}} = \exp\left[\frac{-0.212\ \text{atm}^{-1}(P - 1)}{T}\right]$$

Note that at $P < 3$ atm, $T > 400°\text{C}$, the right-hand side of this expression can be set equal to unity with no greater than about 0.1% error. We can now write for the equilibrium constant of the decomposition reaction

$$K = (Pe^{0.520P/T})\left\{\exp\left[-\frac{0.212(P - 1)}{T}\right]\right\} \qquad (10\text{-}29)$$

where P is the equilibrium CO_2 pressure in atmospheres and T is the temperature in degrees kelvin. We note that for $P = 1$ atm, the factor in the braces becomes unity. The equilibrium value of P reaches 1.000 atm at 1167.7 K, and thus at this temperature K is equal to 1.0005; whereas at 3.400 atm (i.e., at 1300 K) $K = 3.403$. These calculations show that it is a good approximation to write

$$K \approx P_{CO_2}$$

for the decomposition reaction, provided $P\bar{V} = RT + 42.7P$ and P is not too large.

Note that if we have an expression for K as a function of T, then we easily can obtain expressions for ΔG°_{rxn}, ΔS°_{rxn}, ΔH°_{rxn}, and $\Delta C^\circ_{P,rxn}$.

10-6 *The "No!" of Thermodynamics Is Emphatic*

As an example of the use of tables of $\Delta \bar{G}^\circ_f$ values in equilibrium calculations, consider the formation of benzene from acetylene gas:

$$3\ \mathrm{H{-}C{\equiv}C{-}H}(g) \rightleftharpoons C_6H_6(\ell)$$

for which

$$\Delta G^\circ_{\mathrm{rxn},298} = \Delta \bar{G}^\circ_{f,298}[C_6H_6(\ell)] - 3\Delta \bar{G}^\circ_{f,298}[C_2H_2(g)]$$

From data in Appendix I we compute at 25°C (298.15 K)

$$\Delta G^\circ_{\mathrm{rxn}} = 124.5 - 3(209.2) = -503.1\ \mathrm{kJ}$$

The negative value of $\Delta G^\circ_{\mathrm{rxn}}$ means that acetylene gas at 1 atm and 25°C is thermodynamically unstable with respect to liquid benzene. In fact, the Gibbs energy change is large and negative, and this means that at equilibrium the vapor pressure of $C_2H_2(g)$ over $C_6H_6(\ell)$ must be very small. We can compute the equilibrium pressure $C_2H_2(g)$ from the equation

$$\Delta G^\circ_{\mathrm{rxn}} = -RT\ln K = -RT\ln\left(\frac{a_{C_6H_6(\ell)}}{a^3_{C_2H_2(g)}}\right) = -503.1\ \mathrm{kJ}$$

At low pressures, $a_{C_6H_6(\ell)} = 1$ and $a_{C_2H_2(g)} = P_{C_2H_2}$, and thus at 298 K

$$P_{C_2H_2(g)} = \exp\left(\frac{-503.1 \times 10^3\ \mathrm{J \cdot mol^{-1}}}{3 \times 8.314\ \mathrm{J \cdot K^{-1} \cdot mol^{-1}} \times 298.2\ \mathrm{K}}\right) = 4.2 \times 10^{-30}\ \mathrm{atm}$$

which is very small indeed. It is interesting to note, however, that one can prepare and store acetylene gas, and acetylene exhibits no detectable rate of reaction to form benzene. Nonetheless, these thermodynamic calculations show that the synthesis of benzene from acetylene is feasible, and if a suitable catalyst could be found, such a synthesis could be carried out.

From thermodynamic data tables we find that $\Delta G^\circ_{\mathrm{rxn}} = -37.66$ kJ at 25°C for the reaction

$$2\ \mathrm{Na}(s) + \tfrac{1}{2}\ O_2(g) \rightleftharpoons Na_2O(s)$$

and the reaction of metallic sodium with oxygen takes place rapidly at room temperature. We also find $\Delta G^\circ_{\mathrm{rxn}} = +163.2$ kJ at 25°C for the reaction

$$2\ \mathrm{Au}(s) + \tfrac{3}{2}\ O_2(g) \rightleftharpoons Au_2O_3(s)$$

and $O_2(g)$ at 1 atm pressure will not react directly with gold metal in the absence of any other substances no matter what we do.

The preceding examples can be summarized qualitatively as follows:

1. The NO! of thermodynamics is emphatic. If thermodynamics says "NO" to a specific process, then that is all there is to it (provided the data are reliable).

2. The YES! of thermodynamics is actually *maybe.* The fact that a specific process is thermodynamically feasible carries no implications whatsoever regarding whether the process will actually occur at a detectable rate. Thermodynamics is silent on the rates of chemical reactions. *Spontaneous* is not synonymous with *immediate.*

It is important to recognize that it is ΔG_{rxn} and not ΔG°_{rxn} that determines whether a reaction is spontaneous. If all reactants and products are pure condensed phases, then

$$\Delta G_{rxn} = \Delta G^\circ_{rxn} + RT \ln Q = \Delta G^\circ_{rxn} \qquad (Q = 1)$$

However, if one or more of the reactants or products is a gas or a solution-phase species, then the distinction between ΔG_{rxn} and ΔG°_{rxn} is very significant indeed. As an example, consider the synthesis of ammonia from the elements

$$N_2(g) + 3\,H_2(g) \rightleftharpoons 2\,NH_3(g)$$

From data in Appendix I we compute for this reaction at 25°C

$$\Delta G^\circ_{rxn,298} = (2\text{ mol})(-16.48\text{ kJ}\cdot\text{mol}^{-1}) - 1(0) - 3(0) = -32.96\text{ kJ}$$
$$\Delta H^\circ_{rxn,298} = (2\text{ mol})(-46.11\text{ kJ}\cdot\text{mol}^{-1}) = -92.22\text{ kJ}$$

The equilibrium constant at 25°C is given by

$$\ln K = \frac{-\Delta G^\circ_{rxn,298}}{RT} = \frac{32.96 \times 10^3\text{ J}}{8.314\text{ J}\cdot\text{K}^{-1} \times 298.2\text{ K}}$$

thus

$$K = 5.94 \times 10^5$$

Assuming ideal gas behavior, we have at 25°C

$$K = \frac{P^2_{NH_3}}{P_{N_2} \cdot P^3_{H_2}} = 5.94 \times 10^5$$

The ammonia synthesis reaction is a Class II reaction. The number of degrees of feeedom for the reaction is

$$f = c - p + 2 = (3 - 1) - 1 + 2 = 3$$

and thus the specification of three intensive variables is necessary to determine an equilibrium state of the system. If T is chosen, then fixing any two of the four variables P_{N_2}, P_{H_2}, P_{NH_3}, or P_{tot} fixes the other two. For example, if we require that $P_{N_2} = 0.010$ atm, $P_{H_2} = 0.030$ atm, then at 25°C

$$\frac{P^2_{NH_3}}{P_{N_2} \cdot P^3_{H_2}} = \frac{P^2_{NH_3}}{(0.010\text{ atm})(0.030\text{ atm})^3} = 5.94 \times 10^5$$

and

$$P_{NH_3} = 0.40\text{ atm}$$

whereas at $P_{N_2} = 1.0$ atm, $P_{H_2} = 3.0$ atm, we compute $P_{NH_3} = 4.0 \times 10^3$ atm.

The high temperature used in the Haber process is necessary to make the reaction proceed at an economically feasible *rate*; as T increases, the value of K and thus the percent conversion to NH_3 decreases. Assuming that ΔH°_{rxn} is independent of T, we can estimate K at 500°C using the van't Hoff equation:

$$\ln\left(\frac{K_{773}}{5.94 \times 10^5}\right) = \frac{-92.22 \times 10^3 \text{ J}}{8.314 \text{ J}\cdot\text{K}^{-1}}\left(\frac{773 - 298}{773 \times 298}\right)\text{K}$$

thus

$$K_{773} = 6.93 \times 10^{-5}$$

Note that the equilibrium constant decreases by a factor of about 10^{10} relative to 25°C. The decrease in K with increase in T is a consequence of the fact that $\Delta H^\circ_{rxn} < 0$; an increase in T favors the reactants because the reaction liberates energy as heat. At 500°C, with $P_{N_2} = 1.00$ atm, $P_{H_2} = 3.00$ atm at equilibrium, we compute that

$$P_{NH_3} = (6.93 \times 10^{-5} P_{N_2} \cdot P^3_{H_2})^{1/2} = (6.93 \times 10^{-5} \times 27.0)^{1/2} = 0.0433 \text{ atm}$$

At $P_{N_2} = 100$ atm and $P_{H_2} = 300$ atm, we have

$$P_{NH_3} = [6.93 \times 10^{-5} \times 100 \times (300)^3]^{1/2} = 433 \text{ atm}$$

Note that the equilibrium value of P_{NH_3} increases as the pressures of P_{H_2} and P_{N_2} increase, and the percent conversion of N_2 or H_2 to NH_3 increases as P_{tot} increases.

10-7 *Thermodynamic Calculations Can Be Carried Out for Spontaneous Reactions*

In Chapters 4 and 5 we considered some spontaneous changes of state, and we now recall that for any such process occurring within an adiabatic enclosure the total entropy change must be positive. The spontaneous processes that we are about to consider differ from those discussed previously only in the sense that they involve a change in phase.

As our first example, consider the calculation of ΔS for the following transformation:

$$H_2O(\ell, 263 \text{ K}, 1 \text{ atm}) \longrightarrow H_2O(s, 263 \text{ K}, 1 \text{ atm})$$

We know from our consideration of the phase diagram of water (Chapter 8) that pure $H_2O(\ell)$ at 263 K and 1 atm is thermodynamically unstable with respect to the solid and is incapable of coexisting in equilibrium with $H_2O(s)$ under these conditions. Because the process is not reversible, we cannot equate the energy transferred as heat to the surroundings divided by the temperature of the surroundings (263 K) *to the entropy change of the system.* We can, however, devise a reversible path between the same initial and final states, compute the total entropy change for the system along the reversible path, and equate this ΔS to that in the irreversible process, because ΔS is independent of path. The reversible path is shown schematically in the following diagram (1 atm):

$$\begin{array}{ccc} H_2O(\ell, 263\text{ K}) & \xrightarrow{\Delta S_{\text{rxn}} = \Delta S^\circ_{\text{rxn}}} & H_2O(s, 263\text{ K}) \\ \text{①}\downarrow \Delta S_1 & (1\text{ atm}) & \text{③}\uparrow \Delta S_3 \\ H_2O(\ell, 273\text{ K}) & \xrightarrow[\text{②}]{\Delta S_2 = \Delta S^\circ_2} & H_2O(s, 273\text{ K}) \end{array}$$

$$\Delta S_{\text{rxn}} = \Delta S_1 + \Delta S_2 + \Delta S_3$$
$$\Delta H_{\text{rxn}} = \Delta H_1 + \Delta H_2 + \Delta H_3$$

We have taken $H_2O(\ell)$ to 273 K and allowed it to freeze at this temperature, because at 1 atm total pressure this is the temperature (more precisely 273.15 K) at which the transformation can take place reversibly, that is, with $\Delta G = 0$. The heat capacities of the $H_2O(\ell)$ and $H_2O(s)$ are $75.3\ \text{J}\cdot\text{K}^{-1}\cdot\text{mol}^{-1}$ and $37.7\ \text{J}\cdot\text{K}^{-1}\cdot\text{mol}^{-1}$, respectively, and are essentially independent of T over the range 263 K–273 K. The enthalpy of fusion of water at 273 K and 1 atm pressure is $\Delta H_{\text{fus}} = 6008\ \text{J}\cdot\text{mol}^{-1}$. Because, for step 2, $\Delta G_2 = 0$, we have

$$\Delta S_2 = \frac{\Delta H_2}{T} = \frac{-6008\ \text{J}\cdot\text{mol}^{-1}}{273\ \text{K}} = -22.01\ \text{J}\cdot\text{K}^{-1}\cdot\text{mol}^{-1}$$

where the minus sign arises from the fact that $\Delta H_2 = -\Delta H_{\text{fus}}$. The values of ΔS_1 and ΔS_2 can be obtained using the equation

$$\left(\frac{\partial S}{\partial T}\right)_P = \frac{C_P}{T}$$

thus

$$\Delta S_1 = \int_{263}^{273} (75.3\ \text{J}\cdot\text{K}^{-1}\cdot\text{mol}^{-1})\, d\ln T$$
$$= 75.3 \ln\frac{273}{263} = +2.81\ \text{J}\cdot\text{K}^{-1}\cdot\text{mol}^{-1}$$

and

$$\Delta S_3 = \int_{273}^{263} (37.7)\, d\ln T = 37.7 \ln\frac{263}{273} = -1.41\ \text{J}\cdot\text{K}^{-1}\cdot\text{mol}^{-1}$$

Hence we compute for the entropy change when 1 mol of $H_2O(\ell)$ is converted to 1 mol of $H_2O(s)$ at 273 K and 1 atm

$$\Delta S_{\text{rxn}} = \Delta S^\circ_{\text{rxn}} = 2.81 - 22.01 - 1.41 = -20.61\ \text{J}\cdot\text{K}^{-1}\cdot\text{mol}^{-1}$$

We know from the second law that the entropy change of the heat bath must be greater than $20.61\ \text{J}\cdot\text{K}^{-1}\cdot\text{mol}^{-1}$, because

$$\Delta S_{H_2O} + \Delta S_{\text{bath}} > 0$$

for an irreversible process taking place within an adiabatic enclosure. We can compute the entropy change of the surroundings in this process as follows. Although the solidification of the water is an irreversible process, we shall assume that the energy is transferred to the heat bath only as heat under isothermal, isobaric conditions; thus the pressure is constant and

$$q_{\text{bath}} = -(q_{H_2O})_P = -\Delta H_{\text{rxn}}$$

Furthermore, because the energy transfer to the bath is an isothermal pure heat transfer, we have

$$\Delta S_{\text{bath}} = \frac{q_{\text{bath}}}{T} = -\frac{\Delta H_{\text{rxn}}}{T}$$

Referring to the above cycle, we have

$$\Delta H_{\text{rxn}} = \Delta H_1 + \Delta H_2 + \Delta H_3$$

and

$$\Delta H_{\text{rxn}} = \bar{C}_{P,1}\,\Delta T_1 - \Delta \bar{H}_{\text{fus}} + \bar{C}_{P,2}\,\Delta T_2$$

Thus

$$\begin{aligned}\Delta H_{\text{rxn}} &= (75.3\ \text{J}\cdot\text{K}^{-1}\cdot\text{mol}^{-1})(273 - 263)\ \text{K} - 6008\ \text{J}\cdot\text{mol}^{-1} \\ &\quad + (37.7\ \text{J}\cdot\text{K}^{-1}\cdot\text{mol}^{-1})(263 - 273)\ \text{K} \\ \Delta H_{\text{rxn}} &= -5632\ \text{J}\cdot\text{mol}^{-1}\end{aligned}$$

Therefore

$$\Delta S_{\text{bath}} = \frac{-(-5632\ \text{J}\cdot\text{mol}^{-1})}{263\ \text{K}} = +21.41\ \text{J}\cdot\text{K}^{-1}\cdot\text{mol}^{-1}$$

hence the net entropy production is $21.41 - 20.61 = 0.80\ \text{J}\cdot\text{K}^{-1}\cdot\text{mol}^{-1}$. We can also calculate the Gibbs energy change for the water in this irreversible isothermal process, a quantity that we know must be negative:

$$\Delta G^\circ_{\text{rxn}} = \Delta G_{\text{rxn}} = \Delta H_{\text{rxn}} - T\,\Delta S_{\text{rxn}} = -5632 - 263(-20.61) = -212\ \text{J}\cdot\text{mol}^{-1}$$

where $\Delta G_{\text{rxn}} = \Delta G^\circ_{\text{rxn}}$, because $RT\ln Q = RT\ln(a_s/a_\ell) = RT\ln 1 = 0$. Note that we cannot say that because $K = (a_s/a_\ell)$, and at 1 atm $a_s = a_\ell = 1$, that $K = 1$, because at 1 atm and 263 K, $\Delta G \neq 0$ for this transition.

Problems

1. For the reaction

$$C(s) + H_2O(g) \rightleftharpoons CO(g) + H_2(g)$$

$\Delta H_{\text{rxn}} = +131$ kJ. Circle the correct answer for the direction in which the above equilibrium will shift as a result of the stated change in conditions

(NE denotes *no* effect).

(a) An increase in the reaction temperature.	$\longleftarrow$	NE	$\longrightarrow$
(b) A decrease in the amount of C(*s*).	$\longleftarrow$	NE	$\longrightarrow$
(c) A decrease in the reactor volume.	$\longleftarrow$	NE	$\longrightarrow$
(d) An increase in P_{H_2O}.	$\longleftarrow$	NE	$\longrightarrow$
(e) Addition of N_2 gas to the reaction mixture.	$\longleftarrow$	NE	$\longrightarrow$

2. Given that $K_c = 400$ for the reaction

$$C(s) + H_2O(g) \rightleftharpoons CO(g) + H_2(g)$$

and also given that when excess C(*s*) was reacted with $H_2O(g)$, the equilibrium concentration of $H_2O(g)$ in the reaction mixture was found to be 1.00×10^{-2} *M*, calculate the equilibrium concentration of CO(*g*).

3. Suppose that we place 5.00 mol of C(*s*) and 2.00 mol of $CO_2(g)$ in a 2.00-L reaction vessel, and the following equilibrium

$$C(s) + CO_2(g) \rightleftharpoons 2\,CO(g)$$

is established under conditions where $K_c = 5.00$. Compute the equilibrium concentration of $CO_2(g)$ in the reaction vessel.

4. Suppose that C(*s*) is brought into contact with a mixture of $CO_2(g)$ at 2.00 atm and CO(*g*) at 0.50 atm. Is the reaction

$$C(s) + CO_2(g) \rightleftharpoons 2\,CO(g) \qquad K = 1.90$$

at equilibrium? If not, then predict the direction in which the reaction will proceed toward equilibrium and compute the equilibrium pressures of CO(*g*) and $CO_2(g)$.

5. The equilibrium constant at 100°C for the reaction

$$N_2O_4(g) \rightleftharpoons 2\,NO_2(g)$$

is $K = 5.0 \times 10^2$ (1-atm ideal gas standard states). Suppose we mix 2.00 mol of NO_2 with 2.00 mol of N_2O_4 in a 0.500-L reaction vessel. Compute Q/K for the reaction mixture and predict the direction in which the reaction proceeds toward equilibrium. Also compute the equilibrium pressures of $N_2O_4(g)$ and $NO_2(g)$ at 100°C.

6. For the phase transition

$$\text{HgS}(s, \text{red}) = \text{HgS}(s, \text{black})$$
$$\Delta G^\circ_{rxn} = (4184 - 5.44T)\ \text{J}$$

(a) Obtain ΔH°_{rxn} and ΔS°_{rxn}.
(b) Estimate $\Delta C^\circ_{P,rxn}$
(c) Which form of HgS is more stable at 25°C, 1 atm?
(d) Compute the transition temperature at 1 atm.

(e) Compute the pressure at which the two forms are in equilibrium at 525°C. (Take ρ(red) = 8.1 g·cm^{-3} and ρ(black) = 7.7 g·cm^{-3}.)
(f) Compute K at 25°C and 525°C.
(g) Construct the P-T coexistence curve over the range 0°C–600°C, 0 atm–1500 atm.

7. An important industrial synthesis of formaldehyde from methanol involves the reaction

$$CH_3OH(g) \rightleftharpoons HCHO(g) + H_2(g)$$

Given the following data, compute the equilibrium mole fraction of formaldehyde at 1000 K and a total pressure of 2.00 atm (assume ideal gas behavior).

Species	$[-(\bar{G}^\circ_{1000} - \bar{H}^\circ_0)/1000]/cal \cdot K^{-1} \cdot mol^{-1}$	$\Delta\bar{H}^\circ_{0,f}/kcal \cdot mol^{-1}$
$H_2(g)$	32.74	0
$HCHO(g)$	55.11	−26.80
$CH_3OH(g)$	61.58	−45.47

8. Anhydrous calcium sulfate is widely used as a drying agent (Drierite):

$$2\,CaSO_4(s) + H_2O(g) \rightleftharpoons (CaSO_4)_2 \cdot H_2O(s)$$

Given the following data at 25°C, compute the maximum pressure of $H_2O(g)$ (in torr) in a sample of air in equilibrium with excess $CaSO_4(s)$ at 25°C: $(CaSO_4)_2 \cdot H_2O(s)$, $\Delta\bar{G}^\circ_{f,298} = -686.04$ kcal·mol^{-1}; $CaSO_4(s)$, $\Delta\bar{G}^\circ_{f,298} = -312.46$ kcal·mol^{-1}; and $H_2O(g)$, $\Delta\bar{G}^\circ_{f,298} = -54.64$ kcal·mol^{-1}.

9. For the reaction

$$2\,Ca(\ell) + ThO_2(s) = 2\,CaO(s) + Th(s)$$

the following data were obtained at 1 atm pressure in an electrochemical cell:

T/K	$-\Delta G/kJ$
1275	28.48
1375	20.92
1475	13.31

(a) Compute K, ΔG°_{rxn}, ΔH°_{rxn}, and ΔS°_{rxn} for the cell reaction at 1375 K. Also calculate q_{rev} and w_{elec} for the isothermal reversible operation of the cell at 1375 K.
(b) Calculate the maximum temperature at which $Ca(\ell)$ will reduce $ThO_2(s)$ at 1 atm pressure.
(c) Compute the number of moles of the various species present at equilibrium when 20 g of Ca is allowed to react with 50 g of thorium oxide at 1375 K.

10. Deoxygenated nitrogen is often prepared in the laboratory by passing tank nitrogen over hot copper gauze. The reaction is

$$2\,Cu(s) + \tfrac{1}{2}\,O_2(g) \rightleftharpoons Cu_2O(s)$$

for which $\Delta G^\circ_{rxn} = (-166{,}732 + 63.01T)$ J. Assuming equilibrium is attained, calculate the maximum number of parts per million O_2 in the nitrogen exiting at 1 atm pressure from a tube at 600°C containing $Cu(s)$.

11. The equilibrium vapor pressure of water over $K_4Fe(CN)_6 \cdot 3\,H_2O(s)$ plus $K_4Fe(CN)_6(s)$ has been reported as 10.0 torr at 25.0°C and 7.20 torr at 20.0°C.
(a) Compute K at 20.0°C and 25.0°C for the reaction

$$K_4Fe(CN)_6 \cdot 3\,H_2O(s) \rightleftharpoons K_4Fe(CN)_6(s) + 3\,H_2O(g)$$

(b) Compute ΔG°_{rxn} at 25.0°C for the reaction.
(c) Compute ΔH°_{rxn}.
(d) Compute ΔS°_{rxn}.

12. Given that $\Delta G^\circ_{rxn} = +37.66$ kJ at 25°C for the reaction

$$Na_2O(s) \rightleftharpoons 2\,Na(s) + \tfrac{1}{2}\,O_2(g)$$

estimate the value of P_{O_2} in equilibrium with $Na_2O(s) + Na(s)$ at 25°C.

13. Given that $\Delta G^\circ_{rxn} = +163.2$ kJ at 25°C for the reaction

$$2\,Au(s) + \tfrac{3}{2}\,O_2(g) \rightleftharpoons Au_2O_3(s)$$

Estimate the value of P_{O_2} required to convert $Au(s)$ to $Au_2O_3(s)$.

14. Given that $\Delta H^\circ = 1880$ J and $\Delta S^\circ = -3.31\ J \cdot K^{-1}$ at 25°C for the reaction

$$C(\text{graphite}) \rightleftharpoons C(\text{diamond})$$

and that the densities of diamond and graphite are $3.51\ g \cdot cm^{-3}$ and $2.22\ g \cdot cm^{-3}$, respectively;
(a) Obtain an expression for $\Delta G^\circ_{rxn} = f(T)$ for the process assuming ΔH°_{rxn} is constant, and use it to compute ΔG°_{rxn} at 200 K, 298 K, and 500 K.

(b) Find the pressure at which the two forms are in equilibrium at 200 K, 298 K, and 500 K. Which is the thermodynamically more stable form of solid carbon at 1 atm and 25°C? Is this form the more stable at all T, 1 atm?

(c) How would you determine $\Delta H^\circ_{\text{rxn}}$ for the above reaction? On the basis of your calculations, what experimental conditions do you recommend for making diamonds from graphite?

15. Show that if $\Delta H^\circ_{\text{rxn}}$ is independent of temperature, then $\Delta S^\circ_{\text{rxn}}$ is also independent of temperature.

16. Starting from the van't Hoff equation, derive the expression

$$\ln K = \frac{-k_1}{RT} + k_2 + \frac{a}{R}\ln T + \frac{bT}{2R}$$

for the temperature dependence of an equilibrium constant; k_1, k_2, a, and b are constants, where

$$\Delta C_P^\circ = a + bT$$

for the reaction.

17. Using data for water, given in Chapter 6, compute $\Delta G^\circ_{\text{rxn}}$ for the reaction

$$H_2O(\ell) \rightleftharpoons H_2O(g)$$

at 25°C, 100°C, and 374°C, assuming

(a) $H_2O(g)$ is an ideal gas.

(b) $H_2O(g)$ obeys the equation of state

$$P\bar{V} = RT\left\{1 + \frac{9T_cP}{128P_cT}\left[1 - 6\left(\frac{T_c}{T}\right)^2\right]\right\}$$

18. Given the following data at 25°C for the reaction

$$C_6H_5COOH(aq) \rightleftharpoons C_6H_5COO^-(aq) + H^+(aq)$$

$$\log_{10}K = -4.213 \qquad \Delta\bar{H}^\circ_{\text{rxn}} = -420\ \text{cal}\cdot\text{mol}^{-1}$$

$$\Delta\bar{C}^\circ_{P,\text{rxn}} = -39\ \text{cal}\cdot\text{K}^{-1}\cdot\text{mol}^{-1}$$

Derive an expression for $\log_{10}K = f(T)$, and use it to compute $\log_{10}K$ for the above reaction at 15°C and 50°C.

19. Consider the reaction

$$3\,Mn_2O_3(s) \rightleftharpoons 2\,Mn_3O_4(s) + \tfrac{1}{2}\,O_2(g)$$

Utilizing the data tabulated below, estimate the temperature at which the equilibrium pressure of $O_2(g)$ over the two manganese oxide phases is 0.20 atm. State any assumptions that you make.

Species	$\Delta\bar{G}^\circ_{f,298}/kJ\cdot mol^{-1}$	$\bar{S}^\circ_{298}/J\cdot K^{-1}\cdot mol^{-1}$
$O_2(g)$	0	205.0
$Mn_2O_3(s)$	−881.1	110.5
$Mn_3O_4(s)$	−1283.2	155.6

Also compute the temperature at which $P_{O_2} = 1.00$ atm.

20. Using data in Appendix I, determine whether or not $Fe_3O_4(s)$ is stable to air oxidation to $Fe_2O_3(s)$ at 298 K and 1000 K.

21. The equilibrium vapor pressure of $CO_2(s)$ at 216.6 K is 5.11 atm. Given that the equation of state of $CO_2(g)$ is

$$P\bar{V} = RT + (42.7 \text{ cm}^3\cdot\text{mol}^{-1})P$$

and that $\bar{V}$ for $CO_2(s)$ is 28.2 $\text{cm}^3\cdot\text{mol}^{-1}$ at 216.6 K and 1 atm, compute ΔG°_{rxn} for the process

$$CO_2(s) \rightleftharpoons CO_2(g)$$

22. The equilibrium vapor pressure of water at 110°C is 1.414 atm and $\Delta\bar{H}^\circ_{vap} =$ 40.150 $\text{kJ}\cdot\text{mol}^{-1}$ at this temperature. Calculate ΔS°_{rxn}, ΔG°_{rxn}, and ΔH°_{rxn} for the process

$$H_2O(\ell, 1 \text{ atm}, 110°C) \rightleftharpoons H_2O(g, 1 \text{ atm}, 110°C)$$

Assume that $H_2O(g)$ is ideal and that $\bar{V}_{H_2O}(\ell) = 18.0 \text{ cm}^3\cdot\text{mol}^{-1}$.

23. The equilibrium vapor pressures of $H_2O(s)$ and $H_2O(\ell)$ at −15°C are 1.241 torr and 1.436 torr, respectively. Calculate ΔG°_{rxn} at −15°C for the process

$$H_2O(s) \rightleftharpoons H_2O(\ell)$$

24. The equilibrium vapor pressure of $CO_2(s)$ at 216.6 K is 5.11 atm. Assume that $CO_2(g)$ behavior can be described by the van der Waals equation of state ($a = 3.59 \text{ L}^2\cdot\text{atm}$, $b = 42.7 \text{ cm}^3\cdot\text{mol}^{-1}$). The value of $\bar{V}$ for $CO_2(s)$ is 28.2 $\text{cm}^3\cdot\text{mol}^{-1}$.

(a) Assume that $\bar{V}$ for $CO_2(s)$ is independent of pressure and calculate ΔG°_{rxn} at 216.6 K for

$$CO_2(s) \rightleftharpoons CO_2(g)$$

(b) Compute ΔG for the process

$$CO_2(s, 216.6 \text{ K}, 1 \text{ atm}) \longrightarrow CO_2(\text{ideal gas}, 216.6 \text{ K}, 1 \text{ atm})$$

25. Consider the equilibrium

$$CaCO_3(s) \rightleftharpoons CaO(s) + CO_2(g)$$

and suppose that the total pressure in the reaction vessel is increased by pumping in argon gas at fixed T. Show that the change in the equilibrium value of $a_{CO_2(g)}$ will be exactly compensated for by the corresponding change in the $a_{CaO(s)}/a_{CaCO_3(s)}$ ratio.

26. Consider the application of the van't Hoff equation to a two-phase equilibrium

$$\alpha \rightleftharpoons \beta \qquad K = \frac{a_\beta}{a_\alpha}$$

Show that the van't Hoff equation leads to the Clapeyron equation in such a case. Do *not* assume that $\Delta \bar{H}^\circ_{rxn} = \Delta \bar{H}_{rxn}$.

27. When we calculate ΔG°_{rxn} from $\Delta \bar{G}^\circ_f$ values in units of $J \cdot mol^{-1}$, the resulting ΔG°_{rxn} value has the units J. If we calculate ΔG°_{rxn} from the relation $\Delta G^\circ_{rxn} = -RT \ln K$, then ΔG°_{rxn} appears to have the units $J \cdot mol^{-1}$, because R has the units $J \cdot K^{-1} \cdot mol^{-1}$. How would you resolve this paradox? (*Hint:* Review the derivation of Eq. 10-13.)

28. The conversion of coal to methane gas is carried out by various schemes that involve several steps. One gasification scheme involves the following sequence of reactions:

(a) The reaction of carbon with steam at about 1200°C

$$C(s) + H_2O(g) \rightleftharpoons CO(g) + H_2(g)$$

(b) The partial conversion of $CO(g)$ to $H_2(g)$ ("water-gas shift reaction")

$$CO(g) + H_2O(g) \rightleftharpoons CO_2(g) + H_2(g)$$

This reaction is controlled to produce a 3:1 ratio of CO to H_2 in the reaction mixture, which is required for the next step.

(c) The "catalytic methanation" reaction

$$CO(g) + 3H_2(g) \rightleftharpoons H_2O(g) + CH_4(g)$$

Finally, the reaction mixture is dehydrated. Using thermodynamic data from thermodynamic tables, derive expressions for $\ln K$ as a function of T for each of the above reactions. Plot the $\ln K$ expressions as a function of T on the same graph over the range 400 K to 1700 K.

29. Reaction (a) in Problem 28 is only a rough approximation for coal, because the approximate composition of coal is $CH_{0.8}$. Modern coal gasification schemes use the "hydrogasification reaction"

(a1) $$2CH_{0.8}(s) + 1.2\,H_2(g) \rightleftharpoons CH_4(g) + C(s)$$

in an initial stage to produce substantial amounts of methane directly from coal. The H_2-rich gas is obtained by running reaction (a) on the C(s) produced in reaction (a1). Obtain an expression for $\ln K$ as a function of T for reaction (a1). Clearly state your approximations. (See tables following the Index for data on coal.)

Multistage flash distillation plant Jakarta, Indonesia. Seawater is heated by steam in heat exchangers and then flows into a series of distillation chambers (each at a slightly lower pressure) where the water is partially vaporized. The vaporization process is very fast and is therefore called flash distillation. The plant shown can produce 6480 cubic meters (1.7 million U.S. gallons) per day of distillate from seawater.

11

Activities of Solution-Phase Species

It was van't Hoff who first applied the powerful methods of thermodynamics to solutions in a systematic manner. His treatment, however, lacked the generality which might have been achieved at that time if the system of thermodynamics developed by Gibbs ten years earlier had been employed. Gibbs's great treatise provides all the essential basic principles required for the thermodynamics of solutions. The most important contribution of thermodynamics has been to reduce all measurements of systems in equilibrium to the determination of a single thermodynamic function. Measurements of the elevation of the boiling point and the lowering of the freezing point and vapor pressure of a solvent, caused by the addition of solute, and measurements of solubility, osmotic pressure, and the electromotive forces of suitable cells may all be used to determine the Gibbs chemical potentials of electrolytes.

H. S. Harned and B. B. Owen,
The Physical Chemistry of Electrolytic Solutions, 3rd ed. New York: Reinhold Publishing Corp., 1958.

11-1 *The Activity of a Substance in a Solution Is Proportional to the Equilibrium Vapor Pressure of the Substance*

To carry out a detailed thermodynamic analysis of a chemical reaction involving a solution phase, it is necessary to characterize the activities of the various species in the solution as a function of the composition of the solution. There is no general molecular theory of solutions that can be used to calculate the activity of solution-phase species. Consequently, in most cases, the activities must be obtained from experimental measurements on the solutions of interest. One of the most basic methods for the determination of the activity of a solution-phase species involves the measurement of the partial pressure of the species in a gas phase that is in equilibrium with the solution containing that species.

Consider a two-component liquid solution phase in equilibrium with a two-component gas phase (Figure 11-1). The gas phase and the liquid solution phase are in equilibrium, and thus the chemical potential of each component must be the same in the two phases:

$$\mu_i(soln) = \mu_i(g) \tag{11-1}$$

For each gas-phase species, we choose as a standard state the hypothetical ideal gas at 1 atm at the temperature of the equilibrium system; recall that the chemical potential of a gas in its standard state is denoted by $\mu_i^\circ(g)$. For each solution-phase species, we choose as a standard state the pure liquid at 1 atm pressure at the temperature of the equilibrium system; the chemical potential of the liquid in its standard state is denoted by $\mu_i^\circ(\ell)$. The activity of any species

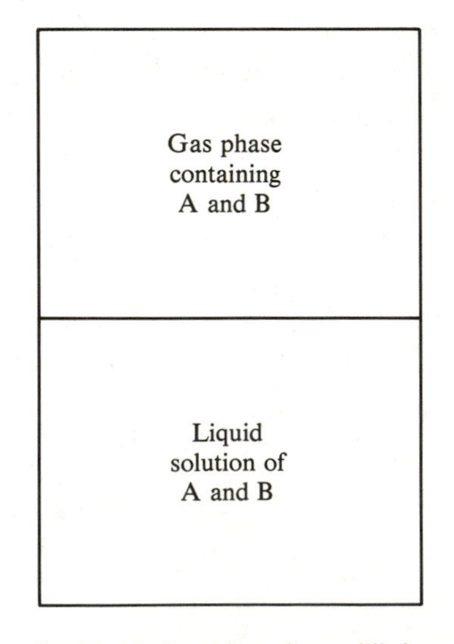

Figure 11-1. Two-component liquid solution phase in equilibrium with a two-component gas phase.

is related to the chemical potential of that species by the equation (see Eq. 9-27)

$$\mu_i = \mu_i^\circ + RT \ln a_i \tag{11-2}$$

Combination of Eqs. 11-2 and 11-1 yields

$$\mu_i^\circ(\ell) + RT \ln a_i(soln) = \mu_i^\circ(g) + RT \ln a_i(g) \tag{11-3}$$

Let $P_i^\bullet$ be the equilibrium vapor pressure of pure liquid i at the same temperature and *total pressure* $P_{tot}(P_{tot} \geqslant P_i^\bullet)$ as in the solution-gas system considered above. At equilibrium

$$\mu_i(\ell) = \mu_i^\bullet(g) \tag{11-4}$$

Using Eq. 11-2 we obtain from Eq. 11-4

$$\mu_i^\circ(\ell) + RT \ln a_i(\ell) = \mu_i^\circ(g) + RT \ln a_i^\bullet(g) \tag{11-5}$$

where $a_i^\bullet(g)$ is the activity of gaseous species in equilibrium with pure liquid i at a pressure of P_{tot}. Subtraction of Eq. 11-5 from Eq. 11-3 yields

$$RT \ln a_i(soln) - RT \ln a_i(\ell) = RT \ln a_i(g) - RT \ln a_i^\bullet(g)$$

or

$$\frac{a_i(soln, T, P_{tot}, X)}{a_i(\ell, T, P_{tot})} = \frac{a_i(g, T, P_i)}{a_i^\bullet(g, T, P_i^\bullet)} \tag{11-6}$$

where X denotes composition variables. Rearranging Eq. 11-6, we obtain

$$a_i(soln, T, P_{tot}, X) = a_i(\ell, T, P_{tot}) \frac{a_i(g, T, P_i)}{a_i^\bullet(g, T, P_i^\bullet)} \tag{11-7}$$

Equation 11-7 tells us that the activity of a solution-phase species i is equal to the activity of pure liquid i times the *ratio* of the equilibrium activities of gaseous i over the solution to the equilibrium activity of gaseous i over *pure* liquid i at the same total pressure as that over the solution.

Equation 11-7 is the basic equation that is used to compute the activity of a solution-phase species from measurements of gas pressures in equilibrium with the solution phase. We are already familiar with $a_i(\ell, T, P_{tot})$; that is, the activity of a pure liquid phase at pressure P_{tot} is (Eq. 9-28)

$$a_i(\ell, T, P_{tot}) = \exp\left[\int_1^{P_{tot}} \frac{\bar{V}_i(\ell)}{RT}\, dP\right] \approx \exp\left[\frac{\bar{V}_i(\ell)}{RT}(P_{tot} - 1)\right] \tag{11-8}$$

As we have already shown (Example 9-4), the value of $a_i(\ell, T, P_{tot})$ is only weakly dependent on P_{tot}, and if P_{tot} is not large (say, <5 atm), then $a_i(\ell, T, P_{tot})$ is equal to unity to within 0.3% error. In any event, if a more precise value of $a_i(\ell, T, P_{tot})$ is needed, then it can be obtained using Eq. 11-8. Taking $a_i(\ell, T, P_{tot}) = 1.000$, Eq. 11-7 becomes

$$a_i(soln, T, P_{tot}, X) = \frac{a_i(g, T, P_i)}{a_i^\bullet(g, T, P_i^\bullet)} \tag{11-9}$$

The activity of a gas is given by Eq. 9-34 as

$$\ln\left[\frac{a_i(g)}{P_i}\right] = \int_0^{P_i}\left(\frac{\bar{V}_i}{RT} - \frac{1}{P}\right)dP \tag{11-10}$$

If the gas phase contains two or more gases, then $\bar{V}_i$ in Eq. 11-10 is the partial molal volume of gas i:

$$\bar{V}_i = \left(\frac{\partial V}{\partial n_i}\right)_{T,P,n_{j(j\neq i)}} \tag{11-11}$$

However, except at high pressure, it is usually a satisfactory approximation for gaseous solutions to take $\bar{V}_i = V/n_i$. If the gas phase is ideal, then

$$\frac{\bar{V}_i}{RT} - \frac{1}{P} = 0$$

and

$$a_i(g) = P_i = Y_i P_{\text{tot}}$$

where P_i is the partial pressure of i in a gas with a total pressure of P_{tot}, and Y_i is the mole fraction of i in the gas phase.

In the case in which the gas phase is ideal, Eq. 11-9 takes the simple form

$$a_i(soln) = \frac{P_i}{P_i^\bullet}$$

where P_i is the equilibrium vapor pressure of i over the solution and $P_i^\bullet$ is the equilibrium vapor pressure of pure liquid i at the same temperature and total pressure as the solution. The equilibrium vapor pressure of a pure liquid is only weakly dependent on total pressure, and thus $P_i^\bullet$ can usually be taken, without significant error, as the equilibrium vapor pressure of pure i; that is, $P_i^\bullet \approx P_i^\circ$. Thus

$$a_i(soln) \approx \frac{P_i}{P_i^\circ} \tag{11-12}$$

Equation 11-12 is remarkably simple to use. To obtain the activity of species i in a solution, all we need to do is to measure the equilibrium vapor pressure of i over the solution and divide by the equilibrium vapor pressure of pure liquid i at the same temperature as that of the solution.

EXAMPLE 11-1

The partial pressures of water and alcohol over a 0.5000-mol fraction alcohol-water solution at 25°C are (1 kPa ≈ 0.010 atm)

$$P_{\text{EtOH}} = 4.914 \text{ kPa} \qquad P_{\text{H}_2\text{O}} = 2.305 \text{ kPa}$$

whereas for the pure liquids at 25°C

$$P^\circ_{\text{EtOH}} = 7.893 \text{ kPa} \qquad P^\circ_{\text{H}_2\text{O}} = 3.169 \text{ kPa}$$

Compute the activities of ethanol and of water in the $X_i = 0.5000$ alcohol-water solution.

Solution: The activities of alcohol and water in the solution are given by Eq. 11-12:

$$a_{\text{EtOH}} = \frac{P_{\text{EtOH}}}{P^\circ_{\text{EtOH}}} = \frac{4.914\text{ kPa}}{7.893\text{ kPa}} = 0.6226$$

$$a_{\text{H}_2\text{O}} = \frac{P_{\text{H}_2\text{O}}}{P^\circ_{\text{H}_2\text{O}}} = \frac{2.305\text{ kPa}}{3.169\text{ kPa}} = 0.7274$$

Note that $a_{\text{H}_2\text{O}} \neq X_{\text{H}_2\text{O}}$ and that $a_{\text{EtOH}} \neq X_{\text{EtOH}}$.

11-2 *Raoult's Law Holds for Ideal Solutions*

Consider a pure liquid that is placed in a closed container with a vapor space; the container is mounted in a thermostat. The surface area of the liquid is constant, and thus the rate of evaporation of molecules from the liquid phase is constant. Initially, the rate of condensation of vapor molecules is zero, because there is no vapor. As the vapor pressure over the liquid increases, the rate of condensation of the vapor increases until the rate of condensation equals the rate of evaporation (Figure 11-2). Equilibrium is attained when the rates

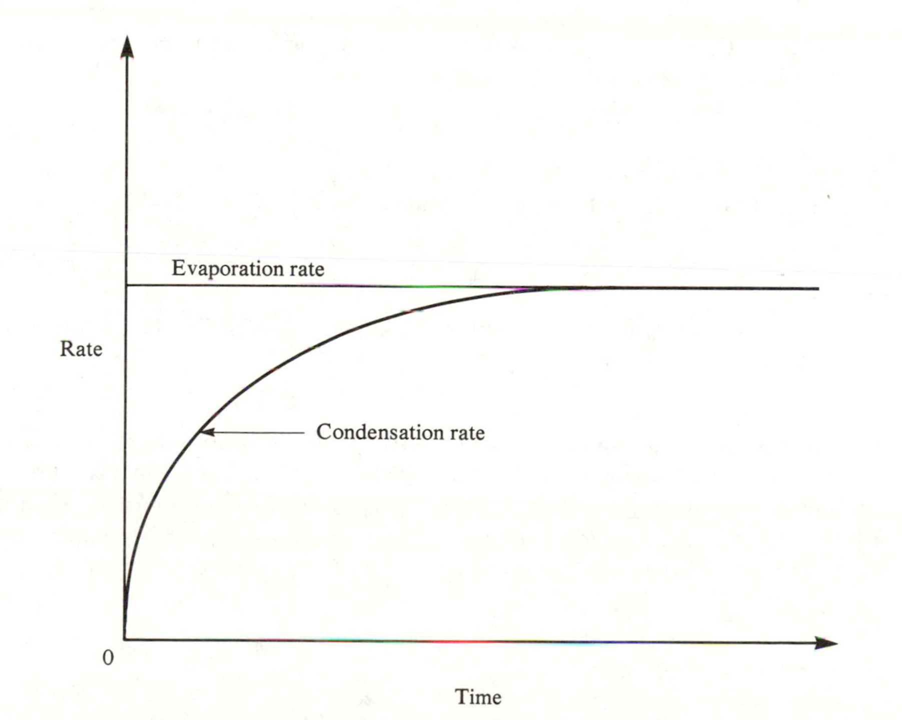

Figure 11-2. Establishment of liquid-vapor equilibrium. At equilibrium, the rate of evaporation of the liquid equals the rate of condensation of the vapor.

of evaporation and condensation are equal. Suppose that we dissolve a nonvolatile solute in the liquid (the solvent) and repeat the experiment. The surface area of the liquid is exactly the same in the two cases; however, the rate of evaporation of the liquid from the solution is less because the number of solvent molecules per unit area of the surface of the liquid is less, owing to the presence of solute molecules at the surface. Because the rate of evaporation of solvent molecules is reduced relative to the pure solvent, the equilibrium vapor pressure of the solvent over the solution is less than the equilibrium vapor pressure of the pure liquid.

If the interactions between A- and B-type molecules in a solution are the same as the interactions between A molecules in pure A and B molecules in pure B, then the A and B molecules in the solution will be randomly distributed. In such a case, the rate of evaporation of the solvent, and hence the equilibrium vapor pressure of the solvent, is directly proportional to the ratio of the number of solvent molecules N_1 to the number of solvent plus solute molecules $N_1 + N_2$. But

$$\frac{N_1}{N_1 + N_2} = \frac{n_1}{n_1 + n_2} = X_1$$

where the n_i is the number of moles of i and X_1 is the mole fraction of the solvent; therefore

$$P_1 \propto X_1 \tag{11-13}$$

The value of $P_1 \to P_1^\circ$ as $X_1 \to 1$ (pure solvent), and thus the proportionality constant in Eq. 11-13 must be P_1°; that is,

$$P_1 = P_1^\circ X_1 \tag{11-14}$$

Equation 11-14 is known as *Raoult's law.*

The activity of the solvent in the solution is given by Eq. 11-12 as

$$a_1 = \frac{P_1}{P_1^\circ} \tag{11-12}$$

If Raoult's law holds for the solvent, then substitution of Eq. 11-14 into Eq. 11-12 yields

$$a_1 = X_1 \tag{11-15}$$

that is, if Raoult's law holds, then the activity of a solution-phase species is equal to the mole fraction of that species in the solution. If Raoult's law holds for all components of a solution, then $a_i = X_i$ for all the components. Such a solution is said to be an *ideal solution.* Ideal solutions are rare, but they are not unknown. Two examples are (1) solutions of the xylenes

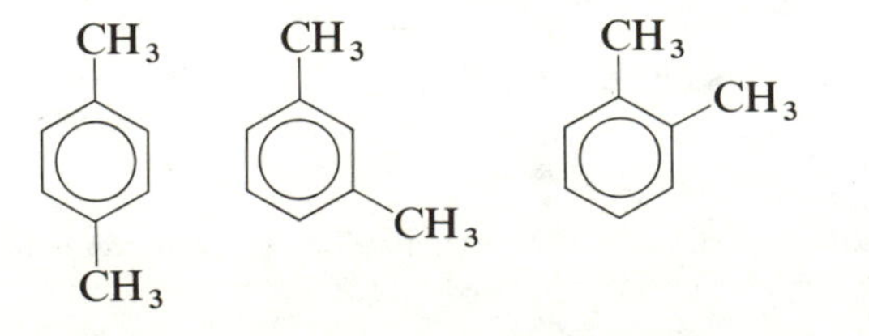

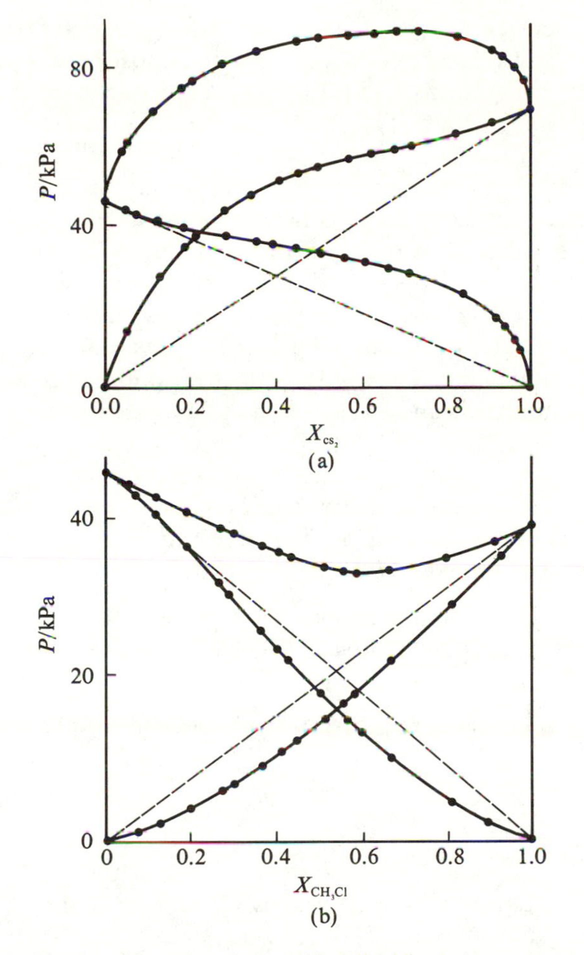

Figure 11-3. Pressure-composition plots for nonideal liquid solutions: (a) $CS_2 + CH_3COCH_3$ and (b) $CS_2 + CH_3Cl$. In case a the deviations from Raoult's law are positive ($P_i > X_iP_i^\circ$), whereas in case b the deviations from Raoult's law are negative ($P_i < X_iP_i^\circ$). The values of the vapor pressure predicted by Raoult's law are shown by the dashed lines in case a and in case b. The deviations from Raoult's law are a consequence of the fact that the intermolecular interactions between unlike molecules are different than those between like molecules.

and (2) solutions of 1-propanol + 2-propanol

$$CH_3CH_2CH_2OH \qquad CH_3\underset{\displaystyle OH}{\underset{|}{C}}HCH_3$$

For an ideal gas, $(\partial U/\partial V)_T = 0$. The corresponding condition for an ideal solution is

$$\left(\frac{\partial U}{\partial V}\right)_{T,1} = \left(\frac{\partial U}{\partial V}\right)_{T,2} \tag{11-16}$$

Most solutions are not ideal because Eq. 11-16 is a very stringent requirement.

If a two-component solution is not ideal, then two cases arise: (1) the attractions between solvent and solute molecules are less than those between the molecules of the two pure components and (2) the attractions between the solvent and solute molecules are greater than those between the molecules of the pure components. Case (1) gives rise to positive deviations from Raoult's law, $P_1 > X_1 P_1^\circ$; case (2) gives rise to negative deviations from Raoult's law, $P_1 < X_1 P_1^\circ$. An example of case (1) is carbon disulfide + acetone solutions, and an example of case (2) is carbon disulfide + chloroform solutions (Figure 11-3). Deviations from Raoult's law are positive when the unlike molecules repel one another to a greater extent than do the like molecules and, in effect, tend to force each other out of the solution. The deviations from Raoult's law are negative when the unlike molecules attract one another to a greater extent than do the like molecules and, in effect, tend to hold each other in the solution.

11-3 *The Vapor over a Liquid with Two Volatile Components Is Richer in the More Volatile Component Than the Liquid Is*

We shall now consider some quantitative aspects of a solution with two volatile components, an example of which is solutions composed of benzene and toluene:

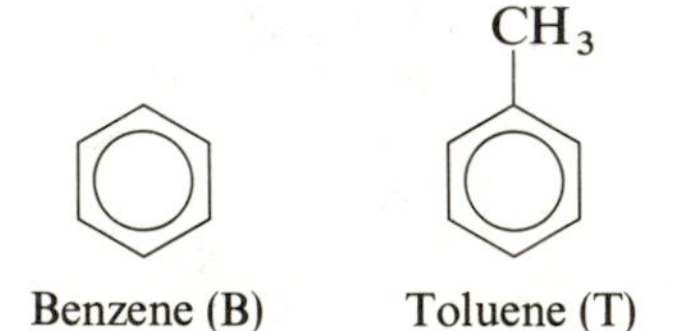

The total equilibrium pressure P_{tot} over the solution (excluding air) is

$$P_{tot} = P_B + P_T \tag{11-17}$$

where P_B is the vapor pressure of benzene and P_T is the vapor pressure of toluene. If Raoult's law holds for both benzene and toluene, then using Eq. 11-14 we have

$$P_B = X_B P_B^\circ \qquad P_T = X_T P_T^\circ \tag{11-18}$$

Substitution of Eqs. 11-18 into Eq. 11-17 yields

$$P_{tot} = X_B P_B^\circ + X_T P_T^\circ \tag{11-19}$$

Further

$$X_T + X_B = 1$$

and thus

$$P_{tot} = X_B P_B^\circ + P_T^\circ - X_B P_T^\circ = P_T^\circ + X_B(P_B^\circ - P_T^\circ) \tag{11-20}$$

Equation 11-20 is the equation for a straight line (P_{tot} versus X_B) because P_B° and P_T° are constants (see Figure 11-4). Equation 11-20 holds nicely for solutions composed of benzene and toluene, as shown in Figure 11-4. Note that a solution of benzene and toluene is an ideal solution.

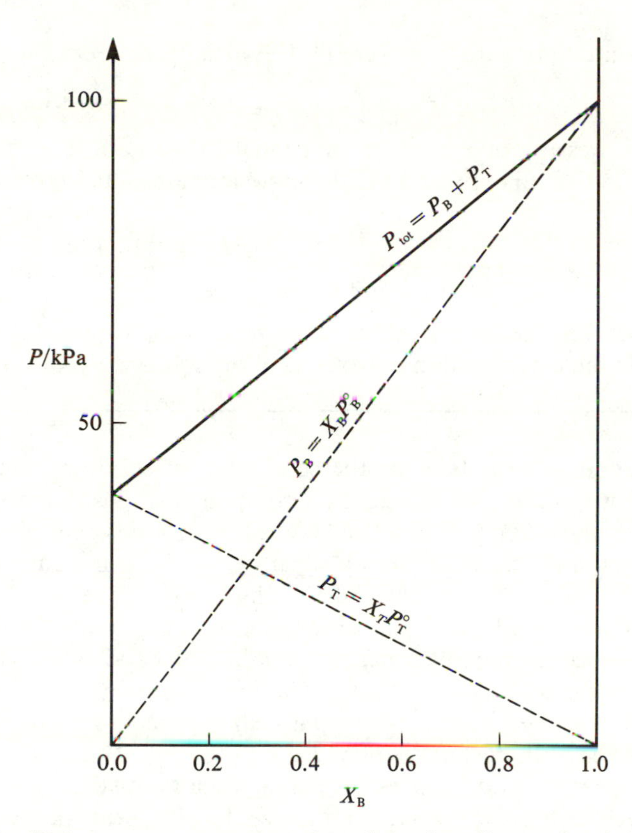

Figure 11-4. Equilibrium vapor pressure versus the mole fraction of benzene in solutions composed of benzene and toluene. These solutions are *ideal*, and the equilibrium vapor pressures of benzene P_B and toluene P_T are given by Raoult's law:

$$P_B = X_B P_B^\circ \qquad P_T = X_T P_T^\circ$$

The total pressure, $P_{tot} = P_B + P_T$, is a straight-line function of X_B when Raoult's law holds for both components, that is, when the solution is ideal.

EXAMPLE 11-2

Given that at 81°C the equilibrium vapor pressures of toluene and benzene are 39.0 kPa and 102.4 kPa, respectively, compute:

(a) The total pressure over the $X_B = 0.500$ solution.

(b) The mole fraction of benzene in the vapor.

Solution:

(a) The partial pressures of benzene and toluene over the solution are computed from Raoult's law:

$$P_B = X_B P_B^\circ = 0.500 \times 102.4 \text{ kPa} = 51.2 \text{ kPa}$$

$$P_T = X_T P_T^\circ = 0.500 \times 39.0 \text{ kPa} = 19.5 \text{ kPa}$$

The total pressure is the sum of the partial pressures:

$$P_{tot} = P_B + P_T = 51.2 + 19.5 = 70.7 \text{ kPa}$$

(b) The pressure of a gas is proportional to the number of moles of the gas in the vapor; therefore, the mole fraction of benzene in the vapor Y_B is given by

$$Y_B = \frac{n_B}{n_B + n_T} = \frac{P_B}{P_B + P_T} = \frac{P_B}{P_{tot}} = \frac{51.2 \text{ kPa}}{70.7 \text{ kPa}} = 0.724$$

Note that the mole fraction of benzene in the vapor (0.724) is greater than the mole fraction of benzene in the solution (0.500).

Distillation is a method for the separation of solution components by preferential evaporation of the solution components. Distillation consists of the thermal evaporation of a liquid and the removal of the vapor phase, followed by condensation of the vapor into a separate vessel. Pure water is obtained from seawater on a commercial scale by distillation. The salts dissolved in seawater are nonvolatile, and thus only the water evaporates; however, as the distallation proceeds, the salt concentration increases and eventually salts precipitate from the concentrated brine. The precipitated salts collect as boiler scale and increase the amount of energy as heat required for the distillation and, therefore, increase the cost of the pure water produced. To avoid this problem, large-scale commercial desalinization plants use a multistage flash distillation process. Such units are designed to minimize the total heat energy input to the process, because the energy input is by far the major cost factor for a desalinization plant. Flash distillation refers to the rapid evaporation of a hot liquid that occurs when the liquid is injected into a chamber where the pressure is much lower than the equilibrium vapor pressure of the hot liquid.

If two or more components of the solution are volatile, then the distillation apparatus must be more elaborate to achieve a separation of the components. In any case it is the more volatile component that is distilled; the less volatile component remains behind in the distillation flask. In essence, the distillation process in such a case utilizes the fact that the vapor is richer than the liquid in the more volatile component; that is, the mole fraction of the more volatile component is higher in the vapor than in the liquid. If this vapor is condensed and partially reevaporated, then the mole fraction of the more volatile component will be increased even further. If the condensation-reevaporation process is repeated a sufficiently large number of times, then a separation can be achieved.

Remarkable separations can be achieved by multistage distillation units. Distillation techniques are used to separate heavy water (D_2O) from light water (H_2O). The heavy water is used on a large scale as a neutron moderator (slows down the neutrons) and coolant in heavy water nuclear power plants. Light water has a normal boiling point of 100.00°C, whereas heavy water has a normal boiling point of 101.42°C. Only 0.015% of the hydrogen atoms in normal water are the deuterium isotope; nonetheless, a modern heavy water distillation plant produces 99.8 atom % D_2O from normal isotopic abundance water at a total cost of around $440 per kilogram of D_2O. The distillation plants have over 300 successive distillation stages and require an input of about 135,000 mol of normal water per mole of D_2O produced. Canada, which uses D_2O as a coolant in its nuclear reactors, has two heavy water plants with a combined D_2O output capability of 1600 tons per year.

Ethyl alcohol + water solutions constitute an especially interesting case of nonideal solution behavior. If we attempt to separate alcohol from water by distillation, an amazing result is obtained. No matter how effective a distillation column we use, the distillate issuing from the column has a maximum alcohol content of 95%; water is 5% of the distillate. If the solution is 50% alcohol, then the distillate is 95% alcohol. If the solution is 12% alcohol, then the distillate is 95% alcohol. If the solution is 98% alcohol, then the distillate is initially 95% alcohol until the liquid phase reaches 100% alcohol. A 95% alcohol-plus-water solution distills *as if it were a pure liquid. A solution that distills without change in composition is called an "azeotrope."* Azeotropes are found only for some nonideal solutions; the composition of the azeotrope corresponds to the composition for which the average intermolecular interactions between the two types of molecules are the same as those in the pure liquids.

Alcohol-water solutions obtained by fermentation have a maximum alcohol content of about 12%. Yeasts that produce alcohol as a waste product cannot survive in solutions with more than 12% alcohol. In effect, the yeasts are poisoned by their own waste products. The *proof* of an alcohol-water solution is defined as two times the percent by volume alcohol in the solution. An 86 proof liquor is 43% alcohol. The proof can be increased by simple distillation up to 190 (95%). *Absolute alcohol* (100% ethanol) can be prepared from 95% alcohol by dehydration with calcium oxide, which removes the water from the solution by the reaction

$$CaO(s) + H_2O(soln) \longrightarrow Ca(OH)_2(s)$$

Absolute alcohol is prepared commercially on a large scale by addition of benzene to 95% alcohol followed by distillation. The benzene, water, and some of the alcohol distill off, leaving absolute alcohol behind.

11-4 *Activity Coefficients Measure the Deviation of a Solution from Ideal Behavior*

In a nonideal solution the ratio a_i/X_i is a measure of the deviation of the solution from ideal behavior. If $a_i/X_i = 1$, then Raoult's law holds and the solution is ideal. If $a_i/X_i \neq 1$, then the solution is nonideal. The *activity coefficient* of species i, γ_i, is defined as

$$\gamma_i = \frac{a_i}{X_i} \tag{11-21}$$

The activity coefficient is a quantitative measure of the deviation of the solution from ideal behavior. Note that if Raoult's law holds for component 1, then

$$\gamma_1 = \frac{a_1}{X_1} = \frac{P_1}{X_1 P_1^\circ} = \frac{X_1 P_1^\circ}{X_1 P_1^\circ} = 1$$

If the solution is not ideal, then $\gamma_i \neq 1$. For any component, call it component 1, capable of existing as a pure liquid at the temperature of the solution, Raoult's law will always be satisfied *in the limit* as $X_1 \rightarrow 1$:

$$\lim_{X_1 \rightarrow 1} \left(\frac{dP_1}{dX_1} \right) = P_1^\circ \tag{11-22}$$

EXAMPLE 11-3

We have shown in Example 11-1 that the activities of water and alcohol in the $X_{H_2O} = 0.5000$ water + alcohol solution at 25°C relative to the pure liquid standard states are

$$a_{H_2O} = 0.7274 \qquad a_{EtOH} = 0.6226$$

Compute the activity coefficients of water and alcohol in this solution.

Solution: The activity coefficients of water and alcohol in the solutions are computed using Eq. 11-21:

$$\gamma_{H_2O} = \frac{a_{H_2O}}{X_{H_2O}} = \frac{0.7274}{0.5000} = 1.455$$

$$\gamma_{EtOH} = \frac{a_{EtOH}}{X_{EtOH}} = \frac{0.6226}{0.5000} = 1.245$$

The $X_{H_2O} = 0.5000$ alcohol + water solution is not ideal, because $\gamma_i \neq 1$.

In Figure 11-5 we have plotted equilibrium vapor pressures versus composition for the water + alcohol solutions at 25°C. Note in Figure 11-5 that Raoult's law holds for alcohol in the solution only over the limited concentration range $0.90 < X_{EtOH} < 1.0$; whereas for water Raoult's law holds over the concentration range $0.92 < X_{H_2O} < 1.0$. Also, the total pressure P_{tot} is not given accurately by Raoult's law at any value of X_{EtOH}, except for those corresponding to the pure liquids. The deviations from Raoult's law are strongly positive. The positive deviations from Raoult's law arise at the molecular level from the hydrocarbon portion of the ethanol molecules that preferentially interact with the hydrocarbon portions of other ethanol molecules rather than with water molecules.

A comparison at 25°C of the equilibrium vapor pressure of water over aqueous sucrose solutions (0 *m to* 2 *m* in sucrose) with the equilibrium vapor

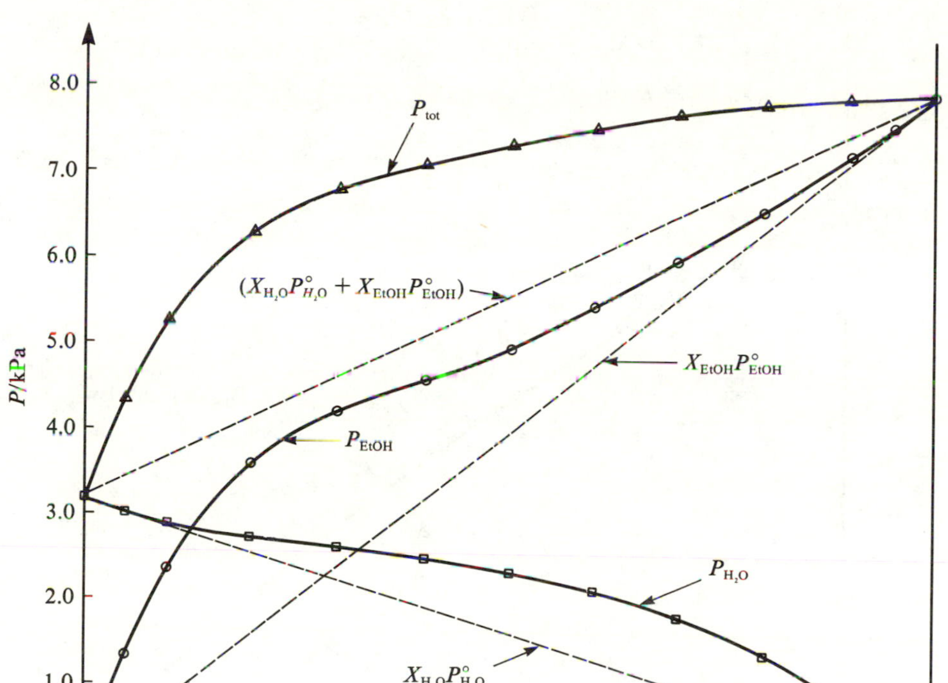

Figure 11-5. Equilibrium vapor pressures of alcohol + water solutions at 25°C. The dashed lines represent the behavior expected if Raoult's law holds.

pressure of water for these solutions predicted by Raoult's law is presented in Figure 11-6. As seen in Figure 11-6, Raoult's law holds for water in these solutions only for $X_{H_2O} > 0.987$, which corresponds to sucrose molalities in the range $0 < m < 0.7$. Above about 0.7 m in sucrose, *negative* deviations from Raoult's law become apparent on the scale of the figure.

We can obtain the activity of water in an aqueous sucrose solution from a determination of the equilibrium vapor pressure of water over the solution. If we know the mole fraction of water in the solution, then we can compute the activity coefficient of water in the solution. Sucrose, $C_{12}H_{22}O_{11}$, is nonvolatile. The vapor pressure of sucrose over an aqueous sucrose solution is effectively zero. How then can we determine the activity of sucrose in aqueous sucrose solutions? A similar problem presents itself for aqueous salt solutions, for example, NaCl(aq); there is no detectable pressure of NaCl(g) over an aqueous sodium chloride solution. To understand how this problem is solved, it is necessary to understand Henry's law and the difference between solute and solvent standard states for solution-phase components.

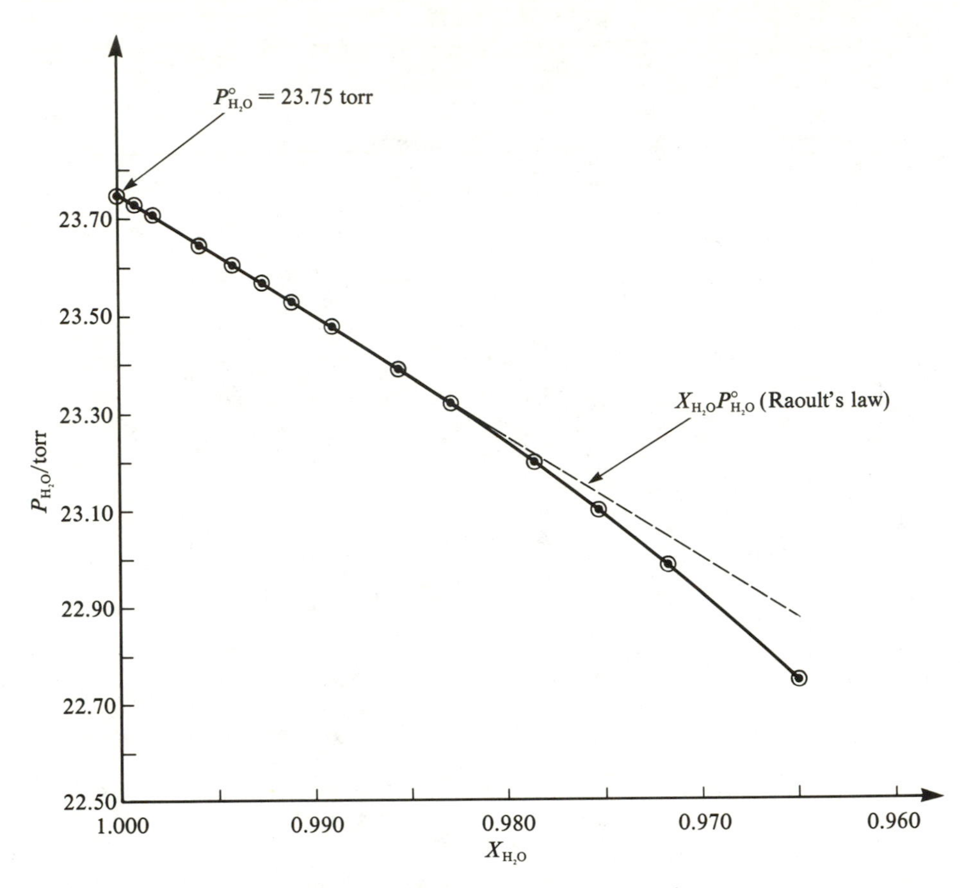

Figure 11-6. Comparison of the actual equilibrium vapor pressure of water over aqueous sucrose solutions at 25°C with that predicted by Raoult's law.

11-5 *Solute Standard States Are Based on Henry's Law*

If the pure liquid exists at the temperature of a solution in which it is a component, then we always choose the pure liquid at 1 atm and the temperature of the solution as the standard state for that component, as done for both water and alcohol in the water + alcohol solutions and for water in the water + sucrose solutions. Pure liquid at 1-atm standard states for solution components are called *solvent or Raoult's law standard states.* Note that the designations *solvent* and *solute* are arbitrary when the solution components both exist as pure liquids. For example, it makes no difference whether we regard water as dissolved in ethanol or the reverse. The activity of a solution component relative to a solvent standard state is given approximately by Eq. 11-12:

$$a_i = \frac{P_i}{P_i^\circ}$$

where P_i° is the equilibrium vapor pressure of pure component i at the temperature of the solution. We cannot establish a solvent standard state for a component of a solution if we cannot determine P_i°, that is, if the pure liquid does not exist at the temperature of the solution. For example, neither sucrose nor oxygen ($T_c = 155$ K) can exist as liquids at 25°C, and thus $P_{O_2}^\circ$ and $P_{C_{12}H_{22}O_{11}}^\circ$ cannot be determined at 25°C. The resolution of this dilemma is based on Henry's law.

Consider the equilibrium between oxygen dissolved in water and oxygen in the vapor space over the aqueous solution; that is,

$$O_2(aq) \rightleftharpoons O_2(g) \tag{11-23}$$

The equilibrium constant expression for the reaction in Eq. 11-23 is

$$K = \frac{a_{O_2}(g)}{a_{O_2}(soln)} \tag{11-24}$$

If the equilibrium pressure of $O_2(g)$ over the solution is low, then $a_{O_2}(g) \approx P_{O_2}$ and

$$K = \frac{P_{O_2}}{a_{O_2}(soln)}$$

Henry's law, which is based on experimental measurements for equilibria of the type shown in the reaction in Eq. 11-23, states that the limiting slope of a plot of the equilibrium gas pressure versus the concentration of dissolved gas is a constant. Denoting the gas pressure of the solute by P_2 and the molality of the dissolved gas by m, we have

$$\lim_{m \to 0} \left(\frac{dP_2}{dm} \right) = P_{2m}^* \tag{11-25}$$

where P_{2m}^* is the *Henry's law constant.* If the slope of a plot of P_2 versus m is a constant at low m, the P_2 is proportional to m at low m, and

$$\frac{dP_2}{dm} = \frac{P_2}{m} = P_{2m}^* \tag{11-26}$$

where P_{2m}^* is the proportionality constant. Thus Henry's law can also be stated in the form

$$P_2 = P_{2m}^* m \qquad \text{(at low } m\text{)} \tag{11-27}$$

or, using mole fraction concentration units,

$$P_2 = P_{2X}^* X_2 \tag{11-28}$$

where $P_{2X}^* \neq P_{2m}^*$.

Henry's law states that at *low concentrations of the dissolved gas the equilibrium partial pressure of the gas over the solution is directly proportional to the concentration of the gas in the solution.* However, we have no way of knowing a priori to how high a concentration Henry's law holds for any particular gas; this question can only be answered by experimental measurements. A plot of P_{O_2} versus m_{O_2} in water at 25°C is given in Figure 11-7. The graph shows that Henry's law holds over the range $0 < m_{O_2} < 5.5 \times 10^{-3}$. Above 5.5×10^{-3} m, positive deviations from Henry's law become evident on the scale of

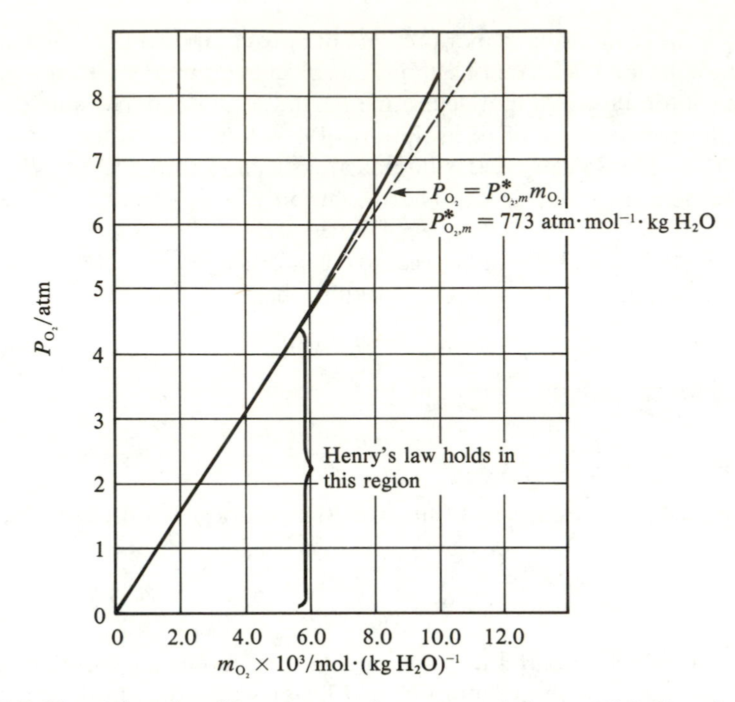

Figure 11-7. Henry's law plot of P_{O_2} versus m_{O_2} for aqueous solutions at 25°C. Note that Henry's law holds (within the resolution of the graph) up to an oxygen pressure of about 4.5 atm.

Figure 11-7. The Henry's law constant for O_2 in water at 25°C is determined from the slope of the P_{O_2} versus m_{O_2} graph in the Henry's law region; the experimental value of the Henry's law constant for O_2 in water is

$$P^*_{O_2,m} = 773 \text{ atm}\cdot\text{mol}^{-1}\cdot\text{kg H}_2\text{O}$$

Henry's law constants for several gases in water are given in Table 11-1.

TABLE 11-1

Henry's Law Constants on the Molality Scale P^*_{2m} for Some Common Gases in Water at 25°C

Gas	*(P^*_{2m}/atm·mol^{-1}·kg H_2O × 10^3)*
CO_2	0.034
NH_3	0.067
O_2	0.773
H_2	1.31
N_2	1.61
He	2.65
H_2S	10

EXAMPLE 11-4

Compute the concentration of dissolved O_2 in water at 25°C in equilibrium with air ($P_{O_2} = 0.20$ atm); take $P^*_{O_2,m} = 773$ atm·mol^{-1}·kg H_2O.

Solution:

$$P_{O_2} = P^*_{O_2,m} m_{O_2}$$

$$P_{O_2} = (773 \text{ atm}\cdot\text{mol}^{-1}\cdot\text{kg H}_2\text{O}) m_{O_2}$$

$$m_{O_2} = \frac{0.20 \text{ atm}}{773 \text{ atm}\cdot\text{mol}^{-1}\cdot\text{kg H}_2\text{O}} = 2.6 \times 10^{-4} \text{ mol}\cdot(\text{kg H}_2\text{O})^{-1}$$

We are now in a position to define a solute (Henry's law) standard state. The solute standard state is defined as *the hypothetical, ideal unit-concentration-of-solute solution at* 1 *atm total pressure and the temperature of interest* (see Figure 11-8).

The general expression for the activity of a solution-phase species relative to a solute standard state is obtained in a manner analogous to the development

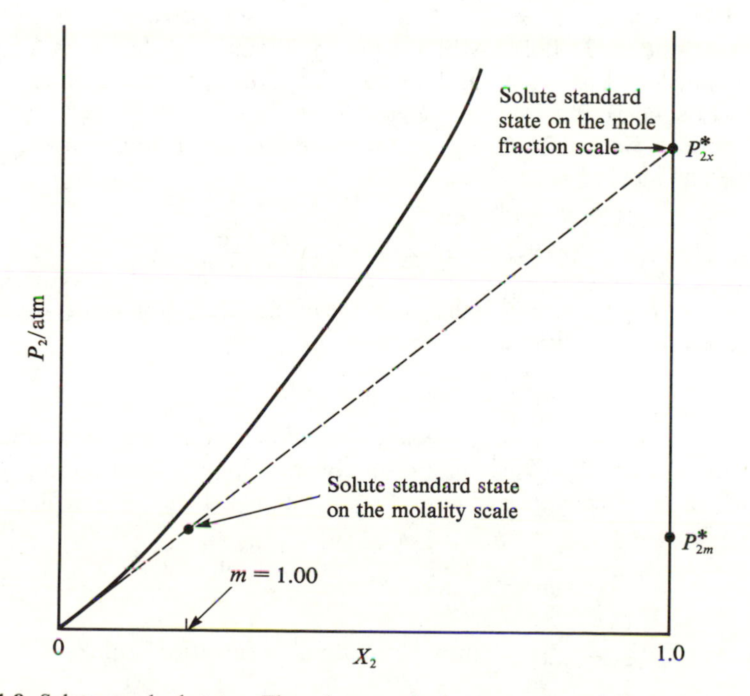

Figure 11-8. Solute standard states. The solute standard state on the molality scale is the *hypothetical* state for which $m = 1$ and $P_2 = P^*_{2m}$; the solute standard state on the mole fraction scale is the *hypothetical* state for which $X_2 = 1$ and $P_2 = P^*_{2X}$ (note that neither of these two states lies on the actual P_2 versus X_2 curve, which is the solid line).

of Eq. 11-7. The results follow:

1. Molality concentration scale:

$$a_2(soln, T, P_{tot}, m) = a_2\begin{pmatrix} soln, T, P_{tot}, \\ \text{hypothetical} \\ \text{ideal } m = 1 \end{pmatrix} \frac{a_2(g, T, P_2)}{a^*_{2m}(g, T, P^*_{2m})} \tag{11-29}$$

2. Mole fraction concentration scale:

$$a_2(soln, T, P_{tot}, X_2) = a_2\begin{pmatrix} soln, T, P_{tot}, \\ \text{hypothetical} \\ \text{ideal } X_2 = 1 \end{pmatrix} \frac{a_2(g, T, P_2)}{a^*_{2X}(g, T, P^*_{2X})} \tag{11-30}$$

If the gas phases are ideal and the total pressure P_{tot} is not too far from 1 atm, then Eqs. 11-29 and 11-30 reduce to the simpler forms

$$a_{2m} = \frac{P_2}{P^*_{2m}} \tag{11-31}$$

$$a_{2X} = \frac{P_2}{P^*_{2X}} \tag{11-32}$$

respectively, where $a_{2m} \neq a_{2X}$. Figure 11-8 was constructed assuming that Eqs. 11-31 and 11-32 hold. Modification of Figure 11-8 to the more general cases defined in Eqs. 11-29 and 11-30 is straightforward. In most cases of interest Eqs. 11-31 and 11-32 are adequate.

Consider Eq. 11-29 at $P_{tot} = 1$ atm. At $P_{tot} = 1$ atm, $a_2(soln, T, P_{tot},$ hypothetical ideal $m = 1) = 1$, and thus

$$a^*_{2m}(g, T, P^*_{2m}) = \frac{a_2(g, T, P_2)}{a_2(soln, T, m)} \tag{11-33}$$

where $a^*_{2m}(g, T, P^*_{2m}) \approx P^*_{2m}$ if the gas is ideal. But Eq. 11-33 is just the equilibrium constant expression for the reaction

$$\text{B}(soln, m) = \text{B}(g) \tag{11-34}$$

where B denotes the chemical species. Therefore, the Henry's law constant is approximately equal to the equilibrium constant for the reaction in Eq. 11-34.

Activity coefficients relative to a solute standard state are defined as

$$\gamma_{2m} = \frac{a_{2m}}{m} \quad \text{(molality concentration scale)} \tag{11-35}$$

$$\gamma_{2X} = \frac{a_{2X}}{X_2} \quad \text{(mole fraction concentration scale)} \tag{11-36}$$

Note that $\gamma_{2m} \neq \gamma_{2X}$; the value of the activity coefficient of a component in solution depends on the choice of standard state, because the value of the activity depends on the choice of standard state.

If Henry's law holds for the solute, and the gas behaves ideally, then

$$a_{2m} = \frac{P_2}{P^*_{2m}} = \frac{mP^*_{2m}}{P^*_{2m}} = m$$

and

$$\gamma_{2m} = \frac{a_{2m}}{m} = 1$$

Thus if Henry's law holds, then the activity coefficient of the solute is equal to one. Further, given the above assumptions but using the mole fraction concentration scale, we have

$$a_{2X} = \frac{P_2}{P^*_{2X}} = \frac{X_2 P^*_{2X}}{P^*_{2X}} = X_2$$

and

$$\gamma_{2X} = \frac{a_{2X}}{X_2} = 1$$

Thus if the concentration of the solute is in the region where Henry's law holds, then the activity is equal to the concentration ($\gamma_2 = 1$). Such a solution is called a *dilute solution.*

EXAMPLE 11-5

The equilibrium vapor pressure of $O_2(g)$ at 25°C over an aqueous solution with an oxygen molality of 1.00×10^{-2} is 8.20 atm. Estimate the activity and the activity coefficient of oxygen in the solution on the molality scale, assuming ideal gas behavior.

Solution: The activity of dissolved oxygen on the molality scale is given by Eq. 11-31 as

$$a_{O_2,m} = \frac{P_{O_2}}{P^*_{O_2,m}} = \frac{8.20 \text{ atm}}{773 \text{ atm}\cdot\text{mol}^{-1}\cdot\text{kg H}_2\text{O}} = 1.06 \times 10^{-2}\ m$$

The activity coefficient of dissolved oxygen on the molality scale is given by Eq. 11-35:

$$\gamma_{O_2,m} = \frac{a_{O_2,m}}{m} = \frac{1.06 \times 10^{-2}\ m}{1.00 \times 10^{-2}\ m} = 1.06$$

An activity coefficient of 1.06 corresponds to a 6% positive deviation from dilute solution behavior of 0.010 m $O_2(aq)$.

Henry's law constants for some common gases in water are given in Table 11-1. The smaller the value of P^*_{2m}, the higher the solubility of the gas. Thus O_2 has a solubility in water twice as great as that of N_2. However, because the partial pressure of N_2 in air is about four times that of O_2, the concentration of N_2 in water in equilibrium with air is twice as great as that for O_2.

TABLE 11-2
Henry's Law Constants P^*_{2m} for O_2 in Water over the Temperature Range 0°C to 40°C

t/°C	P^*_{2m}/atm·mol^{-1}·kg H_2O
0	453
5	511
10	575
15	640
20	707
25	773
30	836
40	962

The solubility of most gases in most liquids *decreases* with increasing temperature; that is, the Henry's law constant increases with increasing temperature. The data for O_2 in water, which illustrate this effect, are given in Table 11-2. The solubility of O_2 in water at 0°C is over twice the solubility of O_2 in water at 40°C.

As noted above, Henry's law constants are equilibrium constants. At 25°C we have

$$O_2(aq) \rightleftharpoons O_2(g) \qquad K = 773 \tag{11-37}$$

The standard Gibbs energy change for the reaction in Eq. 11-37 at 25°C is

$$\Delta G^\circ_{\text{rxn}} = -RT\ln K = \frac{-(8.314\ \text{J}\cdot\text{K}^{-1})(298.15\ \text{K})}{(1000\ \text{J}\cdot\text{kJ}^{-1})}\ln(773) = -16.48\ \text{kJ}$$

Further,

$$\Delta G^\circ_{\text{rxn}} = \Delta G^\circ_f[O_2(g)] - \Delta G^\circ_f[O_2(aq)] = -16.48\ \text{kJ}$$

but $\Delta G^\circ_f[O_2(g)] = 0$, and thus $\Delta G^\circ_f[O_2(aq)] = 16.48$ kJ at 25°C. The value of $\Delta H^\circ_{\text{rxn}}$ at 25°C for the reaction in Eq. 11-37 can be obtained from the data in Table 11-2, together with the van't Hoff equation. The data in Table 11-2 are plotted in the form $\log K$ versus T^{-1} in Figure 11-9. A tangent to the curve at 298.15 K shows that the $\log K$ versus T^{-1} plot is slightly curved, which means that $\Delta H^\circ_{\text{rxn}}$ is slightly temperature dependent ($\Delta C^\circ_{P,\text{rxn}} \neq 0$). The value of $\Delta H^\circ_{\text{rxn}}$ at 25°C can be computed from the van't Hoff equation using the slope of the plot of $\log K$ versus $1/T$ at 25°C. Thus

$$\Delta H^\circ_{\text{rxn}} = \frac{R\ln(K_2/K_1)}{[(1/T_1) - (1/T_2)]}$$

and

$$\Delta H^\circ_{\text{rxn}} = \frac{2.303 \times 8.314\ \text{J}\cdot\text{K}^{-1}[\log(\frac{836}{707})]}{1000\ \text{J}\cdot\text{kJ}^{-1}[(1/293.15) - (1/303.15)]} = 12.38\ \text{kJ}$$

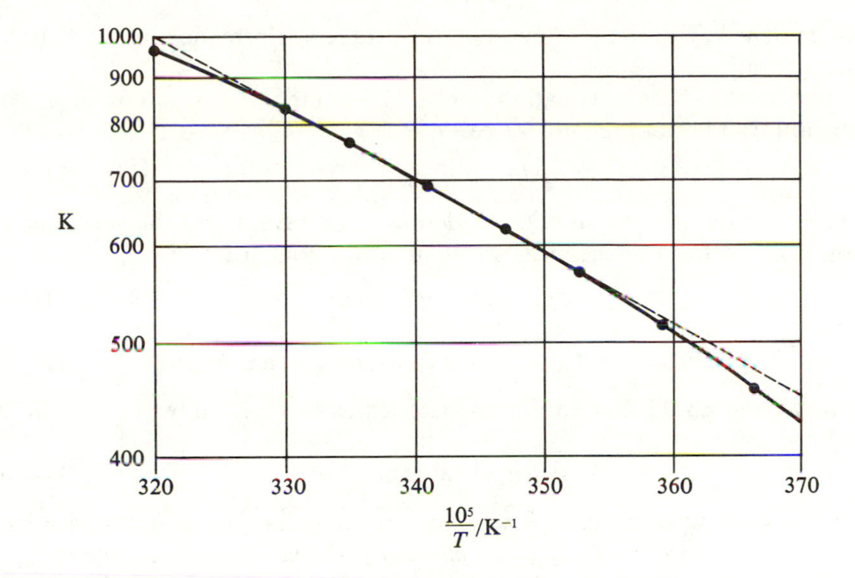

Figure 11-9. Plot of $\log_{10} K$ versus $1/T$ for the reaction

$$O_2(aq) \rightleftharpoons O_2(g)$$

over the range 0°C to 40°C. Note that the plot shows some curvature (compare with dashed straight line that is tangent to the curve at 25°C).

The entropy change for the reaction in Eq. 11-37 can be computed from the equation

$$\Delta G^\circ_{\text{rxn}} = \Delta H^\circ_{\text{rxn}} - T\Delta S^\circ_{\text{rxn}}$$

from which we obtain

$$\Delta S^\circ_{\text{rxn}} = \frac{\Delta H^\circ_{\text{rxn}} - \Delta G^\circ_{\text{rxn}}}{T}$$

and at 25°C we compute

$$\Delta S^\circ_{\text{rxn}} = \frac{12.39 \times 10^3 \text{ J} - (-16.48 \times 10^3 \text{ J})}{298.15 \text{ K}} = 96.8 \text{ J}\cdot\text{K}^{-1}$$

where the large positive $\Delta S^\circ_{\text{rxn}}$ value reflects the formation of a gaseous species from a condensed-phase species [$O_2(aq) \rightarrow O_2(g)$].

11-6 *The Gibbs-Duhem Equation Gives the Activity of a Nonvolatile Solute in Terms of the Activity of the Solvent*

In Section 11-5 we have seen how to establish a solute (Henry's-law) standard state for a volatile solute that does not exist as a pure liquid at the temperature of the solution. We now turn to the problem of establishing a solute standard state for a nonvolatile solute such as sucrose or sodium chloride in water. The key to the resolution of this problem is provided by the Gibbs-Duhem equation.

The Gibbs-Duhem equation for a two-component solution at fixed temperature and total pressure is (see Eq. 8-26)

$$n_1\,d\mu_1 + n_2\,d\mu_2 = 0 \tag{11-38}$$

where by convention the subscript 1 denotes the solvent and the subscript 2 denotes the solute. The chemical potential of component i is

$$\mu_i = \mu_i^\circ + RT\ln a_i \tag{11-39}$$

and thus

$$d\mu_i = RT\,d\ln a_i \qquad \text{(at constant temperature)} \tag{11-40}$$

Substitution of Eq. 11-40 into Eq. 11-38, followed by division by $(n_1 + n_2)RT$, yields

$$X_1\,d\ln a_1 + X_2\,d\ln a_2 = 0 \tag{11-41}$$

Equation 11-41 tells us that, *if we know a_1, the activity of the solvent, as a function of composition, then we can obtain a_2, the activity of the solute, as a function of composition by integrating the Gibbs-Duhem equation.* For example, suppose that Raoult's law holds for the solvent; then $a_1 = X_1$, and

$$X_2\,d\ln a_2 + X_1\,d\ln X_1 = X_2\,d\ln a_2 + dX_1 = 0$$

Because $X_1 + X_2 = 1$, and thus $dX_1 = -dX_2$, we have

$$d\ln a_2 = \frac{-dX_1}{X_2} = \frac{dX_2}{X_2} = d\ln X_2$$

or

$$d\ln\left(\frac{a_2}{X_2}\right) = 0 \tag{11-42}$$

Integration of Eq. 11-42 yields the result that

$$\frac{a_2}{X_2} = k \tag{11-43}$$

where k is a constant. Thus we conclude that if Raoult's law holds for the solvent, then the activity of the solute is proportional to the mole fraction of the solute, and this result holds even if the solute is nonvolatile. In the Henry's law region we have

$$a_2 = \frac{P_2}{P_{2X}^*}$$

and using Eq. 11-43 we obtain

$$P_2 = kP_{2X}^*X_2$$

or $k \equiv 1$. Thus we have shown that in the composition range for which Raoult's law holds for the solvent, Henry's law holds for the solute and *vice versa* (see Figure 11-10).

The activity of a nonvolatile solute species can be determined by measuring the vapor pressure of the solvent as a function of composition and then

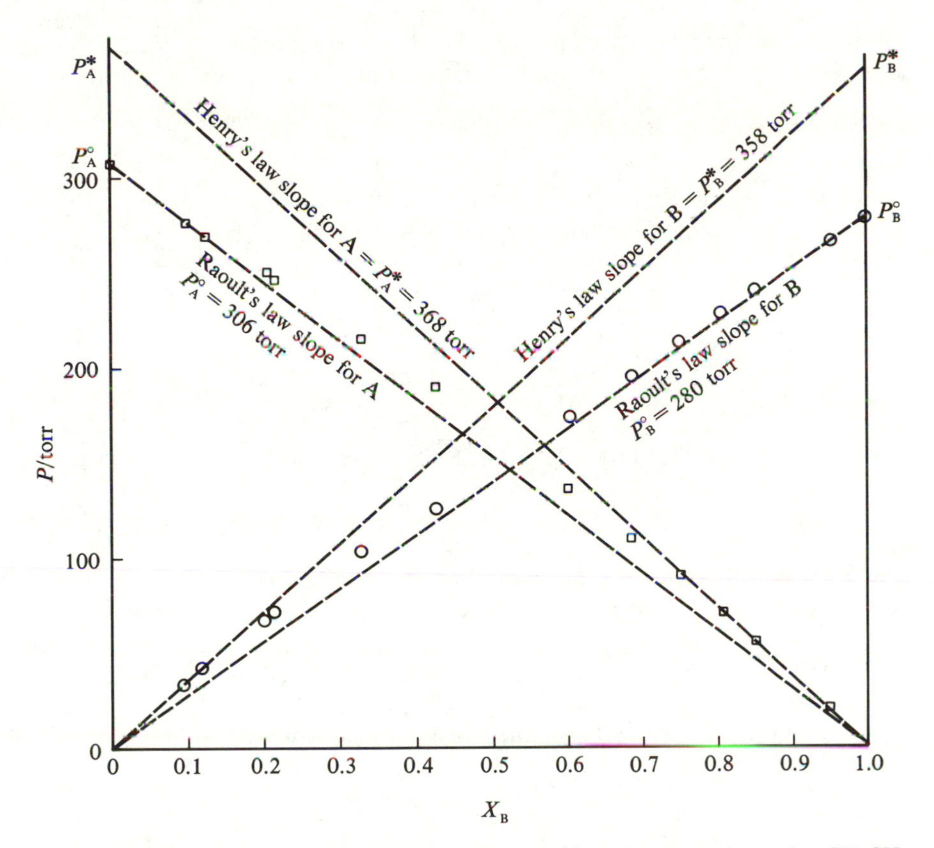

Figure 11-10. Equilibrium vapor pressure versus composition plot for CCl_4(A) plus CH_3CH_2-$OCOCH_3$(B) solutions at 25°C. Note that in the region where Raoult's law holds for the major component (solvent), Henry's law holds for the solute and vice versa.

X_B	P_A	P_B
0.0	306	0
0.096	277	34.4
0.120	269	42.6
0.198	250	67.0
0.215	246	72.3
0.326	215	104
0.425	190	126
0.598	136	175
0.684	110	196
0.748	90.1	213
0.806	70.7	228
0.849	56.2	240
0.950	18.6	266
1.00	0	280

combining these data with the Gibbs-Duhem equation. Consider a dilute solution of sucrose in water; the mole fraction of water X_1 in the solution is

$$X_1 = 1 - X_2 = 1 - \frac{n_2}{n_1 + n_2} = 1 - \frac{m}{(1000/M_1) + m}$$

where M_1 is the molecular mass of the solvent. Note that by definition there are m moles of solute in 1000 grams of solvent. If the solution is dilute, then

$$m \ll \frac{1000}{M_1} = 55.506$$

and

$$X_1 \approx 1 - \frac{m}{55.506}$$

If the solution is sufficiently dilute that Raoult's law holds, then $a_1 = X_1$, and

$$\ln a_1 = \ln X_1 \approx \ln\left(1 - \frac{m}{55.506}\right) \approx -\frac{m}{55.506} \tag{11-44}$$

where we have used the fact that $\ln(1 - x) \approx -x$ for small values of x. Equation 11-44 is the Raoult's law approximation for the activity of water in a solution of molality m.

We shall now derive an equation for the activity coefficient of the solute as a function of composition from the vapor pressure of the solvent as a function of composition. We define the *osmotic coefficient* ϕ of the solvent in the solution by the equation

$$\ln a_1 = -\frac{m\phi}{55.506} \tag{11-45}$$

If Raoult's law holds, then $\phi = 1$; whereas, if Raoult's law does not hold, then $\phi \neq 1$. A measurement of the equilibrium vapor pressure of water over the solution gives $a_1 = P_1/P_1^\circ$, from which ϕ can be calculated using Eq. 11-45. The Gibbs-Duhem equation for the solution is obtained from Eq. 11-41

$$55.506\, d\ln a_1 + m\, d\ln a_2 = 0 \tag{11-46}$$

Substitution of Eq. 11-45 into Eq. 11-46 yields

$$m\, d\ln(m\gamma_2) = d(\phi m)$$

where we have set a_2 equal to $m\gamma_2$. Thus

$$m\, d\ln m + m\, d\ln\gamma_2 = m\, d\phi + \phi\, dm$$

and therefore

$$d\ln\gamma_2 = d\phi + \frac{(\phi - 1)}{m}\, dm \tag{11-47}$$

Integration of Eq. 11-47 from $m = 0$ to m yields (note that $\gamma_2 = 1$ at $m = 0$ and that $\phi = 1$ at $m = 0$)

$$\int_1^{\gamma_2} d\ln\gamma_2 = \int_1^{\phi} d\phi + \int_0^m \left(\frac{\phi - 1}{m}\right) dm$$

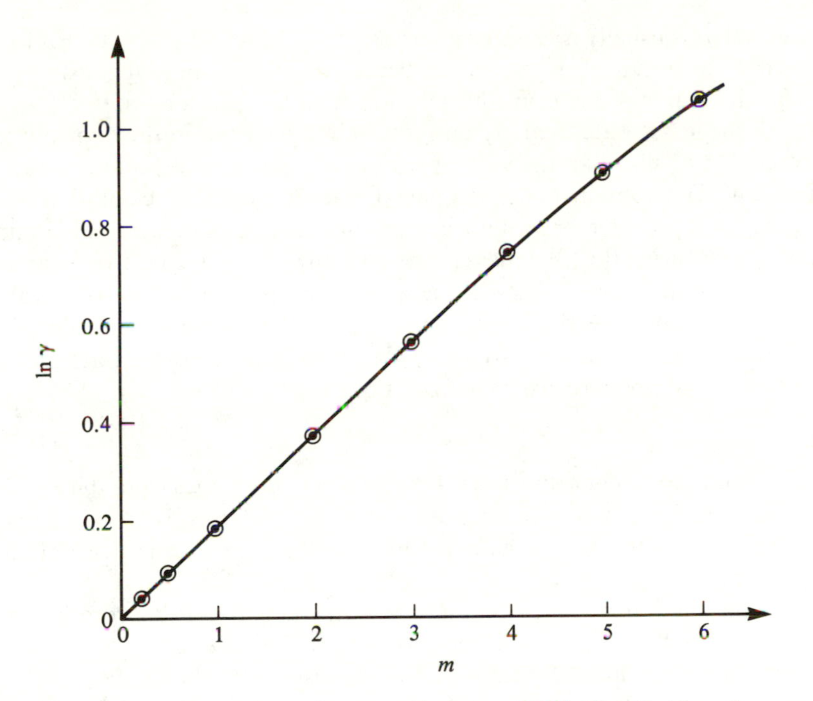

Figure 11-11. A plot of $\ln \gamma_2$ versus m for sucrose in water at 25°C. Note that $\ln \gamma_2$ for sucrose is roughly proportional to m. At $m = 0.10$, $\gamma_2 = 1.017$; whereas, at $m = 1.00$, $\gamma_2 = 1.188$.

or

$$\ln \gamma_2 = \phi - 1 + \int_0^m \left(\frac{\phi - 1}{m}\right) dm \qquad (11\text{-}48)$$

The integral in Eq. 11-48 is the area under the curve for a plot of $[(\phi - 1)/m]$ versus m. A plot of $[(\phi - 1)/m]$ versus m from 0 to m yields the value of the integral in Eq. 11-48. A plot of $\ln \gamma_2$ versus m for sucrose in water at 25°C is shown in Figure 11-11.

The method outlined above requires actual vapor pressure measurements on the equilibrium vapor phase. Such measurements are not easy to carry out at the high accuracies needed for the determination of solute activities. However, once the equilibrium vapor pressure of water is known accurately as a function of the solute concentration, say, for sucrose in water, then these data can be used in a *relative method* to obtain nonvolatile solute activities. The relative method is called the *isopiestic method.*

In the isopiestic method a given amount of reference solute (say, sucrose) is weighed in a boat (usually platinum) and a roughly comparable number of moles of the solute of interest is also weighed in a similar container. Solvent is then added to both containers in an amount that will roughly produce the desired value of m (the exact amount of solvent need not be known), and the solutes are dissolved. Both containers are then placed in a thermostated closed container that is usually evacuated after sealing. The apparatus is then allowed to reach equilibrium at a known temperature. Water spontaneously passes through the vapor phase from the solution with the greater water activity to that with the lesser water activity until the water activities in the two solutions are equal. At equilibrium the vapor pressure of solvent over the two solutions is identical. The container is then opened, and analysis of the solutions gives m_R (molality of the reference solute) and m. In many cases m can be determined simply by measuring the total mass of the dish plus contents and then evaporating the solvent and reweighing. For a particular equilibrium pair of m values, the two solutions have the same solvent activity at the temperature of the experiment. Because the two isopiestic solutions have the same water activity at the same total pressure, we have from Eq. 11-45

$$m_R\phi_R = m\phi \tag{11-49}$$

All the quantities except ϕ in Eq. 11-49 are known (ϕ_R comes from the absolute vapor pressure data on the reference salt solutions), and thus ϕ can be calculated from this expression. From the experimental values of ϕ one can obtain γ_2 using Eq. 11-48.

Any thermodynamic method that yields the activity of one component in a solution as a function of the composition of the solution can be used in conjunction with Gibbs-Duhem equation to obtain the activities of all the other components of the solution as a function of composition. In addition to the isopiestic method described above, some other methods that yield the solvent activity are osmotic pressure, freezing point depression, and boiling point elevation measurements (Chapter 13).

11-7 *Activities of Electrolytes Are Always Based on Henry's Law Solute Standard States*

The determination of activities for electrolytes presents a problem that we have not yet considered in our discussion of solute activities; namely, the electrolyte dissociates into ions on dissolution. Furthermore, the interactions between the ions and a solvent like water are very strong, and deviations from Raoult's law show up at low electrolyte concentrations.

How do we establish the solute standard state for an electrolyte like $HCl(aq)$ that on dissolution in water yields $H^+(aq)$ and $Cl^-(aq)$? The equilibrium between the solute $HCl(aq)$ and $HCl(g)$ is

$$H^+(aq) + Cl^-(aq) \rightleftharpoons HCl(g) \tag{11-50}$$

The equilibrium constant for the reaction in Eq. 11-50 is

$$K = \frac{a_{HCl(g)}}{[a_{H^+(aq)}] \cdot [a_{Cl^-(aq)}]} \tag{11-51}$$

At low $HCl(g)$ pressure, $a_{HCl(g)} = P_{HCl}$, and thus Eq. 11-51 becomes

$$K = \frac{P_{HCl}}{[a_{H^+(aq)}] \cdot [a_{Cl^-(aq)}]} \tag{11-52}$$

We *define* the activity coefficient of an ion, by analogy to that for a neutral solute species, as

$$\gamma_{ion} = \frac{a_{ion}}{m_{ion}} \tag{11-53}$$

Combination of Eqs. 11-53 and 11-52 yields

$$K = \frac{P_{\mathrm{HCl}}}{[m_{\mathrm{H^+}(aq)}][m_{\mathrm{Cl^-}(aq)}][\gamma_{\mathrm{H^+}} \cdot \gamma_{\mathrm{Cl^-}}]} \tag{11-54}$$

But

$$m_{\mathrm{H^+}(aq)} = m_{\mathrm{Cl^-}(aq)} = m$$

and thus

$$K = \frac{P_{\mathrm{HCl}}}{m^2 \gamma_{\mathrm{H^+}} \cdot \gamma_{\mathrm{Cl^-}}} \tag{11-55}$$

Interpreting K, by analogy with neutral solutes such as O_2, as the Henry's law constant for HCl(*aq*), we see that Henry's law for HCl(*aq*) takes the form ($\gamma_{\mathrm{H^+}}\gamma_{\mathrm{Cl^-}} \to 1$ as $m \to 0$)

$$P_{\mathrm{HCl}} = P^*_{\mathrm{HCl},m} m^2_{\mathrm{HCl}} \tag{11-56}$$

The second-order dependence of P_{HCl} on the molality of the solute is a direct consequence of the fact that each HCl molecule on dissolution in water yields *two* ions. Thus in the region where Henry's law holds for HCl(*aq*), a plot of P_{HCl} versus m^2_{HCl} will be linear with

$$\lim_{m^2 \to 0} \left(\frac{dP_{\mathrm{HCl}}}{dm^2}\right) = P^*_{\mathrm{HCl},m}$$

whereas

$$\lim_{m \to 0} \left(\frac{dP_{\mathrm{HCl}}}{dm}\right) = \lim_{m \to 0} 2mP^*_{\mathrm{HCl},m} = 0$$

and thus a plot of P_{HCl} versus m is not linear at low m. The distinction between P_{HCl} versus m_{HCl} and P_{HCl} versus m^2_{HCl} is shown in Figure 11-12, where the establishment of the solute standard state for HCl(*aq*) is illustrated.

The activity of HCl(*aq*) is given by

$$a_{\mathrm{HCl}} = \frac{P_{\mathrm{HCl}}}{P^*_{\mathrm{HCl},m}} = a_{\mathrm{H^+}} a_{\mathrm{Cl^-}} = m_{\mathrm{H^+}} m_{\mathrm{Cl^-}} \gamma_{\mathrm{H^+}} \gamma_{\mathrm{Cl^-}} \tag{11-57}$$

Single-ion activities and single-ion activity coefficients are not measurable, because it is impossible to prepare a solution that contains only a single ion in appreciable concentrations or which contains an excess of, say, total positive charges per unit volume over total negative charges. The *electroneutrality condition* for electrolytes holds to a very high order of accuracy. Consequently, it is conventional to define the *mean ionic activity coefficient* $\gamma_\pm$ *of an electrolyte.* For HCl(*aq*), the mean ionic activity coefficient, $\gamma_{\pm(\mathrm{HCl})}$, is defined as

$$\gamma_{\pm(\mathrm{HCl})} = (\gamma_{\mathrm{H^+}} \gamma_{\mathrm{Cl^-}})^{1/2}$$

whereas for $CaCl_2$(*aq*), which yields three ions per $CaCl_2$ unit, the mean ionic activity coefficient is defined as

$$\gamma_{\pm(\mathrm{CaCl_2})} = (\gamma_{\mathrm{Ca^{2+}}} \gamma^2_{\mathrm{Cl^-}})^{1/3}$$

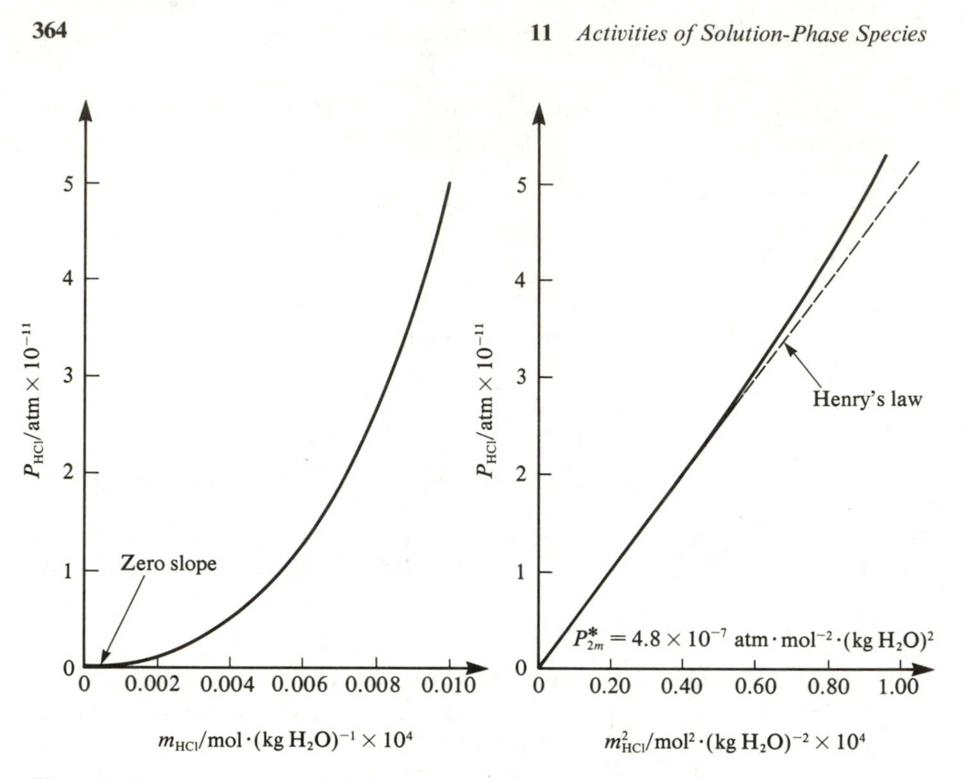

Figure 11-12. Partial pressure of HCl(g) over HCl(aq) solutions at 25°C as a function of m_{HCl} and as a function of m^2_{HCl} at 25°C. Note that Henry's law holds for the m^2_{HCl} case but not for the m_{HCl} case.

The general expression for the mean ionic activity coefficient of an electrolyte is

$$\gamma_\pm = (\gamma_+^{v_+}\gamma_-^{v_-})^{1/v} \tag{11-58}$$

where v_+ is the number of cations per formula unit, v_- is the number of anions per formula unit, and $v = v_+ + v_-$ is the total number of ions produced per formula unit. Mean ionic activity coefficients are measurable quantities.

The relationships between the activity and the mean ionic activity coefficient, $\gamma_\pm$, for $CaCl_2$ are obtained as follows. The Henry's law equilibrium is

$$Ca^{2+}(aq) + 2\,Cl^-(aq) \rightleftharpoons CaCl_2(g) \tag{11-59}$$

and thus

$$\frac{P_{CaCl_2}}{P^*_{m,CaCl_2}} = a_{CaCl_2(aq)} = a_{Ca^{2+}(aq)} \cdot a^2_{Cl^-(aq)} = m_{Ca^{2+}}\gamma_{Ca^{2+}} \cdot m^2_{Cl^-}\gamma^2_{Cl^-}$$

Hence

$$a_{CaCl_2(aq)} = m_{Ca^{2+}}m^2_{Cl^-}(\gamma_{Ca^{2+}}\gamma^2_{Cl^-}) = m_{Ca^{2+}}m^2_{Cl^-}\gamma^3_{\pm(CaCl_2)}$$

If $CaCl_2$ is the only source of $Ca^{2+}(aq)$ and $Cl^-(aq)$ in the solution, then $m_{Cl^-} = 2m_{Ca^{2+}} = 2m$ and

$$a_{CaCl_2(aq)} = 4m^3\gamma^3_\pm = \frac{P_{CaCl_2}}{P^*_{m,CaCl_2}}$$

Note that Henry's law for $CaCl_2$ (aq) takes the form

$$P_{CaCl_2} = 4m^3 P^*_{m,CaCl_2}$$

The equilibrium vapor pressure of $CaCl_2(g)$ over an aqueous solution is incredibly small. The value of ΔG°_{rxn} at 25°C for the reaction

$$Ca^{2+}(aq) + 2\,Cl^-(aq) \rightleftharpoons CaCl_2(g)$$

is 80.54 kJ. Further, for the reaction in Eq. 11-59 we have

$$\Delta G^\circ_{rxn} = -RT\ln K = -RT\ln \frac{P_{CaCl_2}}{a_{CaCl_2(aq)}}$$

The activity of $CaCl_2(aq)$ in a 0.10-*m* $CaCl_2(aq)$ solution at 25°C is

$$a_{CaCl_2} = 4(0.10)^3(0.518)^3 = 5.56 \times 10^{-4}$$

and

$$\frac{P_{CaCl_2}}{a_{CaCl_2(aq)}} = e^{-\Delta G^\circ_{rxn}/RT}$$

Thus at 25°C we have

$$P_{CaCl_2} = 5.56 \times 10^{-4} \exp\left(-\frac{80.54 \times 10^3\ \text{J}}{8.314\ \text{J}\cdot\text{K}^{-1} \times 298.15}\right) = 4.31 \times 10^{-18}\ \text{atm}$$

It is obvious that, with such a low equilibrium vapor pressure of $CaCl_2(g)$, we cannot determine $a_{CaCl_2(aq)}$ by direct $CaCl_2(g)$ vapor pressure measurements. The primary sources of activities and activity coefficients for electrolytes in solution are isopiestic measurements and electrochemical cell measurements.

The application of the isopiestic measurements to electrolytes requires a modification of Eq. 11-45 to account for the dissociation of the electrolyte; namely,

$$\ln a_1 = -\left(\frac{vmM_1}{1000}\right)\phi = -\frac{vm\phi}{55.506} \qquad (11\text{-}60)$$

where v is the total number of ions per formula unit. For $HCl(aq)$, $v = 2$; whereas, for $CaCl_2$, $v = 3$.

TABLE 11-3
Activity and Activity Coefficient Data for HCl(*aq*) at 25°C, 1 atm

m	$\gamma_\pm(expt)$	$\gamma_\pm(calc)^a$	$a_{HCl(aq)}$
0.0001	0.988	0.988	9.8×10^{-9}
0.001	0.965	0.965	9.3×10^{-7}
0.01	0.904	0.899	8.2×10^{-5}
0.1	0.796	0.754	6.3×10^{-3}
1	0.809	0.555	0.66
10	10.44	0.409	1.1×10^4

[a] Calculated using the expression

$$\log \gamma_\pm = \frac{-0.511m^{1/2}}{1 + m^{1/2}}$$

which is an extended form of the Debye-Hückel limiting law.

TABLE 11-4
Osmotic and Activity Coefficient Data for Aqueous Electrolytes at 25°C[a]

m	0.1	0.3	0.5	0.7	1.0	2.0	3.0
$HCl(aq)$	$\gamma_{\pm}$0.796	0.756	0.757	0.772	0.809	1.009	1.316
	ϕ0.943	0.952	0.974	0.998	1.039	1.188	1.348
$HClO_4(aq)$	$\gamma_{\pm}$0.803	0.768	0.769	0.785	0.823	1.055	1.448
	ϕ0.947	0.958	0.976	1.000	1.041	1.210	1.406
$KCl(aq)$	$\gamma_{\pm}$0.769	0.687	0.649	0.626	0.603	0.572	0.568
	ϕ0.929	0.908	0.901	0.899	0.899	0.914	0.939
$KNO_3(aq)$	$\gamma_{\pm}$0.739	0.614	0.545	0.496	0.443	0.333	0.269
	ϕ0.906	0.851	0.817	0.790	0.756	0.669	0.602
$LiClO_4(aq)$	$\gamma_{\pm}$0.812	0.792	0.808	0.834	0.887	1.158	1.582
	ϕ0.951	0.971	0.999	1.027	1.072	1.238	1.419
$NaClO_4(aq)$	$\gamma_{\pm}$0.775	0.701	0.668	0.648	0.629	0.609	0.611
	ϕ0.930	0.915	0.910	0.910	0.913	0.934	0.960
$NaOH(aq)$	$\gamma_{\pm}$0.764	0.706	0.688	0.680	0.677	0.707	0.782
	ϕ0.925	0.929	0.937	0.945	0.958	1.015	1.094
p-toluene–sulfonic acid(*aq*)	$\gamma_{\pm}$0.759	0.660	0.608	0.573	0.535	0.459	0.427
	ϕ0.922	0.887	0.869	0.854	0.838	0.809	0.816
$CaCl_2(aq)$	$\gamma_{\pm}$0.518	0.455	0.448	0.460	0.500	0.792	1.483
	ϕ0.854	0.876	0.917	0.963	1.046	1.376	1.779
$Ca(ClO_4)_2(aq)$	$\gamma_{\pm}$0.557	0.532	0.564	0.618	0.743	1.634	4.21
	ϕ0.883	0.942	1.014	1.089	1.219	1.710	2.261
$K_2SO_4(aq)$	$\gamma_{\pm}$0.436	0.313	0.261	0.229	—	—	—
	ϕ0.779	0.721	0.691	0.670	—	—	—
$H_2SO_4(aq)$	$\gamma_{\pm}$0.266	0.183	0.156	0.142	0.132	0.128	0.142
	ϕ0.680	0.668	0.676	0.689	0.721	0.846	0.991
$K_4Mo(CN)_8(aq)$	$\gamma_{\pm}$0.145	0.0831	0.0632	0.0521	0.0436	—	—
	ϕ0.603	0.529	0.497	0.485	0.485	—	—
$K_3Fe(CN)_6(aq)$	$\gamma_{\pm}$(0.268) est	0.184	0.155	0.140	0.128	—	—
	ϕ0.727	0.682	0.676	0.679	0.705	—	—

[a] Data from Lewis and Randall (1961), Robinson and Stokes (1959), and Harned and Owen (1958).

Activities and activity coefficients for $HCl(aq)$ at 25°C are given in Table 11-3. Note that a_{HCl} ranges over 12 powers of 10 for a 5-powers-in-10 range of molality; this is primarily a consequence of the fact that $a_{HCl} \propto m^2$. The variation of $\gamma_{\pm}$ with m is much less pronounced, except at high m where the activity of $HCl(g)$ increases much more rapidly than m. At 15 m $HCl(aq)$, the equilibrium vapor pressure of $HCl(g)$ over the solution is about 180 torr. Mean ionic activity coefficients for some representative electrolytes in water at 25°C are given in Table 11-4. In Figure 11-13 we have plotted $\log \gamma_{\pm}$ versus $m^{1/2}$ for $KCl(aq)$ and $CaCl_2(aq)$ at 25°C.

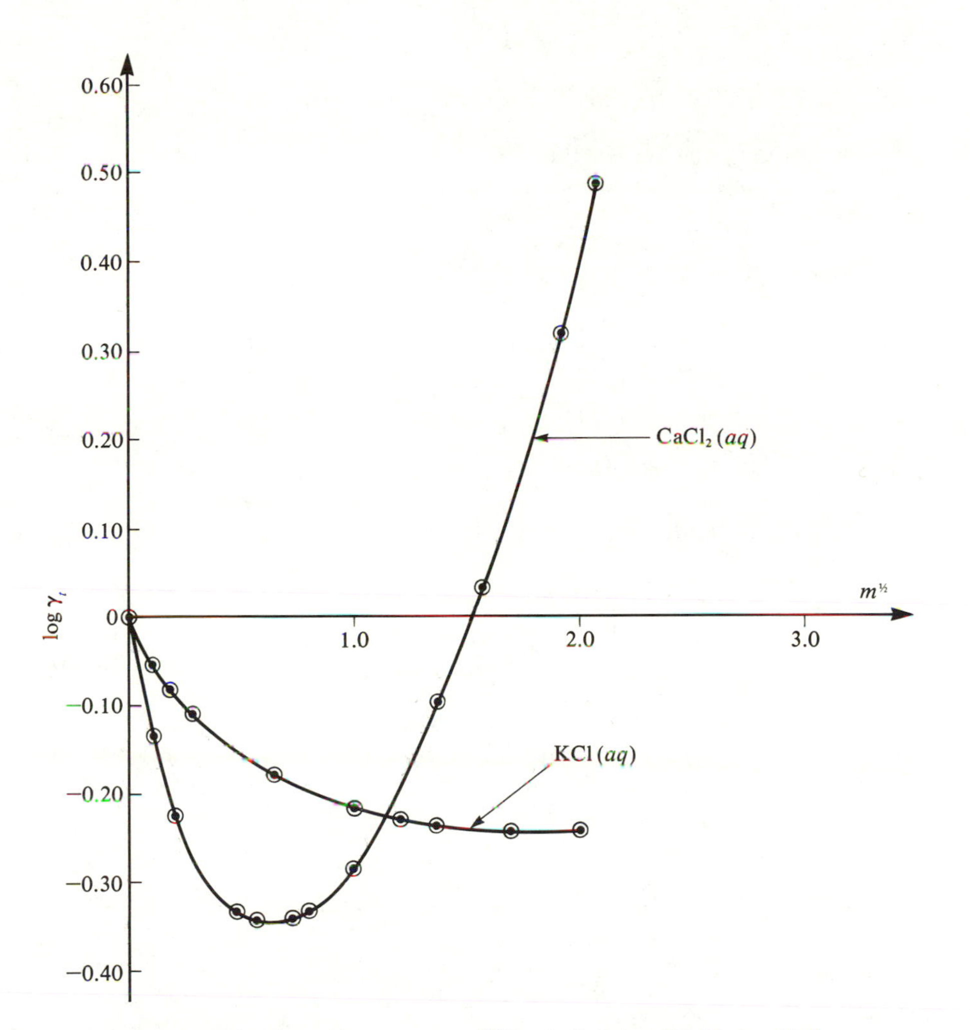

Figure 11-13. Plots of $\log \gamma_{\pm}$ versus $m^{1/2}$ for $CaCl_2(aq)$ and $KCl(aq)$ at 25°C.

11-8 *The Magnitude of the Equilibrium Constant for a Reaction Depends on the Choice of Standard States*

We are now in a position to discuss equilibrium constants for reactions involving a solution phase; this is most conveniently done by means of examples.

The hydrolysis of the ester ethyl acetate has the following stoichiometry:

$$\underset{(1)}{H_2O} + \underset{(2)}{CH_3CH_2OCOCH_3} \longrightarrow \underset{(3)}{CH_3COOH} + \underset{(4)}{CH_3CH_2OH}$$

For notational simplicity, we shall number the substances involved in the manner indicated. Let us now suppose that the equilibrium reaction is established in aqueous solution:

$$H_2O(\ell) + CH_3CH_2OCOCH_3(aq) \rightleftharpoons CH_3COOH(aq) + CH_3CH_2OH(aq)$$

The equilibrium constant has the form

$$K = \frac{a_3a_4}{a_1a_2} \tag{11-61}$$

If we choose our standard states as follows,

H_2O — The pure liquid at 1 atm (solvent standard state). This choice is indicated by writing $H_2O(\ell)$ rather that $H_2O(aq)$. Note that $a_{H_2O} = \gamma_{H_2O}X_{H_2O}$.

$CH_3CH_2OCOCH_3$, CH_3COOH, CH_3CH_2OH — The hypothetical ideal unit molality solute standard states in H_2O at 1 atm. Note that $a_i = \gamma_i m_i$.

then Eq. 11-61 becomes

$$K = \left(\frac{m_3m_4}{X_1m_2}\right)\left(\frac{\gamma_3\gamma_4}{\gamma_1\gamma_2}\right) \tag{11-62}$$

If Raoult's law holds for H_2O (the major component), then $\gamma_1 = 1.00$ and Henry's law holds for components, 2, 3, and 4; thus $\gamma_2 = \gamma_3 = \gamma_4 = 1.00$. Hence if Raoult's law holds for H_2O, then Eq. 11-62 can be written as

$$K = \frac{m_3m_4}{X_1m_2} \tag{11-63}$$

In very dilute solution, $X_1 \approx 1$ (H_2O is the solvent), and Eq. 11-63 becomes

$$K \approx \frac{m_3m_4}{m_2}$$

Therefore, K can be computed from the equilibrium values of m_2, m_3, and m_4 in dilute aqueous solution. Once the numerical value of K has been determined correctly from measurements in dilute solutions, then the resulting value of K can be equated to a_3a_4/a_1a_2, even though the concentrations of the various species involved are such that $\gamma_i \neq 1$, because K is independent of concentration.

In the above case (as in many others) there are numerous possible combinations of choices of standard states for the species involved in a reaction, and each different set of choices leads to a different numerical value of K for the reaction. Consequently, it is necessary to indicate just what standard states have been chosen in the thermodynamic analysis of the reaction; otherwise, the reported value of the equilibrium constant is meaningless. For example, the above reaction could be carried out with ethanol as the solvent; then

$$\underset{(1)}{H_2O(al)} + \underset{(2)}{CH_3CH_2OCOCH_3(aq)} \rightleftharpoons \underset{(3)}{CH_3COOH(al)} + \underset{(4)}{CH_3CH_2OH(\ell)}$$

where *al* denotes a solute in ethyl alcohol. In dilute ethyl alcohol solution, $a_4 \approx X_4 \approx 1$ and $\gamma_3 = \gamma_2 = \gamma_1 \approx 1$, thus

$$K' = \frac{m_3}{m_1 m_2}$$

where $K' \neq K$, because the standard states, and thus the $\Delta \bar{G}_f^\circ$ values for the products and reactants are different in the two cases.

The value of the equilibrium constant for a reaction involving one or more solution species for which a solute standard state has been chosen depends on the solvent in which the reaction is run. This is true even if the solvent is neither a reactant or a product but merely serves as the medium in which the reaction is carried out. The value of the standard chemical potential of a species μ_i° in a solute standard state depends on the Henry's law constant for the solute in the given solvent, and the value of the Henry's law constant for a solute depends on the solvent.

Consider the weak-acid-dissociation equilibrium

$$CH_3COOH(aq) \rightleftharpoons CH_3COO^-(aq) + H^+(aq)$$

The equilibrium constant is given by (where $B \equiv CH_3COO$)

$$K = \frac{a_{B^-} \cdot a_{H^+}}{a_{HB}}$$

Thus

$$K = \frac{(m_{B^-})(\gamma_{B^-})(m_{H^+})(\gamma_{H^+})}{(m_{HB})(\gamma_{HB})} = \frac{(m_{B^-})(m_{H^+})\gamma^2_{\pm(HB)}}{(m_{HB})\gamma_{HB}}$$

where $\gamma_{\pm(BX)}$ is the mean *ionic* activity coefficient of $H^+(aq) + CH_3COO^-(aq)$ and γ_{HB} is the activity coefficient of the *neutral* species $CH_3COOH(aq)$. Note that $\gamma_{\pm(HB)} \neq \gamma_{HB}$, because γ_{HB} refers to undissociated $HB(aq)$, whereas $\gamma_{\pm(HB)}$ refers to the ions $H^+(aq)$, $B^-(aq)$. Generally speaking, *neutral* solutes do not deviate as greatly from dilute solution ($\gamma = 1$) behavior as do electrolytes, and it is usually a satisfactory approximation, in all but the most precise work, to take $\gamma = 1$ for neutral solutes at moderate to low concentrations. Thus, in dilute aqueous solution, $\gamma_{HB} \approx 1$, and we have

$$K \approx \frac{(m_{B^-})(m_{H^+})\gamma^2_{\pm(HB)}}{m_{HB}}$$

If the only significant source of $H^+(aq)$ and $X^-(aq)$ is HB, then $m_{B^-} = m_{H^+}$ and thus

$$K \approx \frac{(m_{H^+})^2\gamma^2_{\pm(HB)}}{m_{HB}}$$

To a first approximation, in dilute aqueous solutions the mean ionic activity coefficient $\gamma_\pm$ of an electrolyte in water at 25°C is given by the equation

$$\log \gamma_\pm = \frac{-0.511|Z_+Z_-|I^{1/2}}{1 + I^{1/2}} \qquad (11\text{-}64)$$

where Z_+ is the charge on the cation and Z_- is the charge on the anion; the vertical rules denote the absolute value of the product Z_+Z_-. The function I is the *ionic strength* of the solution. The ionic strength of the solution is defined (G. N. Lewis) by the equation

$$I = \frac{1}{2}\sum_i m_i Z_i^2 \tag{11-65}$$

where the sum in Eq. 11-65 is over *all* the ion types in the solution. Note from Eq. 11-65 that for a 1:1 electrolyte like NaCl(*aq*)

$$I = \tfrac{1}{2}[m_{Na^+}(1)^2 + m_{Cl^-}(-1)^2] = \tfrac{1}{2}(2m) = m$$

EXAMPLE 11-6

Compute the ionic strength of an aqueous solution that is 0.020 *m* in NaCl(*aq*) and 0.010 *m* in $CaCl_2$(*aq*).

Solution: From Eq. 11-65 we obtain

$$I = \frac{1}{2}\sum_i m_i Z_i^2 = \frac{1}{2}[m_{Na^+}(1)^2 + m_{Ca^{2+}}(2)^2 + m_{Cl^-}(-1)^2]$$

thus

$$I = \tfrac{1}{2}\{(0.020) + 4(0.010) + [2(0.010) + 0.020]\} = 0.050\ m$$

EXAMPLE 11-7

Use the result in Example 11-6, together with Eq. 11-64, to estimate $\gamma_\pm$ for NaCl(*aq*) and for $CaCl_2$(*aq*) in an aqueous solution that is 0.020 *m* in NaCl and 0.010 *m* in $CaCl_2$(*aq*).

Solution: For NaCl(*aq*), $Z_+ = 1$, $Z_- = -1$, and thus from Eq. 11-64 we have for the mean ionic activity coefficient of NaCl(*aq*) in the solution

$$\log \gamma_\pm = \frac{-0.511|(1)(-1)|(0.050)^{1/2}}{1 + (0.050)^{1/2}} = -0.0934$$

or $\gamma_\pm = 0.807$. For $CaCl_2$(*aq*), $|Z_+Z_-| = |(2)(-1)| = 2$; thus $\log \gamma_\pm = -0.1868$, and $\gamma_\pm = 0.650$. Equation 11-64 is a fair approximation for solutions with $I < 0.10$, which is more accurate the lower the value of I. Note from Eq. 11-64 that as $I \to 0$, $(1 + I^{1/2}) \to 1$ and $\log \gamma_\pm$ approaches

$$\log \gamma_\pm \approx -0.511|Z_+Z_1|I^{1/2} \tag{11-66}$$

Equation 11-66 is known as the *Debye-Hückel limiting law* for $\gamma_\pm$. The Debye-Hückel limiting law becomes exact in the limit as $I \to 0$.

It is common practice in the study of equilibria involving ionic species to attempt to hold the $\gamma_\pm$ terms for the electrolytes involved in the reaction at a constant value by keeping the total ionic strength I constant by using a large excess of a suitable "inert" supporting electrolyte (e.g., $LiClO_4$ or KNO_3 for aqueous solution equilibria). Such a procedure is not necessarily effective, because, in general, $\gamma_\pm$ depends not only on the total ionic strength but also on the nature of the various ions present in the solution. Equation 11-64 is only an *approximation.* The value of the so-called equilibrium constant determined in this way is strictly applicable only at the particular value of the ionic strength at which it was determined. Consider once again the dissociation of the weak acid HB(*aq*) discussed above:

$$K = \frac{[H^+][B^-]\gamma^2_{\pm(HB)}}{[HB]\gamma_{HB}}$$

Assuming that $\gamma_{\pm(HB)}$ and γ_{HB} are constant at a given total I, we have

$$K = \frac{[H^+][B^-]}{[HB]} \cdot f(I)$$

where the value of $f(I)$ depends on the ionic strength I. Often $K_{app} = K/f(I)$ ("app" denotes apparent) is the quantity determined experimentally and not K. It is worth keeping in mind that $K_{app} \neq K$.

EXAMPLE 11-8

The equilibrium constant at 25°C for the reaction

$$CH_3COOH(aq) \rightleftharpoons H^+(aq) + CH_3COO^-(aq)$$

is 1.74×10^{-5}. Compute the equilibrium molalities of

$$H^+(aq), \quad CH_3COO^-(aq), \quad \text{and} \quad CH_3COOH(aq)$$

in a 0.100-*m* solution of $CH_3COOH(aq)$, assuming $\gamma_{HB} = 1.00$, and also assuming that $\gamma_{\pm(HB)}$ (where $B^- \equiv CH_3COO^-$) is given by Eq. 11-64.

Solution: To compute $\gamma_\pm$ from Eq. 11-64, we need to know the ionic strength of the solution. The equilibrium constant expression for the reaction is

$$K_a = 1.74 \times 10^{-5} = \frac{[H^+][B^-]}{[HB]}\left(\frac{\gamma^2_{\pm(HB)}}{\gamma_{HB}}\right)$$

where the brackets denote molalities. In dilute solution γ_{HB}, the activity coefficient of the neutral species $CH_3COOH(aq)$, is approximately equal to one; thus

$$\frac{1.74 \times 10^{-5}}{\gamma^2_{\pm(HB)}} = \frac{[H^+][B^-]}{[HB]}$$

We now assume that $\gamma_{\pm(HB)} \approx 1$ for the purpose of estimating I. Because $K_w = 1.01 \times 10^{-14}$ for the dissociation of water at 25°C,

$$H_2O(\ell) \rightleftharpoons H^+(aq) + OH^-(aq)$$

we can neglect the K_w reaction as a significant source of $H^+(aq)$ in the solution ($K_w \ll K_a$). Thus from the reaction stoichiometry we have

$$[H^+] = [B^-]$$

and from the material balance on acetate we have

$$[HB] = 0.100 - [B^-]$$

Thus

$$\frac{[H^+]^2}{0.100 - [H^+]} = 1.74 \times 10^{-5}$$

and $[H^+] = [B^-] = 1.32 \times 10^{-3}$ m and $[HB] = 0.099$ m. The ionic strength of the solution is

$$I = \tfrac{1}{2}\{[H^+] + [B^-]\} = 1.32 \times 10^{-3}$$

and from Eq. 11-64 we compute

$$\log \gamma_\pm = \frac{-0.511(1.32 \times 10^{-3})^{1/2}}{1 + (1.32 \times 10^{-3})^{1/2}}$$

or $\gamma_\pm = 0.960$. Thus

$$\frac{1.74 \times 10^{-5}}{(0.960)^2} = \frac{[H^+][B^-]}{[HB]} = \frac{[H^+]^2}{0.100 - [H^+]}$$

and $[H^+] = [B^-] = 1.37 \times 10^{-3}$ m and $[HB] = 0.099$ m.

11-9 *Reaction Spontaneity Is Determined by ΔG and Not by ΔG°*

The value of ΔG°_{rxn} at 25°C for the reaction

$$Pb(s) + 2\,AgCl(s) \rightleftharpoons 2\,Ag(s) + PbCl_2(s)$$

is -92.30 kJ. From the Lewis equation we have

$$\Delta G_{rxn} = \Delta G^\circ_{rxn} + RT \ln Q$$

and at 25°C

$$\Delta G_{rxn} = -92.30 \times 10^3\ \text{J} + (8.314\ \text{J}\cdot\text{K}^{-1})(298.15) \ln\left[\frac{a^2_{Ag(s)} a_{PbCl_2(s)}}{a_{Pb(s)} a^2_{AgCl(s)}}\right]$$

At 1 atm all the $a_i = 1$, and hence $Q = 1$ and $\Delta G_{rxn} = -92.30$ kJ. In other words, the reaction is spontaneous ($\Delta G_{rxn} < 0$) in the direction written (left to right) at 1 atm and 298 K. Because the value of ΔG_{rxn} in this case *does not change* as reactants are converted to products, the reaction will proceed until whichever of the two reactants [Pb(s) or AgCl(s)] that is stoichiometrically limiting is *totally* consumed. Conversely, if $PbCl_2(s)$ and Ag(s) are brought in contact at 1 atm and 298 K, no Pb(s) or AgCl(s) *whatsoever* will form, because

$\Delta G_{rxn} = +92.30$ kJ. Although $\Delta G_{rxn} \ll 0$, the Pb(*s*) + AgCl(*s*) reaction is extremely slow, because it involves bringing together two solid phases, the reaction rate can be increased greatly, however, by providing a more favorable reaction pathway. For example, the reactants could be mixed in the presence of $H_2O(\ell)$. A small amount of AgCl(*s*) dissolves in the water and reacts with Pb(*s*).

The distinction between ΔG_{rxn} and ΔG°_{rxn} is especially important to keep in mind when dealing with reactions that involve a solution phase. For example, consider the reaction

$$CH_3COOH(aq) \rightleftharpoons H^+(aq) + CH_3COO^-(aq) \qquad K_{298} = 1.74 \times 10^{-5}$$

(*K* on the molality composition scale, solute standard states)

for which, at 25°C, ΔG°_{rxn} is

$$\Delta G^\circ_{rxn} = -(8.314\ \text{J}\cdot\text{K}^{-1})(298.15\ \text{K})\ln(1.74 \times 10^{-5}) = +27.17\ \text{kJ}$$

In this case (or in any case for that matter) we cannot conclude that, owing to the positive value of ΔG°_{rxn}, no products will form spontaneously when we dissolve $CH_3COOH(\ell)$ in water at 25°C. *It is the value of ΔG_{rxn}, and not ΔG°_{rxn}, that determines whether or not the reaction is spontaneous at a particular temperature and pressure.* In the above case, even at 1 atm total pressure, $\Delta G_{rxn} \neq \Delta G^\circ_{rxn}$. The value of ΔG_{rxn} (in kilojoules) at 25°C is given by

$$\Delta G_{rxn}/\text{kJ} = 27.17 + 2.478\ln\left[\frac{a_{H^+(aq)}a_{CH_3COO^-(aq)}}{a_{CH_3COOH(aq)}}\right]$$

and with $a_{CH_3COOH(aq)} = 0.10$ and $a_{H^+(aq)} \approx a_{CH_3COO^-(aq)} \approx 0$ the value of ΔG_{rxn} is very large and *negative*. Therefore, the dissociation of $CH_3COOH(aq)$ takes place *spontaneously* with $a_{CH_3COOH(aq)}$ decreasing and $a_{H^+(aq)}$ and $a_{CH_3COO^-(aq)}$ increasing, until

$$(2.48\ \text{kJ})\ln\left[\frac{a_{H^+(aq)}a_{CH_3COO^-(aq)}}{a_{CH_3COOH(aq)}}\right] = -27.17\ \text{kJ}$$

which corresponds to the equilibrium state, $\Delta G_{rxn} = 0$. Because of the positive value of ΔG°_{rxn}, the values of $a_{H^+(aq)}$ and $a_{CH_3COO^-(aq)}$ at equilibrium will be small ($\sim 1.3 \times 10^{-3}$, see Example 11-8) compared to the equilibrium value of $a_{CH_3COOH(aq)}$ (~ 0.10), but this is not the same as saying that because ΔG°_{rxn} is positive, the reaction is not spontaneous in the direction written. Spontaneity at constant temperature and pressure is governed by ΔG_{rxn} (and not by ΔG°_{rxn}). The value of ΔG_{rxn} depends on the actual value of the activities of the reactants and products, whereas the value of ΔG°_{rxn} for a given reaction depends (for a given choice of standard states) only on the temperature.

The temperature dependence of the equilibrium constants for ionization reactions is unusual because of the large negative $\Delta C^\circ_{P,rxn}$ values for ionization reactions, which for monoprotic acids are typically around $-150\ \text{J}\cdot\text{K}^{-1}$ to $-200\ \text{J}\cdot\text{K}^{-1}$. Such large negative $\Delta C^\circ_{P,rxn}$ values coupled with small ΔH°_{rxn} values can produce the type of *log K versus* $1/T$ behavior found for the aqueous dissociation of benzoic acid shown in Figure 11-14. Note that ΔH°_{rxn} changes sign in the range 0°C–100°C.

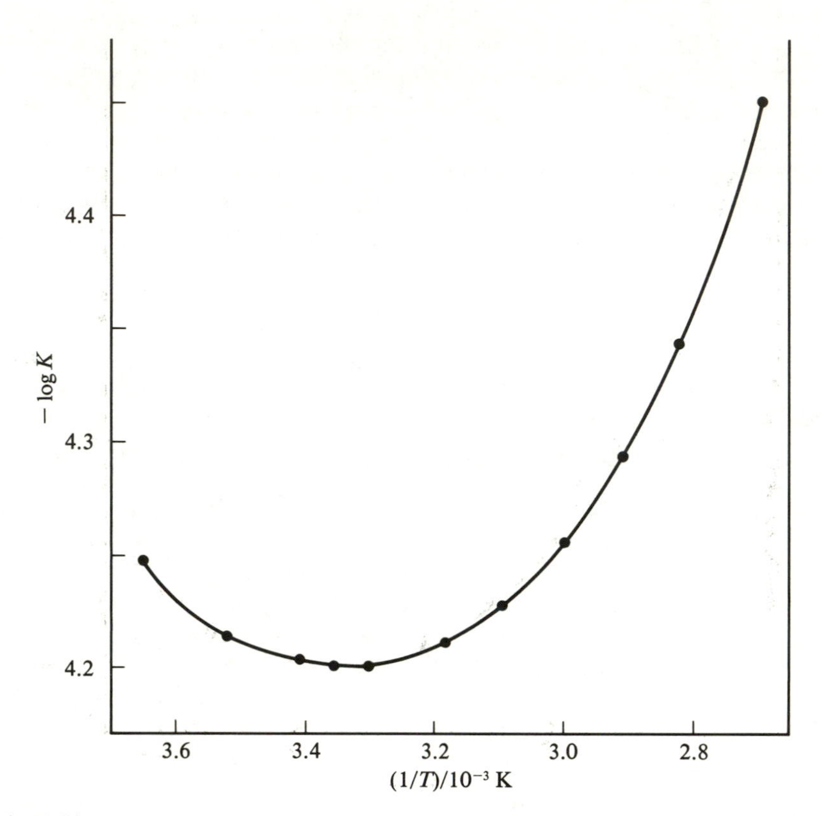

Figure 11-14. Plot of $-\log K$ versus $1/T$ for the aqueous ionization of benzoic acid over the range 0°C to 100°C:

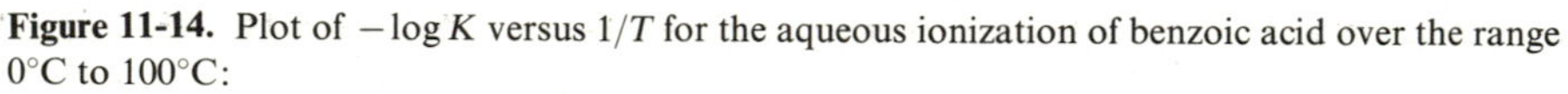

$$C_6H_5COOH(aq) \rightleftharpoons H^+(aq) + C_6H_5COO^-(aq)$$

Note that $-\log K$ passes through a minimum.

11-10 *The Activity Is a Measure of the Escaping Tendency of a Component from a Phase*

On first encounter, the activity function comes across as a somewhat mysterious concept, and there is a widespread tendency to avoid a discussion of equilibrium constants in terms of activities. This is unfortunate, because without the activity function the equilibrium constant expression does not have the particularly simple form given in Eq. 10-14. More importantly, if we express K in the form of Eq. 10-14, but with concentrations (or pressures) in place of activities, then we no longer have a quantity that is a constant, except in the limit of low concentrations and pressures. The activity function was *invented* to preserve the simple algebraic form of K (which otherwise prevails only at low concentrations and pressures) over the entire possible range of concentrations and pressures.

The physical basis for the activity function is that the concentration of a solution species (or pressure of a gaseous species) is not, in general, a reliable quantitative measure of the escaping tendency of a species from a phase, because of the existence of strong, specific interactions between the species. By way of example, consider a solution of NaCl in water. We expect that the escaping tendency of water from the solution is less than that from pure water simply on the basis that the number of water molecules per unit area of solution surface is less than that for pure water. The dilution effect is measured by Raoult's law, which says that the equilibrium vapor pressure (which measures the escaping tendency) of water over the solution is directly proportional to the mole fraction of water in the solution ($P_{H_2O} = X_{H_2O}P^{\circ}_{H_2O}$). The greater the concentration of NaCl, the lower X_{H_2O} and thus the lower P_{H_2O}. However, Raoult's law holds for H_2O in a NaCl(*aq*) solution only at very low concentrations of NaCl. At higher concentrations the equilibrium partial pressure of water over the solution is less than that computed from Raoult's law. Deviations from Raoult's law are a consequence of the fact that the ions in solution interact electrostatically with the dipoles on water molecules and thereby bind some of the water molecules to themselves. As a consequence, the average Gibbs energy of the water molecules in the solution is reduced relative to pure water, which leads to a reduced escaping tendency of water (a lower activity). Note that because

$$a_{H_2O(\ell)} = \frac{P_{H_2O}}{P^{\circ}_{H_2O}}$$

if Raoult's law holds, then $a_{H_2O(\ell)} = X_{H_2O}$. Whereas if

$$\frac{P_{H_2O}}{P^{\circ}_{H_2O}} \neq X_{H_2O}$$

then $a_{H_2O(\ell)} \neq X_{H_2O}$; that is, $a_{H_2O} = \gamma_{H_2O}X_{H_2O}$ ($\gamma_{H_2O} \neq 1$).

Similar considerations apply to the NaCl(*aq*) in the solution. If the solution is very dilute, then Henry's law holds ($P_{NaCl} = m^2_{NaCl}P^*_{NaCl}$) and the escaping tendency of NaCl from the solution (i.e., the activity) is equal to

$$a_{NaCl} = \frac{P_{NaCl}}{P^*_{NaCl}} = m^2_{NaCl}$$

However, Henry's law holds only at very low concentrations for electrolytes, because the electrostatic forces between ions are long-range forces. Consequently, $a_{NaCl} \neq m^2_{NaCl}$ and we have:

$$a_{NaCl} = \frac{P_{NaCl}}{P^*_{NaCl}} = m^2_{NaCl}\gamma^2_{\pm(NaCl)}$$

The activity can be thought of as an effective or *active* concentration of a species. In other words, strong specific interactions between species in solution, which manifest themselves as deviations from either Raoult's law (solvent) or

Henry's law (solute), make the species behave as if it were present at a concentration smaller (or larger) than the actual concentration. This effective or active concentration of a species is called the *activity*.

Problems

1. Compute the ionic strength for each of the following solutions at 25°C.
   (a) 0.020 m $NaCl(aq)$ + 0.030 m $NaOH(aq)$
   (b) 0.010 m $K_2SO_4(aq)$ + 0.020 m $KBr(aq)$
   (c) 0.010 m $CH_3COOH(aq)$; $K_a = 1.74 \times 10^{-5}$
   (d) 0.010 m $H_2SO_4(aq)$; note that $K_a = 0.012$ for

$$HSO_4^-(aq) \rightleftharpoons H^+(aq) + SO_4^{2-}(aq)$$

   (e) A solution prepared by mixing 50 mL of 0.100 m $CH_3COOH(aq)$ with 50 mL of 0.200 m $NaOH(aq)$.

2. Using data in Table 11-4, compute the activities of the electrolytes in the following solutions:

(a) 0.30 m	$HCl(aq)$	(b) 0.10 m	$KCl(aq)$
(c) 0.50 m	$CaCl_2(aq)$	(d) 3.00 m	$H_2SO_4(aq)$
(e) 0.70 m	$K_3Fe(CN)_6$	(f) 0.50 m	$K_4Mo(CN)_8$

3. The average partial pressure of $CO_2(g)$ in air is 3×10^{-4} atm. Estimate the concentration of $CO_2(aq)$ in water in equilibrium with air at 25°C.

4. Plot $\log \gamma_\pm$ versus $I^{1/2}/(1 + I^{1/2})$ and $\log \gamma_\pm$ versus $I^{1/2}$ for $HCl(aq)$ over the range 0 m–10 m. Compute the difference in $\log \gamma_\pm$ values at 0.100 m for the two expressions. Using data in Table 11-3, compute $\Delta \bar{G}_{NaCl}$ for the process at 25°C:

$$NaCl(aq, 0.0010\ m) \longrightarrow NaCl(aq, 10.0\ m)$$

5. (a) Show that if $PbCl_2(s)$ is equilibrated with an aqueous solution and the following equilibrium

$$PbCl_2(s) \rightleftharpoons \underset{(1)}{Pb^{2+}(aq)} + \underset{(2)}{2\,Cl^-(aq)}$$

   is established, then ($m_2 = m_{Cl^-}$)

$$K = m_1 m_2^2 \gamma_\pm^3$$

   (b) Suppose that prior to equilibration with $PbCl_2(s)$ the molality of $Cl^-(aq)$ in the solution was m. Show that

$$K = m_1(m_2 + m)^2 \gamma_\pm^3$$

   (c) Explain why it is not rigorously true that $m_2 = 2m_1$ in either case (a) or (b).

6. Show that if Raoult's law holds, then the activity of water in an aqueous solution containing a strong electrolyte that dissociates into ν ions is given by

$$a_{H_2O} = 1 - \frac{\nu m}{55.506 + \nu m}$$

Use this expression to compute a_{H_2O} for 0.300 m aqueous solutions of HCl, NaOH, and $CaCl_2$. Compare your results for a_{H_2O} to the value calculated using osmotic coefficient data in Table 11-4.

7. The value of K_a for $CH_3COOH(aq)$ at 25°C is 1.74×10^{-5} (molality scale).

(a) Calculate the equilibrium concentrations of all the species in a 0.200-m solution assuming that all $\gamma_i = 1.00$.

(b) Repeat the calculation in part a assuming that

$$\log \gamma_{\pm} = -\frac{0.511 I^{1/2}}{1 + I^{1/2}}$$

and that $\gamma = 1$ for undissociated CH_3COOH.

(c) Compute the pH in part (a) and part (b) where

$$\text{pH} \equiv -\log_{10}\{[H^+]\gamma_{\pm}\}$$

(d) Repeat the calculation in part (b) for a solution that is also 0.100 m in NaCl(aq).

(e) Repeat the calculation in part (b) taking $\gamma = 0.993$ for $CH_3COOH(aq)$.

8. At 800 K for the reaction

$$3\,A(g) \rightleftharpoons A_3(g)$$

we have

$$K_{800} = 0.90 \qquad \Delta H^\circ_{rxn} = -420 \text{ kJ}$$

estimate the mole fraction of trimer at 800 K and 1000 K when

(a) $P_{tot} = 0.010$ atm (b) $P_{tot} = 10$ atm

(Assume ideal gas behavior.)

9. For the reaction

$$3\,Si(s) + 2\,N_2(g) \rightleftharpoons Si_3N_4(g)$$
$$\Delta G^\circ_{rxn}/\text{cal} = -1.77 \times 10^5 - 5.76T \log_{10} T + 96.3T$$

where the ΔG°_{rxn} expression is valid over the range 500 K–1670 K.

(a) Calculate $\Delta \bar{G}^\circ_f$ and $\Delta \bar{H}^\circ_f$ of $Si_3N_4(g)$ at 1000 K.

(b) Derive expressions for ΔS°_{rxn} and $\Delta C^\circ_{P,rxn}$.

(c) What additional data are needed to compute $\bar{S}^\circ$ for $Si_3N_4(g)$ at 1000 K?

(d) Suppose that excess Si(s) is subjected to 1 atm of $N_2(g)$ in a closed vessel at 1000 K and that the system is allowed to come to equilibrium; compute the partial pressures of Si_3N_4 and N_2 gas at equilibrium.

10. Compute $\Delta \bar{G}_f^\circ$ of He(*aq*) at 25°C using data in this chapter.

11. Estimate the moles of O_2 dissolved in 10 gallons of water in contact with air at 25°C.

12. Saturated aqueous salt solutions are used to provide an atmosphere with a fixed vapor pressure of water. Use the phase rule to show why such a procedure is effective.

13. The activity of component 1 in a two-component solution is given by

$$a_1 = X_1 e^{BX_2^2/RT}$$

(a) Derive an expression for γ_1.
(b) Derive an expression for a_2.

14. The solubility of the solid *Z*(*s*) in benzene at 25°C is $X_2 = 0.275$. The value of γ_2 relative to a solute standard state is $\gamma_2 = 0.676$ for the saturated solution. Compute K and $\Delta G_{\text{rxn}}^\circ$ for the reaction

$$Z(s) = Z(soln)$$

15. A reaction of importance in various coal-gasification schemes is the water-gas reaction

$$\text{C(graphite)} + H_2O(g) \rightleftharpoons CO(g) + H_2(g)$$

For this reaction at 25°C, $\Delta G_{\text{rxn}}^\circ = +21.853$ kcal and $\Delta H_{\text{rxn}}^\circ = 31.381$ kcal, whereas $\Delta C_{P,\text{rxn}}^\circ = 1.55\ \text{cal}\cdot\text{K}^{-1}$ (and independent of T) over the range 25°C to 700°C.
(a) Compute the value of K at 25°C.
(b) Compute the value of $\Delta S_{\text{rxn}}^\circ$ at 25°C.
(c) Compute the value of $\Delta H_{\text{rxn}}^\circ$ at 700°C.
(d) Compute the value of $\Delta S_{\text{rxn}}^\circ$ at 700°C.
(e) Compute the value of K for the water-gas reaction at 700°C assuming that $\Delta C_{P,\text{rxn}}^\circ = 0$.
(f) Compute the value of K for the water-gas reaction at 700°C taking $\Delta C_{P,\text{rxn}}^\circ = 1.55\ \text{cal}\cdot\text{K}^{-1}$.

16. Excess graphite is placed in a vessel with a 2.00-L vapor space, and 2.00 mol of water is added; the vessel is closed, the temperature is increased to 700°C, and the water-gas reaction is allowed to reach equilibrium:

$$C(s) + H_2O(g) \rightleftharpoons CO(g) + H_2(g) \qquad K = 1.27$$

Denote the number of moles of H_2 at equilibrium by X and
(a) Express K in terms of the total pressure P_{tot}, X, and constants.
(b) Express P_{tot} in terms of X assuming ideal gas behavior.
(c) Compute the moles of H_2, CO, and H_2O present at equilibrium.

17. The equilibrium constant for the reaction

$$AgOCOCH_3(s) \rightleftharpoons Ag^+(aq) + CH_3COO^-(aq)$$

at 25°C is $K_{sp} = 4.05 \times 10^{-3}$ (molality scale).

(a) Calculate the solubility of $AgOCOCH_3(s)$ in $H_2O(\ell)$ assuming $\gamma_\pm = 1.00$.
(b) Repeat the calculation in part (a) assuming

$$\log_{10}\gamma_\pm = \frac{-0.511 I^{1/2}}{1 + I^{1/2}}$$

(c) Repeat the calculation in part (b) for an aqueous solution 0.100 m in $KNO_3(aq)$.

18. The equilibrium vapor pressure at 25°C of component 1 over a binary liquid solution is given in torr by the expression

$$P_1 = 500X_1(1 + 10X_2^2)$$

(a) Determine the Raoult's law constant (P_1°) and Henry's law constant (P_1^*) for component 1 at 25°C.
(b) Determine the activity of component 1 relative to a solvent standard state for the $X_1 = 0.50$ solution at 25°C.
(c) Determine the activity coefficient of component 1 relative to a solute standard state for the $X_1 = 0.50$ solution at 25°C.
(d) The equilibrium vapor pressure of pure component 2 at 25°C is 600 torr; the total pressure over the two-component solution at $X_2 = 0.500$ is 1275 torr; compute the activity coefficient of component 2.

19. (a) The solubility of $CaSO_4$ in water at 25°C is 0.01535 m, whereas its solubility in $MgSO_4(aq)$ is as follows:

$m(MgSO_4)$	$m(CaSO_4)$
0.00502	0.01441
0.01012	0.01362
0.01528	0.01310

Taking into account activity coefficients, estimate K at 25°C for the reaction

$$CaSO_4(s) \rightleftharpoons Ca^{2+}(aq) + SO_4^{2-}(aq)$$

(b) The solubility of $CaSO_4(s)$ in 0.212 m $CuSO_4(aq)$ at 25°C is 0.01329, and in 0.977 $CuSO_4(aq)$ it is 0.01654. Suggest a possible explanation for the increased solubility of $CaSO_4$ in $CuSO_4(aq)$ at high $CuSO_4$ concentration. How would you check your explanation experimentally?

20. Given that the equilibrium vapor pressure of component 2 in a binary solution over part of the composition range is given by

$$P_2 = \tfrac{1}{3}(4X_2 - X_2^2)P_2^\circ$$

Derive an expression for the equilibrium vapor pressure of component 1 valid over the same composition range. Use your results to calculate γ_1 at $X_1 = 0.50$ relative to both a solvent and solute standard state. Plot P_2 and P_1 in terms of P_2° over the range $0 \leqslant X_1 \leqslant 1$. Take $P_2^\circ = 3P_1^\circ$.

21. The equilibrium vapor pressure at 25°C of component 1 over a binary liquid solution is given in torr by the expansion

$$P_1 = 200X_1(1 + 2X_2^2 + 6X_2^3)$$

(a) Determine the Raoult's law constant (P_1°) and Henry's law constant (P_1^*) for component 1 at 25°C.
(b) Determine the activity of component 1 relative to a solvent standard state for the $X_1 = 0.50$ solution at 25°C.
(c) Determine the activity coefficient of component 1 relative to a solute standard state for the $X_1 = 0.50$ solution at 25°C.

22. Oxygen-free water can be prepared in the lab in either of two ways: (i) By vigorous boiling of the water followed by cooling in the absence of O_2. (ii) By bubbling oxygen-free $N_2(g)$ or $Ar(g)$ into the solution through a narrow-necked flask. Discuss why these procedures are effective.

23. Show that
(a) The equilibrium constant for a reaction written in reverse is the reciprocal of the equilibrium constant for the forward reaction.
(b) The equilibrium constant for a reaction obtained by adding two reactions is the product of the equilibrium constants for the reactions that are added together.

24. Consider the equilibrium between $Ag^+(aq)$ and $Ag(NH_3)_2^+(aq)$ in a $AgNO_3(aq)$ solution containing $NH_3(aq)$; that is

$$\underset{(1)}{Ag^+(aq)} + \underset{(2)}{2\,NH_3(aq)} \rightleftharpoons \underset{(3)}{Ag(NH_3)_2^+(aq)}$$

Show that

$$K = \left\{\frac{[Ag(NH_3)_2^+]}{[Ag^+][NH_3]^2}\right\} \cdot \left\{\frac{\gamma_{\pm(3)}}{\gamma_{\pm(1)}\gamma_2^2}\right\}$$

Explain the significance of $\gamma_{\pm(3)}$ and $\gamma_{\pm(1)}$. Assume that Eq. 11-64 holds for $\gamma_\pm$, and show that in dilute solution

$$K \approx \frac{[Ag(NH_3)_2^+]}{[Ag^+][NH_3]^2}$$

25. Under what assumption can we say that $\gamma_\pm$ for a solution of $AgNO_3(aq)$ is equal to $\gamma_\pm$ for a solution of $Ag(NH_3)_2NO_3(aq)$ at the same values of the ionic strength?

26. Show that

$$P_{2X}^{*} = \left(\frac{1000}{M_1}\right) P_{2m}^{*}$$

where M_1 is the molecular weight of the solvent.

27. Show that the activity coefficients of a substance in solution relative to the solvent and the solute standard states are related by the expression

$$\frac{\gamma_{i(\text{solvent std state})}}{\gamma_{i(\text{solute std state})}} = \frac{P_i^{*}}{P_i^{\circ}} = \text{constant}$$

28. Show that solute activity coefficients on the molality γ_{im} and mole fraction γ_{iX} concentration scales are related as follows:

$$\gamma_{iX} = \gamma_{im}\left(1 + \frac{M_1 m}{1000}\right)$$

where M_1 is the molecular mass of the solvent.

29. Given the following data on ethyl ether (1) + acetone (2) solutions at 30°C:

$X_2(\ell)$	P_2/torr	P_1/torr
0	0	646
0.050	29.4	614
0.100	58.8	581
0.200	90.3	535
0.500	168	391
0.800	235	202
1.000	283	0

(a) Compute the activity coefficients of ether and acetone relative to solvent standard states at $X_2 = 0.50$.
(b) Compute the activity coefficient of acetone relative to a solute standard state at $X_2 = 0.50$.
(c) Compute P_2 and P_1 at $X_2 = 0.080$.
(d) Compute $\Delta G_{\text{rxn}}^{\circ}$ values at 30°C for
  (i) the dissolution of 1 mol of acetone(ℓ) in 1 mol of ether(ℓ), relative to a solvent standard state for acetone in the solution.
  (ii) the dissolution of 1 mol of acetone(ℓ) in 1 mol of ether(ℓ), relative to a solute standard state for acetone in the solution.

30. Given the following data (J. A. Larkin and R. C. Pemberton, *NPL Report Chem.*, 43, January 1976) on alcohol + water solutions at 25°C, compute the activities and activity coefficients of alcohol and of water relative to both solvent and solute standard states, and plot γ versus X for the four cases.

Pressures in kPa

$X_{C_2H_5OH}$	$P_{C_2H_5OH}$	P_{H_2O}
0.00	0	3.169
0.02	0.571	3.108
0.05	1.328	3.022
0.08	1.978	2.942
0.10	2.353	2.893
0.20	3.603	2.700
0.30	4.164	2.578
0.40	4.524	2.466
0.50	4.914	2.305
0.60	5.363	2.070
0.70	5.858	1.755
0.80	6.432	1.318
0.90	7.126	0.717
0.93	7.352	0.510
0.96	7.582	0.297
0.98	7.736	0.151
1.00	7.893	0

31. Consider the following variant on the isopiestic method: Two different solutions of the same electrolyte in water are held at different temperatures (T_1, T_2) and are allowed to equilibrate through the vapor phase, and the resulting equilibrium solutions are analyzed. Suppose that a_1 at T_1, m_1 is known. Show that such measurements (i.e., m_1, m_2, T_1, and T_2 data) yield $(\partial \ln a/\partial T)_{m,P}$.

32. Show that if the activity coefficient of the solute is given by

$$\log \gamma_\pm = -A|Z_+Z_-|I^{1/2}$$

then the osmotic coefficient of the solvent is given by

$$\phi = 1 - \frac{2.303}{3} A|Z_+Z_-|I^{1/2}$$

where Z_+ is the cationic charge and Z_- is the anionic charge.

33. The dissociation constants of weak acids as a function of temperature can be fit to an equation of the form

$$\log_{10} K_a = -\frac{A_1}{T} + A_2 - A_3 T$$

where A_1, A_2, and A_3 are constants. Derive general expressions for $\Delta G^\circ_{\text{rxn}}$, $\Delta H^\circ_{\text{rxn}}$, $\Delta S^\circ_{\text{rxn}}$, and $\Delta C^\circ_{P,\text{rxn}}$ in terms of A_1, A_2, A_3, and T for the dissociation reaction, and apply them to the calculation of these quantities at 25°C for the acetic acid and ammonium ion in water:

Acid	$-\log K_a$ (25°C)	A_1	A_2	A_3
acetic acid	4.756	1170.48	3.1649	0.013399
ammonium ion	9.245	2835.76	0.6322	0.001225

Also plot $\log K_a$ versus $1/T$ in the region 0°C–75°C for the above acids. [Data taken from Robinson and Stokes (1959), where additional data can be found].

34. Using methods analogous to those in the text, obtain an expression for the activity of a component in a solid solution relative to a solvent standard state.

35. Given that the equilibrium vapor pressure of component 2 in a binary solution is given by

$$P_2 = (2X_2 - X_2^2)P_2^+ \qquad 0 \leqslant X_2 < 0.80$$

where P_2^+ is a constant; show that this equation cannot hold as $X_2 \to 1$. Obtain P_2^* in terms of P_2^+ and compute γ_2 relative to a solute standard state at $X_2 = 0.01$, 0.10, and 0.25. Using the above expression for P_2, derive an expression for P_1, and compute γ_1 at $X = 0.01$, 0.10, and 0.25. In what range of X_1 is this P_1 expression valid?

36. In a dilute binary solid solution, the activity coefficient of the solute is given by

$$\gamma_2 = e^{-kX_2}$$

(a) Derive an expression for the activity coefficient of the solvent.
(b) Is your result in part (a) valid over the range $0 \leqslant X_1 < 1$?

37. Show that a catalyst can affect the equilibrium distribution of species but not the equilibrium constant for a reaction.

38. Suppose that the partial molar volumes of the constituents in a gaseous solution are equal to the molar volumes of the pure gases at the same temperature and total pressure as the gaseous solution (Amagat's law), and show that

$$a_i(\text{gas soln}) = X_i a_i(\text{pure gas})$$

where a_i(pure gas) is the activity of pure gaseous i at the same temperature and total pressure as the gas mixture.

39. Determine the value of the limit

$$\lim_{m \to 0} \left(\frac{\phi - 1}{m} \right)$$

where ϕ is the osmotic coefficient and m is the molality of the solute for (a) sucrose(aq), and (b) NaCl(aq).

40. When tap water is heated from, say, 18°C to 60°C, bubbles usually form in the liquid. Explain this phenomenon and estimate the composition of the gas in the bubbles at 60°C. Assume that dP^*_{2m}/dT for N_2 is equal to that for O_2 and neglect argon.

41. Consider the reaction

$$\text{H}X(aq) \rightleftharpoons \text{H}^+(aq) + X^-(aq)$$

Derive relationships between the equilibrium constant for this reaction expressed on the molality, molarity, and mole fraction concentration scales. Apply your results to acetic acid in water at 25°C for which $\log K = -4.756$ at 25°C on the molality scale.

42. Consider the reaction

$$\text{Ag}X(s) \rightleftharpoons \text{Ag}^+(aq) + X^-(aq)$$

Derive relationships between the equilibrium constant for this reaction expressed on the molality, molarity, and mole fraction concentration scales. Apply your results to silver acetate in water at 25°C. Take $K_{sp} = 4.052 \times 10^{-3}$ at 25°C on the molality scale.

43. Use the data in Problem 33 to obtain an expression for $\log K$ as a function of T for the reaction

$$\text{NH}_4^+(aq) + \text{CH}_3\text{COO}^-(aq) \rightleftharpoons \text{NH}_3(aq) + \text{CH}_3\text{COOH}(aq)$$

Use your equation to compute $\Delta H^\circ_{\text{rxn}}$ and $\Delta S^\circ_{\text{rxn}}$ at 25°C for this reaction.

44. Use the data in Table 11-4 to compute $\Delta\bar{G}_{\text{CaCl}_2(aq)}$ at 25°C for the process

$$\text{CaCl}_2(aq, 0.100\ m) \longrightarrow \text{CaCl}_2(aq, 0.200\ m)$$

Also compute $\Delta\bar{G}_{\text{H}_2\text{O}(\ell)}$ for this process. Use your results to compute ΔG_{total} for the process.

45. State all of the assumptions necessary to arrive at the following approximate expressions

(a) $\text{NH}_3(aq) + \text{H}_2\text{O}(\ell) \rightleftharpoons \text{NH}_4^+(aq) + \text{OH}^-(aq)$

$$K \simeq \frac{[NH_4^+][OH^-]}{[NH_3]}$$

where the brackets denote molarities.

(b) $Ag_2SO_4(s) \rightleftharpoons 2Ag^+(aq) + SO_4^{2-}(aq)$

$$s \simeq \left(\frac{K_{sp}}{4}\right)^{1/3}$$

where s is the solubility in moles per liter, and K_{sp} is the solubility product constant on the molality scale.

(c) $Ca^{2+}(aq) + 2HCO_3^-(aq) \rightleftharpoons CaCO_3(s) + H_2O(\ell) + CO_2(g)$

$$K \simeq \frac{P_{CO_2}}{[Ca^{2+}][HCO_3^-]^2}$$

46. Given that $K = 2.399 \times 10^{-14}$ at 37°C for the reaction

$$H_2O(\ell) \rightleftharpoons H^+(aq) + OH^-(aq)$$

Compute the $pH = -\log([H^+]\gamma_\pm)$ of neutral aqueous solution at 37°C for which $I = 0.30\ m$. Assume only 1:1 electrolytes are present. Note that human body temperature is about 37°C.

47. Given that human blood is buffered at pH = 7.4 and that the partial pressure of $CO_2(g)$ in exhaled air is about 0.04 atm, estimate the concentration of $HCO_3^-(aq)$ in blood at 37°C. Take $K_{a1} = 5 \times 10^{-7}$ and $K_{a2} = 1 \times 10^{-11}$ for $CO_2(aq)$ and $K = 0.02 = [CO_2]/P_{CO_2}$. Take $I = 0.30\ m$.

48. If we treat $HCl(aq)$ as a strong electrolyte, what can we say about the value of the equilibrium constant for the reaction

$$HCl(aq) \rightleftharpoons H^+(aq) + Cl^-(aq)$$

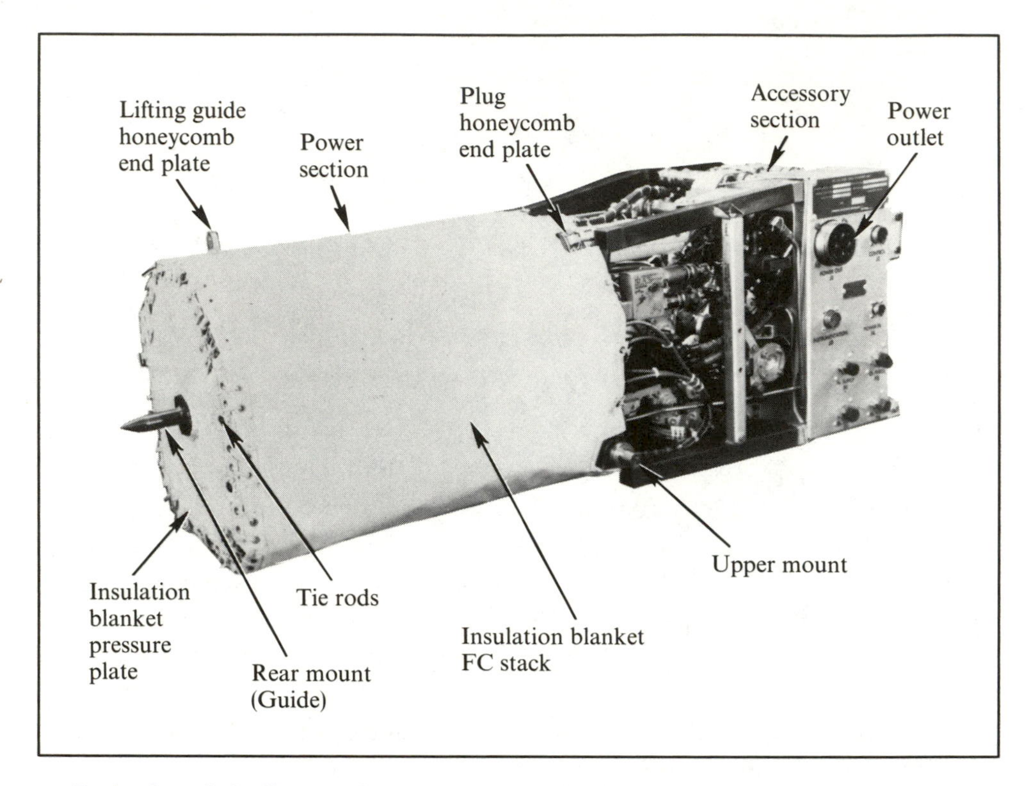

Each orbiter fuel cell power plant contains 64 *single cells divided electrically into two stacks of* 32 *cells each. Each cell consists of an electrolyte sandwiched between two electrodes. Hydrogen passing over one electrode reacts with hydroxyl ions in the electrolyte to produce water and release electrons. The electrons move through an external circuit toward the oxygen electrode. Oxygen passing over the hydrogen electrode reacts with the water and returning electrons to produce hydroxyl ions. The hydroxyl ions then migrate to the hydrogen electrode and enter the hydrogen reaction cycle. The movement of the electrons through the external circuit creates direct current for the spacecraft's electrical system.*

The space shuttle is equipped with three fuel cell power plants located under the payload bay in the forward portion of the orbiter's mid-fuselage.

Each fuel cell power plant is capable of supplying 12 kilowatts at peak and 7 kilowatts average power to the spacecraft's electrical system. At 2 kilowatts power output, each power plant provides 32.5 volts and 61.5 amperes of direct current, and at 12 kilowatts provides 27.5 volts and 436 amperes of direct current.

12

Thermodynamics of Ions in Solution

> *. . . the emf of a galvanic cell is a measure of the affinity of the current-generating chemical process, and, . . . conversely, electromotive forces can be calculated thermodynamically from a chemical equilibrium. [The] equation for the metal-electrolyte potential difference [is]:*
>
> $$E = \frac{RT}{nF} \ln \frac{C}{c}$$
>
> *(where R is the gas constant, n is the chemical valency of the ion concerned, C is a constant specific to the electrode, and c is the ion concentration) . . .*
>
> Walther H. Nernst
> Nobel Prize Lecture, 1920

THE PRINCIPAL sources of thermodynamic data for ions in solution are electrochemical cells, direct equilibrium measurements (determination of K and its temperature dependence), and thermochemical measurements, $\Delta H^\circ_{\text{rxn}}$. In this chapter we shall discuss electrochemical cells as a source of thermodynamic data. Our objective will be to show how $\Delta \bar{G}^\circ_f$, $\Delta \bar{H}^\circ_f$, and $\bar{S}^\circ$ values are determined for ions in solution; these quantities, for any given ion, depend on temperature and solvent but not on the concentration of the ion in solution or on the total pressure.

12-1 *An Electric Current Can Be Obtained from a Chemical Reaction*

Electrochemical cells are a particularly valuable source of thermodynamic data on ions because, at least in principle, any chemical reaction can be made to occur in an electrochemical cell, whether it be a so-called oxidation-reduction reaction, a metathetical reaction, ions combining to form an insoluble salt or a neutral molecule, a phase change, or even a simple dilution reaction. The principal advantages that cells have over direct equilibrium methods for studying chemical reactions are (1) the reaction need not be at equilibrium in the usual chemical sense (i.e., ΔG for the reaction need not be equal to zero), and hence the values of the composition variables of the reactants and products can be chosen for convenience, and (2) because we (and not Mother Nature) choose the values of the composition variables, it is not necessary, in most cases, to analyze for the equilibrium composition of the system.

For a chemical reaction occurring in an electrochemical cell, the Gibbs phase rule takes the form

$$f = c - p + 3 \qquad \text{(electrochemical cell)} \tag{12-1}$$

rather than $f = c - p + 2$ for the same reaction run in, say, a flask. The extra degree of freedom arises from the existence of an additional independent intensive thermodynamic coordinate, namely, the cell voltage. Thus, for a reaction involving only pure condensed phases, for example,

$$Pb(s) + 2\,AgCl(s) \rightleftharpoons PbCl_2(s) + 2\,Ag(s)$$

we have

$$f = (4 - 1) - 4 + 3 = 2$$

when the reaction is run inside the cell, as opposed to

$$f = (4 - 1) - 4 + 2 = 1$$

when the reaction is run outside the cell. Therefore, when the reaction is run in a cell, we can pick both the equilibrium pressure and temperature, and this choice of temperature and pressure determines the cell voltage. When the reaction is run in a flask, we can choose the equilibrium temperature *or* the equilibrium pressure *but not both of them.* If we choose 1 atm pressure, then there is one, and only one, value of the temperature at which equilibrium can be attained, and this equilibrium value of the temperature may be an experimentally inconvenient value.

Consider the chemical reaction

$$H_2(g) + Cl_2(g) \rightleftharpoons 2\,HCl(aq) \tag{12-2}$$

The reaction in Eq. 12-2 can be separated into two half-reactions, an *oxidation half-reaction* and a *reduction half-reaction:*

$$H_2(g) \rightleftharpoons 2\,H^+(aq) + 2e^- \qquad \text{oxidation}$$
$$Cl_2(g) + 2e^- \rightleftharpoons 2\,Cl^-(aq) \qquad \text{reduction}$$

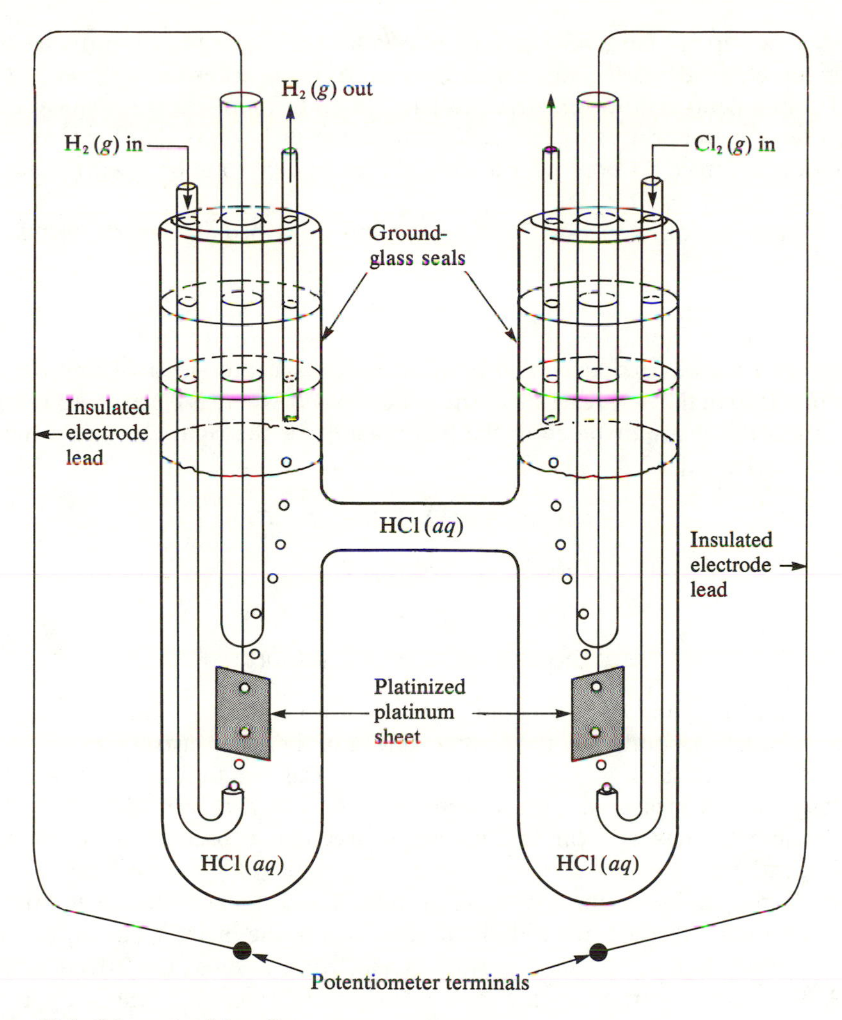

Figure 12-1. Schematic of the cell

$$Pt(s)|H_2(g)|HCl(aq)|Cl_2(g)|Pt(s)$$

The cell electrolyte solution is $HCl(aq)$. Note that the H-tube design of the cell physically separates the $H_2(g)$ and $Cl_2(g)$ and thereby prevents the reaction

$$H_2(g) + Cl_2(g) \longrightarrow 2\,HCl(aq)$$

from occurring directly (i.e., via H_2, Cl_2 collisions). The cell forces the reaction to occur by electron transfer through the external circuit.

Each of these two half-reactions can be made to take place at one of the electrodes in an electrochemical cell (Figure 12-1). The *cell diagram* for the reaction in Eq. 12-2 is

$$Pt(s)|H_2(g)|HCl(aq)|Cl_2(g)|Pt(s) \qquad (12\text{-}3)$$

where the vertical bars indicate phase boundaries. Platinum, Pt, provides the metallic electrode surface on which the electrons are transferred between the cell components and the external electrical circuit. In an electrochemical cell, the reactant and product species are separated in such a way that the chemical reaction is forced to proceed via the transfer of electrons between reactants *through the external circuit.*

An oxidation-reduction reaction can be described in general terms as follows:

$$\text{donor-L} + \text{acceptor-R} \rightleftharpoons \text{donor-R} + \text{acceptor-L}$$

Electron donor-L (reducing agent) and electron acceptor-L (oxidizing agent) are involved in the half-reaction at the left electrode, whereas electron acceptor-R and donor-R are involved in the half-reaction at the right electrode. Thus for the cell in Eq. 12-3 we have

$$\underset{\text{donor-L}}{H_2(g)} \rightleftharpoons \underset{\text{acceptor-L}}{2\,H^+(aq)} + 2e^-$$

$$\underset{\text{acceptor-R}}{Cl_2(g)} + 2e^- \rightleftharpoons \underset{\text{donor-R}}{2\,Cl^-(aq)}$$

Note that $Cl_2(g)$ and $H_2(g)$ do not come into direct contact within the cell; they are physically separated by the H-tube nature of the cell construction.

The reaction driving force ΔG_{rxn} for the cell reaction is manifest as a voltage that is measured with a potentiometer and a current detector. The potentiometer acts as a variable voltage source that is used to determine the voltage that must be placed in opposition to the cell voltage to prevent any *net* current flow in the circuit consisting of the cell and the potentiometer. When a current flows through the cell, the current within the metallic electrodes and the external electrical circuit is carried by electrons, whereas the current in the cell electrolyte is carried by the ions $H^+(aq)$ and $Cl^-(aq)$ moving in opposite directions through the solution of $HCl(aq)$ (Figure 12-2). The electrochemical cell reaction is the source of the electric current obtainable from the cell. An electrochemical cell thus enables us to obtain electricity directly from a chemical reaction.

In the thermodynamic analysis of cell data, it is conventional to *assume that oxidation takes place at the left electrode* and, consequently, that reduction takes place at the right electrode *in the cell diagram as written.* (Compare the half-reactions given with the cell diagram.) The "oxidation-at-the-left-electrode" convention is equivalent to the assumption that *negative current* flows *through the cell* electrolyte from right to left and through the external circuit (electrode leads) from left to right. Electrons flow spontaneously from a region of negative potential to a region of positive potential, whereas positive current flows spontaneously from a region of positive potential to a region of negative potential. A negative charge moving in one direction is equivalent to a positive charge moving in the opposite direction. Electrode polarities serve to define

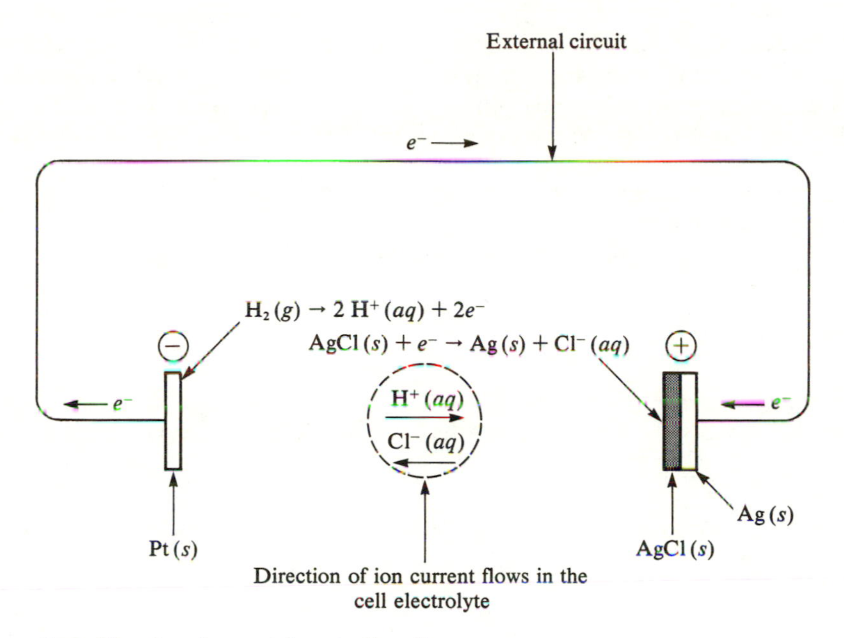

Figure 12-2. Direction of current flows in the cell

$$Pt(s)|H_2(g)|HCl(aq)|AgCl(s)|Ag(s)$$

when the Pt(*s*) electrode is the negative electrode. The negative electrode is the *anode*, and the positive electrode is the *cathode*. Oxidation takes place at the anode and reduction takes place at the cathode. In the cell electrolyte, the current is carried by cations moving toward the cathode and anions moving toward the anode. These distinctions are easily remembered using the following mnemonics:

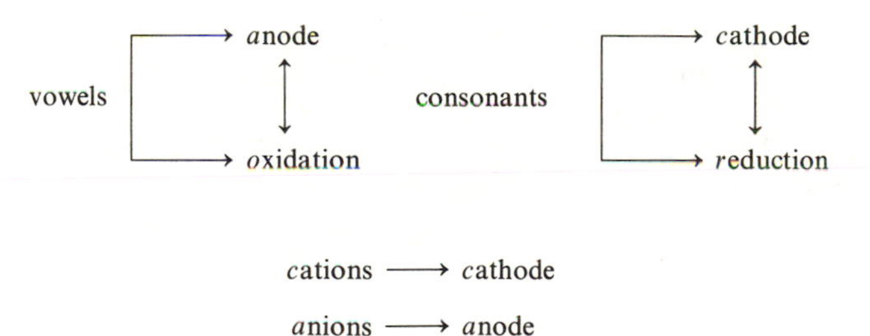

cations ⟶ cathode

anions ⟶ anode

the direction of *external* spontaneous current flow. A negative electrode polarity does not mean that the electrode carries a negative charge. During the spontaneous discharge of a cell, positive ions are produced at the negative electrode and negative *ions* in the solution will move spontaneously *toward* the negative electrode, where the positive ions are produced, in order to restore electroneutrality.

If the *cell reaction* as written proceeds spontaneously from left to right under the conditions prevailing in the cell, then the left electrode will be the negative electrode and the right electrode, the positive electrode; whereas if the cell reaction as written proceeds spontaneously from right to left, then the right electrode will be the negative electrode. The experimentally determined cell polarity is unambiguous and independent of any arbitrary assumptions about the direction of current in the cell or the spatial orientation of the cell in the laboratory with respect to the experimenter.

EXAMPLE 12-1

Write the balanced cell reaction for the following electrochemical cell:

$$Cd(s)|CdSO_4(aq)|Hg_2SO_4(s)|Hg(\ell)$$

Solution: We note at the left electrode that Cd(*s*) is oxidized to $Cd^{2+}(aq)$; thus

$$Cd(s) \longrightarrow Cd^{2+}(aq) + 2e^- \qquad \text{(oxidation at left electrode)}$$

At the right electrode Hg_2^{2+}, in $Hg_2SO_4(s)$, is reduced to $Hg(\ell)$:

$$Hg_2SO_4(s) + 2e^- \longrightarrow 2\,Hg(\ell) + SO_4^{2-}(aq) \qquad \text{(reduction at right electrode)}$$

The electrons supplied by the oxidation half-reaction are consumed by the reduction half-reaction (conservation of electrons); thus the balanced cell reaction is

$$Cd(s) + Hg_2SO_4(s) \longrightarrow CdSO_4(aq) + 2\,Hg(\ell)$$

12-2 *The Nernst Equation Gives the Dependence of the Cell Voltage on Product and Reactant Activities*

Application of the First and Second Laws of Thermodynamics to an electrochemical cell yields

$$dG = -S\,dT + V dP + \sum_i \mu_i\,dn_i - E\,dZ \qquad (12\text{-}4)$$

where E is the cell voltage (electromotive force, *emf*) and dZ is the charge transferred. At constant temperature, constant pressure, and constant composition, Eq. 12-4 becomes

$$dG = -E\,dZ \qquad (12\text{-}5)$$

For the transfer of a finite charge Z through the cell at fixed cell voltage, as a result of the occurrence of the cell reaction, we obtain from Eq. 12-5

$$\Delta G = -EZ = -Z(V^+ - V^-)$$

where $V^+ - V^-$ is the *Volta potential difference* of the electrons at the potentiometer terminals to which the cell terminal wires are connected. If the cell diagram is written such that the left electrode, where oxidation is *assumed* to occur, is negative for the conditions prevailing in the cell, then the cell voltage is positive, $E > 0$; and the cell reaction is spontaneous in the direction left to right. If, on the other hand, the left electrode is positive, then $E < 0$; and the reaction is spontaneous in the direction opposite to that written. Our current convention then is that *negative* current passes from right to left through the cell and from left to right in the external circuit. If left to right in the external circuit is indeed the direction of spontaneous current flow, then $E > 0$; if the reverse is true, then $E < 0$.

For the occurrence of the cell reaction *as written*, without change in the composition of any of the phases in the cell (a miniscule current is sufficient to measure E), the quantity of charge that is transferred is

$$Z = nF$$

where F is *Faraday's constant:*

$$F = 96{,}484.6 \frac{\text{coulombs}}{\text{mole of electrons}} \tag{12-6}$$

and n is the number of moles of electrons transferred from the negative to the positive electrode for the cell reaction as written. That is, n is the number of moles of electrons transferred from the reducing agent to the oxidizing agent in the balanced cell reaction as written. Therefore

$$\Delta G = -nFE \tag{12-7}$$

If all reactants and products are in their standard states, then Eq. 12-7 becomes

$$\Delta G^\circ = -nFE^\circ \tag{12-8}$$

Combination of Eqs. 12-7 and 12-8 with the Lewis equation

$$\Delta G = \Delta G^\circ + RT \ln Q \tag{12-9}$$

yields

$$-nFE = -nFE^\circ + RT \ln Q$$

or

$$E = E^\circ - \frac{RT}{nF} \ln Q \tag{12-10}$$

Equation 12-10 is known as the *Nernst equation* after W. H. Nernst, who discovered the analog of Eq. 12-10 with concentrations in place of activities in the Q expression. The general form of the Nernst equation, Eq. 12-10, is due to G. N. Lewis.

In the Nernst equation, E is the *observed cell voltage*, and E° is the *standard cell voltage;* that is, E° is the voltage that the cell would have if all the reactants and products were in their unit activity standard states (whence $Q = 1$). The standard cell voltage is related to the equilibrium constant for the cell reaction via the equation

$$\Delta G^\circ = -nFE^\circ = -RT \ln K$$

or

$$E^\circ = \frac{RT}{nF} \ln K \tag{12-11}$$

Equation 12-11 shows that the equilibrium constant K for the cell reaction can be calculated from E° for the cell reaction.

Substitution of Eq. 12-11 into the Nernst equation, Eq. 12-10, yields

$$E = -\frac{RT}{nF}\ln\left(\frac{Q}{K}\right) \tag{12-12}$$

Equation 12-12 is an alternate formulation of the Nernst equation that shows that the cell voltage is a quantitative measure of the *driving force* (emf) for the cell reaction under the conditions prevailing in the cell; that is, E is a measure of reaction driving force for a particular value of the ratio Q/K, as shown in the following table:

Value of Q/K	*Direction in which the cell reaction is spontaneous*	*Sign of the cell voltage*	*Sign of ΔG_{rxn}*
$Q/K < 1$	Left to right, $\longrightarrow$	$E > 0$	$\Delta G < 0$
$Q/K > 1$	Right to left, $\longleftarrow$	$E < 0$	$\Delta G > 0$
$Q/K = 1$	Equilibrium, $\rightleftharpoons$	$E = 0$	$\Delta G = 0$

Note that a measurement of the cell voltage yields a numerical value of the Gibbs energy change ΔG_{rxn} for the cell reaction ($\Delta G_{rxn} = -nFE$) for a particular value of $Q \neq K$, whereas equilibrium measurements ($Q = K$) yield ΔG°_{rxn}. However, as we shall see in the next section, ΔG°_{rxn} values also can be obtained from cell measurements.

12-3 *Equilibrium Constants and Activities Can Be Obtained from Electrochemical Cell Measurements*

Application of the Nernst equation, Eq. 12-10, to the cell reaction

$$H_2(g) + Cl_2(g) \rightleftharpoons 2\,HCl(aq)$$

yields

$$E = E^\circ - \frac{RT}{2F}\ln\left[\frac{a^2_{HCl(aq)}}{a_{H_2(g)}a_{Cl_2(g)}}\right] \tag{12-13}$$

Note that $n = 2$ for the cell reaction because 2 mol of electrons are transferred from H_2 to Cl_2 for the reaction as written. The activities of the gas-phase species at known gas pressures can be obtained from the respective equations of state for the gases via the equation

$$\ln\frac{a}{P} = \int_0^P\left(\frac{\bar{V}}{RT} - \frac{1}{P}\right)dP$$

For simplicity we assume in the present context that the gases behave ideally.

In such a case, Eq. 12-13 becomes

$$E = E^\circ - \frac{RT}{2F}\ln\left(\frac{a^2_{\mathrm{HCl}(aq)}}{P_{\mathrm{H_2}}P_{\mathrm{Cl_2}}}\right) \tag{12-14}$$

The activity of HCl(*aq*) is given by (see Eq. 11-57)

$$a_{\mathrm{HCl}(aq)} = a_{\mathrm{H^+}}a_{\mathrm{Cl^-}} = m_{\mathrm{H^+}}\gamma_{\mathrm{H^+}}m_{\mathrm{Cl^-}}\gamma_{\mathrm{Cl^-}} = m^2\gamma_\pm^2 \tag{12-15}$$

where the last equality assumes that HCl(*aq*) is the sole source of $H^+(aq)$ and $Cl^-(aq)$ in the solution. Substitution of Eq. 12-15 into Eq. 12-14 yields

$$E = E^\circ - \frac{RT}{2F}\ln\left(\frac{m^4\gamma_\pm^4}{P_{\mathrm{H_2}}P_{\mathrm{Cl_2}}}\right) \tag{12-16}$$

The cell is generally set up with known values of m, $P_{\mathrm{H_2}}$, $P_{\mathrm{Cl_2}}$ at a known temperature T. The value of RT/F at 25.00°C is

$$\frac{RT}{F} = \frac{8.3144\ \mathrm{J\cdot K^{-1}\cdot mol^{-1}} \times 298.15\ \mathrm{K}}{96{,}484.6\ \mathrm{C\cdot mol^{-1}}} = 0.025693\ \mathrm{V}$$

Thus (1000 mV = 1 V)

$$\frac{RT}{F} = 25.693\ \mathrm{mV} \qquad \text{at } 25.00^\circ\mathrm{C} \tag{12-17}$$

Combination of Eq. 12-17 with Eq. 12-16 yields

$$E = E^\circ + \frac{25.693}{2}\ln(P_{\mathrm{H_2}}P_{\mathrm{Cl_2}}) - 25.693\ln m^2 - 25.693\ln\gamma_\pm^2 \tag{12-18}$$

Rearrangement of Eq. 12-18 with all the known or measured quantities on the left-hand side yields

$$E + 25.693\left[\ln m^2 - \tfrac{1}{2}\ln(P_{\mathrm{H_2}}P_{\mathrm{Cl_2}})\right] = E^\circ - 25.693\ln\gamma_\pm^2 \tag{12-19}$$

The extended Debye-Hückel limiting law for the mean ionic activity coefficient at 25°C of a 1:1 electrolyte like HCl(*aq*) at 25°C (see Eq. 11-64) is ($2.3026\log\gamma_\pm = \ln\gamma_\pm$)

$$\ln\gamma_\pm = \frac{-1.177I^{1/2}}{1 + I^{1/2}} \tag{12-20}$$

where I is the *ionic strength* (see Eq. 11-65)

$$I = \frac{1}{2}\sum_i m_i Z_i^2 \tag{12-21}$$

In Eq. 12-21, m_i is the molality of ion i with charge Z_i. The ionic strength of an HCl(*aq*) solution of molality m is

$$I = \tfrac{1}{2}[m_{\mathrm{H^+}}(1)^2 + m_{\mathrm{Cl^-}}(-1)^2] = m$$

and thus Eq. 12-20 can be rewritten for HCl(*aq*) as

$$\ln\gamma_\pm = \frac{-1.177m^{1/2}}{1 + m^{1/2}} \tag{12-22}$$

The Debye-Hückel limiting law value of $\ln(\gamma_\pm)_{DH}$ given by Eq. 12-22 approaches the *actual* value of $\ln \gamma_\pm$ at low concentrations. Thus for HCl(*aq*)

$$\ln \gamma_\pm = \ln(\gamma_\pm)_{DH} + f(m) = \frac{-1.177m^{1/2}}{1 + m^{1/2}} + f(m) \qquad (12\text{-}23)$$

where $f(m)$ is some as yet unspecified function of the molality of HCl(*aq*). Substitution of Eq. 12-23 with $f(m) = -bm/51.386$ into Eq. 12-19 gives

$$E + 25.693\left[\frac{1}{2}\ln\left(\frac{m^4}{P_{H_2}P_{Cl_2}}\right) - 2\left(\frac{1.177m^{1/2}}{1 + m^{1/2}}\right)\right] = E^\circ + bm \qquad (12\text{-}24)$$

where b is a constant.

Note that all the variables on the left-hand side of Eq. 12-24 are known from experiment. Provided the various assumptions that we have made hold, a plot of the left-hand side of Eq. 12-24, denoted for simplicity as E', versus m should become linear at low m with a slope of b and an intercept of E°. Such a plot is called a *Hitchcock extrapolation.* Once the value of E° is obtained from the E' versus m plot, it can be used in conjunction with Eq. 12-16 to obtain values of $\gamma_\pm$, including $\gamma_\pm$ values at molalities where the Debye-Hückel limiting law does not hold. The Hitchcock extrapolation is necessary to obtain E°, because it was assumed that we do not know either E° *or* $\gamma_\pm(m)$ at the start of the experiments. The cell measurements are used to establish the solute standard state for HCl(*aq*). The attainment of a linear dependence of E' on m as $m \to 0$ substantiates Eq. 12-24 and, in effect, establishes the solute standard state for HCl(*aq*).

A Hitchcock plot for the cell reaction

$$H_2(g) + Cl_2(g) \rightleftharpoons 2\,HCl(aq) \qquad (12\text{-}25)$$

is shown in Figure 12-3. The data used to construct Figure 12-3 are given below:

$$T = 298.15\text{ K} \qquad P_{H_2} = P_{Cl_2} = 1.000\text{ atm}$$

m/mol · kg^{-1}	E/mV	E'/mV	$\gamma_\pm$(calculated using E° obtained from Figure 12-3)
0.0001000	1834.31	1360.41	0.988
0.001000	1717.20	1360.38	0.965
0.01000	1602.24	1360.10	0.904
0.1000	1490.45	1357.60	0.796
0.3000	1436.65	1353.38	0.756
0.5000	1410.33	1349.66	0.757
0.7000	1392.04	1346.16	0.772
1.0000	1371.30	1341.06	0.809

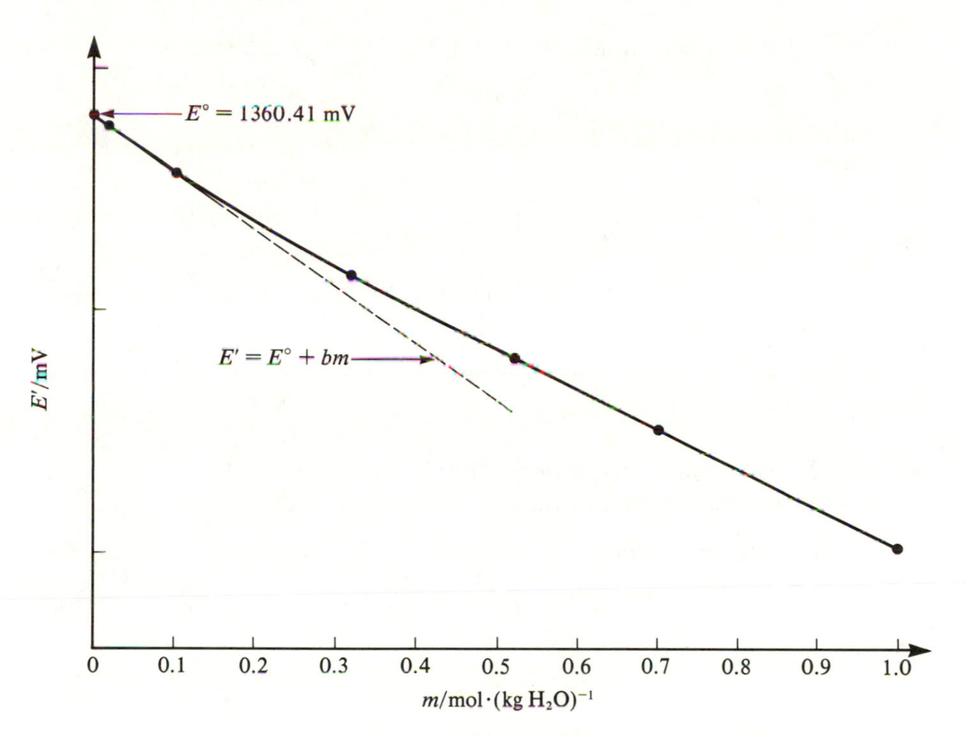

Figure 12-3. Determination of E° for the reaction $H_2(g) + Cl_2(g) \rightleftharpoons 2\ HCl(aq)$, $t = 25.00°C$, $P_{H_2} = P_{Cl_2} = 1.000$ atm. Note from Eq. 12-24 that we have

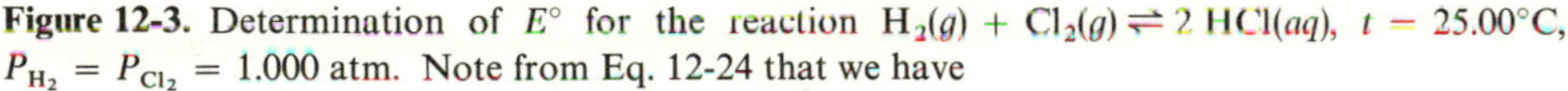

$$E' = E + 25.693\left[\frac{1}{2}\ln\left(\frac{m^4}{P_{H_2}P_{Cl_2}}\right) - 2\left(\frac{1.177m^{1/2}}{1 + m^{1/2}}\right)\right]$$

The $m = 0$ intercept on the E' axis is E°; the value obtained is $E^\circ = 1360.4$ mV at 25°C. This value of E° can be used together with Eq. 12-16 to compute the $\gamma_\pm$ values for HCl(*aq*) given in the table above.

EXAMPLE 12-2

Given that $E^\circ = 1.3604$ V at 298.15 K for the reaction

$$H_2(g) + Cl_2(g) \rightleftharpoons 2\ HCl(aq)$$

Compute ΔG°_{rxn} and the equilibrium constant K at 298.15 K for the reaction.

Solution: From Eq. 12-8 we have

$$\Delta G^\circ_{rxn} = -nFE^\circ = -(2\ \text{mol})(96{,}484.6\ \text{C}\cdot\text{mol}^{-1})(1.3604\ \text{V})$$

One volt·coulomb = 1 joule, and 1000 J = 1 kJ; thus

$$\Delta G^\circ_{rxn} = -262.52\ \text{kJ}$$

The value of K is computed using Eq. 12-11:

$$\ln K = \frac{nFE^\circ}{RT} = \frac{-\Delta G^\circ_{\text{rxn}}}{RT}$$

thus

$$\ln K = \frac{-(-262{,}520 \text{ J})}{(8.3144 \text{ J}\cdot\text{K}^{-1})(298.15 \text{ K})} = 1.0590 \times 10^2$$

and

$$K = 9.814 \times 10^{45}$$

at 25°C.

12-4 *Thermodynamic Properties of Ions in Solution Can Be Obtained from Cell Data*

For the cell reaction

$$H_2(g) + Cl_2(g) \rightleftharpoons 2\,HCl(aq) \tag{12-25}$$

we have at 25.00°C (see Example 12-2)

$$\Delta G^\circ_{\text{rxn}} = -262.52 \text{ kJ}$$

At any particular temperature, the $\Delta G^\circ_{\text{rxn}}$ value for the reaction in Eq. 12-25 is equal to

$$\Delta G^\circ_{\text{rxn},T} = 2\Delta\bar{G}^\circ_{f,T}[HCl(aq)] - \Delta\bar{G}^\circ_{f,T}[H_2(g)] - \Delta\bar{G}_{f,T}[Cl_2(g)]$$

But with our choices of standard states we have

$$\Delta\bar{G}^\circ_{f,T}[H_2(g)] = 0 = \Delta\bar{G}^\circ_{f,T}[Cl_2(g)]$$

and thus at 25°C

$$\Delta\bar{G}^\circ_f[HCl(aq), 298.15\text{ K}] = \frac{-262.52 \text{ kJ}}{2 \text{ mol}} = -131.26 \text{ kJ}\cdot\text{mol}^{-1}$$

We have treated HCl(aq) as a strong (completely dissociated) electrolyte in the analysis of the cell data; therefore, we have for the process

$$HCl(aq) \longrightarrow H^+(aq) + Cl^-(aq) \tag{12-26}$$

$$\Delta G^\circ_{\text{rxn}} = 0 \qquad \text{(at all } T\text{)}$$

and thus

$$0 = \Delta\bar{G}^\circ_{f,T}[H^+(aq)] + \Delta\bar{G}^\circ_{f,T}[Cl^-(aq)] - \Delta\bar{G}^\circ_{f,T}[HCl(aq)] \tag{12-27}$$

At 25°C,

$$0 = \Delta\bar{G}^\circ_{f,T}[H^+(aq)] + \Delta\bar{G}^\circ_{f,T}[Cl^-(aq)] - (-131.26 \text{ kJ}\cdot\text{mol}^{-1})$$

Setting $\Delta G^\circ_{\text{rxn}} = 0$ at all T for the reaction in Eq. 12-26 requires $\Delta H^\circ_{\text{rxn}} = \Delta S^\circ_{\text{rxn}} = 0$ at all T for the reaction in Eq. 12-26, because

$$\Delta G^\circ_{\text{rxn}} = \Delta H^\circ_{\text{rxn}} - T\Delta S^\circ_{\text{rxn}}$$

As we have noted previously (Chapter 11) *single-ion thermodynamic properties are not measurable;* therefore, if we are to set up a scale of $\Delta \bar{G}_f^\circ$ values for ions, it is necessary (and sufficient) to assign arbitrarily a value of $\Delta \bar{G}_f^\circ$ to one (and only one) ion in the particular solvent at each temperature. The purely arbitrary conventional choice for aqueous ions is

$$\Delta \bar{G}_f^\circ[H^+(aq)] \equiv 0 \qquad \text{(at all } T) \tag{12-28}$$

Combination of Eq. 12-28 with Eq. 12-27 yields

$$0 = 0 + \Delta \bar{G}_{f,298.15}^\circ[Cl^-(aq)] + 131.26 \text{ kJ}\cdot\text{mol}^{-1}$$

or

$$\Delta \bar{G}_f^\circ[Cl^-(aq), 298.15 \text{ K}] = -131.26 \text{ kJ}\cdot\text{mol}^{-1}$$

To obtain ΔH_{rxn}° and ΔS_{rxn}° for a cell reaction we need to know dE°/dT, as well as E°, as seen from the following: The temperature dependence of ΔG_{rxn} is given by

$$\left(\frac{\partial \Delta G_{rxn}}{\partial T}\right)_P = -\Delta S_{rxn} \tag{12-29}$$

but $\Delta G_{rxn} = -nFE$, and thus

$$\left(\frac{\partial E}{\partial T}\right)_P = \frac{\Delta S_{rxn}}{nF} \tag{12-30}$$

Note that the voltage of an electrochemical cell will be independent of temperature if, and only if, $\Delta S_{rxn} = 0$. For the reversible operation of the cell we have (from the second law and Eq. 12-30)

$$q_{rev} = T\Delta S_{rxn} = nFT\left(\frac{\partial E}{\partial T}\right)_P \tag{12-31}$$

Thus if $(\partial E/\partial T)_P < 0$, then the cell evolves energy as heat when the cell reaction takes place. With all reactants and products in their standard states, we have from Eq. 12-30

$$\frac{dE^\circ}{dT} = \frac{\Delta S_{rxn}^\circ}{nF} \tag{12-32}$$

where the subscript P has been dropped because with fixed-pressure standard states E° is independent of pressure.

The value of ΔH_{rxn} for the cell reaction is given by

$$\Delta H_{rxn} = \Delta G_{rxn} + T\Delta S_{rxn} \tag{12-33}$$

and thus, using Eqs. 12-7 and 12-30, we obtain

$$\Delta H_{rxn} = -nFE + nFT\left(\frac{\partial E}{\partial T}\right)_P$$

or

$$\Delta H_{rxn} = nF\left[T\left(\frac{\partial E}{\partial T}\right)_P - E\right] \tag{12-34}$$

Equation 12-34 is known as the *Gibbs-Helmholtz equation.* If all reactants and products are in their standard states, then the Gibbs-Helmholtz equation becomes

$$\Delta H^\circ_{\text{rxn}} = nF\left(T\frac{dE^\circ}{dT} - E^\circ\right) \tag{12-35}$$

For the reaction

$$\text{H}_2(g) + \text{Cl}_2(g) \rightleftharpoons 2\,\text{HCl}(aq)$$

at 25°C

$$\frac{dE^\circ}{dT} = -0.00125\ \text{V}\cdot\text{K}^{-1}$$

Thus at 25°C, using Eq. 12-35,

$$\Delta H^\circ_{\text{rxn}} = (2\ \text{mol})(96{,}484.6\ \text{C}\cdot\text{mol}^{-1})[298.15\ \text{K}(-0.00125\ \text{V}\cdot\text{K}^{-1}) - 1.3604\ \text{V}]$$

and

$$\Delta H^\circ_{\text{rxn}} = -334.4\ \text{kJ}$$

Further, the value of $\Delta H^\circ_{\text{rxn}}$ is equal to

$$\Delta H^\circ_{\text{rxn}} = 2\,\Delta\bar{H}^\circ_f[\text{HCl}(aq)] - \Delta\bar{H}^\circ_f[\text{H}_2(g)] - \Delta\bar{H}^\circ_f[\text{Cl}_2(g)]$$

and with our choices of standard states

$$\Delta\bar{H}^\circ_f[\text{H}_2(g)] = \Delta\bar{H}^\circ_f[\text{Cl}_2(g)] = 0$$

and thus at 25°C

$$\Delta\bar{H}^\circ_f[\text{HCl}(aq)] = \frac{-334.4\ \text{kJ}}{2\ \text{mol}} = -167.2\ \text{kJ}\cdot\text{mol}^{-1}$$

The value of $\Delta H^\circ_{\text{rxn}}$ is zero for the process

$$\text{HCl}(aq) \longrightarrow \text{H}^+(aq) + \text{Cl}^-(aq)$$

because HCl(*aq*) is a strong electrolyte $[\bar{H}^\circ_{\text{HCl}(aq)} = \bar{H}^\circ_{\text{H}^+(aq)} + \bar{H}^\circ_{\text{Cl}^-(aq)}]$; thus

$$0 = \Delta\bar{H}^\circ_f[\text{H}^+(aq)] + \Delta\bar{H}^\circ_f[\text{Cl}^-(aq)] - \Delta\bar{H}^\circ_f[\text{HCl}(aq)]$$

Our convention that establishes a numerical ionic enthalpy scale is

$$\Delta\bar{H}^\circ_f[\text{H}^+(aq)] \equiv 0 \qquad (\text{at all } T) \tag{12-36}$$

and thus we compute

$$\Delta\bar{H}^\circ_f[\text{Cl}^-(aq)] = \Delta\bar{H}^\circ_f[\text{HCl}(aq)] = -167.2\ \text{kJ}\cdot\text{mol}^{-1}$$

The standard entropy change $\Delta S^\circ_{\text{rxn}}$ for reaction in Eq. 12-25 is given by

$$\Delta S^\circ_{\text{rxn}} = nF\frac{dE^\circ}{dT} = (2\ \text{mol})(96{,}485\ \text{C}\cdot\text{mol}^{-1})(-0.00125\ \text{V}\cdot\text{K}^{-1}) = -241\ \text{J}\cdot\text{K}^{-1}$$

The value of $\Delta S^\circ_{\text{rxn}}$ is also given by

$$\Delta S^\circ_{\text{rxn}} = 2\bar{S}^\circ[\text{H}^+(aq)] + 2\bar{S}^\circ[\text{Cl}^-(aq)] - \bar{S}^\circ[\text{H}_2(g)] - \bar{S}^\circ[\text{Cl}_2(g)] = -241\ \text{J}\cdot\text{K}^{-1}$$

where $\bar{S}^\circ$ for $\text{H}_2(g)$ and $\text{Cl}_2(g)$ can be obtained from the third-law data. To set up a scale of standard partial molar entropies of ions in solutions we must arbitrarily pick a value for one (and only one) ion in the particular solvent at each T, as done for the $\Delta\bar{G}_f^\circ$ and $\Delta\bar{H}_f^\circ$ scales. Our convention is to set arbitrarily*

$$\bar{S}^\circ[H^+(aq)] = 0 \qquad \text{(at all } T) \tag{12-37}$$

At 25°C we have $\bar{S}^\circ = 130.57\ J\cdot K^{-1}\cdot mol^{-1}$ for $H_2(g)$ and $\bar{S}^\circ = 222.96\ J\cdot K^{-1}\cdot mol^{-1}$ for $Cl_2(g)$, and hence we compute for $\bar{S}^\circ[Cl^-(aq)]$ at 25°C

$$-241\ J\cdot K^{-1} = 2\bar{S}^\circ[Cl^-(aq)] - 130.57\ J\cdot K^{-1}\cdot mol^{-1} - 222.96\ J\cdot K^{-1}\cdot mol^{-1}$$

or

$$\bar{S}^\circ[Cl^-(aq), 298.15\ K] = 56.3\ J\cdot K^{-1}\cdot mol^{-1}$$

Proceeding in the manner outlined above, tables of $\Delta\bar{G}_f^\circ$, $\Delta\bar{H}_f^\circ$, and $\bar{S}^\circ$ can be built up for ions in solution. Of course, just as with data on neutral species, the real value of such tables is that they enable one to calculate thermodynamic equilibrium constants at various temperatures for all possible chemical reactions among the species for which the data have been tabulated.

EXAMPLE 12-3

Consider the cell

$$Pt(s)|H_2(g)|HCl(aq)|Hg_2Cl_2(s)|Hg(\ell)$$

(a) Determine the cell reaction for the cell.

(b) Given that $E^\circ = 0.2680$ V at 298.15 K for the above cell and that $\Delta\bar{G}_f^\circ = -131.26$ kJ · mol for $Cl^-(aq)$, compute $\Delta\bar{G}_f^\circ$ for $Hg_2Cl_2(s)$ at 25°C.

(c) Compute the equilibrium constant at 25°C for the reaction

$$Hg_2Cl_2(s) \rightleftharpoons Hg_2^{2+}(aq) + 2\,Cl^-(aq)$$

given that at 25°C

$$\Delta\bar{G}_f^\circ[Hg_2^{2+}(aq)] = 152.09\ kJ\cdot mol^{-1}$$
$$\Delta\bar{G}_f^\circ[Cl^-(aq)] = -131.26\ kJ\cdot mol^{-1}$$

Solution:

(a) Assuming that oxidation takes place at the left electrode we have

$$H_2(g) \rightleftharpoons 2\,H^+(aq) + 2e^- \qquad \text{oxidation}$$

The reduction half-reaction at the right electrode is

$$Hg_2Cl_2(s) + 2e^- \rightleftharpoons 2\,Hg(\ell) + 2\,Cl^-(aq) \qquad \text{reduction}$$

* Extrathermodynamic theoretical considerations lead to $\bar{S}^\circ = -23.0\ J\cdot K^{-1}\cdot mol^{-1}$ for $H^+(aq)$ at 25°C on the practical absolute entropy scale. This value is, however, not known with certainty and is subject to change, which means that if it were used to set up an $\bar{S}^\circ$ scale for ions in water, then it would be necessary to change all the ionic $\bar{S}^\circ$ data whenever $\bar{S}^\circ[H^+(aq)]$ changed. For this reason it is preferable to employ the above conventional relative scale of $\bar{S}^\circ$ values for ions.

The net reaction is the sum of the above two half-reactions:

$$H_2(g) + Hg_2Cl_2(s) \rightleftharpoons 2\,Hg(\ell) + 2\,HCl(aq)$$

(b) The value of ΔG°_{rxn} is given by $-nFE^\circ$ where $n = 2$; thus

$$\Delta G^\circ_{rxn} = (-2\text{ mol})(96{,}485\text{ C}\cdot\text{mol}^{-1})(0.2680\text{ V}) = -51{,}716\text{ J}$$

Further

$$\Delta G^\circ_{rxn} = -51{,}716\text{ J} = 2\Delta\bar{G}^\circ_f[H^+(aq)] + 2\Delta\bar{G}^\circ_f[Cl^-(aq)] + 2\Delta\bar{G}^\circ_f[Hg(\ell)] - \Delta\bar{G}^\circ_f[H_2(g)] - \Delta\bar{G}^\circ_f[Hg_2Cl_2(s)]$$

Because the first, third, and fourth terms on the right-hand side of the above equation are zero by convention, we have for $\Delta\bar{G}^\circ_f$ of $Hg_2Cl_2(s)$

$$-51{,}716\text{ J} = 2(0) + 2(-131{,}260\text{ J}) + 2(0) - (0) - \Delta\bar{G}^\circ_f[Hg_2Cl_2(s)]$$

Thus

$$\Delta\bar{G}^\circ_f[Hg_2Cl_2(s), 298.15\text{ K}] = -210.80\text{ kJ}\cdot\text{mol}^{-1}$$

(c) The value of ΔG°_{rxn} is computed from the $\Delta\bar{G}^\circ_f$ values:

$$\Delta G^\circ_{rxn} = \Delta\bar{G}^\circ_f[Hg_2^{2+}(aq)] + 2\Delta\bar{G}^\circ_f[Cl^-(aq)] - \Delta\bar{G}^\circ_f[Hg_2Cl_2(s)]$$

At 25°C we have

$$\Delta G^\circ_{rxn} = 152.09\text{ kJ} + 2(-131.260\text{ kJ}) - (-210.80\text{ kJ}) = 100.37\text{ kJ}$$

For the equilibrium constant at 25°C of the above reaction, we compute

$$\Delta G^\circ_{rxn} = 100.37 \times 10^3\text{ J} = -RT\ln K = -(8.314)(298.15)\ln K$$

Thus at 25°C

$$K = \frac{(a_{Hg_2^{2+}})(a_{Cl^-})^2}{a_{Hg_2Cl_2(s)}} = \frac{\gamma_\pm^3(m_{Hg_2^{2+}})(m_{Cl^-})^2}{a_{Hg_2Cl_2(s)}} = 2.60 \times 10^{-18}$$

Because at 1 atm total pressure Hg_2Cl_2 is in its standard state $a_{Hg_2Cl_2(s)} = 1$ and the K expression becomes

$$K = \gamma_\pm^3(m_{Hg^{2+}})(m_{Cl^-})^2 = 2.60 \times 10^{-18}$$

If the only source of $Cl^-(aq)$ and $Hg_2^{2+}(aq)$ in the solution is $Hg_2Cl_2(s)$, then we can write

$$K = 4m^3\gamma_\pm^3 = 2.60 \times 10^{-18}$$

We note from the Debye-Hückel expression for $\gamma_\pm$ (Eq. 11-64) that $\gamma_\pm$ decreases as the ionic strength increases, and hence the solubility of Hg_2Cl_2 will be greater the total ionic strength of the solution, assuming no other sources of $Hg_2^{2+}(aq)$ and $Cl^-(aq)$.

12-5 *The E° Values For Cells Can Be Separated into E° Values for Half-Reactions*

Every reaction can be separated into two half-reactions or redox couples; in these two half-reactions, electrons appear either as a product (oxidation half-reaction) or a reactant (reduction half-reaction). Any two half-reactions can be combined to yield a complete balanced reaction.

A set of n half-reactions can be combined in $n(n-1)$ possible ways, and thus a table of n half-reactions, each of which is associated with a particular E° value, can be used to calculate the E° values and thus the equilibrium constants for $n(n-1) \approx n^2$ reactions.

Single-electrode potentials (i.e., half-reaction E° values) cannot be obtained by thermodynamic methods because single-ion thermodynamic properties are not measurable. Nonetheless, we can set up a table of relative single-electrode potentials, if we choose arbitrarily a value of E° for one electrode reaction at each T in each solvent. Such a choice provides a reference point to which all other E° values for half-reactions in the same solvent are referred. The convention adopted for aqueous solutions is to set $E^\circ \equiv 0$ at all T for the $H_2(g)/H^+(aq)$ electrode reaction; that is,

$$\tfrac{1}{2}\,H_2(g) \rightleftharpoons H^+(aq) + e^- \qquad E^\circ = 0 \quad (\text{at all } T) \tag{12-38}$$

For the reaction in Eq. 12-38 we have

$$\Delta G^\circ_{\text{rxn}} = -nFE^\circ = -F(0) = 0 = \Delta\bar{G}^\circ_f[e^-] + \Delta\bar{G}^\circ_f[H^+(aq)] - \tfrac{1}{2}\Delta\bar{G}^\circ_f[H_2(g)]$$

and by convention

$$\Delta\bar{G}^\circ_f[H^+(aq)] = \Delta\bar{G}^\circ_f[H_2(g)] = 0$$

Thus we have

$$\Delta\bar{G}^\circ_f[e^-] = 0 \qquad (\text{at all } T) \tag{12-39}$$

Note that we have not unambiguously specified the standard state of e^-, in that no phase is indicated. All we require is that the standard state of e^- be the same for all half-reactions at a given T.

From the Gibbs-Helmholtz equation (Eq. 12-35), together with the convention that E° for the $H_2(g)/H^+(aq)$ couple is zero at all T, we have for the half-reaction

$$\tfrac{1}{2}\,H_2(g) \rightleftharpoons H^+(aq) + e^-$$

$$\Delta H^\circ_{\text{rxn}} = 0 = \Delta\bar{H}^\circ_f[H^+(aq)] + \Delta\bar{H}^\circ_f[e^-] - \tfrac{1}{2}\Delta\bar{H}^\circ_f[H_2(g)]$$

Because

$$\Delta\bar{H}^\circ_f[H^+(aq)] = \Delta\bar{H}^\circ_f[H_2(g)] = 0$$

we obtain

$$\Delta\bar{H}^\circ_f[e^-] = 0 \tag{12-40}$$

The value of $\Delta S^\circ_{\text{rxn}}$ for the above half-reaction is given by

$$\Delta S^\circ_{\text{rxn}} = nF\frac{dE^\circ}{dT}$$

and therefore

$$\Delta S^\circ_{\text{rxn}} = 0 = \bar{S}^\circ[\text{H}^+(aq)] + \bar{S}^\circ[e^-] - \tfrac{1}{2}\bar{S}^\circ[\text{H}_2(g)]$$

But $\bar{S}^\circ[\text{H}^+(aq)] = 0$ at all T, and thus we obtain the result that

$$\bar{S}^\circ[e^-] = \tfrac{1}{2}\bar{S}^\circ[\text{H}_2(g)] \qquad \text{(at all } T) \tag{12-41}$$

We know from the third law that $\bar{S}^\circ$ for $H_2(g)$ at any particular temperature has a well-defined value {e.g., $\bar{S}^\circ[\text{H}_2(g)] = 130.57\ \text{J}\cdot\text{K}^{-1}$ at 25°C}; therefore, Eq. (12-41) can be said to define the entropy of e^- in the standard state adopted for use in electrochemical calculations. Of course, this result is a consequence of our arbitrary conventions regarding E° for the hydrogen couple and $\Delta\bar{G}^\circ_f$, $\Delta\bar{H}^\circ_f$, and $\bar{S}^\circ$ for $H^+(aq)$ and is not to be taken as the third-law value of e^-. What we are really interested in is what we can measure; therefore, only $\Delta G^\circ_{\text{rxn}}$, $\Delta H^\circ_{\text{rxn}}$, and $\Delta S^\circ_{\text{rxn}}$ values for complete balanced reactions have experimental significance. In forming a complete balanced reaction from half-reactions, the electrons cancel, and therefore any arbitrarily chosen value of $\bar{S}^\circ(e^-)$ whatsoever will not affect our results for balanced reactions. However, we can carry out thermodynamic calculations directly on half-reactions if we take $\Delta\bar{G}^\circ_{f,T}[e^-] = 0 = \Delta\bar{H}^\circ_{f,T}[e^-]$ and $\bar{S}^\circ_T[e^-] = \frac{1}{2}\bar{S}^\circ_T[\text{H}_2(g)]$.

There remains one additional point to be considered before we proceed, and it concerns the sign of E° for half-reactions. This problem arises because all we can measure is the algebraic sum of two half-reaction E° values; that is, all we can measure is E° for the whole cell. Suppose we choose to tabulate our half-reactions with electrons on the left-hand side (in which case we refer to them as *reduction potentials*); then, because our cell current convention is equivalent to the assumption that oxidation takes place at the left electrode, we have

$$E^\circ_{\text{cell}} = E^\circ_{\text{right}} - E^\circ_{\text{left}} = E^\circ_r - E^\circ_l \tag{12-42}$$

where r stands for the right electrode and l stands for the left electrode. For the cell

$$\text{Pt}(s)|\text{H}_2(g)|\text{HCl}(aq)|\text{Cl}_2(g)|\text{Pt}(s)$$

at 25°C, $E^\circ_{\text{cell}} = 1.3595$ V. For the half-reaction

$$\text{H}^+(aq) + e^- \rightleftharpoons \tfrac{1}{2}\text{H}_2(g) \qquad E^\circ_l = 0$$

and we compute the E°_r value of the half-reaction

$$\tfrac{1}{2}\text{Cl}_2(g) + e^- \rightleftharpoons \text{Cl}^-(aq)$$

from the E°_{cell} value

$$E^\circ_{\text{cell}} = E^\circ_r - E^\circ_l$$

and

$$E^\circ_r = E^\circ_l + E^\circ_{\text{cell}} = 0 + 1.3604\ \text{V} = +1.3604\ \text{V}$$

Thus we have at 25°C for the standard reduction potential of the half-reaction

$$\tfrac{1}{2}\,Cl_2(g) + e^- \rightleftharpoons Cl^-(aq)$$

the value

$$E^\circ = +1.3604 \text{ V}$$

Note that for $n\,Cl^-(aq) = (n/2)\,Cl_2(g) + ne^-$, $E^\circ = +1.3604$ (25°C), because multiplying through a whole or half-reaction by a constant does not change E°. Voltage is an intensive quantity and does not depend on the number of moles consumed or produced. The value of ΔG°_{rxn}, however, is extensive and is directly proportional to the number of moles consumed: $\Delta G^\circ = -nFE^\circ$. A 2-V lead storage battery puts out 2 V whether it weighs a pound or a ton. A 1-ton lead storage battery, however, is capable of performing 2000 times as much work as a 1-lb lead storage battery, all other factors being equal.

EXAMPLE 12-4

For the cell

$$Pt(s)|H_2(g)|HCl(aq)|Hg_2Cl_2(s)|Hg(\ell)$$

at 25°C, $E^\circ_{cell} = 0.2680$ V. Compute the standard reduction potential at 25°C of the right electrode in this cell.

Solution: By convention, reduction occurs at the right electrode; thus

$$Hg_2Cl_2(s) + 2e^- \rightleftharpoons 2\,Hg(\ell) + 2\,Cl^-(aq)$$

Using Eq. 12-42 we obtain

$$E^\circ_r = E^\circ_{cell} + E^\circ_l = 0.2680 \text{ V} + 0 = 0.2680 \text{ V}$$

for the standard reduction potential of the above half-reaction.

An abbreviated list of standard reduction potentials in water at 25°C and their temperature coefficients is presented in Table 12-1. Using reduction potentials, stronger reducing agents than $H_2(g)$ have negative E° values, whereas weaker reducing agents than $H_2(g)$ have positive E° values. The stronger the oxidizing agent in a half-reaction, the more positive the E° value. Referring to Table 12-1 we note that the data are classified into one of two groups, namely, acidic or basic solutions. The reason for the subdivision is that it is desirable to formulate our reactions in terms of the principal species that exist in the solution. Thus $OH^-(aq)$ rather than $H^+(aq)$ is used in basic solution, and we write in basic solution

$$2\,H_2O(\ell) + 2e^- \rightleftharpoons H_2(g) + 2\,OH^-(aq) \qquad E^\circ = -0.82803 \text{ V (at 298.15 K)}$$

rather than

$$2\,H^+(aq) + 2e^- \rightleftharpoons H_2(g) \qquad E^\circ = 0$$

TABLE 12-1

Standard Reduction Potentials and Their Temperature Coefficients for Various Electrodes in Water at 25°C[a,b]

Acidic solutions	$E°/V$	$(dE°/dT)/mV \cdot K^{-1}$
$\frac{3}{2} N_2(g) + H^+(aq) + e^- \rightleftharpoons HN_3(g)$	−3.40	−1.193
$Li^+(aq) + e^- \rightleftharpoons Li(s)$	−3.045	−1.57
$K^+(aq) + e^- \rightleftharpoons K(s)$	−2.925	−1.080
$Ca^{2+}(aq) + 2e^- \rightleftharpoons Ca(s)$	−2.866	−0.175
$Na^+(aq) + e^- \rightleftharpoons Na(s)$	−2.714	−0.772
$Mg^{2+}(aq) + 2e^- \rightleftharpoons Mg(s)$	−2.363	+0.103
$\frac{1}{2} H_2(g) + e^- \rightleftharpoons H^-(aq)$	−2.25	−1.57
$Pu^{3+}(aq) + 3e^- \rightleftharpoons Pu(s)$	−2.031	+0.06
$n\, H_2O + e^- \rightleftharpoons e^-(aq)$	−(1.7)	—
$Zn^{2+}(aq) + 2e^- \rightleftharpoons Zn(s)$	−0.7628	+0.091
$Fe^{2+}(aq) + 2e^- \rightleftharpoons Fe(s)$	−0.4402	+0.052
$Eu^{3+}(aq) + e^- \rightleftharpoons Eu^{2+}(aq)$	−0.429	—
$Cr^{3+}(aq) + e^- \rightleftharpoons Cr^{2+}(aq)$	−0.408	—
$Cd^{2+}(aq) + 2e^- \rightleftharpoons Cd(s)$	−0.403	−0.093
$PbSO_4(s) + 2e^- \rightleftharpoons Pb(s) + SO_4^{2-}(aq)$	−0.3588	−1.015
$Cd^{2+}(aq) + 2e^- \rightleftharpoons Cd(Hg)[\text{2-phase } 11\%]$	−0.3516	−0.250
$CO_2(g) + 2\,H^+(aq) + 2e^- \rightleftharpoons HCOOH(aq)$	−0.199	−0.936
$Pb^{2+}(aq) + 2e^- \rightleftharpoons Pb(s)$	−0.126	−0.451
$HgI_4^{2-}(aq) + 2e^- \rightleftharpoons Hg(\ell) + 4\,I^-(aq)$	−0.038	+0.04

$2D^+(\text{in } D_2O) + 2e^- \rightleftharpoons D_2(g)$	−0.0098	—
$2\,H^+(aq) + 2e^- \rightleftharpoons H_2(g)$	0	0
$HCOOH(aq) + 2\,H^+(aq) + 2e^- \rightleftharpoons HCHO(aq) + H_2O(\ell)$	0.056	—
$Cu^{2+}(aq) + e^- \rightleftharpoons Cu^+(aq)$	0.153	+0.073
$AgCl(s) + e^- \rightleftharpoons Ag(s) + Cl^-(aq)$	0.2222	−0.658
$Hg_2Cl_2(s) + 2e^- \rightleftharpoons 2\,Hg(\ell) + 2\,Cl^-(aq)$	0.2680	−0.317
$Cu^{2+}(aq) + 2e^- \rightleftharpoons Cu(s)$	0.337	+0.008
$Fe(CN)_6^{3-}(aq) + e^- \rightleftharpoons Fe(CN)_6^{4-}(aq)$	0.361	−0.058
$Hg_2SO_4(s) + 2e^- \rightleftharpoons 2\,Hg(\ell) + SO_4^{2-}(aq)$	0.6151	−0.826
$O_2(g) + 2\,H^+(aq) + 2e^- \rightleftharpoons H_2O_2(aq)$	0.6824	−1.033
$Fe^{3+}(aq) + e^- \rightleftharpoons Fe^{2+}(aq)$	0.771	+1.188
$Hg_2^{2+}(aq) + 2e^- \rightleftharpoons 2\,Hg(\ell)$	0.788	—
$Ag^+(aq) + e^- \rightleftharpoons Ag(s)$	0.7991	−1.000
$2\,Hg^{2+}(aq) + 2e^- \rightleftharpoons Hg_2^{2+}(aq)$	0.920	—
$Pd^{2+}(aq) + 2e^- \rightleftharpoons Pd(s)$	0.987	—
$O_2(g) + 4\,H^+(aq) + 4e^- \rightleftharpoons 2\,H_2O(\ell)$	1.229	−0.846
$MnO_2(s) + 4\,H^-(aq) + 2e^- \rightleftharpoons Mn^{2+}(aq) + 2\,H_2O(\ell)$	1.23	−0.661
$Cl_2(g) + 2e^- \rightleftharpoons 2\,Cl^-(aq)$	1.3604	−1.250
$PbO_2(s) + 4\,H^+(aq) + 2e^- \rightleftharpoons Pb^{2+}(aq) + 2\,H_2O(\ell)$	1.455	−0.238
$MnO_4^-(aq) + 8\,H^+(aq) + 5e^- \rightleftharpoons Mn^{2+}(aq) + 4\,H_2O(\ell)$	1.51	−0.66
$PbO_2(s) + SO_4^{2-}(aq) + 4\,H^+(aq) + 2e^- \rightleftharpoons PbSO_4(s) + 2\,H_2O(\ell)$	1.682	+0.326
$MnO_4^-(aq) - 4\,H^+(aq) + 3e^- \rightleftharpoons MnO_2(s) + 2\,H_2O(\ell)$	1.695	−0.666

TABLE 12-1 (continued)

Acidic solutions	$E°/V$	$(dE°/dT)/mV \cdot K^{-1}$
$H_2O_2(aq) + 2\,H^+(aq) + 2e^- \rightleftharpoons 2\,H_2O(\ell)$	1.776	−0.658
$XeO_3(s) + 6\,H^+(aq) + 6e^- \rightleftharpoons Xe(g) + 3\,H_2O(\ell)$	1.8	—
$Co^{3+}(aq) + e^- \rightleftharpoons Co^{2+}(aq)$	1.808	—
$S_2O_8^{2-}(aq) + 2e^- \rightleftharpoons 2\,SO_4^{2-}(aq)$	2.01	−1.26
$O_3(g) + 2\,H^+(aq) + 2e^- \rightleftharpoons O_2(g) + H_2O(\ell)$	2.07	−0.483
$OH(aq) + H^+(aq) + e^- \rightleftharpoons H_2O(\ell)$	2.85	−1.855
$F_2(g) + 2e^- \rightleftharpoons 2\,F^-(aq)$	2.87	−1.830
$H_4XeO_6(aq) + 2\,H^+(aq) + 2e^- \rightleftharpoons XeO_3(s) + 3\,H_2O(\ell)$	3.0	—
$F_2(g) + 2\,H^+(aq) + 2e^- \rightleftharpoons 2\,HF(aq)$	3.06	−0.60
Basic solutions	$E°/V$	$(dE°/dT)/mV \cdot K^{-1}$
$Ca(OH)_2(s) + 2e^- \rightleftharpoons Ca(s) + 2\,OH^-(aq)$	−3.02	−0.965
$Mg(OH)_2(s) + 2e^- \rightleftharpoons Mg(s) + 2\,OH^-(aq)$	−2.690	−0.945
$Pu(OH)_3(s) + 3e^- \rightleftharpoons Pu(s) + 3\,OH^-(aq)$	−2.42	—
$ZnS(s, \text{wurtzite}) + 2e^- \rightleftharpoons Zn(s) + S^{2-}(aq)$	−1.405	−0.85
$2\,SO_3^{2-}(aq) + 2\,H_2O + 2e^- \rightleftharpoons S_2O_4^{2-}(aq) + 4\,OH^-(aq)$	−1.12	−0.71
$Cd(CN)_4^{2-}(aq) + 2e^- \rightleftharpoons Cd(s) + 4\,CN^-(aq)$	−1.028	—
$2H_2O(\ell) + 2e^- \leftrightharpoons H_2(g) + 2\,OH^-(aq)$	−0.82803	−0.8342
$Cd(OH)_2(s) + 2e^- \rightleftharpoons Cd(s) + 2\,OH^-(aq)$	−0.809	−1.014

$HgS(s, \text{black}) + 2e^- \rightleftharpoons Hg(s) + S^{2-}(aq)$	−0.69	−0.79
$Fe(OH)_3(s) + e^- \rightleftharpoons Fe(OH)_2(s) + OH^-(aq)$	−0.56	−0.96
$O_2(g) + e^- \rightleftharpoons O_2^-(aq)$	−0.563	—
$Hg(CN)_4^{2-}(aq) + 2e^- \rightleftharpoons Hg(\ell) + 4\,CN^-(aq)$	−0.37	+0.78
$Cu(NH_3)_2^+(aq) + e^- \rightleftharpoons Cu(s) + 2\,NH_3(aq)$	−0.12	−0.78
$2\,Cu(OH)_2(s) + 2e^- \rightleftharpoons Cu_2O(s) + 2\,OH^-(aq) + H_2O(\ell)$	−0.080	−0.725
$HgO(\text{red}) + H_2O + 2e^- \rightleftharpoons Hg(\ell) + 2\,OH^-(aq)$	+0.098	−1.120
$O_2(g) + 2\,H_2O(\ell) + 4e^- \rightleftharpoons 4\,OH^-(aq)$	+0.401	−1.680
$2\,AgO(s) + H_2O(\ell) + 2e^- \rightleftharpoons Ag_2O(s) + 2\,OH^-(aq)$	+0.607	−1.117
$ClO^-(aq) + H_2O(\ell) + 2e^- \rightleftharpoons Cl^-(aq) + 2\,OH^-(aq)$	+0.891	−1.079
$O_3(g) + H_2O(\ell) + 2e^- \rightleftharpoons O_2(g) + 2\,OH^-(aq)$	+1.24	−1.318
$OH(g) + e^- \rightleftharpoons OH^-(aq)$	+2.02	−2.689

[a] Note that for the whole cell $E^\circ_{cell} = E^\circ_r - E^\circ_l$ and

$$\frac{dE^\circ_{cell}}{dT} = \frac{dE^\circ_r}{dT} - \frac{dE^\circ_l}{dT}$$

[b] Data taken primarily from A. J. de Bethune and N. A. Swendeman Loud, *Standard Aqueous Electrode Potentials and Temperature Coefficients at* 25°*C* (C. A. Hampel Pub. Co., Skokie, Ill., 1964). See this reference for a much more extended list and additional data. See also W. M. Latimer, *Oxidation Potentials* (Prentice-Hall, Englewood Cliffs, N.J., 1952), for extensive E° data.

Consider the problem of combining two half-reactions to obtain a complete balanced reaction. In the combination of half-reactions to obtain the whole reaction, the electrons cancel, and thus we can simply combine the two half-reaction E° values directly:

$$E^\circ_{\text{cell}} = E^\circ_r - E^\circ_l$$

For example, suppose that we want E° and dE°/dT at 25°C for the reaction

$$\text{(1)}\quad 5\,\text{Pb}^{2+}(aq) + 2\,\text{MnO}_4^-(aq) + 2\,\text{H}_2\text{O}(\ell) \rightleftharpoons 5\,\text{PbO}_2(s) + 2\,\text{Mn}^{2+}(aq) + 4\,\text{H}^+(aq)$$

From Table 12-1 we find at 25°C

	E°/V	(dE°/dT)/mV·K^{-1}
(2) $PbO_2(s) + 4\,H^+(aq) + 2e^- \rightleftharpoons Pb^{2+}(aq) + 2\,H_2O(\ell)$	+1.455	−0.238
(3) $MnO_4^-(aq) + 8\,H^+(aq) + 5e^- \rightleftharpoons Mn^{2+}(aq) + 4\,H_2O(\ell)$	+1.51	−0.66

Multiplying reaction 2 by 5 and reaction 3 by 2 we obtain

	E°/V	(dE°/dT)/mV·K^{-1}
(4) $5\,PbO_2(s) + 20\,H^+(aq) + 10e^- \rightleftharpoons 5\,Pb^{2+}(aq) + 10\,H_2O(\ell)$	+1.455	−0.238
(5) $2\,MnO_4^-(aq) + 16\,H^+(aq) + 10e^- \rightleftharpoons 2\,Mn^{2+}(aq) + 8\,H_2O(\ell)$	+1.51	−0.66

Subtraction of reaction 4 from 5 yields reaction 1. Thus we have for $\Delta G^\circ_{\text{rxn}}$ for reaction 1

$$\Delta G^\circ_{\text{rxn},1} = -10FE^\circ_1 = \Delta G^\circ_5 - \Delta G^\circ_4 = -10FE^\circ_5 + 10FE^\circ_4$$

Whereas $\Delta S^\circ_{\text{rxn}}$ for reaction 1 is given by

$$\Delta S^\circ_{\text{rxn},1} = 10F\left(\frac{dE^\circ_1}{dT}\right) = \Delta S^\circ_5 - \Delta S^\circ_4 = 10F\frac{dE^\circ_5}{dT} - 10F\frac{dE^\circ_4}{dT}$$

but

$$E^\circ_4 = E^\circ_2 \quad \text{and} \quad E^\circ_5 = E^\circ_3$$

because E° is independent of n. Thus at 25°C we obtain

$$E^\circ_1 = E^\circ_3 - E^\circ_2 = 1.51 - 1.455 = 0.06\ \text{V}$$

and

$$\frac{dE^\circ_1}{dT} = \frac{dE^\circ_3}{dT} - \frac{dE^\circ_2}{dT} = -0.66 + 0.238 = -0.42\ \text{mV}\cdot\text{K}^{-1}$$

If we combine two half-reactions to yield a complete reaction, then we can work directly with the E° values for the half-reactions rather than Gibbs energies, because the number of electrons in the two half-reactions is always made the same before combining them to obtain a complete reaction.

EXAMPLE 12-5

Consider the problem of calculating an E° value for a half-reaction from the E° values for two other half-reactions. Given the following data at 25°C:

	E°/V	$(dE^\circ/dT)/\mathrm{mV \cdot K^{-1}}$
(a) $Cu^{2+}(aq) + 2e^- \rightleftharpoons Cu(s)$	+0.337	+0.008
(b) $Cu^{+}(aq) + e^- \rightleftharpoons Cu(s)$	+0.521	−0.058

Compute E° and dE°/dT at 25°C for the half-reaction

(c) $Cu^{2+}(aq) + e^- \rightleftharpoons Cu^{+}(aq)$

Solution: Subtraction of electrode reaction (b) from reaction (a) yields reaction (c); thus we have for ΔG°_c

$$\Delta G^\circ_c = -n_c F E^\circ_c = \Delta G^\circ_a - \Delta G^\circ_b = -n_a F E^\circ_a + n_b F E^\circ_b$$

and for ΔS°_c we have

$$\Delta S^\circ_c = n_c F \frac{dE^\circ_c}{dT} = \Delta S^\circ_a - \Delta S^\circ_b = n_a F \frac{dE^\circ_a}{dT} - n_b F \frac{dE^\circ_b}{dT}$$

Solving for E°_c from the ΔG°_c expression yields

$$E^\circ_c = \frac{1}{n_c}(n_a E^\circ_a - n_b E^\circ_b) \tag{12-43}$$

Solving for dE°_c/dT from the ΔS°_c expression yields

$$\frac{dE^\circ_c}{dT} = \frac{1}{n_c}\left(n_a \frac{dE^\circ_a}{dT} - n_b \frac{dE^\circ_b}{dT}\right) \tag{12-44}$$

Application of Eq. 12-43 to the above case, where $n_c = n_b = 1$ and $n_a = 2$, yields for electrode reaction c at 25°C

$$E^\circ_c = 2(0.337) - (0.521) = +0.153 \text{ V}$$

Application of Eq. 12-44 to the above case yields at 25°C

$$\frac{dE^\circ_c}{dT} = 2(0.008) - (-0.058) = +0.074 \text{ mV} \cdot \text{K}^{-1}$$

EXAMPLE 12-6

Standard reduction potentials for half-reactions can be combined with E° values for complete reactions to calculate E° values for other half-reactions. Thus we have for the $H^+(aq)/H_2(g)$ electrode reaction

(a) $2\,H^+(aq) + 2e^- \rightleftharpoons H_2(g) \qquad E^\circ = 0$

and at 25°C we have for the reaction

(b) $H_2O(\ell) \rightleftharpoons H^+(aq) + OH^-(aq) \qquad K_w = 1.008 \times 10^{-14}$

Use these data to compute E° for the half-reaction

(c) $2\,H_2O(\ell) + 2e^- \rightleftharpoons H_2(g) + 2\,OH^-(aq)$

Solution: Reaction (c) is obtained by multiplying reaction (b) by 2 and adding the result to reaction (a). Thus

$$\Delta G^\circ_c = -2FE^\circ_c = \Delta G^\circ_a + 2\,\Delta G^\circ_b = -2FE^\circ_a + 2(-RT\ln K_w)$$

where we have used the relation $\Delta G^\circ_{rxn} = -RT\ln K$. Solving for E°_c we obtain

$$E^\circ_c = E^\circ_a + \frac{RT}{F}\ln K_w$$

Thus at 25°C we compute

$$E^\circ_c = 0 + 0.025693\ln(1.008 \times 10^{-14}) = -0.82803\text{ V}$$

which is the desired result.

12-6 *Fuel Cells Are Used to Obtain Electric Power Directly from the Oxidation of a Fuel*

A fuel cell is an electrochemical cell in which the substance oxidized is capable of being used as a fuel in a heat engine, that is, capable of being "burned" (e.g., hydrogen, methane, and kerosene). A fuel cell, then, is an electrochemical cell whose cell reaction is the reaction of a fuel with an oxidizer.

Consider the following electrochemical fuel cell:

$$Pt(s)|H_2(g)|NaOH(aq)|O_2(g)|Pt(s)$$

The electrode reactions of the above cell are

$$4\,OH^-(aq) + 2\,H_2(g) \rightleftharpoons 4\,H_2O(\ell) + 4e^- \qquad \text{oxidation}$$
$$O_2(g) + 2\,H_2O(\ell) + 4e^- \rightleftharpoons 4\,OH^-(aq) \qquad \text{reduction}$$

and the complete cell reaction is

$$2\,H_2(g) + O_2(g) \rightleftharpoons 2\,H_2O(\ell)$$

At 25°C, $E^\circ_{cell} = 1.229$ V and $dE_{cell}/dT = -0.846\text{ mV}\cdot\text{K}^{-1}$. Application of the Nernst equation to the fuel cell reaction yields

$$E = E^\circ - \frac{RT}{4F}\ln\left[\frac{a^2_{H_2O(\ell)}}{a^2_{H_2(g)}a_{O_2(g)}}\right]$$

At moderate gas pressures, $a_{H_2} \approx P_{H_2}$ and $a_{O_2} \approx P_{O_2}$; therefore,

$$E = E^\circ + \frac{RT}{4F}\ln P^2_{H_2}P_{O_2} - \frac{RT}{4F}[2\ln a_{H_2O(\ell)}]$$

The activity of water can be expressed in terms of the osmotic coefficient of water (Eq. 11-60):

$$\ln a_{H_2O(\ell)} = \left(\frac{-2m\phi}{55.506}\right)$$

where m is the molality of NaOH(aq) and ϕ is the osmotic coefficient of water in NaOH(aq, m) at the temperature of the cell electrolyte. Thus

$$E \approx E^\circ + \frac{RT}{4F}\left(\ln P_{H_2}^2 P_{O_2} + \frac{4m\phi}{55.506}\right)$$

From this expression we can estimate the reversible cell voltage for any set of cell conditions. Thus, at 25°C, with $P_{H_2} = P_{O_2} = 1.00$ atm and $m_{NaOH} = 1.00$, we compute ($\phi = 0.958$ from Table 11-4)

$$E = 1.229 + \frac{0.02569}{4}\left[\frac{4(1.00)(0.958)}{55.506}\right] = 1.229 \text{ V}$$

Although increasing the NaOH(aq) concentration does not affect E much, a high NaOH(aq) is advantageous if large currents are desired, because Na^+ and OH^- ions carry the current through the cell electrolyte, and a low internal cell resistance gives a low voltage drop ($V = IR$ drop) within the cell.

The principal advantage of consuming a fuel electrochemically as opposed to burning the fuel is that work can be extracted directly from the energy released without passing through a Carnot-type heat engine cycle. We can thus avoid the requirement of the discharge of energy as heat into a low-temperature reservoir. For a cell operating reversibly at constant T and P, the electrical work done *on the surroundings* ($-w_{elec}$) is

$$-w_{elec} = -\Delta G = nFE = -\Delta H + T\Delta S$$

For a heat engine operating reversibly with the high-temperature heat source produced by the combustion of the same fuel as in the fuel cell, we have

$$\frac{-w_{eng}}{q_2} = \frac{-w_{eng}}{-\Delta H_{rxn}} = \left(\frac{T - T'}{T}\right)$$

where ΔH_{rxn} is the enthalpy of combustion of the fuel. Thus

$$-w_{eng} = -\Delta H_{rxn}\left(\frac{T - T'}{T}\right)$$

For the reaction

$$2\,H_2(g) + O_2(g) \rightleftharpoons 2\,H_2O(\ell)$$

run reversibly in a fuel cell, we compute

$$-w_{elec} = 4FE = \frac{(4 \text{ mol})(96{,}500 \text{ C}\cdot\text{mol}^{-1})(1.229 \text{ V})}{(1000 \text{ J}\cdot\text{kJ}^{-1})} = 474 \text{ kJ}$$

Whereas for the reaction

$$2\,H_2(g) + O_2(g) \longrightarrow 2\,H_2O(g)$$

run in a heat engine at 700°C, with a low-temperature reservoir at 27°C, we have

$$-w_{\text{eng}} = (483 \text{ kJ})(\tfrac{673}{973}) = 334 \text{ kJ}$$

or 69% of the fuel cell value for the work done. Fuel cells are not competitive on an economic basis at the present time with heat engines because of problems arising from irreversibility of the electrode reactions. However, for remote low-power applications, such as on spaceships, fuel cells are superior to heat engines.

12-7 *Electrochemical Cells Are a Major Source of Equilibrium Constants for Chemical Reactions*

Electrochemical cells can be used to obtain equilibrium constants for all types of chemical reactions. As an example, we shall outline the electrochemical cell method for the determination of the equilibrium constant K_a for a weak-acid–dissociation reaction in water. The reaction of interest is

$$\text{HB}(aq) \rightleftharpoons \text{H}^+(aq) + \text{B}^-(aq) \tag{12-45}$$

for which

$$K_a = \frac{a_{\text{H}^+}a_{\text{B}^-}}{a_{\text{HB}}} \tag{12-46}$$

Consider the cell

$$\text{Pt}(s)|\text{H}_2(g)|\text{HB}(m_1), \text{NaB}(m_2), \text{NaCl}(m_3)|\text{AgCl}(s)|\text{Ag}(s) \tag{12-47}$$

for which the electrode reactions are

$$\tfrac{1}{2}\text{H}_2(g) \rightleftharpoons \text{H}^+(aq) + e^- \qquad \text{(oxidation, left electrode)}$$
$$e^- + \text{AgCl}(s) \rightleftharpoons \text{Ag}(s) + \text{Cl}^-(aq) \qquad \text{(reduction, right electrode)}$$

The sum of these two electrode reactions yields the net cell reaction

$$\tfrac{1}{2}\text{H}_2(g) + \text{AgCl}(s) \rightleftharpoons \text{Ag}(s) + \text{H}^+(aq) + \text{Cl}^-(aq) \tag{12-48}$$

Application of the Nernst equation to the cell reaction yields

$$E = E^\circ - \frac{RT}{F}\ln\left[\frac{(a_{\text{H}^+})(a_{\text{Cl}^-})}{a^{1/2}_{\text{H}_2(g)}}\right] \tag{12-49}$$

where we have taken $a_{\text{Ag}(s)} = a_{\text{AgCl}(s)} = 1.000$. At a fixed value of P_{H_2} we have

$$E = E^* - \frac{RT}{F}\ln(a_{\text{H}^+})(a_{\text{Cl}^-}) \tag{12-50}$$

where

$$E^* = E^\circ + \frac{RT}{2F}\ln a_{\text{H}_2} \tag{12-51}$$

From the K_a expression (Eq. 12-46) we obtain

$$a_{\text{H}^+} = \frac{K_a a_{\text{HB}}}{a_{\text{B}^-}}$$

Substitution of the expression for a_{H^+} into Eq. 12-50 yields

$$E = E^* - \frac{RT}{F}\ln K_a - \frac{RT}{F}\ln\frac{(a_{Cl^-})(a_{HB})}{a_{B^-}} \tag{12-52}$$

From the reaction stoichiometry, we can write (see the cell diagram, Eq. 12-47)

Species:	$HB(aq)$	$\rightleftharpoons$	$H^+(aq)$	$B^-(aq)$
Molality:	$m_1 - m_H$		m_H	$m_2 + m_H$

and, therefore,

$$\frac{(a_{Cl^-})(a_{HB})}{a_{B^-}} = \left[\frac{(\gamma_{Cl^-})(\gamma_{HB})}{\gamma_{B^-}}\right]\left[\frac{m_3(m_1 - m_H)}{(m_2 + m_H)}\right] \tag{12-53}$$

Substitution of Eq. 12-53 into Eq. 12-52 yields

$$E = E^* - \frac{RT}{F}\ln K_a - \frac{RT}{F}\ln\left[\frac{(\gamma_{Cl^-})(\gamma_{HB})}{\gamma_{B^-}}\right] - \frac{RT}{F}\ln\left\{\frac{m_3(m_1 - m_H)}{(m_2 + m_H)}\right\} \tag{12-54}$$

The value of m_H, which is small compared to m_1 [the stoichiometry molality of $HB(aq)$] and m_2 [the stoichiometric molality of $B^-(aq)$], can be estimated from the K_a expression by assuming that $a_i \approx m_i$; thus

$$m_H \approx K_a\left(\frac{m_1}{m_2}\right)$$

If K_a is small ($<10^{-4}$), then a rough value of K_a suffices to estimate the magnitude of the small correction term m_H. A plot of

$$E + \frac{RT}{F}\ln\left[\frac{m_3(m_1 - m_H)}{(m_2 + m_H)}\right] \quad \text{versus} \quad m(=m_1 = m_2 = m_3)$$

gives an intercept at $m = m_1 = m_2 = m_3 = 0$ of $E^* - (RT/F)\ln K_a$, from which K_a can be calculated. Note that the activity coefficient term in Eq. 12-54 is equal to

$$\frac{RT}{F}\ln\frac{\gamma_{Cl^-}\gamma_{HB}}{\gamma_{B^-}} = \frac{RT}{F}\ln\frac{\gamma^2_{\pm(NaCl)}\gamma_{HB}}{\gamma^2_{\pm(NaB)}} \tag{12-55}$$

Thus, the activity coefficient term is proportional to m as $m \to 0$, because the $I^{1/2}/(1 + I^{1/2})$ terms arising from the $\gamma_\pm$ terms (Debye-Hückel theory) in the numerator and denominator of Eq. 12-54 cancel at low m, whereas for nonelectrolytes $\ln\gamma$ is proportional to m at low m. Using cells of this type, a large number of K_a values for weak acids have been determined at various temperatures. Note also that if $B^- \equiv OH^-$, then $HB \equiv HOH$, and thus a cell like that in Eq. 12-47 can be used to determine K_a for $H_2O(\ell)$, that is, K_w.

12-8 *The pH of a Solution Can Be Determined Using an Electrochemical Cell*

The electrochemical determination of pH is essentially the evaluation of a quasi-thermodynamic quantity (pH), which was designed to be a measure of the acidity of the solution, from a measurement of the electromotive force of a suitable galvanic cell.

Consider the cell

$$\mathrm{Pt}(s)|\mathrm{H_2}(g)|\text{soln X}[\text{containing } \mathrm{H^+}(aq)]\,\vdots\,\mathrm{KCl}(satd)|\mathrm{Hg_2Cl_2}(s)|\mathrm{Hg}(\ell) \qquad (12\text{-}56)$$

↑ Liquid junction

The electrode reactions of the cell in Eq. 12-56 are

$$\tfrac{1}{2}\,\mathrm{H_2}(g) \rightleftharpoons \mathrm{H^+}(aq, \text{soln X}) + e^- \qquad \text{(oxidation)}$$

$$e^- + \tfrac{1}{2}\,\mathrm{Hg_2Cl_2}(s) \rightleftharpoons \mathrm{Hg}(\ell) + \mathrm{Cl^-}\,(aq, satd\ \mathrm{KCl}) \qquad \text{(reduction)}$$

The complete cell reaction is

$$\tfrac{1}{2}\,\mathrm{H_2}(g) + \tfrac{1}{2}\,\mathrm{Hg_2Cl_2}(s) \rightleftharpoons \mathrm{Hg}(\ell) + \mathrm{H^+}(aq, \mathrm{X}) + \mathrm{Cl^-}\,(satd) \pm \text{ion transfer} \qquad (12\text{-}57)$$

where "$\pm$ ion transfer" represents the contribution to the cell voltage arising from the transfer of ions across the liquid junction (two miscible electrolyte solutions in contact). At $P_{\mathrm{H_2}} = 1.00$ atm $\approx a_{\mathrm{H_2}(g)}$, application of the Nernst equation to the reaction in Eq. 12-57 yields

$$E = E^\circ + E_J - \frac{RT}{F}\ln a_{\mathrm{H^+}\,(\text{soln X})} - \frac{RT}{F}\ln a_{\mathrm{Cl^-}\,(\text{satd})} \qquad (12\text{-}58)$$

where E_J is the *liquid junction potential* that arises in the boundary zone between two different electrolyte solutions that are allowed to interdiffuse. Note that the product $(a_{\mathrm{H^+}})(a_{\mathrm{Cl^-}})$ is not measurable because it involves single-ion activities *in different solutions.* We obtain from Eq. 12-58

$$-\log a_{\mathrm{H^+}\,(\text{soln X})} = F\left[\frac{E - E^\circ - E_J}{2.303RT}\right] + \log a_{\mathrm{Cl^-}\,(\text{KCl satd})} \qquad (12\text{-}59)$$

Because $a_{\mathrm{Cl^-}}$ in saturated KCl(aq) is a constant at a given T, we can set up an operational scale of pH using the equation

$$\mathrm{pH}(X) - \mathrm{pH}(S) = -\log\left[\frac{a_{\mathrm{H^+}\,(X)}}{a_{\mathrm{H^+}\,(S)}}\right] \qquad (12\text{-}60)$$

where S denotes a standard reference solution of known pH = pH(S). Substitution of Eq. 12-59 twice (once for solution X and once for solution S), into Eq. 12-60 yields

$$\mathrm{pH}(X) = \mathrm{pH}(S) + \left(\frac{E_X - E_S}{2.303RT}\right)F - \left(\frac{E_{JX} - E_{JS}}{2.303RT}\right)F \qquad (12\text{-}61)$$

Meaningful pH measurements are possible if, and only if, $E_{JX} - E_{JS} = 0$, because the E_J values are not known, nor can they be estimated reliably. Using the above approach, pH measurements become possible when pH values are assigned to standard solutions. However, such a procedure does not guarantee that it is possible to calculate unambiguously the concentration of $H^+(aq)$ from a measured pH. Suppose we define $pH = -\log a_{H^+}$, where $a_{H^+} = m_{H^+} \cdot \gamma_?$. The interpretation given to $\gamma_?$ is necesarily conventional and depends on how the pH(S) values for the standards are assigned. We shall assume that the U.S. National Bureau of Standards buffer standardization procedure was used to assign pH(S) to the standard buffer solutions (see Bates, 1974).

If the composition and ionic strength of the electrolyte solution whose pH is to be determined are not much different from that of the solution used for the standardization of the instrument, then it is usually possible to measure the pH of the solution to within 0.02 pH units [assuming that the scale reading can be made to better than this precision, which is not the case for some commercial pH meters, and that $pH(S) - pH(X) < 2$ pH units]. For fairly concentrated solutions ($I > 1$), especially those involving ions with $|Z| > 2$, it is pure fancy to place a reliability of better than ± 0.1 pH unit on the measured pH no matter how expensive the measurement package, owing to the uncertainties surrounding the liquid junction potentials and the activity coefficients. We distinguish between the measurement of a voltage for the cell and the assignment of a pH to the solution. The fact that one may measure a voltage to better than 0.1 mV in no way implies that the pH has a reliability of $0.1\ \text{mV}/(59\ \text{mV} \cdot \text{pH}^{-1}) \approx 0.002$ pH units.

12-9 *An Ion-Selective Electrode Uses a Semipermeable Membrane That Ideally Is Permeable Only to the Ion of Interest*

The hydrogen electrode

$$\text{Pt}(s)|\text{H}_2(g)|\text{H}^+(aq)$$

is not suitable for routine pH measurements because the platinized platinum electrode interacts chemically with a wide variety of possible solution constituents that either are catalytically decomposed on the electrode surface or are irreversibly adsorbed on the electrode surface and thereby "poison" the the catalytic surface. In addition, a conventional hydrogen electrode requires a continuous supply of hydrogen gas and the exiting H_2 poses a serious laboratory explosion hazard.

Routine laboratory pH measurements are made with a hydrogen-ion–sensitive glass electrode. A hydrogen-ion–sensitive glass electrode consists of a silver-silver chloride internal electrode element dipping into an internal solution containing $HCl(aq) + NaCl(aq)$ enclosed by a thin-walled bulb of sodium-containing glass (Figure 12-4). The current is carried through the glass

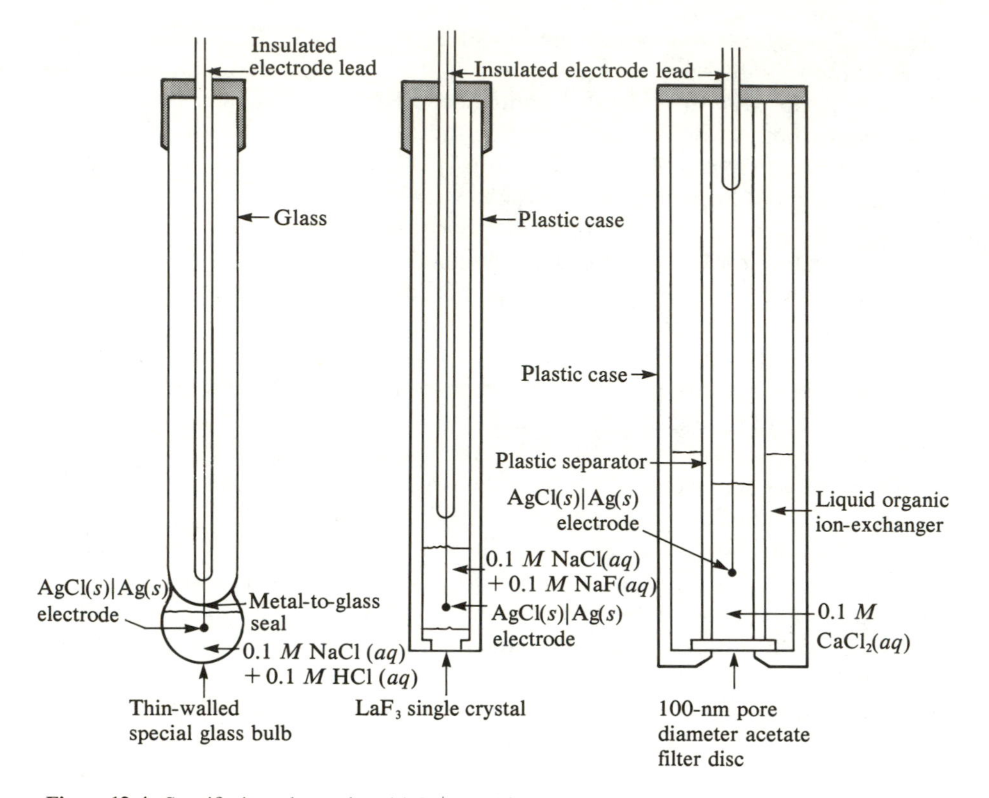

Figure 12-4. Specific-ion electrodes: (a) H^+-sensitive glass electrode; (b) solid-state F^--sensitive electrode; (c) Ca^{2+}-sensitive liquid ion-exchanger electrode.

membrane primarily by sodium ions. A cell involving a hydrogen glass electrode is

$$\overbrace{Ag(s)|AgCl(s)|HCl(aq),NaCl(aq)}^{\text{Hydrogen glass electrode}}\underset{\substack{\text{Glass}\\ \text{membrane}}}{\underset{\uparrow}{\vdots}}\text{soln containing } H^+(aq)\underset{\substack{\text{Liquid}\\ \text{junction}}}{\underset{\uparrow}{\vdots}}KCl(aq)(satd)|Hg_2Cl_2(s)|Hg(\ell) \tag{12-62}$$

The detailed analysis of the operation of the glass electrode is complicated and involves ion-exchange equilibria of the type

$$Na^+(aq) + H^+(glass) \rightleftharpoons Na^+(glass) + H^+(aq)$$

at the two glass surfaces, and also involves the movement of Na^+ ions through the dry glass layer within the glass bulb membrane. For thermodynamic purposes we can ignore these complexities and represent the glass electrode simply as H(*glass*). That is, we can represent the above cell as

$$M(s)|H(glass)|H^+(aq)\,\|\,KCl(aq,\ satd)|Hg_2Cl_2(s)|Hg(\ell)$$

where the glass electrode can be thought of as functioning like a conventional hydrogen electrode; hence

$$H(glass) \rightleftharpoons H^+(aq) + e^-$$

It is important to recognize, however, that this is merely a mnemonic device that simplifies the thermodynamic analysis. The thermodynamic analysis of the cell in Eq. 12-62 is the same as that of the cell in Eq. 12-47 with $P_{H_2} =$ 1.00 atm. The result of the analysis is the same as Eq. 12-61.

The development of the hydrogen-ion–sensitive glass electrode has made the measurement of pH a routine matter, and this fact has greatly facilitated numerous significant scientific discoveries in the chemical and biological sciences. There has been a tremendous upsurge of interest in membrane electrode technology and function in recent years, and special glass electrodes that respond to potassium- or sodium-ion activities, as well as a more complex type of liquid ion-exchange membrane electrodes, that respond, for example, to the activity of calcium ion, have been developed. A variety of solid-state electrodes that respond to the activities of ions such as $F^-(aq)$ and $S^{2-}(aq)$ have also been developed.

The basic idea involved in the development of an ion-specific electrode is to find a membrane (solid, liquid, or glass) that (ideally) is permeably only to the ion of interest. In practice, no ion-selective electrode is ever perfectly selective; if ions x and y can both pass through the membrane, then the electrode voltage usually depends on the activities of ions x and y in the following way:

$$E = E^* - \frac{RT}{z_x F}\ln(a_x + pa_y)$$

where p is the permeability ratio for ions x and y in the membrane; $p = p_y/p_x$. For a Nernstian electrode response to ion x in the presence of ion y, we require $pa_y \ll a_x$. In other words, a desirable membrane for ion x has $p \ll 1$ for all other ions.

Different types of membrane involve different types of ion transport through the membrane. Liquid ion-exchanger membranes involve the formation of a mobile complex between the ion being measured and a carrier (complexing agent). The calcium ion-selective electrode has a liquid ion exchanger consisting of an organic solvent (e.g., *n*-octanol) in which calcium complex of the composition CaX_2 is dissolved. The ion X^- is a phosphate diester anion of the type

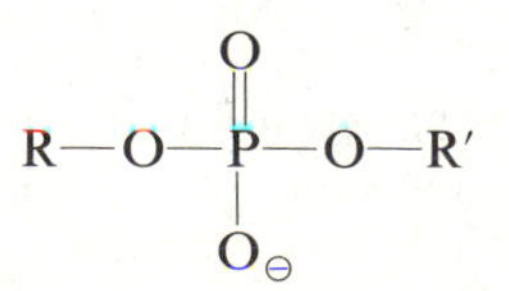

A sequence of reactions consistent with the electrode response (see Figure 12-4), where *ex* refers to the external electrolyte solution of interest, and *in* refers to the internal $CaCl_2(aq)$ solution, is

$$\begin{aligned} Ca^{2+}(ex) + 2\,X^-(\text{octanol}) &\rightleftharpoons CaX_2(\text{octanol}) \\ CaX_2(\text{octanol}) &\rightleftharpoons Ca^{2+}(in) + 2\,X^-(\text{octanol}) \\ 2\,AgCl(s) + 2e^- &\rightleftharpoons 2\,Ag(s) + 2\,Cl^-(in) \\ \hline \text{net: } Ca^{2+}(ex) + 2\,AgCl(s) + 2e^- &\rightleftharpoons Ca^{2+}(in) + 2\,Cl^-(in) + 2\,Ag(s) \end{aligned}$$

All the terms in the above net reaction are *constant for a given electrode* except for $Ca^{2+}(ex)$, which is the ion of interest in the external media. Divalent ions such as Zn^{2+} and Mg^{2+} interfere with the response of the electrode to Ca^{2+} by competing with Ca^{2+} for the X^- (octanol) carriers.

The fluoride ion-specific electrode has a single crystal of lanthanum fluoride, LaF_3, that acts as the ion-selective membrane (Figure 12-4). Lanthanum fluoride is an electrolytic (ionic) conductor. The ionic charge carrier in the crystal lattice is the fluoride ion. The operation of the electrode is consistent with the following reaction sequence:

$$\begin{aligned} F^-(LaF_3) &\rightleftharpoons F^-(ex) \\ F^-(in) &\rightleftharpoons F^-(LaF_3) \\ AgCl(s) + e^- &\rightleftharpoons Ag(s) + Cl^-(in) \\ \hline \text{net: } AgCl(s) + F^-(in) + e^- &\rightleftharpoons F^-(ex) + Ag(s) + Cl^-(in) \end{aligned}$$

All the terms in the above net reaction are constant for a given F^- electrode, except for $F^-(ex)$. The fluoride-ion–sensitive solid-state electrode is thus equivalent to an electrode with the reaction

$$\tfrac{1}{2}\,F_2(g) + e^- \rightleftharpoons F^-(aq)$$

Electrodes for measuring gases such as O_2 or CO_2 are also available. These electrodes utilize a thin Teflon membrane as the semipermeable barrier. The Teflon membrane allows the passage of small molecules like CO_2 or O_2 but not ions or large neutral molecules. Once the CO_2 or O_2 has passed through the barrier into an internal solution, its concentration is sensed *either* by measuring the concomitant change in the pH of a dilute $NaHCO_3(aq)$ solution [$CO_2 + OH^-(aq) \rightarrow HCO_3^-(aq)$] as in the CO_2 gas electrode *or* by measuring the electrode current arising from the reduction of oxygen on a rhodium coil under a constant, externally applied voltage:

$$\begin{aligned} 4e^- + O_2(aq) + 2\,H_2O(\ell) &\longrightarrow 4\,OH^-(aq) \qquad &\text{(rhodium cathode)} \\ 4\,Ag(s) + 4\,Cl^-(aq) &\longrightarrow 4\,AgCl(s) + 4e^- \qquad &\text{(silver anode)} \end{aligned}$$

The higher the concentration of oxygen, the greater the electrode current. These gas-sensitive electrodes are capable of monitoring O_2 and CO_2 directly in the gas phase as well as in solution.

12-10 *Redox Potentials Are Determined in Electrochemical Cells*

The designation *redox potential* is used to denote the voltage of an electrode half-reaction for which both the oxidized and reduced species in the electrode reaction are soluble species. For example,

$$Fe(CN)_6^{3-}(aq) + e^- \rightleftharpoons Fe(CN)_6^{4-}(aq) \tag{12-63}$$

Because both the oxidized and reduced species are soluble, the design of cells used to study reactions of this type requires special techniques involving the use of ion-selective electrodes.

A suitable cell for the study of the half-reaction in Eq. 12-63 is

$$\underbrace{Hg(\ell)|Hg_2Cl_2(s)|KCl(aq, m_3)|\text{K-glass}|}_{\text{K -ion–selective electrode}}$$

$$K_4Fe(CN)_6(aq, m_2), K_3Fe(CN)_6(aq, m_1)|Au(s) \tag{12-64}$$

where K-glass represents a K^+-selective membrane for the transfer of K^+ ions from the left-hand cell electrolyte solution to the right-hand cell electrolyte solution. The net cell reaction for the cell in Eq. 12-64 is

$$2\,Hg(\ell) + 2\,KCl(m_3) + 2\,K_3Fe(CN)_6(m_1) \rightleftharpoons 2\,K_4Fe(CN)_6(m_2) + Hg_2Cl_2(s) \tag{12-65}$$

Note that this reaction does not involve any quantities that refer to the particular glass electrode used.

Application of the Nernst equation to the reaction in Eq. 12-65 yields (at 1 atm and 25°C)

$$E = E^\circ - 0.059158 \log\left[\frac{a_{K_4Fe(CN)_6}}{a_{KCl}a_{K_3Fe(CN)_6}}\right] \tag{12-66}$$

and thus from Eq. 12-66

$$E = E^\circ - 0.059158 \log\left[\frac{(4m_2 + 3m_1)^4 m_2 \gamma_{\pm(0)}^5}{(4m_2 + 3m_1)^3 m_1 \gamma_{\pm(i)}^4 m_3^2 \gamma_{\pm(KCl)}^2}\right] \tag{12-67}$$

where we have used the relations

$$m_2 = m_{Fe(CN)_6^{4-}} \qquad m_1 = m_{Fe(CN)_6^{3-}} \qquad m_{K^+} = 4m_2 + 3m_1$$

$$\gamma_{\pm(0)}^5 = \gamma_{K^+}^4 \gamma_{Fe(CN)_6^{4-}} \qquad \gamma_{\pm(i)}^4 = \gamma_{K^+}^3 \gamma_{Fe(CN)_6^{3-}}$$

Equation 12-67 can be rearranged to

$$E + 0.059158\left\{\log\left[\frac{(4m_2 + 3m_1)m_2}{m_1 m_3^2}\right] - 2\log\gamma_{\pm(KCl)}\right\} = E^\circ - 0.059158[5\log\gamma_{\pm(0)} - 4\log\gamma_{\pm(i)}] \tag{12-68}$$

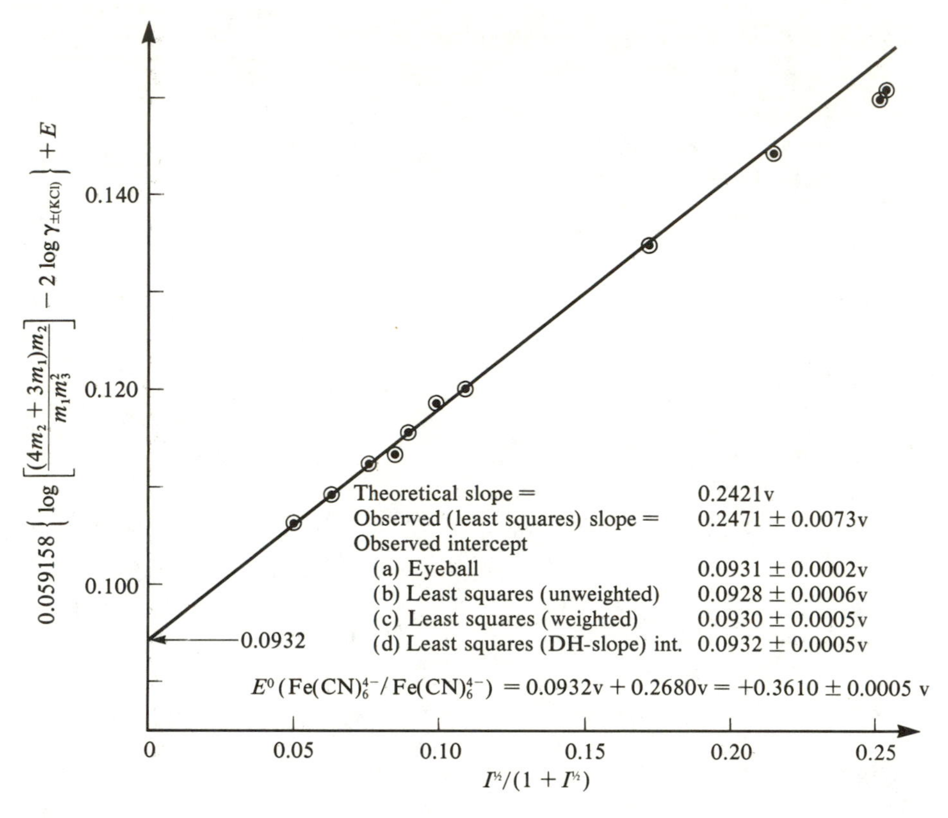

Figure 12-5. Determination of E° for the electrode

$$Fe(CN)_6^{3-}(aq) + e^- \rightleftharpoons Fe(CN)_6^{4-}(aq)$$

Hence, utilizing the available activity coefficient data for KCl(aq), a plot of the left-hand side of Eq. 12-68, call it E', versus $I^{1/2}/(1 + I^{1/2})$ can be constructed and used to obtain E° for the cell (Figure 12-5). The theoretical Debye-Hückel slope of the plot in Figure 12-5 is (see Eq. 12-68)

$$0.059158 \times 0.5115 \times [5|(-4)(+1)| - 4|(-3)(+1)|] = 0.2421 \text{ V}$$

From the intercept of Figure 12-5 we obtain for the redox potential of the half-reaction

$$Fe(CN)_6^{3-}(aq) + e^- \rightleftharpoons Fe(CN)_6^{4-}(aq)$$

a value of $E_r^\circ = E_{cell}^\circ + E_\ell^\circ = 0.0930 \text{ V} + 0.2680 \text{ V} = 0.3610 \text{ V}$ at 25°C, where 0.2680 V is the E° value for the $Hg_2Cl_2(s)|Hg(\ell)$ electrode.

We have presented a few examples that show how electrochemical cells can be used as a source of thermodynamic data for chemical reactions. Many additional examples are illustrated in the problems. Electrochemical cells are the major source of ΔG_{rxn}° values for chemical reactions.

Problems

1. Give the cell reaction for the following cells:
   (a) $Zn(s)|ZnSO_4(aq)|Hg_2SO_4(s)|Hg(\ell)$
   (b) $Pt(s)|H_2(g)|HBr(aq)|AgBr(s)|Ag(s)$
   (c) $Li(s)|LiBr$(propylene carbonate)$|TlBr(s)|Tl(s)$
   (d) $Pt(s)|D_2(g)|DCl(D_2O)|Hg_2Cl_2(s)|Hg(\ell)$

2. Give cell diagrams for cells in which the following reactions are the cell reactions:
   (a) $Pb(s) + 2\,AgCl(s) \rightleftharpoons PbCl_2(s) + 2\,Ag(s)$
   (b) $Zn(s) + CuSO_4(aq) \rightleftharpoons Cu(s) + ZnSO_4(aq)$
   (c) $AgCl(s) \rightleftharpoons Ag^+(aq) + Cl^-(aq)$
   (d) $H_2(g) + HgO(s) \rightleftharpoons Hg(\ell) + H_2O(\ell)$

3. Give the cell diagrams for electrochemical cells in which the following reactions occur:
   (a) $PbSO_4(s) + Zn(s) \rightleftharpoons ZnSO_4(aq) + Pb(s)$
   (b) $H_2(g) + I_2(s) \rightleftharpoons 2\,HI(aq)$
   (c) $Cd(s) + Hg_2SO_4(s) \rightleftharpoons CdSO_4(aq) + 2\,Hg(\ell)$

4. The standard reduction potential at 25°C of the electrode reaction

$$Ag^+(aq) + e^- \rightleftharpoons Ag(s)$$

   is $E^\circ = +0.799$ V. The equilibrium constant for the reaction

$$Ag_3Co(CN)_6(s) \rightleftharpoons 3\,Ag^+(aq) + Co(CN)_6^{3-}(aq)$$

   is $K = 8.5 \times 10^{-21}$ at 25°C. Compute E° for the half-reaction

$$Ag_3Co(CN)_6(s) + 3e^- \rightleftharpoons 3\,Ag(s) + Co(CN)_6^{3-}(aq)$$

5. The cell

$$Hg(\ell)\left|\begin{matrix}\text{mercurous nitrate}\\ 0.100\text{ g/L in }0.1\ M\ HNO_3(aq)\end{matrix}\right\|\begin{matrix}\text{mercurous nitrate}\\ 1.00\text{ g/L in }0.1\ M\ HNO_3(aq)\end{matrix}\left|Hg(\ell)\right.$$

   has an emf of 0.029 V at 25°C. Compute the charge on the cation in a mercurous salt. (State your assumptions.)

6. Compute E° and dE°/dT at 25°C for the half-reactions using $\Delta\bar{G}_f^\circ$ data (Appendix I):
   (a) $Ag^+(aq) + e^- \rightleftharpoons Ag(s)$
   (b) $Ag(NH_3)_2^+(aq) + e^- \rightleftharpoons Ag(s) + 2\,NH_3(aq)$

7. Explain why the following cells cannot be employed to determine $\gamma_\pm$ for $NaCl(aq)$ and $CuSO_4(aq)$, respectively.

$$Na(s)|NaCl(aq)|Hg_2Cl_2(s)|Hg(\ell)$$

$$Cu(s)|CuSO_4(aq)|PbSO_4(s)|Pb(Hg)\ (2\text{ phase})$$

8. Use the data in Table 12-1 to compute ΔG°_{rxn}, ΔH°_{rxn} and ΔS°_{rxn} for the following reaction at 25°C:

$$H_2(g) + CO_2(g) \rightleftharpoons HCOOH(aq)$$

9. Using the data in Appendix I, compute ΔG°_{rxn}, ΔH°_{rxn}, ΔS°_{rxn}, and K for the following reactions (at 25°C):

$$H_2S(g) \rightleftharpoons H_2S(aq)$$
$$H_2S(aq) \rightleftharpoons H^+(aq) + HS^-(aq)$$
$$HS^-(aq) \rightleftharpoons H^+(aq) + S^{2-}(aq)$$

Use your results to compute the m of $H_2S(aq)$ for an aqueous solution in equilibrium with $H_2S(g)$ at 0.250 atm. Also, compute the molalities of the ions $H^+(aq)$, $HS^-(aq)$, $S^{2-}(aq)$ in the above solution using Debye-Hückel theory to estimate $\gamma_\pm$ values.

10. Calculate the percentage error in ΔG°_{rxn} at 25°C that results when there is a 1% error and a 10% error, respectively, in K at 25°C. Do the calculation for K values of 1×10^{10} and 1.

11. Given the following data (25°C):

	E°	dE°/dT
$Hg^{2+}(aq) + 2e^- \rightleftharpoons Hg(\ell)$	+0.0854 V	—
$HgI_4^{2-}(aq) + 2e^- \rightleftharpoons Hg(\ell) + 4\,I^-(aq)$	−0.038 V	+0.04 mV·K^{-1}

$\bar{S}^\circ[Hg(\ell)] = 77.4\ J\cdot K^{-1}\cdot mol^{-1}$; $\bar{S}^\circ[Hg^{2+}(aq)] = -22.6\ J\cdot K^{-1}\cdot mol^{-1}$; and $\bar{S}^\circ[H_2(g)] = 130.587\ J\cdot K^{-1}\cdot mol^{-1}$. Compute the equilibrium constant at 25°C for the reaction

$$Hg^{2+}(aq) + 4\,I^-(aq) \rightleftharpoons HgI_4^{2-}(aq)$$

Also, compute ΔH°_{rxn} and ΔS°_{rxn} for the reaction, and estimate K at 0°C and 50°C. Take $\bar{S}^\circ[I^-(aq)] = 109.37\ J\cdot K^{-1}\cdot mol^{-1}$(25°C), and calculate $\bar{S}^\circ$ at 25°C for $HgI_4^{2-}(aq)$.

12. Given the following standard electrode potentials for iron at 25°C:

$$Fe^{2+}(aq) + 2e^- \rightleftharpoons Fe(s) \qquad E^\circ = -0.440\ V$$
$$Fe^{3+}(aq) + e^- \rightleftharpoons Fe^{2+}(aq) \qquad E^\circ = +0.771\ V$$

Compute E° for the couple

$$Fe^{3+}(aq) + 3e^- \rightleftharpoons Fe(s)$$

Using the above data, show that there are three possible E° values for the disproportionation reaction

$$3\,Fe^{2+}(aq) \rightleftharpoons 2\,Fe^{3+}(aq) + Fe(s)$$

Also show that all three E° values lead to the same equilibrium constant, and compute its value at 25°C. Give the cell diagrams to which each of the E° values applies. Note that E° values for disproportionation reactions are not unique.

13. The following data were obtained for the cell

$$Pb(s)|PbCl_2(s)|NaCl(aq)|AgCl(s)|Ag(s)$$

T/K	P/atm	E/V
280.0	1.00	0.49364
300.0	1.00	0.49004
320.0	1.00	0.48644
300.0	1000.	0.49093
300.0	2000.	0.49182

(a) Write the cell reaction.
(b) Compute E° and ΔG°_{rxn} at 300 K for the cell reaction.
(c) Compute ΔS°_{rxn} and ΔH°_{rxn} at 300 K for the cell reaction.
(d) Compute ΔV°_{rxn} and ΔU°_{rxn} at 300 K for the cell reaction.

14. The observed voltage of the cell

$$Na(s)|NaI(H_2NCH_2CH_2NH_2)|Na(Hg, X_{Na})$$

at 25°C, $X_{Na} = 0.0176$ is 0.8453 V. Compute γ_{Na} in the amalgam relative to Na(s) at 1 atm, 25°C as the standard state for Na. Comment on the magnitude of γ_{Na}.

15. Consider the cell

$$Cd(s)|CdSO_4(m_1), K_2SO_4(m_2) \text{ in } H_2O|Hg_2SO_4(s)|Hg(\ell)$$

for which $E = 1.1647$ V at 298 K and 1 atm when $m_1 = 1.25 \times 10^{-3}$ and $m_2 = 2.50 \times 10^{-2}$.
(a) Write the cell reaction.
(b) Obtain a general expression for the cell emf.
(c) Using the above data, together with the Debye-Hückel expression for $\log \gamma_\pm \propto I^{1/2}/(1 + I^{1/2})$, compute E° and the equilibrium constant for the cell reaction at 25°C.
(d) What additional data is needed to compute ΔS°_{rxn} for the cell reaction?

16. Construct plots at 25°C of $\log \gamma_\pm$ versus $I^{1/2}$ and $I^{1/2}/(1 + I^{1/2})$ for HCl(aq), KCl(aq), $CaCl_2(aq)$, $K_3Fe(CN)_6(aq)$, and $K_4Fe(CN)_6(aq)$ using data in Table 11-4. For each case, also plot the theoretical Debye-Hückel curves for $\log \gamma_\pm \propto [(I^{1/2})/(1 + I^{1/2})]$ on the same graph. Try plotting $\log \gamma_\pm$ versus $\log m$ for the above electrolytes, and note the behavior in the region $m = 0.1$ to $m = 1.0$.

17. The voltage of the saturated Weston standard cell as a function of t (°C) at 1 atm is given by the expression (see Problem 29 for cell reaction)

$$E = 1.01845 - 4.05 \times 10^{-5}(t - 20) - 9.5 \times 10^{-7}(t - 20)^2 \text{ V}$$

Obtain expressions for ΔG_{rxn}, ΔS_{rxn}, ΔH_{rxn}, and $\Delta C_{P,rxn}$ for the cell reaction. Use your results to compute w_{elec} and q_{rev} per mole of cadmium consumed when the cell operates reversibly at 25°C and 50°C. What additional data are needed to convert the above $E = f(t)$ expression to $E^\circ = f(t)$?

18. Given the following thermodynamic data at 25°C:

Species	$\bar{S}^\circ/cal \cdot K^{-1} \cdot mol^{-1}$	$\Delta\bar{G}^\circ_f/kcal \cdot mol^{-1}$	$\Delta\bar{H}^\circ_f/kcal \cdot mol^{-1}$
$H_2(g)$	31.2	—	—
$H^+(aq)$	—	—	—
$Fe^{2+}(aq)$	−27.1	−20.30	−21.0
$Fe^{3+}(aq)$	−70.1	−2.53	−11.4
$CN^-(aq)$	28.2	39.6	36.1
$Fe(CN)_6^{4-}(aq)$	—	—	—
$Fe(CN)_6^{3-}(aq)$	64.8	—	—
$HFe(CN)_6^{3-}(aq)$	—	—	—
$Fe^{3+}(aq) + 6\,CN^-(aq) \rightleftharpoons Fe(CN)_6^{3-}(aq)$		$\Delta H^\circ_{rxn} = -70.14$ kcal	
$Fe^{2+}(aq) + 6\,CN^-(aq) \rightleftharpoons Fe(CN)_6^{4-}(aq)$		$\Delta H^\circ_{rxn} = -85.77$ kcal	
$Fe(CN)_6^{3-}(aq) + e^- \rightleftharpoons Fe(CN)_6^{4-}(aq)$		$E^\circ = +0.361$ V	
$HFe(CN)_6^{3-}(aq) \rightleftharpoons H^+(aq) + Fe(CN)_6^{4-}(aq)$		$K = 6.7 \times 10^{-5}$	$\Delta H^\circ_{rxn} = 0.0$

Complete the table and calculate K at 25°C for the reaction

$$Fe^{2+}(aq) + Fe(CN)_6^{3-}(aq) \rightleftharpoons Fe^{3+}(aq) + Fe(CN)_6^{4-}(aq)$$

19. For the cell

$$Na(s)|NaI \text{ in } C_2H_5NH_2\left|\begin{matrix}Na(Hg)\\ \sim 0.1\%\end{matrix}\right|NaCl(m = 1.022, aq)|Hg_2Cl_2(s)|Hg(\ell)$$

the observed cell voltage at 25°C and 1 atm is 3.0035 V. Take $\gamma_\pm = 0.650$ for 1.022 m NaCl(aq). (Na(Hg) represents a dilute sodium amalgam.):

(a) Write the cell reaction. Will changing the concentration of NaI in $C_2H_5NH_2$ affect the observed cell voltage? Is it necessary to know the concentration of Na in the amalgam to analyze the cell data?
(b) Compute E° for the above cell.
(c) Given that $\Delta\bar{G}^\circ_f[Hg_2Cl_2(s)] = -210.618$ kJ and $\Delta\bar{G}^\circ_f[Cl^-(aq)] = -131.26$ kJ both at 25°C, compute $\Delta\bar{G}^\circ_f$ at 25°C for NaCl(aq) and $Na^+(aq)$.

(d) Given E° (25°C) = +0.2680 V for the half-reaction

$$Hg_2Cl_2(s) + 2e^- \rightleftharpoons 2\,Hg(\ell) + 2\,Cl^-(aq)$$

Compute E° (25°C) for the half-reaction

$$Na^+(aq) + e^- \rightleftharpoons Na(s)$$

(e) Given that $dE^\circ/dT = 0.455$ mV/K at 25°C for this cell, compute ΔH°_{rxn} and ΔS°_{rxn} at 25°C for the cell reaction. Using these results together with the data (25°C)

Species	$\Delta\bar{H}^\circ_f/kJ$	$\bar{S}^\circ/J\cdot K^{-1}$
$Na(s)$	0	51.0
$Hg_2Cl_2(s)$	−264.931	195.8
$Cl^-(aq)$	−167.16	56.5
$Hg(\ell)$	0	77.4

compute $\Delta\bar{H}^\circ_f$ and $\bar{S}^\circ$ of $Na^+(aq)$ at 25°C.

20. Some reversible cells that are used as rechargeable batteries are listed below. In each case: (i) write the cell reaction; (ii) calculate the standard cell voltage and its temperature coefficient at 25°C; (iii) obtain a general expression for the cell emf, converting activities to activity coefficients and composition variables, and estimate the cell voltage at 25°C for the electrolyte compositions given (which are not necessarily those for the fully charged cell). (Parts ii and iii will require looking up auxiliary data in the library in certain cases.)

(a) The lead storage cell ($E \approx 2.0$ V):

$$Pb(s)|PbSO_4(s)|H_2SO_4(aq,\ 10\ m)|PbO_2(s),\ PbSO_4(s)|Pb(s)$$

(b) The Edison cell ($E \approx 1.3$ V):

$$\text{steel}|Fe(s)|Fe(OH)_2(s)|KOH(aq),\ LiOH(aq)(I_{tot} = 2.0)$$
$$|NiOOH(s),\ Ni(OH)_2(s)|\text{steel}$$

(c) The nickel-cadmium cell, which is the same as the Edison cell except for the replacement of $Fe(s)|Fe(OH)_2(s)$ with $Cd(s)|(Cd(OH)_2(s)$: This cell is widely used at present in all sorts of cordless rechargeable convenience items like electric toothbrushes and knives ($E \approx 1.3$ V).

(d) The silver-zinc cell ($E \approx 1.9$ V) (much less weight per ampere·hour than the lead storage cell; it is used in some space vehicles):

$$Zn(s)|K_2ZnO_2(s)|KOH(aq)\ (m = 3)|Ag_2O_2(s)|Ag(s)$$

(e) The mercury battery ($E \approx 1.35$ V): This cell has a constant voltage under discharge until exhausted; then E drops rapidly to zero (explain).

It is widely used in hearing aids, electric watches, heart pacemakers, and so forth.

$$\mathrm{Zn}(s)|\mathrm{ZnO}(s)|\mathrm{KOH}(aq)|\mathrm{HgO}(s)|\mathrm{Hg}(\ell)$$

21. Consider the cells

$$\mathrm{Cd}(s)|\mathrm{CdSO_4}(aq)|\mathrm{Hg_2SO_4}(s)|\mathrm{Hg}(\ell)|\mathrm{Hg_2SO_4}(s)|\mathrm{ZnSO_4}(aq)|\mathrm{Zn}(s)$$
$$\mathrm{Cd}(s)|\mathrm{CdCl_2}(aq)|\mathrm{Hg_2Cl_2}(s)|\mathrm{Hg}(\ell)|\mathrm{Hg_2Cl_2}(s)|\mathrm{ZnCl_2}(aq)|\mathrm{Zn}(s)$$

Write the cell reactions, and show that the anion must be included in the cell reaction if consistent E° values are to be obtained from the measured E° values. (This problem demonstrates the importance of writing complete cell reactions rather than so-called "net" reactions, which in this case would amount to neglecting the anion in the reaction.)

22. The $\mathrm{p}K_m(=-\log_{10} K_m)$ and $\mathrm{p}K_X(=-\log_{10} K_X)$ values for the ionization of H_2O and D_2O at various temperatures are

	H_2O		D_2O	
t(°C)	$\mathrm{p}K_m$	$\mathrm{p}K_X$	$\mathrm{p}K_m$	$\mathrm{p}K_X$
10	14.535	16.279	15.526	17.224
25	13.997	15.741	14.955	16.653
40	13.542	15.279	14.468	16.166

Compute $\Delta H^\circ_{\mathrm{rxn}}$ and $\Delta S^\circ_{\mathrm{rxn}}$ at 25°C on the molality and mole fraction composition scales for the reactions

$$\mathrm{H_2O}(\ell) \rightleftharpoons \mathrm{H^+(in\ H_2O)} + \mathrm{OH^-(in\ H_2O)}$$
$$\mathrm{D_2O}(\ell) \rightleftharpoons \mathrm{D^+(in\ D_2O)} + \mathrm{OD^-(in\ D_2O)}$$

Suggest an explanation for why the reaction

$$\mathrm{D_2O}(\ell) \rightleftharpoons \mathrm{D^+(in\ H_2O)} + \mathrm{OD^-(in\ H_2O)}$$

cannot be directly investigated. Suggest an explanation for why the pK values for D_2O are larger than for H_2O. (*Hint:* The heavier isotope forms the stronger bond to a given atom because its vibrational ground state lies lower in the same potential well as the lighter isotope—see Chapter 14.)

23. Devise a cell without liquid junction in which the reaction

$$\mathrm{CH_4}(g) + 2\,\mathrm{O_2}(g) \rightleftharpoons \mathrm{CO_2}(g) + 2\,\mathrm{H_2O}(\ell)$$

takes place. Use the data in Appendix I to compute E° and dE°/dT for this cell at 298 K and 500 K. Set up the general expression for the emf of your cell, and estimate E at 500 K when $P_{\mathrm{CH_4}} = 2.50$ atm and $P_{\mathrm{O_2}} =$ 1.50 atm. Take $P_{\mathrm{CO_2}} = 1.00$ atm.

24. Consider the cell

$$\mathrm{Pt}(s)|\mathrm{H_2}(g)|\mathrm{HCl}(aq)|\mathrm{AgCl}(s)|\mathrm{Ag}(s)$$

Obtain a relationship between the E° values for this cell based on the molality, molarity, and mole fraction composition scales for HCl(aq).

25. Show that the van't Hoff equation in the form $d \ln K/dT = \Delta H^\circ_{\mathrm{rxn}}/RT^2$ is not applicable to equilibrium constants involving solution species whose concentrations are expressed in molarities.

26. Show that the activity of a metal in a two-phase amalgam is independent of the relative proportions of the two phases.

27. A cell suitable for the determination of E° for the redox couple

$$\mathrm{Co(en)_3^{3+}}(aq) + e^- \rightleftharpoons \mathrm{Co(en)_3^{2+}}(aq)$$

where en $\equiv$ $\mathrm{H_2NCH_2CH_2NH_2}$ is

$$\mathrm{Au}(s)|\mathrm{Co(en)_3Cl_3}(m_1),\ \mathrm{Co(en)_3Cl_2}(m_2),\ \mathrm{NaCl}(m_3),\ \mathrm{en}(m)|\mathrm{Na^+}\text{-glass}|$$
$$\mathrm{NaCl}(m_4)|\mathrm{Hg_2Cl_2}(s)|\mathrm{Hg}(\ell)$$

(a) Show that the net cell reaction for the above cell is

$$\begin{aligned}&[\mathrm{Co(en)_3^{2+}}(m_2) + 2\,\mathrm{Cl^-}(3m_1 + 2m_2 + m_3)]\\ &\quad + [\mathrm{Na^+}(m_3) + \mathrm{Cl^-}(3m_1 + 2m_2 + m_3)]\\ &\quad + \tfrac{1}{2}\,\mathrm{Hg_2Cl_2}(s) \rightleftharpoons [\mathrm{Co(en)_3^{3+}}(m_1) + 3\,\mathrm{Cl^-}(3m_1 + 2m_2 + m_3)]\\ &\quad + [\mathrm{Na^+}(m_4) + \mathrm{Cl^-}(m_4)] + \mathrm{Hg}(\ell)\end{aligned}$$

(b) Apply the Nernst equation to the above cell reaction, and show that at 25°C

$$E + 0.059158\left[\log\left(\frac{m_4^2 m_1}{m_3 m_2}\right) + 2\log\gamma_{\pm(\mathrm{NaCl\ at\ }m_4)}\right] =$$
$$E^\circ - 0.059158\log\left\{\frac{\gamma^4_{\pm[\mathrm{Co(en)_3Cl_3}]}}{\gamma^2_{\pm[\mathrm{NaCl}]}\gamma^3_{\pm[\mathrm{Co(en)_3Cl_2}]}}\right\}$$

(c) Show that the right-hand side of the equation in part (b) becomes proportional to $I^{1/2}/(1 + I^{1/2})$ as $I \to 0$.

(d) Show that the theoretical Debye-Hückel slope of a plot of the left-hand side of the equation in part (b) versus I is 0.121 V.

28. Consider the cell

$$\mathrm{Cd}(s)|\mathrm{CdSO_4}(aq)|\mathrm{Cd(Hg,\ 2\text{-}phase)}$$

The two-phase cadmium amalgam involves a phase with $X_{\mathrm{Hg}} = 0.90$ (β phase) and a phase with $X_{\mathrm{Hg}} = 0.77$ (α phase). Show that the reaction of the above cell is

$$\mathrm{Cd}(s) + (Z - 1)\,\mathrm{Cd(Hg,}\ \beta) \rightleftharpoons Z\,\mathrm{Cd(Hg,}\ \alpha)$$

where

$$Z = \frac{X_{\mathrm{Hg}\beta}}{(X_{\mathrm{Hg}\beta} - X_{\mathrm{Hg}\alpha})}$$

The voltage of the above cell is (t in degrees Celsius, E in volts):

$$10^5 E = 5538 - 14.8t - 0.385t^2 + 0.0075t^3$$

Compute ΔG_{rxn}, ΔH_{rxn}, ΔS_{rxn}, and $\Delta C_{P,\mathrm{rxn}}$ for the cell reaction.

29. The cell diagrams for the unsaturated Weston standard cell is

$$\begin{array}{r|l} \mathrm{Cd(Hg)} & \mathrm{CdSO_4}(aq)|\mathrm{Hg_2SO_4}(s)|\mathrm{Hg}(\ell) \\ \text{2-phase } \alpha, \beta & \\ \sim 10\%\text{–}11\%\ \mathrm{Cd} & \end{array}$$

whereas the cell diagram for the saturated Weston standard cell (which is used by the National Bureau of Standards to maintain the volt) is

$$\begin{array}{r|l} \mathrm{Cd(Hg)} & \mathrm{CdSO_4} \cdot \frac{8}{3}\,\mathrm{H_2O}(s)|\mathrm{CdSO_4}(aq,\ satd\ soln)|\mathrm{Hg_2SO_4}(s)|\mathrm{Hg}(\ell) \\ \text{2-phase } \alpha, \beta & \\ 10\%\text{–}11\%\ \mathrm{Cd} & \end{array}$$

Show that the cell reactions of these cells are

$$Z\,\mathrm{Cd}(\alpha - \mathrm{Hg}) + \mathrm{Hg_2SO_4}(s) \rightleftharpoons \mathrm{CdSO_4}(aq) + 2\,\mathrm{Hg}(\ell) + (Z-1)\,\mathrm{Cd}(\beta - \mathrm{Hg})$$

$$Z\,\mathrm{Cd}(\alpha - \mathrm{Hg}) + \mathrm{Hg_2SO_4}(s) + y\,\mathrm{CdSO_4}\cdot A\,\mathrm{H_2O}(\text{satd soln}) \rightleftharpoons$$
$$(Z-1)\,\mathrm{Cd}(\beta - \mathrm{Hg}) + (y+1)\,\mathrm{CdSO_4}\cdot\tfrac{8}{3}\,\mathrm{H_2O}(s) + 2\,\mathrm{Hg}(\ell)$$

where

$$Z = \frac{X_{\mathrm{Hg},\beta}}{X_{\mathrm{Hg},\beta} - X_{\mathrm{Hg},\alpha}} \quad \text{and} \quad y = \frac{8}{3A - 8}$$

30. Consider the following cell:

$$\mathrm{Zn}(s)|\mathrm{ZnSO_4}\cdot\mathrm{Na_2SO_4}\cdot 4\,\mathrm{H_2O}(s)|\mathrm{ZnSO_4}\cdot A\,\mathrm{Na_2SO_4}\cdot B\,\mathrm{H_2O}(aq,\ satd\ soln)|\mathrm{Hg_2SO_4}(s)|\mathrm{Hg}(\ell)$$

where A and B vary with temperature and pressure. Write the cell reaction, and obtain an expression for the cell emf at 1 atm total pressure.

31. Devise electrochemical cells without liquid junction for the determination of the equilibrium constants for the following reactions:
(a) $\mathrm{H_2}(g) + \mathrm{HgO}(s) \rightleftharpoons \mathrm{H_2O}(\ell) + \mathrm{Hg}(\ell)$
(b) $\mathrm{D_2}(g) + 2\,\mathrm{HCl(H_2O)} \rightleftharpoons \mathrm{H_2}(g) + 2\,\mathrm{DCl(D_2O)}$
(c) $^7\mathrm{Li}(s) + {}^6\mathrm{LiBr(DMF)} \rightleftharpoons {}^6\mathrm{Li}(s) + {}^7\mathrm{LiBr(DMF)}$
where DMF denotes N,N-dimethylformamide.

32. Find the relationship between the voltages of the following two cells
(a) $\mathrm{Zn}(s)|\mathrm{ZnSO_4}(aq, m)|\mathrm{SB}|\mathrm{CuSO_4}(aq)|\mathrm{Cu}(s)$
(b) $\mathrm{Zn}(s)|\mathrm{ZnSO_4}(aq, m)|\mathrm{Hg_2SO_4}(s)|\mathrm{Hg}(\ell)|\mathrm{Hg_2SO_4}(s)|\mathrm{CuSO_4}(aq)|\mathrm{Cu}(s)$
where SB denotes a salt bridge.

33. Outline a set of experiments for the determination of $\Delta \bar{G}_f^\circ$, $\Delta \bar{H}_f^\circ$ and $\bar{S}^\circ$ at 25°C for the ion $Co(NH_3)_6^{3+}(aq)$.

34. Design an electrochemical cell for the determination of ΔG_{rxn}° for the reaction

$$H_2(g) + Cl_2(g) \rightleftharpoons 2HCl(g)$$

35. Batteries are rated by energy density. For example, the lead storage battery has a theoretical energy density rating of 175 watt·hour·kg^{-1}. Batteries with theoretical energy densities higher than 3500 W·h·kg^{-1} have been investigated. Suggest possible anodes and cathodes for high-energy-density battery systems. After formulating your suggestions see "Advanced Electrochemical Energy Systems" by L. R. McCoy in *Special Topics in Electrochemistry*, P. A. Rock, Ed., Elsevier Scientific Publishing Co., Amsterdam, 1977.

36. Derive relationships between the E° values on the molality, molarity, and mole fraction concentration scales [for $HCl(aq)$] for the reaction

$$H_2(g) + 2\,AgCl(s) \rightleftharpoons 2\,Ag(s) + 2\,HCl(aq)$$

Reverse osmosis desalination system. One RO unit produces 2271 cubic meters (600,000 U.S. gallons) per day of high-quality drinking water from deep brackish water wells. The net driving pressure is 400 psi. The solution temperature is 25°C, and the recovery rate is 10% at a maximum flow rate of 60 gallons per minute through an element.

13

Phase Equilibria Involving Solutions

Now we may without violence to the general laws of gases which are embodied in our equations suppose other gases to exist than such as actually do exist, and there does not appear to be any limit to the resemblance which there might be between two such kinds of gas. But the increase of entropy due to the mixing of given volumes of the gases at a given temperature and pressure would be independent of the degree of similarity or dissimilarity between them. We might also imagine the case of two gases which should be absolutely identical in all the properties (sensible and molecular) which come into play while they exist as gases either pure or mixed with each other, but which should differ in respect to the attractions between their atoms and the atoms of some other substances, and therefore in their tendency to combine with such substances. In the mixture of such gases by diffusion an increase of entropy would take place, although the process of mixture, dynamically considered, might be absolutely identical in its minutest details (even with respect to the precise path of each atom) with processes which might take place without any increase of entropy. In such respects, entropy stands strongly contrasted with energy. Again, when such gases have been mixed, there is no more impossibility of the separation of the two kinds of molecules in virtue of their ordinary motions in the gaseous mass without any especial external influence, than there is of the separation of a homogeneous gas into the same two parts into which it has once been divided, after these have once been mixed. In other words, the impossibility of an uncompensated decrease of entropy seems to be reduced to improbability.

The Scientific Papers of J. Willard Gibbs, Vol. I. New York: Dover Publications.

13-1 $\Delta G_{mix} = \sum_i n_i(\mu_i - \mu_i')$

The value of the Gibbs energy change on isothermally mixing two or more components to form a solution ΔG_{mix} depends on the change in the chemical potentials of each component on going from the initial (before mixing) to the final (after mixing) state. The total Gibbs energy change on isothermally mixing $n_1, n_2, \ldots, n_c$ moles of the c components is given by

$$\Delta G_{mix} = \sum_{i=1}^{c} n_i(\mu_i - \mu_i') \tag{13-1}$$

where μ_i is the chemical potential of i in the final state and μ_i' is the chemical potential of i in the initial state. Using the relation

$$\mu_i = \mu_i^\circ + RT \ln a_i$$

we obtain

$$\Delta G_{mix} = \sum_{i=1}^{c} n_i(\mu_i^\circ - \mu_i^{\circ\prime}) + RT \sum_{i=1}^{c} n_i \ln \frac{a_i}{a_i'} \tag{13-2}$$

If we choose the same standard states for the various components in the final and initial states, then $\sum_i n_i(\mu_i^\circ - \mu_i^{\circ\prime}) = 0$, and in such a case Eq. 13-2 becomes

$$\Delta G_{mix} = RT \sum_{i=1}^{c} n_i \ln \frac{a_i}{a_i'} \tag{13-3}$$

If the components in the initial state are the pure components, if we choose the pure components at 1 atm as standard states, and if the mixing is carried out at fixed total pressure, then

$$\frac{a_i}{a_i'} = \frac{a_i}{1} = X_i \gamma_i$$

Thus from Eq. 13-3 we obtain

$$\Delta G_{mix} = RT \sum_{i=1}^{c} n_i \ln X_i \gamma_i \tag{13-4}$$

where γ_i is the activity coefficient of i in the solution at T, P, X_i.

We can obtain expressions for ΔS_{mix}, ΔH_{mix}, ΔV_{mix}, ΔU_{mix}, and $\Delta C_{P,mix}$ from Eq. 13-4. The expression for ΔS_{mix} is obtained as follows. Because

$$\Delta S_{mix} = -\left(\frac{\partial \Delta G_{mix}}{\partial T}\right)_{P,comp} = \sum_{i=1}^{c} n_i(\bar{S}_i - \bar{S}_i') \tag{13-5}$$

we have from Eq. 13-4

$$\Delta S_{mix} = -R \sum_{i=1}^{c} n_i \ln X_i \gamma_i - RT \sum_{i=1}^{c} n_i \left(\frac{\partial \ln \gamma_i}{\partial T}\right)_{P,comp} \tag{13-6}$$

The value of ΔH_{mix} is given by

$$\Delta H_{mix} = \Delta G_{mix} + T \Delta S_{mix} = \sum_{i=1}^{c} n_i(\bar{H}_i - \bar{H}_i') \tag{13-7}$$

Substitution of Eqs. 13-4 and 13-6 into Eq. 13-7 yields for ΔH_{mix}

$$\Delta H_{\text{mix}} = -RT^2 \sum_{i=1}^{c} n_i \left(\frac{\partial \ln \gamma_i}{\partial T}\right)_{P,\text{comp}} \tag{13-8}$$

An expression for ΔV_{mix} is obtained from the relation

$$\Delta V_{\text{mix}} = \left(\frac{\partial \Delta G_{\text{mix}}}{\partial P}\right)_{T,\text{comp}} = \sum_{i=1}^{c} n_i(\bar{V}_i - \bar{V}_i') \tag{13-9}$$

Using Eq. 13-4 we obtain

$$\Delta V_{\text{mix}} = RT \sum_{i=1}^{c} n_i \left(\frac{\partial \ln \gamma_i}{\partial P}\right)_{T,\text{comp}} \tag{13-10}$$

The change in internal energy on mixing at fixed T and P is given by

$$\Delta U_{\text{mix}} = \Delta H_{\text{mix}} - P\,\Delta V_{\text{mix}} = \sum_{i=1}^{c} n_i(\bar{U}_i - \bar{U}_i')$$

and thus

$$\Delta U_{\text{mix}} = -RT^2 \sum_{i=1}^{c} n_i \left(\frac{\partial \ln \gamma_i}{\partial T}\right)_{P,\text{comp}} - PRT \sum_{i=1}^{c} n_i \left(\frac{\partial \ln \gamma_i}{\partial P}\right)_{T,\text{comp}} \tag{13-11}$$

The value of $\Delta C_{P,\text{mix}}$ is given by

$$\Delta C_{P,\text{mix}} = \left(\frac{\partial \Delta H_{\text{mix}}}{\partial T}\right)_{P,\text{comp}} = \sum_{i=1}^{c} n_i(\bar{C}_{P_i} - \bar{C}_{P_i}')$$

Thus from Eq. 13-8 we obtain ∂T^2

$$\Delta C_{P,\text{mix}} = -2RT \sum_{i=1}^{c} n_i \left(\frac{\partial \ln \gamma_i}{\partial T}\right)_{P,\text{comp}} - RT^2 \sum_{i=1}^{c} n_i \left(\frac{\partial^2 \ln \gamma_i}{\partial T^2}\right)_{P,\text{comp}}$$

or

$$\Delta C_{P,\text{mix}} = -RT \sum_{i=1}^{c} n_i \left[2\left(\frac{\partial \ln \gamma_i}{\partial T}\right)_{P,\text{comp}} + T\left(\frac{\partial^2 \ln \gamma_i}{\partial T^2}\right)_{P,\text{comp}}\right] \tag{13-12}$$

EXAMPLE 13-1

An *ideal solution* is defined as a solution for which Raoult's law $a_i(g) = X_i a^\circ_{i(g)}$ holds for all components over the entire composition range. Recall from Chapter 11 that $\gamma_i = 1$ at all T, X_i, and P for all components of an ideal solution. Obtain expressions for ΔG_{mix}, ΔS_{mix}, ΔH_{mix}, ΔV_{mix}, ΔU_{mix}, and $\Delta C_{P,\text{mix}}$ for the isothermal, 1-atm mixing of pure components to form an ideal solution.

Solution: From Eq. 13-4 with $\gamma_i = 1$, we obtain for ΔG_{mix}

$$\Delta G_{\text{mix}} = RT \sum_{i=1}^{c} n_i \ln X_i \tag{13-13}$$

From Eq. 13-6 with $\gamma_i = 1$, we obtain for ΔS_{mix}

$$\Delta S_{\text{mix}} = -R \sum_{i=1}^{c} n_i \ln X_i \tag{13-14}$$

From Eq. 13-8 we obtain

$$\Delta H_{\text{mix}} = 0 \tag{13-15}$$

From Eq. 13-10 we obtain

$$\Delta V_{\text{mix}} = 0 \tag{13-16}$$

From Eq. 13-11 we obtain

$$\Delta U_{\text{mix}} = 0 \tag{13-17}$$

From Eq. 13-12 we obtain

$$\Delta C_{P,\text{mix}} = 0 \tag{13-18}$$

Note that, because all the X_i values are necessarily less than unity, we have

$$\Delta G_{\text{mix}} < 0$$

and

$$\Delta S_{\text{mix}} > 0$$

as required by the Second Law of Thermodynamics.

EXAMPLE 13-2

Use the results in Example 13-1 to compute ΔG_{mix} and ΔS_{mix} when 2.00 mol of *m*-xylene is mixed with 1.00 mol of *p*-xylene at 300 K to form an ideal solution.

Solution: The value of ΔG_{mix} is computed using Eq. 13-13:

$$\Delta G_{\text{mix}} = RT \sum_i n_i \ln X_i$$

Thus

$$\Delta G_{\text{mix}} = (8.314\ \text{J}\cdot\text{K}^{-1}\cdot\text{mol}^{-1})(300\ \text{K})\left[(2.00\ \text{mol})\ln\left(\frac{2.00}{3.00}\right)\right.$$

$$\left. + (1.00\ \text{mol})\ln\left(\frac{1.00}{3.00}\right)\right] = -4763\ \text{J}$$

The value of ΔS_{mix} is computed from Eq. 13-14:

$$\Delta S_{\text{mix}} = -R \sum_i n_i \ln X_i$$

Thus

$$\Delta S_{mix} = -(8.314\ \mathrm{J \cdot K^{-1} \cdot mol^{-1}})\left[(2.00\ \mathrm{mol})\ln\left(\frac{2.00}{3.00}\right) + (1.00\ \mathrm{mol})\ln\left(\frac{1.00}{3.00}\right)\right]$$

and

$$\Delta S_{mix} = +15.9\ \mathrm{J \cdot K^{-1}}$$

Note that because $\Delta H_{mix} = 0$, $\Delta S_{mix} = -\Delta G_{mix}/T$. Note also that 4.76 kJ is the minimum amount of work necessary at 300 K to separate an ideal solution composed of 2.00 mol of *m*-xylene and 1.00 mol of *p*-xylene into the pure components.

If two or more substances are mixed to produce a homogeneous solution, then the entropy of the solution is higher than the sum of the entropies of the individual substances before mixing, because the particles in the solution are mixed over a large number of positions in space; thus, $\Delta S_{mix} > 0$ (which, of course, we know already from the second law). Likewise, the spontaneous expansion of a gas into a vacuum leads to an increase in the entropy of the gas, because the gas molecules in the final state have a larger configurational randomness (due to the larger volume available) than in the initial state. The spontaneous process of temperature equalization that occurs when two bodies of differing temperature are brought in contact can be visualized as the mixing of the available energy more completely over the combined system. In some cases configurational and thermal randomness factors oppose one another as, for example, with the incomplete miscibility of two liquids (benzene and water as opposed to ethanol and water). Incomplete mixing arises when the intermolecular interaction energy between unlike molecules is quite different than between like molecules; thus although complete mixing would lead to increased configurational randomness, it could be accomplished only at the expense of a decrease in the thermal randomness of the system. That is to say, some internal kinetic energy would have to be converted to internal potential energy if the particles are to be placed in energetically (potential) less favorable positions.

Classical thermodynamics can be used to predict the various thermodynamic properties of solutions once a theoretical expression for the activity coefficient γ_i has been obtained from considerations outside the realm of classical thermodynamics (see Problems 21 and 23). Solution theory is an active area of physicochemical research (both theoretical and experimental), and although the method of theoretical attack is primarily statistical thermodynamics in nature, classical thermodynamics provides a powerful tool for use in the assessment of the validity of any statistical thermodynamic theory of solutions.

13-2 *Colligative Properties of Solutions Depend Primarily on the Ratio of the Number of Solute to Solvent Particles*

Colligative properties (osmotic pressure, freezing point depression, and boiling point elevation) of *dilute* solutions in a given solvent depend only on the colligative molality of the solute. The colligative molality is the stoichiometric molality times the number of solute particles produced per formula unit of solute on dissolution. Thus, for glucose in water, $m_c = m$; whereas, for $CaCl_2$ in water, $m_c = 3m$ ($CaCl_2$ dissociates into three ions per formula unit); for benzoic acid in benzene, $m_c = m/2$ (benzoic acid dimerizes in benzene). If the solution is not dilute in a thermodynamic sense, that is, if $\gamma \neq 1$, then the magnitude of the colligative effects depend on the nature of the solute as well as on the solute concentration.

13-3 *The Activity of a Solvent Can Be Increased by an Increase in the Pressure on the Solution*

Consider the system depicted in Figure 13-1. The dissolution of a solute in a solvent lowers the chemical potential of the solvent in the solution below that of the pure solvent at the same temperature. If we were to fill the right-hand side of the apparatus in Figure 13-1 with the solution to the height 0, then the system is not in equilibrium, and pure solvent passes through the semipermeable membrane from the left (higher chemical potential) to the right (lower chemical potential) until equilibrium is attained when the column of solution reaches the height h. The spontaneous transfer of solvent through a semipermeable membrane leading to a pressure buildup is known as *osmosis*. The column of liquid rises until the hydrostatic head of the solution in the column is sufficient to increase the chemical potential of the solvent in the solution near the semipermeable membrane to that of the pure solvent.

Thus two opposing effects are involved in osmotic pressure, namely, the decrease in chemical potential of the solvent in the solution due to the presence of solute and the increase in chemical potential of the solvent in the solution due to the increase in pressure on the solution. We can take a solution of known composition and find the height h by experiment, not permitting a significant amount of dilution to take place; or we could design the apparatus with a piston that would permit us to increase the pressure on the solution without using a hydrostatic head of solution. If the solution is added to the right-hand side of the apparatus shown in Figure 13-1 to such an extent that it exceeds the height at which equilibrium prevails, then solvent will pass spontaneously from right to left (*reverse osmosis*) through the semipermeable membrane. Reverse osmosis can be used to obtain fresh water from brines (Figure 13-2).

We define the osmotic pressure π as the increase in pressure above one atmosphere that is necessary to restore equilibrium between the solvent in the solution and pure solvent at one atmosphere; that is,

$$\pi = P - 1 \tag{13-19}$$

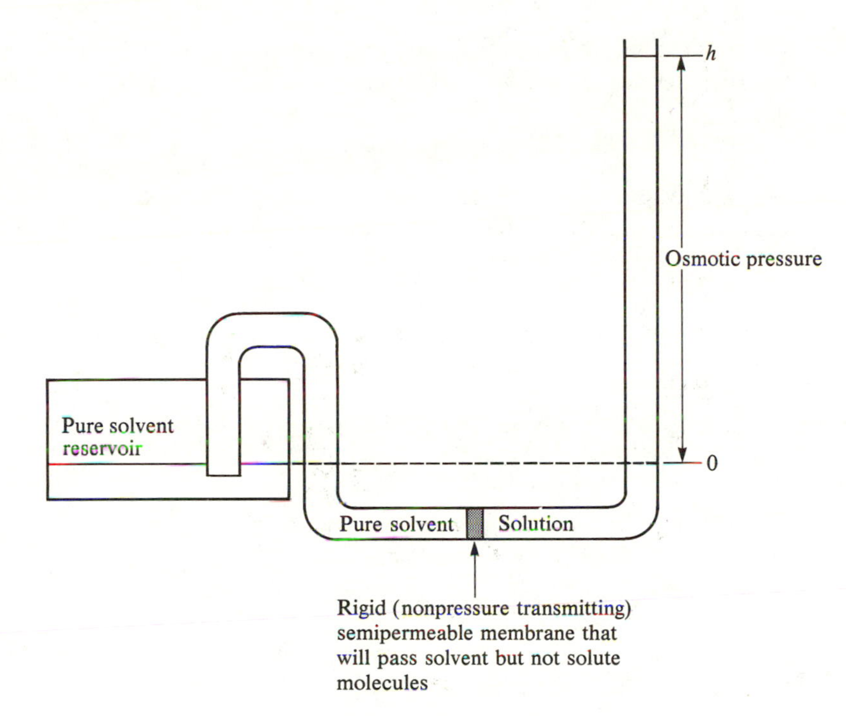

Figure 13-1. The osmotic pressure phenomenon.

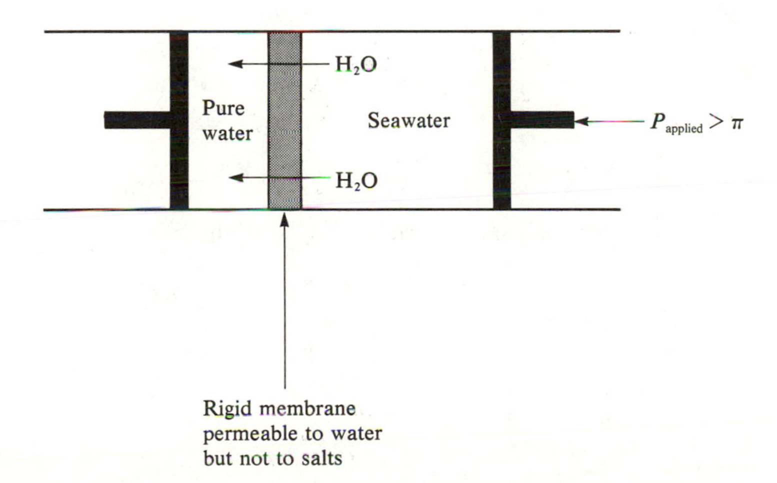

Figure 13-2. Reverse osmosis. A rigid semipermeable membrane separates pure water from seawater. A pressure in excess of the osmotic pressure of seawater (26 atm at 15°C, see Problem 3) is applied to the seawater. The applied pressure increases the activity of water in the seawater to a value in excess of that for pure water at the same temperature. Under these conditions, the net flow of water is from the seawater side through the semipermeable membrane to the pure-water side. The net effect is the production of fresh water from seawater by the application of pressure. Note that the pressure gradient is not transmitted through the rigid semipermeable membrane.

At osmotic equilibrium, the chemical potential of the solvent in the solution must equal the chemical potential of the pure solvent in the 1-atm standard state; thus

$$\mu_1^\circ(l, T, P = 1) = \mu_1(soln, T, P, X) \tag{13-20}$$

At osmotic equilibrium, the decrease in the chemical potential of the solvent arising from the dissolution of the solute, that is,

$$\mu_1(T, X, P = 1) - \mu_1^0 = RT \ln a_1(T, X, P = 1) \tag{13-21}$$

is exactly offset by the increase in the chemical potential of the solvent arising from the increase in the pressure on the solution:

$$\mu_1(T, X, P) - \mu_1(T, X, P = 1) = \int_1^P \bar{V}_1 \, dP \tag{13-22}$$

Thus from Eqs. 13-21 and 13-22 we have

$$\mu_1(T, X, P = 1) - \mu_1^\circ + \mu_1(T, X, P) - \mu_1(T, X, P = 1) = 0$$

or

$$RT \ln a_1(T, X, P = 1) + \int_1^P \bar{V}_1 \, dP = 0 \tag{13-23}$$

If we assume that $\bar{V}_1$, the partial molar volume of the solvent, is independent of pressure, then we obtain from Eq. 13-23

$$RT \ln a_1(T, X, P = 1) + \bar{V}_1(P - 1) = 0 \tag{13-24}$$

or $(\pi = P - 1)$

$$\pi = -\frac{RT}{\bar{V}_1} \ln a_1 \tag{13-25}$$

If the solution is ideal as well as incompressible, then we obtain from Eq. 13-25 with $a_1 = X_1$

$$\pi = -\frac{RT}{\bar{V}_1} \ln X_1 \tag{13-26}$$

For a solute that dissociates into v particles per formula unit,

$$X_1 = 1 - vX_2$$

If $vX_2 \ll 1$, then

$$\ln X_1 = \ln(1 - vX_2) \approx -vX_2$$

and Eq. 13-26 becomes

$$\pi \approx \frac{vRTX_2}{\bar{V}_1} \tag{13-27}$$

Equation 13-27 shows that osmotic pressure measurements can be used to obtain a value of vX_2, which is more reliable the more dilute the solution. From the value of X_2 and a known mass of solute in the solution, a value of the molecular mass M_2 of the solute can be computed.

In the following table we have presented some osmotic pressure data on aqueous sucrose solutions at 25°C.

m	1.651	2.373	3.273	4.120
π (atm)	43.84	67.68	100.43	134.71

From these data it is evident that the osmotic pressure is a large effect indeed, and for this reason it is often employed for molecular weight determinations on dilute solutions of macromolecules. By way of comparison, consider a 0.010-m solution of a nonvolatile, nondissociating solute in water at 23°C. The presence of this solute will produce a lowering of the equilibrium vapor pressure of water of (assuming Raoult's law for the purposes of estimation)

$$P_1^\circ - P_1 = P_1^\circ - X_1 P_1^\circ = X_2 P_1^\circ = \left[\frac{m}{m + (1000/18.02)}\right] P_1^\circ$$

$$= \left(\frac{0.010}{0.010 + 55.5}\right)(25.0 \text{ torr})$$

or

$$P_1^\circ - P_1 = 1.80 \times 10^{-4} \times 25.0 = 4.5 \times 10^{-3} \text{ torr}$$

The osmotic pressure for the above solution is

$$\pi \approx \frac{RT}{\bar{V}_1^\circ} X_2$$

$$\pi = \frac{(82.06 \text{ cm}^3\cdot\text{atm}\cdot\text{K}^{-1}\cdot\text{mol}^{-1})(296.2 \text{ K})(1.80 \times 10^{-4})(760 \text{ torr}\cdot\text{atm}^{-1})}{(18.02 \text{ cm}^3\cdot\text{mol}^{-1})}$$

$$= 185 \text{ torr}$$

The magnitude of the ratio of the osmotic pressure effect to the vapor pressure lowering effect in this case is

$$\frac{185}{4.5 \times 10^{-3}} = 41{,}000$$

EXAMPLE 13-3

The colligative molality in a typical biological cell is about 0.30. Estimate the osmotic pressure in a biological cell at 37°C.

Solution: The mole fraction of the solutes in the biological cell is

$$X_2 = \frac{\nu m}{(1000 \text{ g}/18.02 \text{ g}\cdot\text{mol}^{-1}) + \nu m}$$

but $\nu m = 0.30$; thus

$$X_2 = \frac{0.30}{55.81}$$

From Eq. 13-27 we have

$$\pi = \frac{RTX_2}{\bar{V}_1} = \frac{(0.0821\ \text{L}\cdot\text{atm}\cdot\text{K}^{-1}\cdot\text{mol}^{-1})(310\ \text{K})(0.30)}{(0.0180\ \text{L}\cdot\text{mol}^{-1})(55.81)} = 7.6\ \text{atm}$$

Osmotic pressure is a large effect, and this fact makes osmotic pressure an especially good method for the determination of the molecular weights of proteins. Proteins have large molecular weights and, therefore, yield a relatively small number of solute particles for a given dissolved mass. The only colligative effect sufficiently sensitive to provide useful molecular information on proteins is osmotic pressure.

EXAMPLE 13-4

A 4.00-g sample of human hemoglobin was dissolved in 100 mL of water, and the osmotic pressure at 7.00°C of the solution was found to be $\pi =$ 0.0132 atm. Estimate the molecular weight of the hemoglobin.

Solution: The concentration of the hemoglobin in the aqueous solution is

$$X_2 = \frac{n_2}{n_1 + n_2} = \frac{M_c}{55.5\ \text{mol}\cdot\text{L}^{-1} + M_c} \approx \frac{M_c}{(55.5\ \text{mol}\cdot\text{L}^{-1})}$$

Thus from Eq. 13-27

$$M_c \approx \frac{(55.5\ \text{mol}\cdot\text{L}^{-1})\pi\bar{V}_1}{RT} = \frac{(55.5\ \text{mol}\cdot\text{L}^{-1})(0.0132\ \text{atm})(0.0180\ \text{L}\cdot\text{mol}^{-1})}{(0.0821\ \text{L}\cdot\text{atm}\cdot\text{K}^{-1}\cdot\text{mol}^{-1})(280\ \text{K})}$$

$$M_c = 5.74 \times 10^{-4}\ \text{mol}\cdot\text{L}^{-1}$$

The molecular mass W_2 of the protein can be computed from the concentration of the protein, because the dissolved mass is known:

$$M_c = \frac{\text{grams of solute}/W_2}{\text{liter of solution}}$$

Thus

$$5.74 \times 10^{-4}\ \text{mol}\cdot\text{L}^{-1} = \frac{4.00\ \text{g}/W_2}{0.100\ \text{L}}$$

Solving for W_2 yields

$$W_2 = \frac{40.0\ \text{g}\cdot\text{L}^{-1}}{5.74 \times 10^{-4}\ \text{mol}\cdot\text{L}^{-1}} = 69.7 \times 10^3\ \text{g}\cdot\text{mol}^{-1}$$

Thus the molecular weight of the hemoglobin is about 69,700.

The colligative molarity of the solution inside a biological cell is approximately 0.30 *M*. Most animal cells have approximately the same internal colligative molality as the intercellular fluids (plasma) with which the cells are bathed. Plant cells differ from animal cells in that they possess a *rigid* cellulose cell wall. In either case, the cell's walls are permeable to water but not, for example, to sucrose.

Water will pass spontaneously through the cell wall from the side with the higher water activity (lower colligative molality) to the side with the lower water activity (higher colligative molality). The entry of water into an animal cell causes the cell to expand (inflation), and the exit of water from the cell causes the cell to contract (deflation). The cell assumes its normal volume when it is placed in a solution with a colligative molarity of 0.30 M. More concentrated solutions cause the cell to contract, and less concentrated solutions cause the cell to expand. When animal cells are placed in distilled water at 37°C, equilibrium would be obtained at an internal cell pressure equal to the osmotic pressure of a 0.30-M solution, that is, 7.6 atm (see Example 13-3). A pressure of 7.6 atm cannot be maintained by the animal cell walls, and the cells burst. Rigid plant cell walls can tolerate a pressure of 7.6 atm, and the entry of water into rigid-walled plant cells gives the nonwoody plants the rigidity required to stand erect. Osmotic phenomena are also involved in the uptake of groundwater by plant roots. The water is drawn from the soil into the plant fluids across water-permeable membranes.

13-4 *Nonvolatile Solutes Increase the Boiling Point of a Liquid*

Consider a two-component solution consisting of a nonvolatile solute B and a volatile solvent A, for example, a solution of sucrose in water. The equilibrium of interest is

$$A(soln, X_A, T, P_A) = A(gas, T, P_A) \qquad (13\text{-}28)$$

and the problem of interest can be stated as follows: If the normal boiling point of pure A(ℓ) is T_b°, and the chemical potential (and hence the equilibrium vapor pressure) of A(ℓ) at T_b° is reduced by the dissolution of a nonvolatile solute in A, then what is the value of the temperature of the solution that makes P_A once again equal to exactly 1 atm?

The chemical potentials of A in the solution and A in the gas phase are

$$\mu_A(soln) = \mu_A^\circ(soln) + RT \ln a_A(soln) \qquad (13\text{-}29)$$

$$\mu_A(gas) = \mu_A^\circ(gas) + RT \ln a_A(gas) \qquad (13\text{-}30)$$

At equilibrium between A(soln) and A(gas) we have

$$\mu_A(soln) = \mu_A(gas)$$

and thus

$$\mu_A^\circ(gas) - \mu_A^\circ(soln) = -RT \ln \frac{a_A(gas)}{a_A(soln)} = -RT \ln K \qquad (13\text{-}31)$$

where

$$K = \frac{a_A(gas)}{a_A(soln)}$$

is the equilibrium constant for the process in Eq. 13-28 at T, and $\mu_A^\circ(\text{gas}) - \mu_A^\circ(\text{soln})$ is the standard Gibbs energy of vaporization at T. From the van't Hoff equation, we have

$$\frac{d(\ln(a_A(gas)/a_A(soln))}{dT} = \frac{\bar{H}_A^\circ(gas) - \bar{H}_A^\circ(soln)}{RT^2} \tag{13-32}$$

Equation 13-32 can be integrated, once the temperature dependence of $\bar{H}_A^\circ(gas) - \bar{H}_A^\circ(soln)$ is known. The simplest case is $\Delta\bar{C}_{P,A}^\circ = \bar{C}_{P,A}^\circ(gas) - \bar{C}_P^\circ(soln) \approx 0$, which gives $\bar{H}_A^\circ(gas) - \bar{H}_A^\circ(soln) = \text{constant}$. Integration of Eq. 13-32 with $\bar{H}_A^\circ(gas) - \bar{H}_A^\circ(soln) = \text{constant}$ from $T = T_b^\circ$, the boiling point of pure A(ℓ), to $T = T_b$, the boiling point of the solution, yields

$$\ln\left[\frac{a_A(gas)}{a_A(soln)}\right]_{T_b} - \ln\left[\frac{a_A(gas)}{a_A(soln)}\right]_{T_b^\circ} = \frac{[\bar{H}_A^\circ(gas) - \bar{H}_A^\circ(soln)]}{R}\left(\frac{T_b - T_b^\circ}{T_bT_b^\circ}\right) \tag{13-33}$$

At T_b°, $a_A(soln) = 1$ [pure A(ℓ) at 1 atm]; further, $a_A(gas, T_b, 1 \text{ atm}) \approx a_A(gas, T_b^\circ, 1 \text{ atm})$, because $T_b \approx T_b^\circ$, and thus we obtain from Eq. 13-33

$$-\ln a_A(soln, T_b, 1 \text{ atm}) = \left[\frac{\bar{H}_A^\circ(gas) - \bar{H}_A^\circ(soln)}{R}\right]\left(\frac{T_b - T_b^\circ}{T_bT_b^\circ}\right) \tag{13-34}$$

If Raoult's law holds, then $a_A(soln) = X_A$. Furthermore, because $X_A < 1$ and $\bar{H}_A^\circ(gas) - \bar{H}_A^\circ(soln) > 0$, we note from Eq. 13-34 that the normal boiling point of a liquid that contains a solute is greater than the normal boiling point of the pure liquid. Because $\ln X_A = \ln(1 - X_B)$, we can use Eq. 13-34 to estimate the molecular weight of B from a measurement of the boiling point elevation $T_b - T_b^\circ$. We only need $\Delta\bar{H}_{\text{vap}}^\circ$ for pure A and the masses of A and B used to prepare the solution, because

$$X_B = \left[\frac{W_B/M_B}{W_B/M_B + W_A/M_A}\right]$$

Of course, it should be evident that such a procedure has several simplifying assumptions built into it and can be expected to give reliable results only when X_B is small (say, <0.01) and $T_b - T_b^\circ$ is small. Because Eq. 13-34 with $a_A = X_A$ is valid only in dilute solution [the more dilute the solution, the more nearly $a_A(soln) = X_A$], we can write

$$\ln X_A = \ln(1 - X_B) \approx -X_B \qquad (X_B \ll 1) \tag{13-35}$$

and

$$X_B = \frac{m_c}{(1000/M_A) + m_c} \approx \frac{m_cM_A}{1000} \qquad \left(\text{if } m_c \ll \frac{1000}{M_A}\right) \tag{13-36}$$

Substitution of Eqs. 13-35 and 13-36 into Eq. 13-34 yields for the boiling point elevation produced by a nonvolatile nondissociating solute of molality $m =$

TABLE 13-1

Boiling Point Elevation Constants k_b and Freezing Point Depression Constants k_f for Various Solvents

Solvent	$t_b/°C$	$k_b \Big/ \left(\frac{K \cdot kg\ solvent}{mol\ solute}\right)$	$t_f/°C$	$k_f \Big/ \left(\frac{K \cdot kg\ solvent}{mol\ solute}\right)$
Water (H_2O)	100.00	0.512	0.00	1.86
Acetic acid (CH_3COOH)	117.9	2.93	16.6	3.90
Benzene (C_6H_6)	80.0	2.53	5.50	5.10
Chloroform ($CHCl_3$)	61.2	3.63	−63.5	4.68
Cyclohexane (C_6H_{12})	80.7	2.79	6.5	20.2
Nitrobenzene ($C_6H_5NO_2$)	210.8	5.24	5.7	6.87
Camphor ($C_{10}H_{16}O$)	208.0	5.95	179.8	40.0
p-Dichlorobenzene ($C_6H_4Cl_2$)	——— sublimes ———		53.1	7.1

W_B/M_B

$$T_b - T_b^\circ \approx \left[\frac{RT_bT_b^\circ M_A}{1000\,\Delta\bar{H}^\circ_{A(vap)}}\right]m_c \approx \left[\frac{RT_b^{\circ 2} M_A}{1000\,\Delta\bar{H}^\circ_{A(vap)}}\right]m_c = k_b m_c \quad (13\text{-}37)$$

In Eq. 13-37, k_b, the *boiling point elevation constant*, is a constant characteristic of the solvent. For water, k_b has the value

$$k_b\ (\text{for } H_2O) = \frac{(1.987\ \text{cal}\cdot\text{K}^{-1}\cdot\text{mol}^{-1})(373.15\ \text{K})^2(18.02\ \text{g}\cdot\text{mol}^{-1})}{(1000\ \text{g}\cdot\text{kg}^{-1})(9730\ \text{cal}\cdot\text{mol}^{-1})}$$

$$k_b = 0.512\ \text{K}\cdot\text{kg}\cdot\text{mol}^{-1}$$

Values of k_b for several solvents are given in Table 13-1. Note that if a solute on dissolution yields ν particles per formula unit, then $X_A = 1 - \nu X_B$; thus Eq. 13-37 becomes $T_b - T_b^\circ = k_b \nu m$, where $\nu = 2$ for NaCl(*aq*) and $\nu = \frac{1}{2}$ for benzoic acid in benzene.

EXAMPLE 13-5

As a rough approximation, seawater can be regarded as an aqueous solution that is 0.50 *m* in NaCl(*aq*) and 0.05 *m* in $MgSO_4$(*aq*). Estimate the normal boiling point of seawater.

Solution: The colligative molality of seawater is

$$m_c = 2 \times 0.50 + 2 \times 0.05 = 1.10\ \text{mol}\cdot\text{kg}^{-1}$$

Assuming $a_{H_2O} \approx X_{H_2O}$ as a rough approximation, we have from Eq. 13-37

$$T_b - T_b^\circ = k_b m_c = (0.51\ \text{K}\cdot\text{kg}\cdot\text{mol}^{-1})(1.10\ \text{mol}\cdot\text{kg}^{-1}) = +0.56\ \text{K}$$

or

$$T_b = 373.15 + 0.56\ \text{K} = 373.71\ \text{K} \qquad (100.56°\text{C})$$

Boiling point elevation is a small effect, unless the solution is concentrated.

13-5 *Solutes Decrease the Freezing Point of a Liquid*

The freezing point of an aqueous solution is the temperature at which the chemical potential of water in the solution equals the chemical potential of water in ice. The origin of the depression of the freezing point of an aqueous solvent by a solute can be understood with the aid of Figure 13-3. When the temperature of an aqueous solution is lowered to the freezing point, the solid

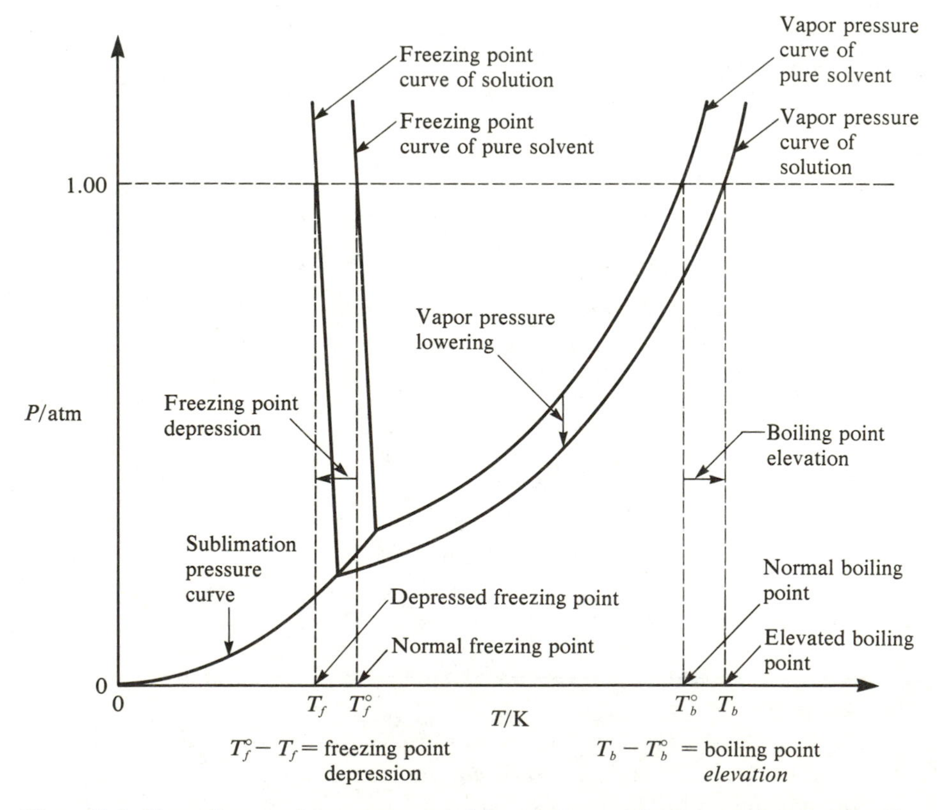

Figure 13-3. Phase diagrams for pure water and for water containing a nonvolatile solute. The presence of the solute lowers the vapor pressure of the solvent. The reduced vapor pressure of the solvent gives rise to an *increase* in the boiling point of the solution relative to the pure liquid, $T_b - T_b^\circ > 0$, and to a *decrease* in the freezing point of the solution relative to the pure solvent, $T_f^\circ - T_f > 0$.

that separates out is almost always pure ice. Icebergs are essentially pure water; impurities may be trapped in occlusions (air or liquid pockets), but the solid phase is pure. The lowering of the vapor pressure of water by the solute causes the equilibrium vapor pressure curve to intersect the equilibrium sublimation pressure curve at a lower temperature than that for the pure water. The freezing point curve begins at the intersections of the sublimation pressure and vapor pressure curves (Figure 13-3), that is, at the triple point. Therefore, the freezing point curve for the solution lies at a *lower* temperature than the freezing point curve of the pure liquid.

Consider the equilibrium between a solid A and a liquid solution containing A; that is,

$$\text{A}(s, T, P) = \text{A}(soln \text{ at } T, P, X_\text{A}) \tag{13-38}$$

For the chemical potential of solid A we have

$$\mu_\text{A}(s) = \mu_\text{A}^\circ(s) + RT \ln a_\text{A}(s) \tag{13-39}$$

whereas for the chemical potential of A in solution we have

$$\mu_\text{A}(soln) = \mu_\text{A}^\circ(soln) + RT \ln a_\text{A}(soln) \tag{13-40}$$

The equilibrium condition for the reaction in Eq. 13-38 is

$$\mu_\text{A}(s) = \mu_\text{A}(soln) \tag{13-41}$$

Substitution of Eqs. 13-39 and 13-40 into Eq. 13-41 yields

$$\mu_\text{A}^\circ(soln) - \mu_\text{A}^\circ(s) = -RT \ln\left[\frac{a_\text{A}(soln)}{a_\text{A}(s)}\right] = -RT \ln K \tag{13-42}$$

where $\mu_\text{A}^\circ(soln) - \mu_\text{A}^\circ(s)$ is the standard Gibbs energy of solution at T, and K, the equilibrium constant for the reaction in Eq. 13-38, is equal to

$$K = \frac{a_\text{A}(soln)}{a_\text{A}(s)} = a_\text{A}(soln) \tag{13-43}$$

where the second equality follows from the fact that the ice is pure and the pressure is close to 1 atm. Substitution of Eq. 13-43 into the van't Hoff equation yields

$$\frac{d \ln a_\text{A}(soln)}{dT} = \frac{\bar{H}_\text{A}^\circ(soln) - \bar{H}_\text{A}^\circ(s)}{RT^2} \tag{13-44}$$

Integration of Eq. 13-44 from pure $\text{A}(a_\text{A} = 1)$, which freezes at T_f°, to the solution with a_A that freezes at T_f, assuming that $\Delta C_{P,\text{rxn}}^\circ = 0$, yields

$$\int_{a_\text{A}=1}^{a_\text{A}} d \ln a_\text{A}(soln) = \left[\frac{\bar{H}_\text{A}^\circ(soln) - \bar{H}_\text{A}^\circ(s)}{R}\right] \int_{T_f^\circ}^{T_f} \frac{dT}{T^2}$$

$$\ln a_\text{A}(soln, T_f, P = 1) = -\left[\frac{\bar{H}_\text{A}^\circ(soln) - \bar{H}_\text{A}^\circ(s)}{R}\right]\left[\frac{T_f^\circ - T_f}{T_f T_f^\circ}\right] \tag{13-45}$$

Note that because $a_\text{A}(soln, T_f, P = 1) < 1$ and $\bar{H}_\text{A}^\circ(soln) - \bar{H}_\text{A}^\circ(s) > 0$ that $T_f^\circ - T_f > 0$; that is, *the freezing point of the solution is less than the freezing point of the pure liquid.*

The boiling point elevation and the freezing point depression effects have the same origin, namely, the reduction of the chemical potential of the solvent on dissolution of the solute. The origin of these effects is shown in Figure 13-3.

EXAMPLE 13-6

Show that as the solute B concentration becomes small, that is, as $X_B \to 0$, we have for the freezing point depression

$$T_f^\circ - T_f = \left(\frac{RT_f^2 M_A}{1000\,\Delta\bar{H}^\circ_{A,\mathrm{fus}}}\right) m_c = k_f m_c \qquad (13\text{-}46)$$

where m_c is the colligative molality of the solute(s), M_A is the molecular mass of the solvent, $\Delta\bar{H}^\circ_{A,\mathrm{fus}}$ is the molar enthalpy of fusion of the solvent, and

$$k_f = \frac{RT_f^2 M_A}{1000\,\Delta\bar{H}^\circ_{A,\mathrm{fus}}} \qquad (13\text{-}47)$$

is the freezing point depression constant of the solvent.

Solution: As $X_B \to 0$, we have

$$\ln a_A = \ln X_A = \ln(1 - X_B) \approx -X_B \approx -\frac{m_c}{1000/M_A}$$

Combination of the above result for $\ln a_A$ with Eq. 13-45 yields

$$\frac{M_A m_c}{1000} = \frac{\Delta\bar{H}^\circ_{A,\mathrm{fus}}}{R}\left(\frac{T_f^\circ - T_f}{T_f T_f^\circ}\right)$$

where we have taken $\bar{H}^\circ_A(soln) = \bar{H}^\circ_A(\ell)$ as $X_B \to 0$. Because $T_f^\circ \approx T_f$, rearrangement of the above equation gives Eq. 13-47. Values of k_f, the freezing point depression constant for several solvents, are given in Table 13-1. Note that the value k_f depends only on solvent properties and is thus independent of solute properties.

EXAMPLE 13-7

For water, $k_f = 1.86\ \mathrm{K \cdot kg \cdot mol^{-1}}$. Estimate the freezing point of seawater given that the colligative molality of seawater is $1.10\ \mathrm{mol \cdot kg^{-1}}$.

Solution: We can estimate $T_f^\circ - T_f$ from Eq. 13-46:

$$T_f^\circ - T_f = k_f m_c = (1.86\ \mathrm{K \cdot kg \cdot mol^{-1}})(1.10\ \mathrm{mol \cdot kg^{-1}}) = 2.05\ \mathrm{K}$$

or

$$T_f = 273.15\ \mathrm{K} - 2.05\ \mathrm{K} = 271.10\ \mathrm{K} \qquad (-2.05^\circ\mathrm{C})$$

The above calculation is only an approximation because we have assumed that the activity coefficients are equal to one.

The freezing point depression effect is the basis of the action of *antifreeze* mixtures. The most commonly used antifreeze is ethylene glycol (boiling point 197°C, freezing point −17.4°C):

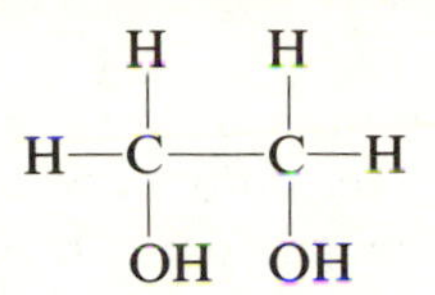

The addition of ethylene glycol to water depresses the freezing point and elevates the boiling point. A 50% by volume (equal volumes of antifreeze and water) ethylene glycol in water solution has a freezing point of −36°C. Above about 55% ethylene glycol in water, the freezing point of the solution *increases*. Pure ethylene glycol freezes at −17.4°C.

The freezing points of ethylene glycol in water solutions are given in Figure 13-4. The effectiveness of ethylene glycol as an antifreeze is a result of several factors, among which are a high boiling point, chemical stability, and the tendency of water to freeze out of the solution to form a slushy mass rather than a solid block of ice. In the absence of antifreeze (or of a sufficient amount

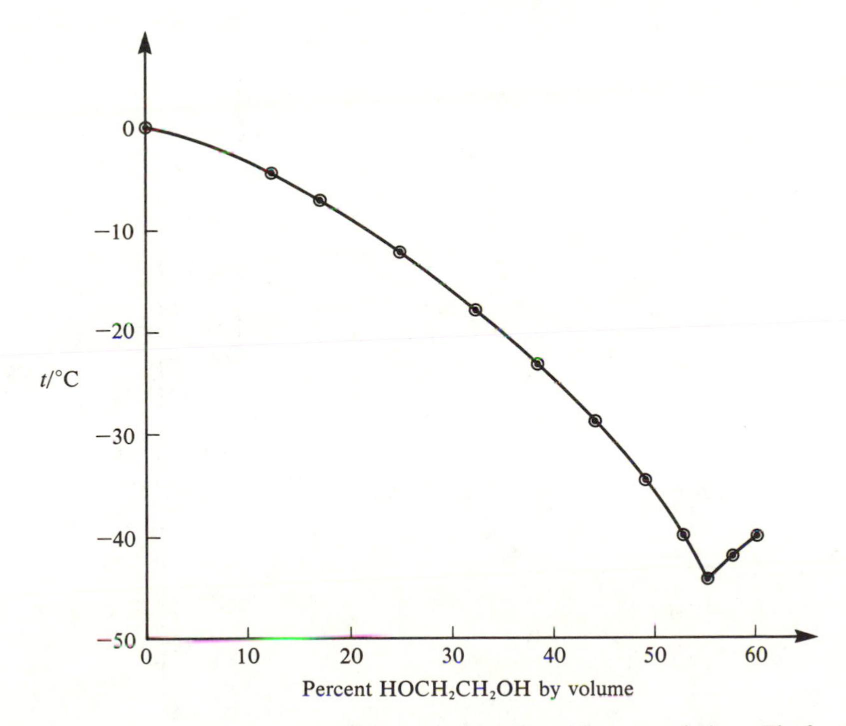

Figure 13-4. Freezing point curve for ethylene glycol (antifreeze) in water solutions. The freezing point of pure water is 0.0°C, and the freezing point of pure ethylene glycol is −17.4°C. Solutions in the range 32% to 55% ethylene glycol in water have a freezing point lower than either of the pure liquids.

of antifreeze), the 11% expansion by volume of water on freezing generates a force (30,000 psi at $-22°C$), which is sufficiently great to rupture a radiator or even a metal engine block.

The freezing point of seawater is $-1.85°C$, and this is the temperature of seawater surrounding an iceberg. Two species of fish called *trematomus* and *dissostichus* live in the cold waters of the Ross Sea of Antarctica near the sea ice. The blood of these fishes is protected from freezing by certain proteins (Figure 13-5). These proteins have an enhanced (i.e., much greater than that predicted by Eq. 13-46 capacity to lower the freezing point of water. The freezing point of the serum from these fishes is $-2.06°C$. The freezing point of the serum after removal of salts (but not the antifreeze proteins) is $-0.60°C$. The concentration of antifreeze–active proteins in the fish blood is about $6\ g \cdot L^{-1}$. The molecular weight of the active proteins averages about $2 \times 10^4\ g \cdot mol^{-1}$, and thus the active protein concentration is about $3 \times 10^{-4}\ M$. The freezing

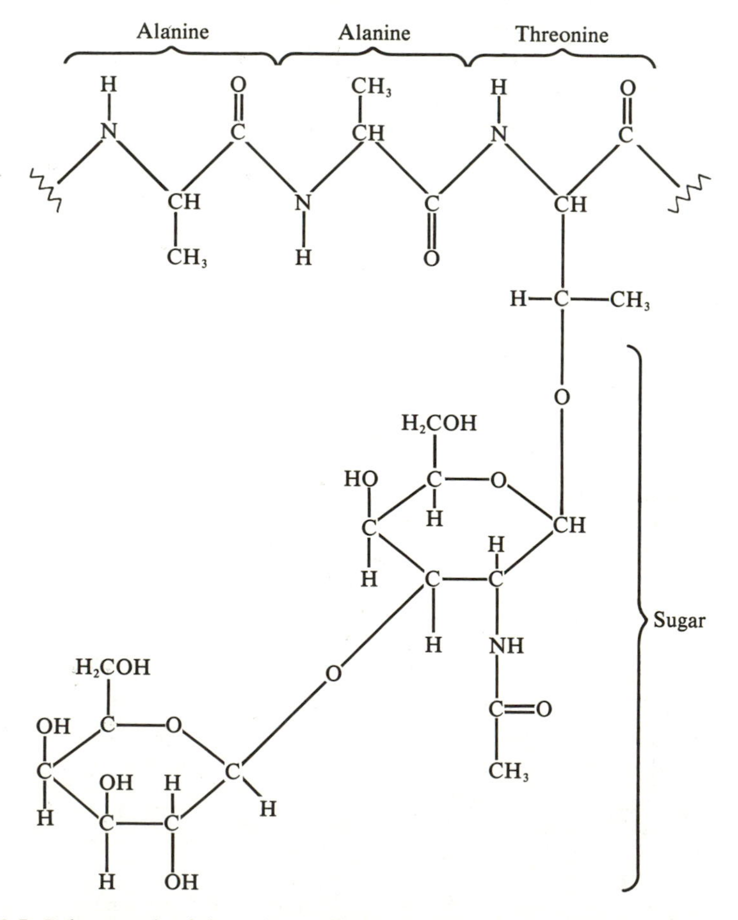

Figure 13-5. Polymer unit of the active antifreeze protein. The polymer unit is repeated about 20 times in the antifreeze protein. The repeat sequence is two alanines followed by a threonine with the sugar units attached to the threonine.

point of the solution predicted using Eq. 13-46 is

$$t_f = -1.86 \times 3 \times 10^{-4} = -0.0006°\text{C}$$

The observed freezing point depression of the antifreeze-protein solution is thus a factor of 1000 times greater $(-0.60/-0.0006)$ than that predicted by simple colligative property considerations. The mode of action of the antifreeze proteins is not entirely clear. The hypothesis is that the proteins adsorb on the surfaces of ice crystal nuclei thereby stopping the growth of the crystals.

13-6 *Two-Component Solid-Liquid–Phase Equilibria*

Although Eq. 13-45 was derived on the assumption that pure A(s) is in equilibrium with the solution (which we shall assume for the purposes of the discussion to follow consists of A and B), we could equally well have considered the problem from the point of view of B(s) in equilibrium with the solution. If this is done, then the resulting expression (at 1 atm) is

$$\ln X_\text{B} = \frac{[\bar{H}_\text{B}^\circ(soln) - \bar{H}_\text{B}^\circ(s)]}{R}\left(\frac{1}{T_{f,\text{B}}^\circ} - \frac{1}{T_f}\right) \tag{13-48}$$

where we have assumed that $a_\text{B} = X_\text{B}$ for simplicity. The analogous expression involving A(s) in equilibrium with the solution is

$$\ln X_\text{A} = \frac{[\bar{H}_\text{A}^\circ(soln) - \bar{H}_\text{A}^\circ(s)]}{R}\left(\frac{1}{T_{f,\text{A}}^\circ} - \frac{1}{T_f}\right) \tag{13-49}$$

If we now plot both Eqs. 13-48 and 13-49 on a T versus X_B diagram, we obtain a result like that shown in Figure 13-6. The temperature at which the two freezing point depression curves intersect is called the *eutectic temperature*; its value, as well as the eutectic composition, is dependent on the total pressure. The eutectic temperature then is the temperature at which both of the above equations (which are valid for $P_\text{tot} = 1$ atm, $\Delta\bar{C}_{P,\text{A}} = \Delta\bar{C}_{P,\text{B}} = 0$, and only pure solids freezing out) are satisfied simultaneously, because at the eutectic temperature both A(s) and B(s) separate simultaneously as a heterogeneous mixture from the solution (see Problem 18).

Referring to Figure 13-6, we note that if a solution having $X_\text{B} = 0.20$ is cooled, then A(s) will begin to separate out at the temperature corresponding to the intersection of $X_\text{B} = 0.20$ dashed line with the freezing point depression curve of A. Further cooling leads to continued separation of A(s) (X_B in solution, therefore, increases) until the eutectic temperature is reached, at which temperature the remaining solution freezes out as a *heterogeneous mixture* of A(s) + B(s). A eutectic mixture melts at a fixed temperature just like a pure compound but is easily distinguishable from a pure compound in several ways, such as by microscopic examination of the crystals.

If we cool a solution with $X_\text{B} = 0.80$ (see Figure 13-6), then the solid that freezes out of the solution is pure B(s). Further cooling of the solution leads to continued separation of B(s), with a decreasing freezing point, until the eutectic temperature is reached.

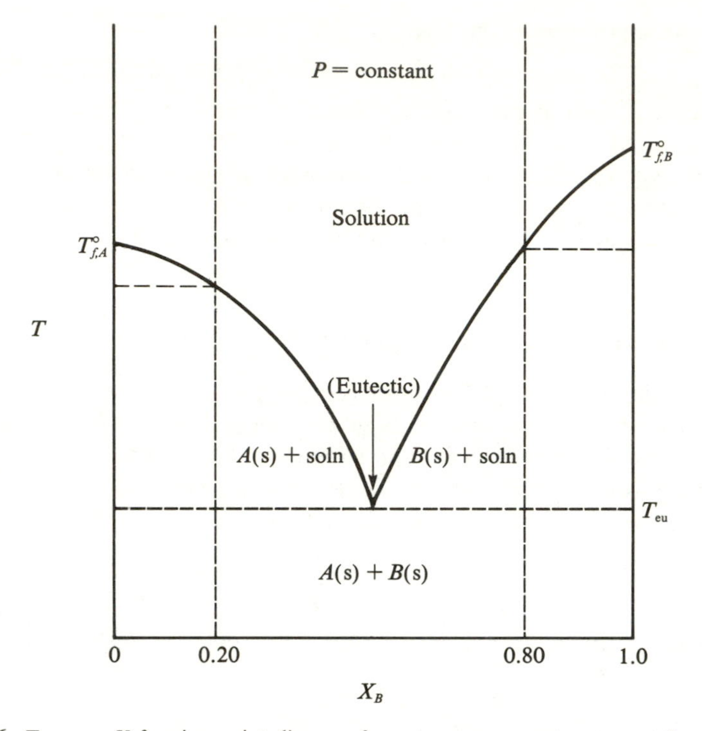

Figure 13-6. T versus X freezing point diagram for a two-component system at fixed pressure. Note that $f = 2 - p + 2 = 4 - p$, or at fixed pressure $f = 3 - p$. Along either freezing point curve, $p = 2$, and thus a specification of T defines the state of the system; whereas, in the "solution" field, $p = 1$, and hence both T and X must be specified to determine the state of the system.

13-7 *Two-Component Liquid-Vapor–Phase Equilibria*

An ideal solution is one for which Raoult's law holds for all components over the entire composition range. For a two-component ideal solution we have

$$P_1 = X_1 P_1^\circ \quad \text{and} \quad P_2 = X_2 P_2^\circ$$

where the X_i values refer to the liquid phase. The total equilibrium pressure P_{tot} over the solution is given by

$$P_{\text{tot}} = P_1 + P_2 = (P_1^\circ - P_2^\circ)X_1 + P_2^\circ = (P_2^\circ - P_1^\circ)X_2 + P_1^\circ \quad (13\text{-}50)$$

where the P_i° values are the equilibrium vapor pressures of the pure liquids at the temperature of the solution. The composition of the ideal vapor over the solution is (where Y_i denotes a mole fraction in the vapor)

$$Y_1 = \frac{P_1}{P_{\text{tot}}} = \frac{X_1 P_1^\circ}{P_{\text{tot}}} \quad \text{and} \quad Y_2 = \frac{P_2}{P_{\text{tot}}} = \frac{X_2 P_2^\circ}{P_{\text{tot}}} \quad (13\text{-}51)$$

In Figure 13-7 we have plotted P_{tot} versus X_2 at fixed T for an ideal solution having $P_1^\circ = 100$ torr and $P_2^\circ = 200$ torr, using Eq. 13-50 to obtain the *liquid-*

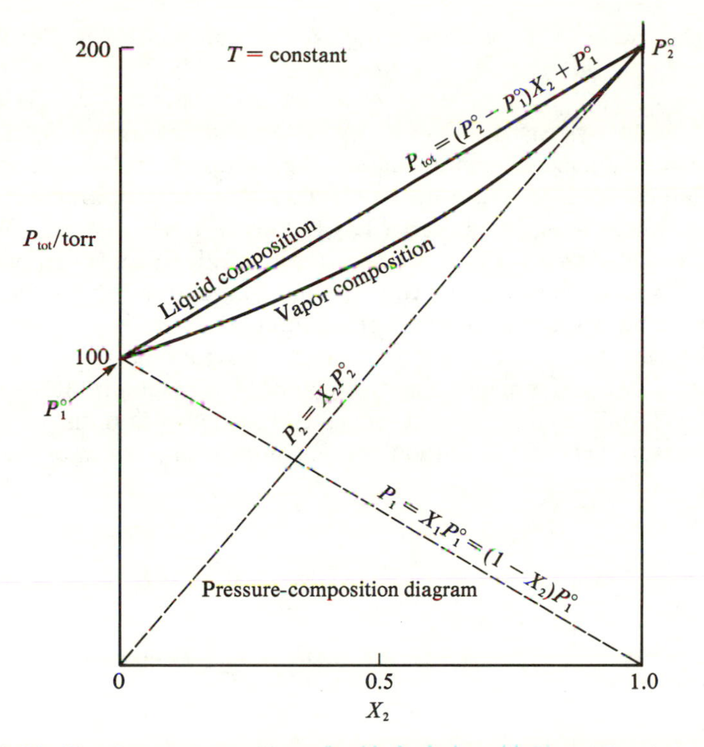

Figure 13-7. Liquid and vapor compositions of an ideal solution with $P_1^\circ = 100$ and $P_2^\circ = 200$ torr.

X_2	P_{tot}	$Y_2 = X_2P_2^\circ/P_{tot}$
0	100	0
0.100	110	0.182
0.200	120	0.333
0.300	130	0.462
0.400	140	0.572
0.500	150	0.667
0.600	160	0.750
0.700	170	0.824
0.800	180	0.889
0.900	190	0.947
1	200	1

composition curve. Equation 13-51 was then used to determine the *vapor-composition curve* given in Figure 13-7. For example, when $X_2 = 0.40$, then $P_{tot} = 140$ torr and $Y_2 = (0.40)(200)/(140) = 0.57$. Diagrams like Figure 13-7 are known as *pressure-composition* (or *P-X*) *diagrams.* We note that the vapor has a higher mole fraction of the more volatile component (higher P_i° value) than does the liquid.

In Figure 13-8 we have constructed the *P-X* diagram for the nonideal alcohol + water system at 343.15 K. An important feature of Figure 13-8 is the existence of a maximum in the liquid-composition curve. At the maximum in the liquid-composition curve the liquid and vapor compositions are identical. A solution whose equilibrium vapor has the same composition as that of the liquid is known as an *azeotrope.* An azeotropic solution distills without change in composition. Such a situation can arise only if the solution is nonideal. The azeotropic composition is a function of the total pressure, because the activities of the components in solution are a function of the total pressure.

Another instructive diagram for two-component systems is the *temperature-composition* (*T-X*) *diagram.* For a temperature-composition diagram we plot the temperature versus the solution composition *for a fixed total pressure.* If

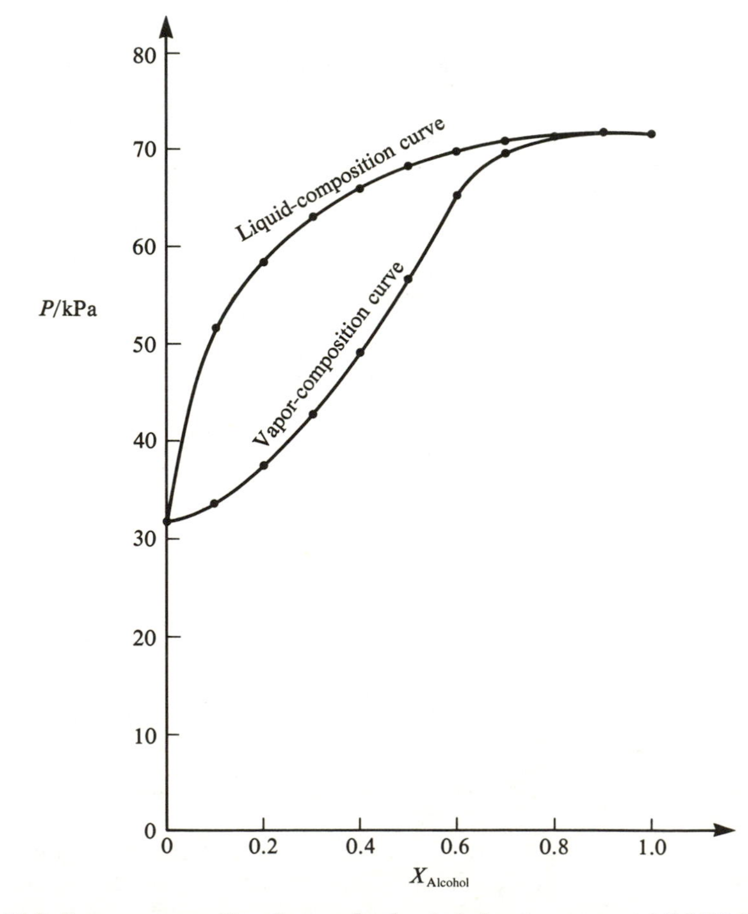

Figure 13-8. Pressure-composition diagram for the alcohol-water system at 343.15 K (data from J. A. Larkin and R. C. Pemberton, *NPL Report Chem.*, 43, January 1976).

the vapor and liquid are ideal, then Eq. 13-50 is applicable; at a fixed total pressure of, say, 760 torr, we can write

$$760 \text{ torr} = (P_2^\circ - P_1^\circ)X_2 + P_1^\circ \tag{13-52}$$

Using Eq. 13-52 we can construct a temperature-composition diagram at 1 atm total pressure, if we have P_2° and P_1° as functions of temperature. We simply take P_2° and P_1° at the same temperature and calculate X_2 and then repeat the calculation at other temperatures obtaining $T = T(X_2)$. It is clear that if the above expression yields a value of X_2 outside the limits $0 < X_2 < 1$, then there exists no solution of the two components at the temperature to which the P_i° values refer, which has the total pressure chosen. At each temperature we can then calculate Y_2 (the mole fraction in the vapor phase) using Eq. 13-51.

Figure 13-9 is a temperature-composition diagram for the (nonideal) alcohol + water system at $P_{\text{tot}} = 101.325$ kPa (1 atm). Figure 13-10 is a T-X

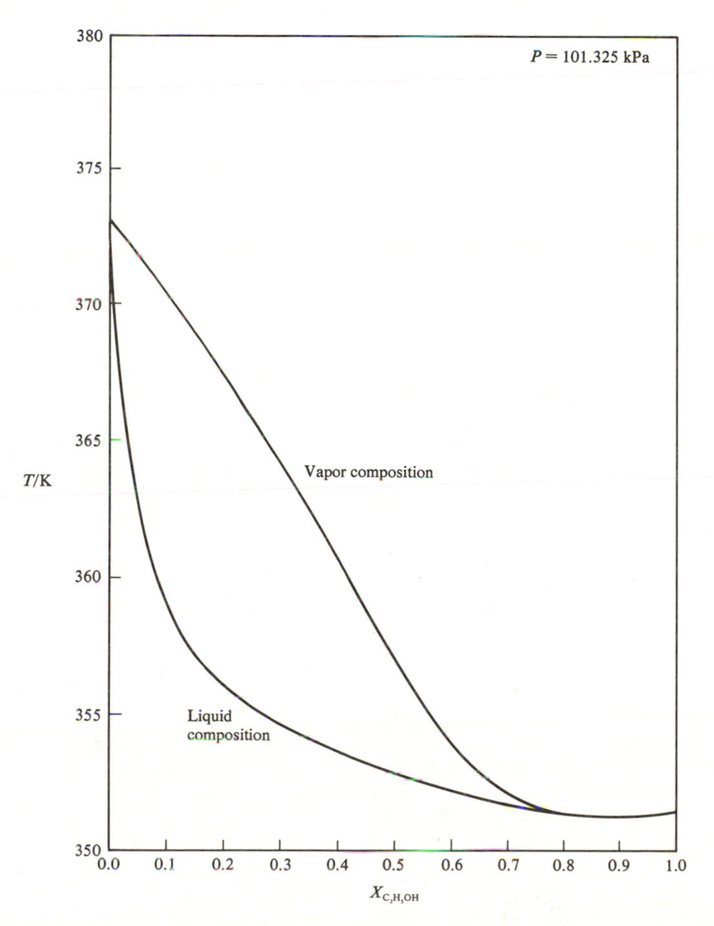

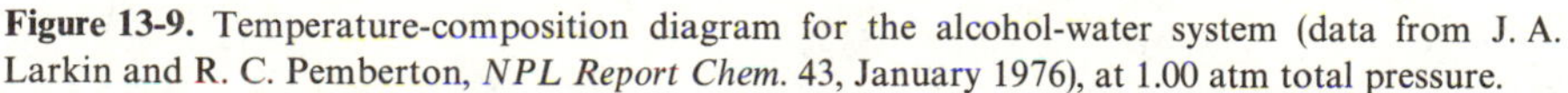

Figure 13-9. Temperature-composition diagram for the alcohol-water system (data from J. A. Larkin and R. C. Pemberton, *NPL Report Chem.* 43, January 1976), at 1.00 atm total pressure.

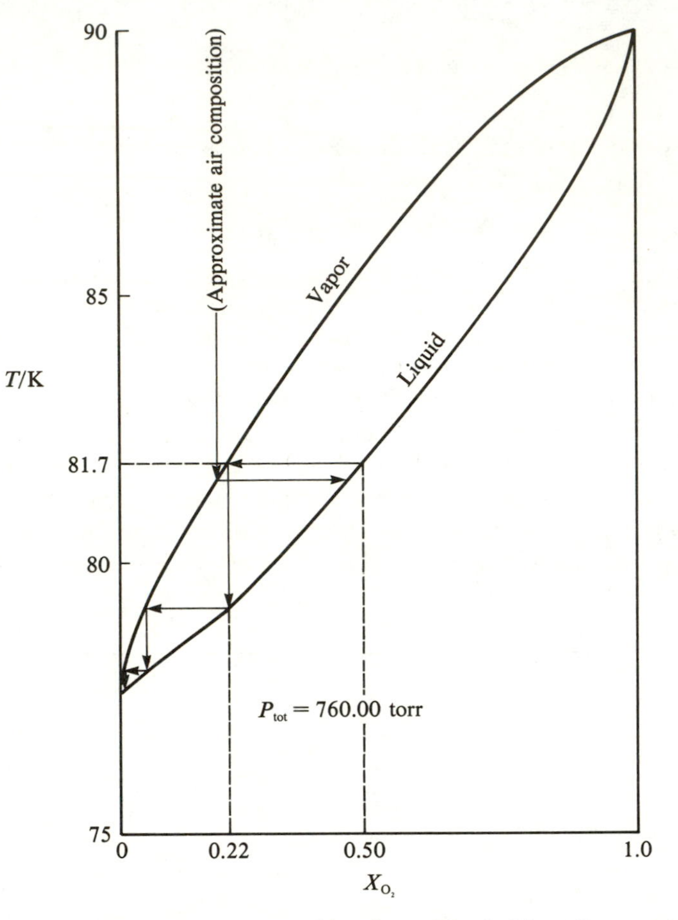

Figure 13-10. Temperature-composition diagram for the $N_2 + O_2$ system.

Data (*International Critical Tables*, Vol. III, 1929)

T/K	$X_{O_2}(\ell)$	X_{O_2} (vap)
77.3	0	0
78.0	0.081	0.022
79.0	0.216	0.068
80.0	0.334	0.120
81.0	0.434	0.177
82.0	0.522	0.236
83.0	0.595	0.299
84.0	0.662	0.369
85.0	0.723	0.443
86.0	0.778	0.522
87.0	0.829	0.603
88.0	0.885	0.696
89.0	0.938	0.798
89.5	0.965	0.870
90.2	1	1

plot for the N_2, O_2 system. Although $N_2 + O_2$ solutions are not ideal, the temperature-composition diagram for an ideal solution is similar to that shown in Figure 13-10; as far as gross features go, it is identical to that for the $N_2 + O_2$ system (as the reader can verify by constructing such a diagram for an ideal solution).

The normal boiling point of pure $N_2(\ell)$ is 77.3 K and that for $O_2(\ell)$ is 90.2 K. Figure 13-10 gives the normal (1-atm) boiling points of all solutions composed of N_2 and O_2, as well as the compositions of the vapor phase in equilibrium with the liquid. From the figure, we read that the boiling point of a solution having $X_{O_2} = 0.50$ is 81.7 K and that the composition of the vapor is $X_{O_2}(vap) = 0.22$. We note that the vapor is richer than the liquid in N_2, which is the more volatile component (i.e., the one with the largest P_i° at the particular temperature). If some of this vapor is in turn condensed and reboiled, and the process is repeated a sufficient number of times, then a separation of the two components can be achieved, provided the T-X diagram has the general form of that in Figure 13-10 (follow the connected arrows). If, however, the system possesses an azeotrope, then this is no longer true, as seen from Figure 13-9.

A *fractionating column* is a distillation column that carries out the repeated evaporation-condensation processes automatically. The essential features of such a column are the existence of a temperature gradient along the column ($dT/dh < 0$) and the presence of a suitable packing material to act as a surface for condensation and to provide a support for heat transfer between rising higher-temperature vapor and descending lower-temperature liquid. With a suitably designed column it is possible for a case like the $N_2 + O_2$ system to achieve a complete separation of the two components with the lower-boiling (more volatile) component (in this case N_2) being first to leave the column. We note also from Figure 13-10 that if we lower the temperature of air (which is essentially 20% O_2, 79% N_2, 1% Ar) with, say, liquid nitrogen at 77.3 K, then the air will begin to condense when its temperature reaches about 81.4 K, and the liquid formed will have $X_{O_2} \approx 0.47$. Such an experiment can be performed easily in the laboratory by pouring $N_2(\ell)$ into a metal cup; the $X_{O_2} \approx 0.47$ solution will form spontaneously at the bottom of the cup and drip off. Commercially available $N_2(g)$ and $O_2(g)$ are obtained by fractional distillation of liquid air.

13-8 *Partial Molar Quantities Are Thermodynamic Properties of a Solution*

A straightforward, but not precise, method for determining partial molar quantities (see Section 5-9) involves plotting the extensive thermodynamic variable $Y = Y(n_i)$ at constant T, P and mole numbers of all components except i versus n_i, the moles of i. The slope of the resulting curve at any point is $\bar{Y}_i = (\partial Y/\partial n_i)_{T,P,n_j(j \neq i)}$ at that particular composition. Such a plot is easily obtained for a two-component system by using the molality composition scale, which fixes the moles of solvent at $1000/M_1$. The slope of the curve at any point gives $\overline{Y}_i$ for the solute at that composition.

A more precise method for the determination of partial molar quantities is the so-called *slope-intercept method*, which we shall illustrate for the special case of a two-component system. With the slope-intercept method we begin by computing the molar value of Y for the solution as a whole, a quantity that we shall designate as $\overline{Y}_{\text{tot}}$ where

$$\overline{Y}_{\text{tot}} = \frac{Y}{\sum_i n_i} \tag{13-53}$$

For those cases in which an absolute value for Y cannot be given, for example, enthalpy and Gibbs energy, we work with ΔY values; thus

$$\Delta\overline{Y}_{\text{tot}} = \frac{\Delta Y}{\sum_i n_i} \tag{13-54}$$

For the special case of a two-component system we have

$$Y = (n_1 + n_2)\overline{Y}_{\text{tot}} \tag{13-55}$$

and, therefore,

$$\overline{Y}_1 = \left(\frac{\partial Y}{\partial n_1}\right)_{n_2} = \overline{Y}_{\text{tot}} + (n_1 + n_2)\left(\frac{\partial \overline{Y}_{\text{tot}}}{\partial n_1}\right)_{n_2} \tag{13-56}$$

where we have omitted T and P as subscripts on the partial derivatives in the interests of simplicity—it is to be understood, however, that the development applies only to a system at fixed T and P. Because

$$\left(\frac{\partial \overline{Y}_{\text{tot}}}{\partial n_1}\right)_{n_2} = \left(\frac{\partial \overline{Y}_{\text{tot}}}{\partial X_2}\right)_{n_2}\left(\frac{\partial X_2}{\partial n_1}\right)_{n_2} \tag{13-57}$$

and

$$X_2 = \frac{n_2}{n_1 + n_2}$$

we have

$$\left(\frac{\partial X_2}{\partial n_1}\right)_{n_2} = -\frac{n_2}{(n_1 + n_2)^2} = -\frac{X_2}{(n_1 + n_2)} \tag{13-58}$$

Furthermore, because an arbitrary change in X_2 gives rise to a definite change in $\overline{Y}_{\text{tot}}$ (for a two-component system), we can write $(\partial \overline{Y}_{\text{tot}}/\partial X_2)_{n_2} = d\overline{Y}_{\text{tot}}/dX_2$ and, therefore,

$$\overline{Y}_{\text{tot}} = \overline{Y}_1 + X_2\frac{d\overline{Y}_{\text{tot}}}{dX_2} \tag{13-59}$$

A similar treatment for $\overline{Y}_2$ yields

$$\overline{Y}_{\text{tot}} = \overline{Y}_2 + X_1\frac{d\overline{Y}_{\text{tot}}}{dX_1} \tag{13-60}$$

A tangent to the curve $\overline{Y}_{\text{tot}} = f(X_2)$ at any particular X_2 has an intercept on the $\overline{Y}_{\text{tot}}$ axis of $\overline{Y}_1$ (Eq. 13-59), whereas a tangent to the curve $\overline{Y}_{\text{tot}} = f(X_1)$ at any

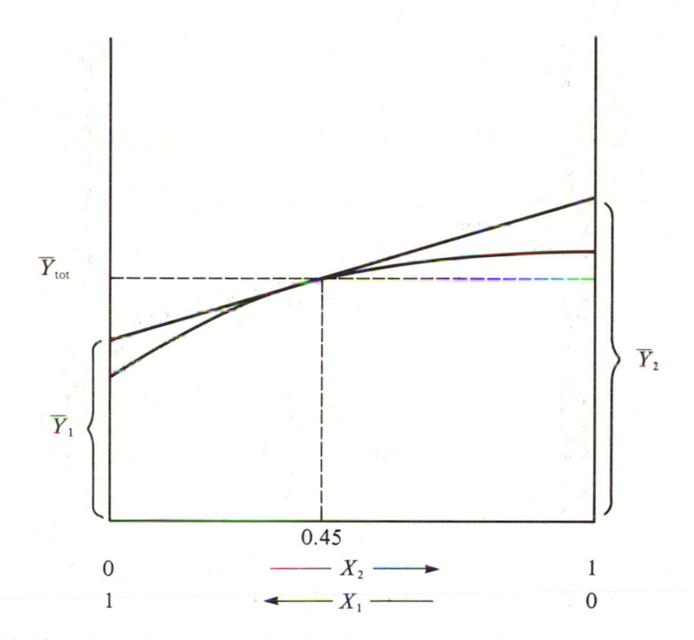

Figure 13-11. Slope-intercept method for the determination of partial molar quantities. The tangent to the curve at $X_2 = 0.45$ yields as intercepts $\overline{Y}_1$ and $\overline{Y}_2$ at $X_2 = 0.45$.

particular X_1 has an intercept on the $\overline{Y}_{tot}$ axis of $\overline{Y}_2$ (Eq. 13-60). The slope-intercept method is depicted graphically in Figure 13-11.

Another useful graphical technique for obtaining partial molar quantities in two-component systems is the following: We define a quantity called the *apparent molar value of Y*, designated ${}^{\phi}Y$, by the expression

$$ {}^{\phi}Y = \frac{Y - n_1 \overline{Y}'_1}{n_2} \tag{13-61}$$

where $\overline{Y}'_1$ is the molar value of Y for pure component 1 at T, P. We shall distinguish partial molar quantities from molar quantities by means of a prime on the latter. Thus $\overline{Y}_i$ and $\overline{Y}'_i$ refer, respectively, to the partial molar value of Y in the solution and the molar value of pure Y for the ith component.

From Eq. 13-61 we obtain

$$\overline{Y}_2 = \left(\frac{\partial Y}{\partial n_2}\right)_{n_1} = {}^{\phi}Y + n_2\left(\frac{\partial {}^{\phi}Y}{\partial n_2}\right)_{n_1} = {}^{\phi}Y + n_2 \frac{d^{\phi}Y}{dn_2} \tag{13-62}$$

and thus the slope of a plot of ${}^{\phi}Y$ versus n_2 at any particular composition can be inserted into Eq. 13-62, together with ${}^{\phi}Y$ and n_2 appropriate to that composition, and $\overline{Y}_2$ results. If Y is a thermodynamic coordinate whose absolute value cannot be specified, then Eq. 13-61 takes the form

$$ {}^{\phi}\Delta Y = \frac{\Delta Y - n_1 \Delta \overline{Y}'_1}{n_2} \tag{13-63}$$

In Figure 13-12 we have plotted the reciprocal of the density of ethanol + water solutions versus the weight percent of ethanol rather than $\overline{V}_{tot}$ versus X.

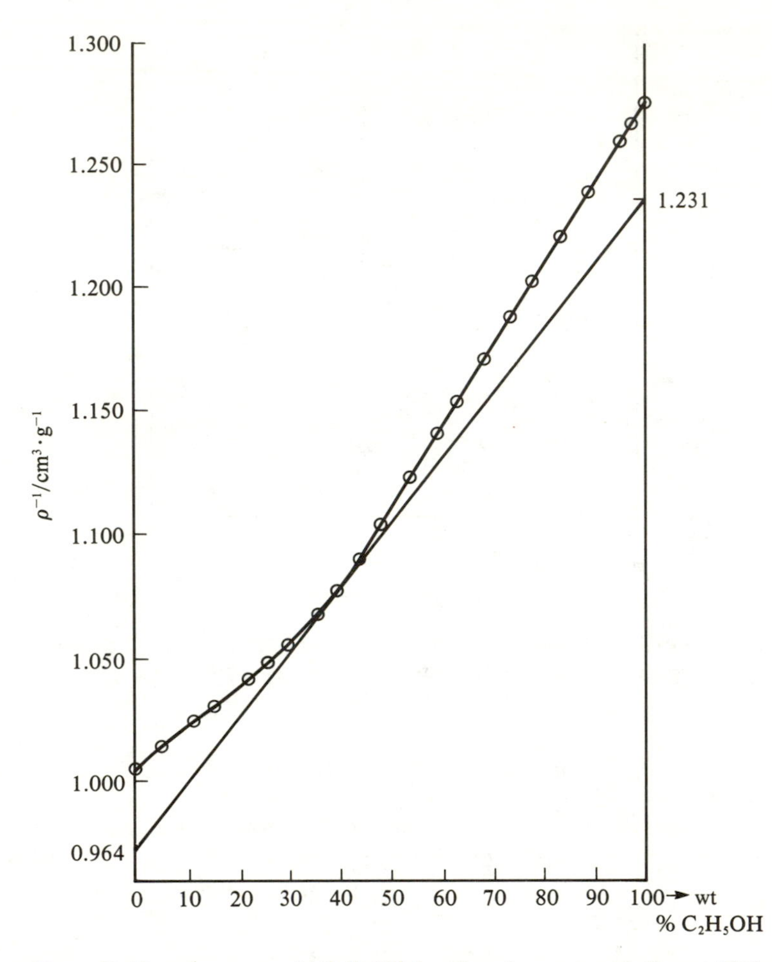

Figure 13-12. ρ^{-1} versus wt % C_2H_5OH for ethanol + water solutions at 25°C.

Inspection of Eqs. 13-59 and 13-60 shows that in such a case the intercepts of a tangent to the curve at a particular weight percent, multiplied by the molecular weights of the respective compounds, yields $\bar{V}_1$ and $\bar{V}_2$ at that composition ($\mathrm{cm^3 \cdot g^{-1} \times g \cdot mol^{-1} = cm^3 \cdot mol^{-1}}$). From Figure 13-12 at 50.0 wt % we find $\bar{V}_{H_2O} = 0.9640 \times 18.015 = 17.37\ \mathrm{cm^3 \cdot mol^{-1}}$, $\bar{V}_{C_2H_5OH} = 1.2321 \times 46.070 = 56.76\ \mathrm{cm^3 \cdot mol^{-1}}$, compared with the values for the pure compounds of

$$\bar{V}'_{H_2O} = 1.0029 \times 18.015 = 18.07\ \mathrm{cm^3 \cdot mol^{-1}}$$

and

$$\bar{V}_{C_2H_5OH} = 1.273 \times 46.07 = 58.65\ \mathrm{cm^3 \cdot mol^{-1}}.$$

Close inspection of the water-ethanol curve in Figure 13-12 reveals some interesting volumetric behavior in the range 0 wt % to 30 wt % ethanol, which is displayed in terms of partial molar volumes in Figure 13-13.

It is worth noting that partial molar volumes can be negative; this happens when the $\bar{V}_{\mathrm{tot}}$ versus X curve has a sufficiently deep downward dip that a

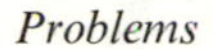

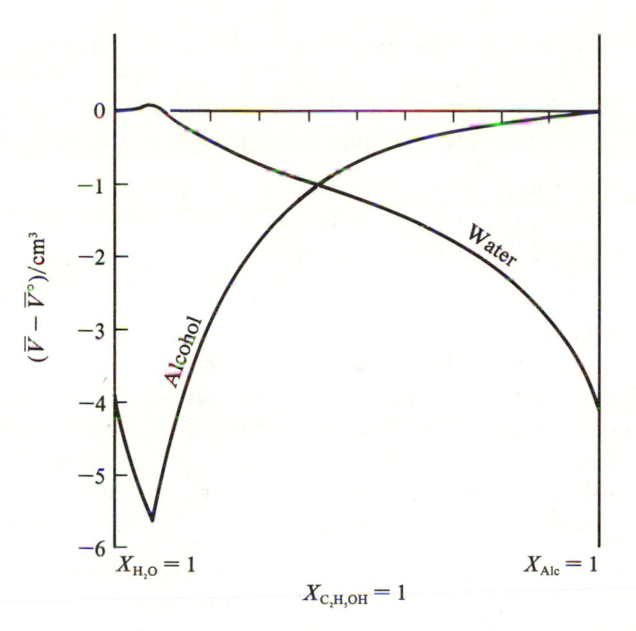

Figure 13-13. Differences between the partial molar volumes and the molar volumes of ethanol and water in ethanol + water solutions at 25°C. (From G. N. Lewis and M. Randall, *Thermodynamics, 2nd ed.*, Revised by K. S. Pitzer and L. Brewer. New York: McGraw-Hill, 1961, p. 209.)

tangent to the curve over part of the composition range intercepts one of the vertical axes below zero over some part of the composition range. One possible physical interpretation of this effect is that if $\bar{V}_i < 0$, then $\Delta V_{\text{mix}} < 0$; hence molecules of component 1 can be thought of either as entering "cavities" in the other liquid in which the volume of a cavity is less than the average volume available to a molecule of component 1 in the pure liquid; or the molecules of component 1 cause the partial collapse of a rather "open" solution structure to a more compact packing arrangement or possibly a combination of both of these effects operate simultaneously.

Problems

1. The equilibrium vapor pressure of water at 25°C is 23.6 torr. Compute the equilibrium vapor pressure of water at 25°C over a solution prepared by mixing 100 g of $H_2O(\ell)$ with 100 g of ethylene glycol, $HOCH_2CH_2OH$ (C = 12.0, O = 16.0, H = 1.0). Assume an ideal solution.

2. Compute the depression of the freezing point of water due to dissolved air. Take air to be 80 mol % N_2 and 20 mol % O_2. See Table 11-1 for Henry's law constants for N_2 and O_2. Show that the normal freezing point of air-saturated water lies 0.0101 K below the triple point of pure water. Assume that Henry's law holds.

3. Seawater contains a variety of cations and anions. The principal ions are $Cl^-(aq)$, $Na^+(aq)$, $SO_4^{2-}(aq)$, and $Mg^{2+}(aq)$. As a rough approximation, seawater can be regarded as a solution that is 0.50 *m* in NaCl and 0.05 *m* in $MgSO_4$. Estimate the osmotic pressure of seawater at 15°C and 37°C.

4. The freezing point constant for water is $k_f = 1.86\ \text{K}\cdot\text{kg}\cdot\text{mol}^{-1}$. Estimate the freezing point in degrees Celsius for an aqueous solution that is 0.10 *m* in NaCl(*aq*) and 0.10 *m* in $CaCl_2(aq)$.

5. One method that is used to estimate the osmotic pressure in biological cells is to place the cells in a series of aqueous salt solutions of different salt concentrations to find the salt concentration at which the cells neither expand nor contract. Given that for a certain type of cell the equilibrium water vapor pressure over the salt solution in which the cell volume remained constant at 37°C was 45.87 torr, compute the osmotic pressure (in atmospheres) in the cell. Take the density of water as $1.00\ \text{g}\cdot\text{cm}^{-3}$ and the vapor pressure of pure water at 37°C as 47.08 torr.

6. The normal isotopic composition of lithium is 92.48 atom % ^{7}Li and 7.52 atom % ^{6}Li. Assume ideal solution behavior, and calculate ΔG_{mix}, ΔH_{mix}, and ΔS_{mix} for the formation of 1.00 mol of normal isotopic composition Li(*s*) from the pure isotopes at 25°C.

7. Calculate ΔG_{mix}, ΔH_{mix}, and ΔS_{mix} when 2.00 mol of *p*-xylene is added to a very large volume of a solution of *p*-xylene and *m*-xylene with $X_p = \frac{1}{3}$. Assume ideal solution behavior and take $T = 298$ K

8. A 1.00 mol sample of benzene ($C_P = 20\ \text{cal}\cdot\text{K}^{-1}\cdot\text{mol}^{-1}$) at 20.0°C is mixed adiabatically with 2.00 mol of toluene ($C_P = 23\ \text{cal}\cdot\text{K}^{-1}\cdot\text{mol}^{-1}$) at 40.0°C. Calculate the total entropy change (assume benzene and toluene form an ideal solution).

9. Compute the entropy change on mixing a solution containing 5.00 mol of benzene and 7.00 mol of toluene with a second solution at the same temperature containing 3.00 mol of benzene, 2.00 mol of toluene, and 4.00 mol of *p*-xylene.

10. The element europium consists of 47.8 atom % Eu^{151} and 52.2 atom % Eu^{153}. Calculate ΔG and ΔS at 300 K for the separation of 1 mol of natural Eu into two equal mass fractions one of which is 90.0 atom % Eu^{151}. Assume that the separation is carried out reversibly.

11. A solution obtained by dissolving a 2.000-g sample of an unknown solid in 875.0 g of $C_6H_6(\ell)$ was observed to begin freezing out $C_6H_6(s)$ at 3.75°C. Estimate the molecular weight of the unknown. Take the melting point of benzene at 5.51°C, and $\Delta\bar{H}_{\text{fus}} = 2366\ \text{cal}\cdot\text{mol}^{-1}$.

12. Given the following data for ethanol + water solutions at 25°C, calculate γ_{H_2O} and $\gamma_{C_2H_5OH}$ relative to the solvent standard states for each component at $X_{H_2O} = 0.1, 0.2, \ldots, 1.0$, and plot γ_{H_2O} and $\gamma_{C_2H_5OH}$ versus X_{H_2O} over the entire composition range.

$X_{C_2H_5OH}$	$P_{C_2H_5OH}$/kPa	P_{H_2O}/kPa
0.10	2.353	2.893
0.20	3.603	2.700
0.30	4.164	2.578
0.40	4.524	2.466
0.50	4.914	2.305
0.60	5.363	2.070
0.70	5.858	1.755
0.80	6.432	1.318
0.90	7.126	0.717
1.00	7.893	0

(Data from J. A. Larkin and B. C. Pemberton, *NPL Report Chem.*, 43, January 1976.)

13. Show that the dependence of the activity of component i in a solution on the total pressure at fixed T and composition (*comp*) is given by

$$\left(\frac{\partial \ln a_i}{\partial P}\right)_{T,\mathrm{comp}} = \frac{\bar{V}_i}{RT}$$

14. Show that the dependence of the activity of component i in a solution on the temperature at fixed total pressure and composition (*comp*) is given by

$$\left(\frac{\partial \ln a_i}{\partial T}\right)_{P,\mathrm{comp}} = -\left(\frac{\bar{H}_i - \bar{H}_i^\circ}{RT^2}\right)$$

where $\bar{H}_i$ is the partial molar enthalpy of i in the solution.

15. (a) A method used to melt ice on streets and sidewalks is to spread rock salt, NaCl(*s*), crystals on the ice. The solubility of NaCl in liquid water around 0°C is 4.8 $\mathrm{mol\cdot kg^{-1}}$. Explain how the rock-salt–ice-melting procedure works, and estimate the lowest temperature at which ice can be melted by this method.

(b) Suppose excess NaCl(*s*) is added to an ice-water mixture and the mixture is stirred; estimate the temperature of the resulting equilibrium mixture. Note this ice-salt-water mixture is used as the cooling medium for the preparation of homemade ice cream.

16. A saturated aqueous solution of NaCl has an NaCl molality at 25°C of 4.8 $\mathrm{mol\cdot kg^{-1}}$. Compute the osmotic pressure in atmospheres at 25°C for a saturated NaCl(*aq*) solution.

17. Show that for the process

$$n_B\mathrm{B}(s) + n_A\mathrm{A}(\ell) \longrightarrow (\text{solution of } n_B \text{ mol of B and } n_A \text{ mol of A})$$

$$\Delta H_{\mathrm{rxn}} = n_A[\bar{H}_A(soln) - \bar{H}_A(\ell)] + n_B[\bar{H}_B(soln) - \bar{H}_B(s)]$$

Also show that if the solution is ideal, then

$$\Delta H_{\text{rxn}} = n_{\text{B}} \Delta \bar{H}_{\text{B,fus}}$$

18. Show, by eliminating T between Eqs. 13-48 and 13-49, that the eutectic composition $X_{\text{B,eu}}$ at 1 atm of a solution of A and B is given by

$$\Delta \bar{H}_{\text{A}}^{\circ} \ln X_{\text{B,eu}} - \Delta \bar{H}_{\text{B}}^{\circ} \ln(1 - X_{\text{B,eu}}) = \frac{\Delta \bar{H}_{\text{A}}^{\circ} \cdot \Delta \bar{H}_{\text{B}}^{\circ}}{R} \left(\frac{1}{T_{f,\text{B}}^{\circ}} - \frac{1}{T_{f,\text{A}}^{\circ}} \right)$$

from which $X_{\text{B,eu}}$ can be obtained by successive approximations if $\Delta \bar{H}_i^{\circ}$ and $T_{f,i}^{\circ}$ are known for A and B. Note that once $X_{\text{B,eu}}$ is known, the value of T_{eu}, the eutectic temperature, can be obtained from either Eq. 13-48 or Eq. 13-49.

19. (a) Show that the solubility of A(s) in mole fraction units is the same in any solvent with which A forms an ideal solution.

(b) Compute the solubility of benzene in mole fraction units at 0°C and 1 atm pressure in any solvent with which benzene forms an ideal solution. For benzene, $T_f^{\circ} = 278.66$ K and $\Delta \bar{H}_{\text{fus}}^{\circ} = 19.67 \text{ kJ} \cdot \text{mol}^{-1}$. Assume $\Delta C_P^{\circ} = 0$.

20. Show that the solubility of a pure solid in an ideal solution always increases with increasing temperature. Does the same result always hold for non-ideal solutions? Explain.

21. The Flory-Huggins theory of polymer solutions yields

$$\Delta G_{\text{mix}} = RT(n_1 \ln \phi_1 + n_2 \ln \phi_2) + A(n_1 \bar{V}_1' + n_2 \bar{V}_2')\phi_1 \phi_2$$

for the mixing of a liquid polymer and a solvent. The ϕ_i values are the volume fractions, A is a function of T alone, and $\bar{V}_i'$ is the molar volume of pure i at the T and P of the solution. Given the above expression for ΔG_{mix}, show that

(a) $\Delta S_{\text{mix}} = -R(n_1 \ln \phi_1 + n_2 \ln \phi_2) - (n_1 \bar{V}_1' + n_2 \bar{V}_2')\phi_1 \phi_2 \dfrac{dA}{dT}$

(b) $\Delta H_{\text{mix}} = (n_1 \bar{V}_1' + n_2 \bar{V}_2')\left(A - T \dfrac{dA}{dT} \right)$

(c) If $\phi_1 \approx 1.0$ ($n_2 \ll n_1$, a dilute polymer solution), then

$$\Delta G_{\text{mix}} = RT(n_1 \ln \phi_1 + n_2 \ln \phi_2) + n_1 B \phi_2$$

$$\Delta S_{\text{mix}} = -R(n_1 \ln \phi_1 + n_2 \ln \phi_2) + n_1 \frac{dB}{dT} \phi_2$$

$$\Delta H_{\text{mix}} = \left(B - T \frac{dB}{dT} \right) n_1 \phi_2$$

where $B = A\bar{V}_1'$.

22. Use the results in Problem 21 to derive an expression for the activities of the solvent and polymer solute in a dilute polymer solution.

23. *Regular solutions* are characterized by an ideal ΔS_{mix}, but have $\Delta H_{\text{mix}} \neq 0$. Regular solutions do not involve strong specific intermolecular interactions like hydrogen bonding and specific solvation. The activity coefficient of component 1 in a two-component regular solution is given by

$$\ln \gamma_1 = \frac{A\bar{V}'_1}{RT} \phi_2^2$$

where

$$\phi_2 = \frac{n_2\bar{V}'_2}{n_2\bar{V}'_2 + n_1\bar{V}'_1}$$

$\bar{V}'_1 = \bar{V}'_1(T, P)$ is the molar volume of pure component 1 at the T and P of the solution; and A is a constant at a given T and P.

(a) Use the Gibbs-Duhem equation to show that

$$\ln \gamma_2 = \frac{A\bar{V}'_2}{RT} \phi_1^2$$

(b) Show for the isothermal, isobaric mixing of two components to form a regular solution that

$$\Delta G_{\text{mix}} = RT(n_1 \ln X_1 + n_2 \ln X_2) + A(n_1\bar{V}'_1\phi_2^2 + n_2\bar{V}'_2\phi_1^2)$$

(c) Given $\bar{V}'_1 = \bar{V}'_2$, show that

$$\Delta G_{\text{mix}} = RT(n_1 \ln X_1 + n_2 \ln X_2) + A\bar{V}'(n_1 + n_2)X_1X_2$$

(d) Given the result in part c, show that

$$\Delta H_{\text{mix}} = (n_1 + n_2)X_1X_2\left[B - T\left(\frac{\partial B}{\partial T}\right)_P\right]$$

where $B = A\bar{V}'$.

(e) Given that, for $C_6H_{12}(\ell) + C_6H_6(\ell)$ solutions, $B = 305\ \text{cal}\cdot\text{mol}^{-1}$ and $(\partial B/\partial T)_P = -1.68\ \text{cal}\cdot\text{K}^{-1}\cdot\text{mol}^{-1}$, compute ΔG_{mix}, ΔS_{mix}, and ΔH_{mix} when 1.00 mol of $C_6H_6(\ell)$ is mixed with 1.00 mol of $C_6H_{12}(\ell)$ at 25.0°C.

24. The solubility of the solid Z(*s*) in benzene at 298 K and 1 atm is $X_2 = 0.275$ and $\gamma_2 = 0.676$ (relative to the solute standard state) for the saturated solution. Compute the standard Gibbs energy of solution of Z(*s*) in benzene at 298 K. Also compute the equilibrium constant at 298 K for the reaction

$$\text{Z}(s) = \text{Z}(soln \text{ in benzene})$$

25. Assuming the activity of water in aqueous solutions is given by Raoult's law, estimate the osmotic pressure of aqueous sucrose solutions at 0°C having the molalities given in the following table:

m	1.651	2.373	3.273	4.120
π(atm)	43.84	67.68	100.43	134.71

Compare your results with those determined experimentally (given in the above table). Use the experimental π data to estimate the activity coefficient of water in the above solutions. To what pressures do these γ values apply?

26. Show that for a two-component regular solution with equal molar volumes

$$\log \gamma_1 = k - k'm$$

where m is the molality of component 2 and k and k' are constants at a given T (see problem 23).

27. Using Figure 13-13, estimate ΔV when 10 g of ethanol is mixed with 90 g of water and when 90 g of ethanol is mixed with 10 g of water. Also estimate the partial molar volumes of ethanol and water in 80 proof vodka at 25°C.

28. List four methods of distinguishing between (a) a pure solid and a eutectic mixture, (b) a solid solution and a heterogeneous mixture, and (c) an azeotrope and a pure liquid.

29. Derive the expression

$$\ln a_A(soln \text{ at } T_b) = \left[\frac{\bar{H}^\circ_A(gas) - \bar{H}^\circ_A(soln)}{R}\right]_{T^\circ_b}\left(\frac{T^\circ_b - T_b}{T_b T^\circ_b}\right) - \left[\frac{\bar{C}^\circ_{P,A}(gas) - \bar{C}^\circ_{P,A}(soln)}{R}\right]\left[\ln\left(\frac{T_b}{T^\circ_b}\right) - \left(\frac{T_b - T^\circ_b}{T_b}\right)\right]$$

for the boiling point elevation when $\Delta\bar{C}_{P,A} \neq 0$.

30. Taking $X_{H_2O} = 0.974$, estimate the freezing point of seawater at 1 atm given that $\gamma_{H_2O} \approx 0.99$ and assuming that $\Delta\bar{C}_{P,fus} = 0$; repeat the calculation with $\Delta\bar{C}_{P,fus} = 9.1 \text{ cal}\cdot\text{K}^{-1}$.

31. Taking $X_{H_2O} = 0.974$ for seawater, and assuming that Raoult's law holds and that water is an incompressible fluid with $\rho = 1.00 \text{ g}\cdot\text{cm}^{-3}$, calculate the minimum pressure (in atmospheres) necessary to obtain fresh water from seawater by reverse osmosis at 17°C. Repeat the calculation given that for water $\beta = 4.67 \times 10^{-5} \text{ atm}^{-1}$, $\rho = 0.999 \text{ g}\cdot\text{cm}^{-3}$ and that for seawater $\gamma_{H_2O} \approx 0.99$.

32. Obtain an expression for the boiling point of a system composed of two immiscible liquids, and discuss your result in reference to steam distillation techniques, as commonly used in organic preparative chemistry. How is your result affected if the liquids are slightly soluble in one another?

33. Offer a qualitative molecular level explanation for the volumetric behavior of water plus ethanol solutions shown in Figure 13-13.

34. In the study of solid-liquid–phase equilibria, a situation similar to that depicted in Figure 13-14 is encountered often. What is involved in such a case is the formation of a compound ($CaZrO_3$ in the case depicted because the maximum comes at $X = 0.50$). Show that the phase diagram in Figure 13-14 can be regarded as two phase diagrams of the type in Figure 13-6 (one involving A + AB and the other involving AB + B) placed beside one another with a common $X_{AB} = 1$, T axis.

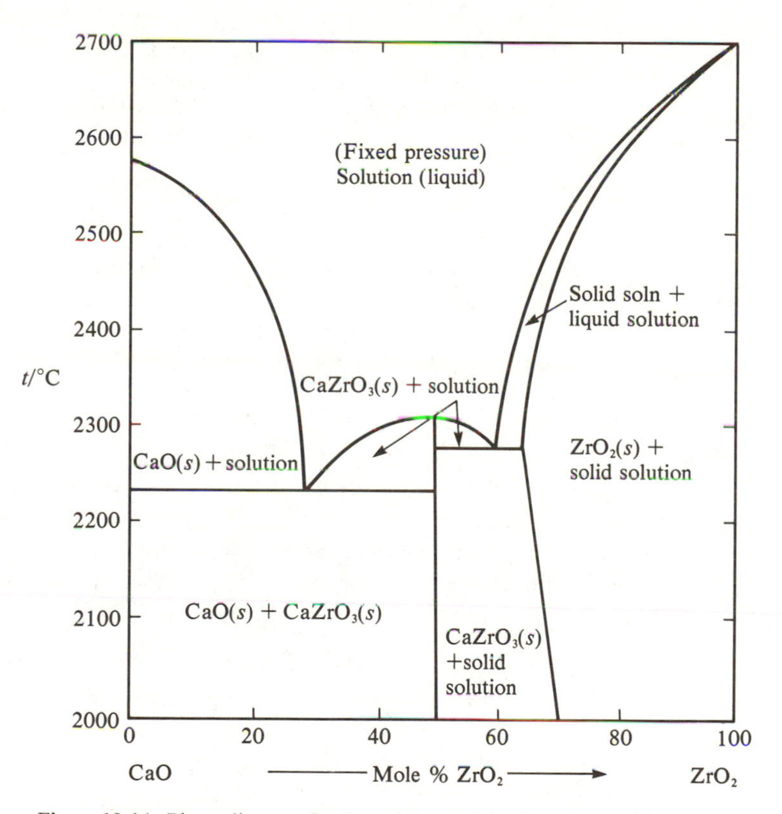

Figure 13-14. Phase diagram for the calcium oxide–zirconium oxide system.

35. In certain systems the solid phase that separates from the melt is not a pure compound but rather is a solid solution. In Figure 13-15 we have sketched a T-X diagram for a system exhibiting solid-solution formation over the entire composition range. Show that if a solution having $X_B = 0.60$ is cooled, then the solid that first separates out from the solution has $X_B = 0.25$. Note that $T^\circ_{f,A}$ and $T^\circ_{f,B}$ are the freezing points of pure A and pure B, respectively.

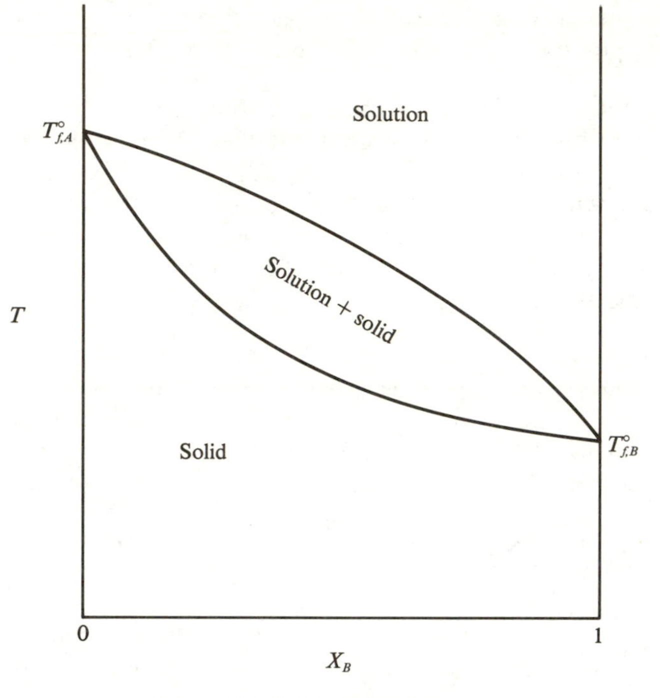

Figure 13-15. Solid solution formation.

36. An osmotic pressure "concentration cell" and an electrolytic concentration cell are opposed and matched thermodynamically by suitable devices such that no solvent or current can flow. Assuming that the solvent in the former is water and that one electron is transferred in the helf-cell reaction of the latter, calculate the osmotic pressure difference when the cell voltage is 0.010 V. State your assumptions.

37. Liquids are capable of existing in a metastable condition known as *negative pressure*, or hydrostatic tension. There is considerable evidence that sap rises in trees under a negative pressure arising from the powerful attractive force generated in capillaries by the evaporation of sap through the stomata (leaf pores). Given that the pressure of water in the roots of a 100-m tree is about 4 atm, and assuming that this root water is in equilibrium with water in the tree all along its height, estimate the pressure of water in the top of the tree. [*Hint:* Take $\bar{G}_{H_2O} = \bar{G}_{H_2O}(h, P)$].

38. Consider the influence of pressure on the solubility of solid A in liquid B. Take pure solid A at 1 atm as the standard state for A(s) and the hypothetical ideal $X_A = 1$ solution at 1 atm as the standard state for A(soln). Show that

$$\left(\frac{\partial \ln X_A}{\partial P}\right)_T = -\left(\frac{\bar{V}_A(soln) - \bar{V}_A(s)}{RT}\right)$$

39. Given the following data on water and *n*-propyl alcohol (denoted *alc*) solutions at 25°C, construct a graph of $\bar{V}_{tot}$ versus X_{alc} and determine $\bar{V}_i$ for H_2O and *n*-propyl alcohol at $X_{alc} = 0.25, 0.50,$ and 0.75.

X_{alc}	$\rho(g \cdot cm^{-3})$	X_{alc}	$\rho(g \cdot cm^{-3})$	X_{alc}	$\rho(g \cdot cm^{-3})$
0.00000	0.99913	0.11391	0.91234	0.72454	0.82179
0.00608	0.99580	0.23075	0.89184	0.87804	0.81730
0.01554	0.99141	0.31031	0.87158	0.90653	0.81490
0.02881	0.98626	0.41173	0.85126	0.93630	0.81282
0.05027	0.97914	0.54542	0.84101	0.96742	0.80982
0.06976	0.95318	0.62960	0.83051	1.00000	0.80733

40. The three possible geometrical isomers of dinitrobenzene are *o*-, *m*-, and *p*-dinitrobenzene. Given the following data (1 atm) and assuming ideal solution behavior and the absence of solid solution formation:

Isomer	$t_f/°C$	$\Delta H_{fus}/cal \cdot g^{-1}$
o	116.93	32.3
m	90.08	24.7
p	173.50	40.0

(a) Compute the eutectic temperatures and eutectic compositions at 1 atm of $o + m$, $o + p$, and $m + p$ solutions.
(b) Construct on graph paper the solid-liquid–phase diagram for the $p + m$ system at 1 atm.
(c) Estimate the freezing point of a liquid composed of equal amounts of the three isomers.

41. Consider a two-component ideal solution at 1 atm and fixed T subject to a gravitational field. Show that

$$d \ln X_1 = -\frac{g}{RT}(M_1 - \rho \bar{V}_1)\, dh$$

where M_1 is the molecular weight of 1, ρ is the density of the solution

$$\rho = \frac{X_1 M_1 + X_2 M_2}{X_1 \bar{V}_1 + X_2 \bar{V}_2} \qquad \text{(ideal solution)}$$

and g is the gravitational acceleration.

(a) Use the above expression to calculate the ratio of mole fractions of CCl_4 and C_6H_6 at the top and bottom of a column of a solution 1.00 m high having a total mole fraction of 0.500 at 300 K. Take $g = 980 \text{ m}\cdot\text{s}^{-1}$
(b) Repeat the calculation for $g = 98{,}000 \text{ m}\cdot\text{s}^{-1}$.

42. Show, with the aid of a phase diagram, that the purer a compound is, the larger its melting point range. How do you reconcile this with the use of sharpness of melting point as a criterion of purity?

43. Show that the freezing point of a pure solid that dissociates in solution is not depressed by the addition of an infinitesimal amount of one of its dissociation products to the melt. Let the formula of the solid be AB(*s*) and let the dissociation reaction be

$$\text{AB}(soln) \rightleftharpoons \text{A}(soln) + \text{B}(soln)$$

44. Derive an expression for the osmotic pressure of a solution for which the partial molar volume of the solvent is not constant but the compressibility of the solution is a constant.

45. Prove that an ideal solution cannot have an azeotrope.

46. Consider the case of a two-component A, B system that involves a continuous series of solid solutions and for which $T^\circ_{f,\text{A}} = T^\circ_{f,\text{B}}$. Sketch the T-X diagram for the following cases:
(a) The liquid and solid solutions are ideal.
(b) The liquid and solid solutions are not ideal, and there is a maximum freezing point $T_{\text{max}} > T^\circ_{f,\text{A}}$.
(c) The same situation as part (b), except that there is a minimum melting point $T_{\text{min}} < T^\circ_{f,\text{A}}$.

47. Suppose that when the solid $A_2(s)$ melts, it becomes monomeric, $A(\ell)$. Derive expressions for the freezing point depression, boiling point elevation, and osmotic pressure of solutions of $A_2(s)$ in $B(\ell)$.

48. One of the requirements for reverse osmosis is the generation of sufficient pressure to increase the activity of water to a value equal to that for pure water. It is proposed to operate a reverse osmosis unit, submerged in the ocean, which utilizes the hydrostatic pressure of the ocean water. The purified water would be pumped to the surface. Given normal seawater at 5°C, how deep would the unit have to be placed? Is such a proposal thermodynamically feasible?

49. Using data in Figure 13-13, estimate the pressure required to increase the activity of water in an $X_{H_2O} = 0.20$ alcohol-water solution at 25°C to that of pure water.

50. Estimate the boiling point of a 55 volume percent solution of ethylene glycol in water.

51. Estimate the vapor pressure in air exhaled from your lungs while you are breathing slowly.

52. How would you produce a high vacuum (say, $<10^{-6}$ torr) in a closed system without using pumps?

53. It is reported that tennis balls, which are normally pressurized at a total pressure of about 2 atm, retain their "bounce" much longer when filled with a 70:30 mixture of SF_6:air. Offer an explanation for this observation.

54. Starting from the Gibbs-Duhem equation for a two-component liquid at fixed T and total P, derive the Duhem-Margules equation

$$\frac{X_1}{P_1}\frac{dP_1}{dX_1} = \frac{X_2}{P_2}\frac{dP_2}{dX_2}$$

Use this equation to show that an azeotrope vaporizes without change in composition.

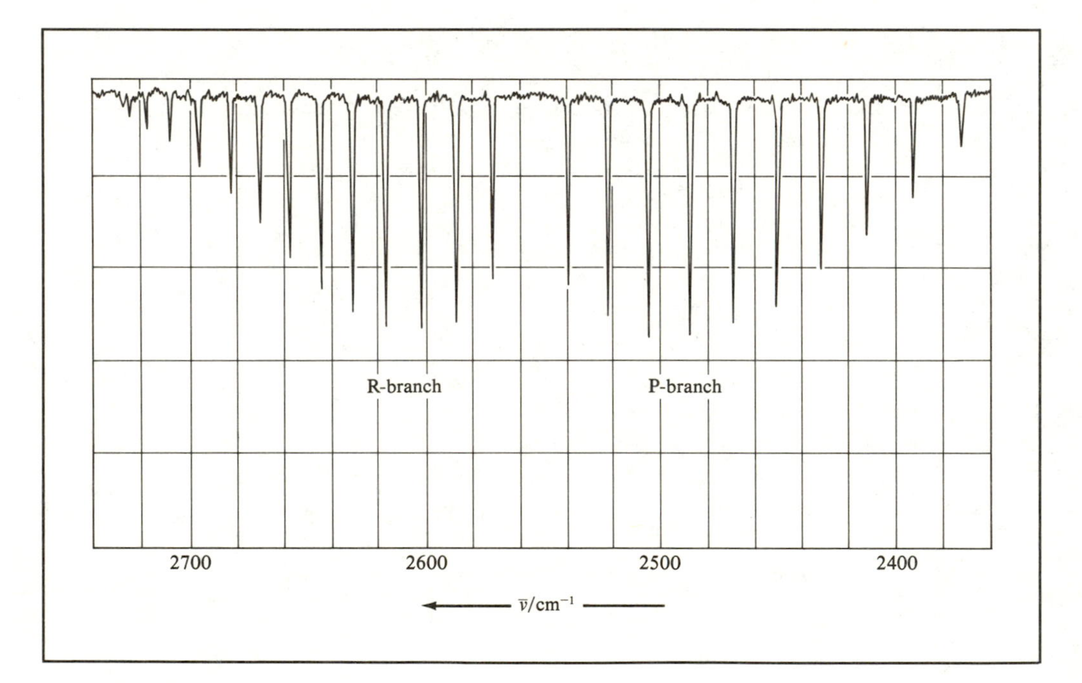

The rotational structure of the $v = 0$ to $v = 1$ transition of HBr gas.

14

Statistical Thermodynamics

There is, essentially, only one problem in statistical thermodynamics: the distribution of a given amount of energy U over N identical systems. Or perhaps better: to determine the distribution of an assembly of N identical systems over the possible states in which this assembly can find itself, given that the energy of the assembly is a constant U. The idea is that there is weak interaction between them, so weak that the energy of interaction can be disregarded, that one can speak of the "private" energy of every one of them and that the sum of their "private" energies has to equal U. The distinguished role of the energy is, therefore, simply that it is a constant of the motion—the one that always exists, and, in general, the only one.

Erwin Schrödinger,
Statistical Thermodynamics.
Cambridge: Cambridge University Press,
1967.

OUR PRIMARY objective in this chapter is to show how equilibrium constants and third-law entropies can be calculated from molecular level properties (namely, vibrational frequencies, bond lengths and bond angles, molecular dissociation energies, and atomic masses). The methods of calculation are based on the theory of equilibrium statistical thermodynamics; the molecular level data used in the calculations comes primarily from spectroscopic measurements.

Recall that classical thermodynamics places certain general restrictions on the allowed values of thermodynamic properties of substances (e.g., at $T > 0$, we have $\bar{S} > 0$ and $\bar{C}_P > 0$). However, classical thermodynamics provides no information whatsoever as to why the molar entropies and heat capacities of various substances at a particular temperature and pressure have the observed values. For example, why is it that at 25.00°C and 1.00 atm we obtain the following values of $\bar{S}°$, $\bar{C}_P^\circ$, and $\bar{H}_{298}^\circ - \bar{H}_0^\circ$?

Species	$\bar{S}°/J \cdot K^{-1} \cdot mol^{-1}$	$\bar{C}_P^\circ/J \cdot K^{-1} \cdot mol^{-1}$	$(\bar{H}_{298}^\circ - \bar{H}_0^\circ)/kJ \cdot mol^{-1}$
$H_2(g)$	130.57	28.82	8.468
$D_2(g)$	144.85	29.20	8.569
$H_2O(g)$	188.72	33.58	9.902
$CO_2(g)$	213.64	37.11	9.363
$CH_4(g)$	186.15	35.31	9.991
$CH_3CH_2OH(g)$	282.59	65.44	14.184

We are not content simply to know that the value of the equilibrium constant at 25°C for the reaction

$$H_2(g) + I_2(g) \rightleftharpoons 2\,HI(g)$$

is $K = 616$ (with 1-atm ideal gas standard states); we want to know *why*, from a molecular point of view, $K = 616$. As we shall see in subsequent sections, the methods of statistical thermodynamics enable us to calculate the equilibrium constant for the above reaction from the molecular properties of $H_2(g)$, $I_2(g)$, and $HI(g)$. We shall confine out statistical thermodynamic analysis primarily to gas-phase species and reactions because they are much simpler to analyze than reactions that involve solids, liquids, and/or solution-phase species.

In our treatment of statistical thermodynamics, we shall take the Planck-Boltzmann equation

$$S = k \ln \Omega \tag{14-1}$$

as our basic postulate. In Eq. 14-1, S is the total entropy of the system, k is Boltzmann's constant ("the gas constant per molecule")

$$k = \frac{R}{N_0} = \frac{8.3144 \text{ J} \cdot \text{K}^{-1} \cdot \text{mol}^{-1}}{6.0221 \times 10^{23} \text{ mol}^{-1}} = 1.3807 \times 10^{-23} \text{ J} \cdot \text{K}^{-1} \tag{14-2}$$

and Ω is the total number of ways (i.e., the total number of *microstates*) in which the particles can be distributed among the accessible energy levels of the system.

Once we have an expression for Ω, then we can calculate all the thermodynamic properties of the system, using Eq. 14-1 as a starting point.

14-1 *Distribution Formulas**

The energy for a system of *identical, noninteracting* particles can be subdivided into intervals labeled i that are sufficiently large that each energy interval includes a large number of *microcells* g_i occupied by a large number of particles n_i. Further, the spread of the energy intervals is sufficiently small compared to the total energy spread of the particles that the energies of the microcells of the ith energy interval can be approximated by a single energy value ε_i. In other words, the number of microcells per unit of energy is very large.

The basic problem in statistical thermodynamics is as follows. Given an *energy distribution* of the particles described by the occupation numbers of the microcells

$$n_1, n_2, n_3, \ldots$$

that is arbitrary except for the restrictions

$$\sum_i n_i = N \qquad (N \text{ is the total number of particles}) \tag{14-3}$$

and

$$\sum_i n_i\varepsilon_i = U \qquad (U \text{ is the total energy of the system of } N \text{ particles}) \tag{14-4}$$

one then computes the total number of ways Ω in which the n_i particles can be distributed among the g_i microcells of the ith interval for all i. When we have obtained an expression for Ω, we then find the most probable value of Ω, call it Ω_{mp}, by varying the n_i subject to the restrictions given by Eqs. 14-3 and 14-4.

Consider the simple case of a single interval i containing three microcells a, b, c among which two particles are distributed. There are three distinct cases for the distribution of the particles:

1. *Bose-Einstein Case:* Indistinguishable particles (x, x), with any number of particles per cell (*bosons*).
2. *Fermi-Dirac Case:* Indistinguishable particles (x, x), with only one particle per cell (*fermions*).
3. *Boltzmann Case:* Distinguishable particles (x, y), with any number of particles per cell (*boltzons*).

The two-particle distributions among the three microcells for the above three cases are given on the next page.

* This section follows closely the treatment of R. D. Cowan, *Amer. J. Phys.*, 25, 463, 1957.

Bose-Einstein			*Fermi-Dirac*			*Boltzmann*		
a	*b*	*c*	*a*	*b*	*c*	*a*	*b*	*c*
x	x	0	x	x	0	x	y	0
x	0	x	x	0	x	y	x	0
0	x	x	0	x	x	x	0	y
x, x	0	0	(3 microstates)			y	0	x
0	x, x	0				0	x	y
0	0	x, x				0	y	x
(6 microstates)						x, y	0	0
						0	x, y	0
						0	0	x, y
						(9 microstates)		

Strictly speaking, Boltzmann statistics can be used only in the case of distinguishable particles, for example, with atoms in a solid that are distinguishable by virtue of their locations at fixed lattice positions. However, Boltzmann statistics may be used for indistinguishable particles *as an approximation* to the correct quantum statistics, if we divide the number of microstates, calculated in the manner outlined above, by $N!$, the number of possible permutations of the N particles. The division by $N!$, in effect, constitutes a correction for the overcounting of microstates that results when we treat indistinguishable particles as distinguishable particles. The result for Ω that follows from division by $N!$ is called *corrected Boltzmann statistics.* In the above example, the "corrected" number of microstates is $9/2! = 4.5$, a result that is intermediate between the Bose-Einstein (BE) value of 6 microstates and the Fermi-Dirac (FD) value of 3 microstates. The intermediate value of Ω for corrected Boltzmann statistics is perfectly general; that is,

$$\Omega_{\mathrm{BE}} \geqslant \frac{\Omega_{\mathrm{B}}}{N!} \geqslant \Omega_{\mathrm{FD}} \tag{14-5}$$

EXAMPLE 14-1

A dilute statistical thermodynamic system is defined as one for which $g_i \gg n_i$, and thus it is very improbable that there will be more than one particle per microcell. Show for the 2-particle, 3-microcell case for a dilute system that

$$\Omega_{\mathrm{BE}} \approx \frac{\Omega_{\mathrm{B}}}{N!} \approx \Omega_{\mathrm{FD}} \tag{14-6}$$

Solution: Reference to the 2-particle, 3-microcell case described above shows that if we discard all microstates with more than one particle per microcell, then

$$\Omega_{\text{BE}} = 3 \qquad \Omega_{\text{FD}} = 3 \qquad \frac{\Omega_{\text{B}}}{N!} = \frac{6}{2!} = 3$$

and thus Eq. 14-6 holds.

The reason for the intermediate character of Boltzmann statistics is easy to understand. The states

a	*b*	*c*
x, *y*	0	0
0	*x*, *y*	0
0	0	*x*, *y*

that were counted in the Boltzmann case are not allowed in the Fermi-Dirac case, whereas the states

a	*b*	*c*
y, *x*	0	0
0	*y*, *x*	0
0	0	*y*, *x*

that were not counted in the Boltzmann case would have to be counted before division by $N!$ to obtain a result for Ω identical to that obtained with the BE case. We shall consider each of the three cases in detail.

I. *Boltzmann Statistics*

The number of distinguishable arrangements of N particles when subdivided into groups with n_1 particles in the first group, n_2 particles in the second group, and so on is

$$\phi = \frac{N!}{\prod_i n_i!} \tag{14-7}$$

EXAMPLE 14-2

Apply Eq. 14-7 to a case for which $N = 4$, $n_1 = 2$, and $n_2 = 2$, and write all the possible arrangements of the four particles.

Solution: From Eq. 14-7 we compute

$$\phi = \frac{N!}{\prod_i n_i!} = \frac{4!}{2!2!} = \frac{4 \times 3 \times 2 \times 1}{(2 \times 1)(2 \times 1)} = 6$$

Distinguishing the particles in the two groups by x and y, we have

$$\begin{array}{ccc} x\,x\,y\,y & y\,y\,x\,x & x\,y\,y\,x \\ y\,x\,x\,y & x\,y\,x\,y & y\,x\,y\,x \end{array}$$

Consider a specific case of the ϕ ways of distributing the N particles. For the ith group of particles, each of the n_i particles can be placed in any one of the g_i microcells of the ith energy level, and thus the n_i particles can be distributed among the g_i cell in $g_i^{n_i}$ different ways. That is, the first particle can be placed in any one of g_i microcells; the second particle can be placed in any one of the g_i microcells; and so forth. Thus

$$g_i \cdot g_i \cdot g_i \cdots = g_i^{n_i} \tag{14-8}$$

EXAMPLE 14-3

Verify Eq. 14-8 for the case $n_i = 2$, $g_i = 2$.

Solution: From Eq. 14-8 we have

$$g_i^{n_i} = 2^2 = 4$$

The four possible distributions are

x_1, x_2	0	x_1	x_2
0	x_1, x_2	x_2	x_1

The number of different ways of distributing the n_i particles of all i groups is

$$\prod_i g_i^{n_i} \tag{14-9}$$

and thus the total number of microstates is given by the product of Eqs. 14-7 and 14-8:

$$\Omega_B = N! \prod_i \frac{g_i^{n_i}}{n_i!} \tag{14-10}$$

Equation 14-10 applies to the case of uncorrected Boltzmann statistics. The value of Ω for corrected Boltzmann statistics, that is, Ω_{BC}, is obtained from Eq. 14-10 on division by $N!$; thus

$$\Omega_{BC} = \prod_i \frac{g_i^{n_i}}{n_i!} \tag{14-11}$$

EXAMPLE 14-4
Evaluate Ω_B and Ω_{BC} for the case in which $n_1 = 2$, $g_1 = 3$ and $n_2 = 2$, $g_2 = 3$.

Solution: From Eq. 14-10 we compute

$$\Omega_B = N! \prod_i \frac{g_i^{n_i}}{n_i!} = 4! \frac{3^2 \cdot 3^2}{2 \cdot 2} = 486$$

For Ω_{BC} we have

$$\Omega_{BC} = \frac{\Omega_B}{N!} = \frac{3^2 \cdot 3^2}{2 \cdot 2} = \frac{81}{4}$$

II. *Bose-Einstein Statistics*

In the BE case the particles are indistinguishable; thus the various groupings leading to the factor $(N!/\prod_i n_i!)$ in the Boltzmann case must be regarded as one and the same. The problem is to determine the number of ways in which n_i particles can be distributed among the g_i cells without regard to which particles are located in a given cell; that is, we are only concerned with how many particles are located in a given cell.

The first of the n_i particles can be placed in any of the g_i cells. The second particle can be placed in any of $g_i + 1$ positions ($g_i - 1$ unoccupied cells, plus to the left or the right of the particle in the occupied cell). The third particle can be placed in any of $g_i + 2$ positions ($g_i - 2$ unoccupied cells, plus left or right of the particles in the two occupied cells, i.e., $g_i - 2 + 4$, or if both particles are in the same cell, then the next particle can be placed in any of the $g_i - 1$ unoccupied cells plus to the left, to the right, or in between the two particles in the occupied cell). Thus for n_i particles we have

$$g_i(g_i + 1)(g_i + 2) \cdots (g_i + n_i - 1) \tag{14-12}$$

possible arrangements. The result in Eq. 14-12 must be divided by $n_i!$ because of the indistinguishability of the particles:

$$\frac{g_i(g_i + 1)(g_i + 2) \cdots (g_i + n_i - 1)}{n_i!} \tag{14-13}$$

The value of Ω_{BE} is obtained by taking the running product of Eq. 14-13 over all i intervals:

$$\Omega_{BE} = \prod_i \frac{g_i(g_i + 1)(g_i + 2) \cdots (g_i + n_i - 1)}{n_i!} \tag{14-14}$$

Equation 14-14 can be rewritten as

$$\Omega_{BE} = \prod_i \frac{(g_i + n_i - 1)!}{(g_i - 1)!n_i!} \tag{14-15}$$

as verified easily by writing the factorials $(g_i + n_i - 1)!$ and $(g_i - 1)!$, and cancelling like terms.

III. *Fermi-Dirac Statistics*

In the FD case the particles are indistinguishable; in addition, at most one particle may be placed in a given cell. Thus for n_i particles of the ith group $(n_i \leqslant g_i)$, the first particle can be placed in any of the g_i cells, the second in any of the $g_i - 1$ cells, and so on. The total number of microstates is

$$\Omega_{\text{FD}} = \prod_i \frac{g_i(g_i - 1)(g_i - 2)\cdots(g_i - n_i + 1)}{n_i!} \tag{14-16}$$

or

$$\Omega_{\text{FD}} = \prod_i \frac{g_i!}{n_i!(g_i - n_i)!} \tag{14-17}$$

IV. *General Treatment*

The BC, BE, and FD cases can be handled simultaneously by defining a quantity δ such that

$\delta = 0$ boltzons

$\delta = 1$ fermions

$\delta = -1$ bosons

The general result for Ω_n is

$$\Omega_n = \prod_i \frac{g_i(g_i - \delta)(g_i - 2\delta)\cdots[g_i - (n_i - 1)\delta]}{n_i!} \tag{14-18}$$

Note that Eq. 14-18 gives Ω_n, that is, the value of Ω for a particular distribution. The value of Ω for all possible distributions is

$$\Omega = \sum_n \Omega_n \tag{14-19}$$

However, only the maximum term in the sum in Eq. 14-19 makes an appreciable contribution to the entropy S

$$S = k \ln\left(\sum_n \Omega_n\right) \tag{14-20}$$

if the distribution is sharply peaked. The entropy S is a maximum at equilibrium; therefore, the value of Ω_n that we want is $\Omega_{\max}$.

$$S = k \ln \Omega = k \ln\left(\sum_n \Omega_n\right) \approx k \ln \Omega_{\max}$$

provided that Ω is very sharply peaked. Because S is of the order of R (the gas constant), the value of $\Omega_{\max}$ is of the order of e^{N_0}.

$$S \approx k \ln \Omega_{\max} \sim k \ln e^{N_0} = kN_0 = R$$

EXAMPLE 14-5

Show that even if there are N_0 (Avogadro's number) terms of magnitude comparable to Ω_{max}, we can take $S = k \ln \Omega_{max}$ without significant error.

Solution: Given that $\Omega = N_0 \Omega_{max}$, we have

$$S = k \ln \Omega = k \ln N_0 + k \ln \Omega_{max}$$

but $N_0 \sim 10^{24}$ and $\Omega_{max} \sim e^{N_0}$; thus

$$S = 55k + N_0 k = N_0 k$$

where $55k$ is totally negligible compared to $6.02 \times 10^{23}\, k$.

14-2 $S = k \ln \Omega_{max}$

Our objective in this section is to find an expression for the distribution of particles among the microcells that corresponds to Ω_{max}. The mathematics is simpler if we find the maximum in $\ln \Omega$ rather than Ω. Thus we seek a maximum in $\ln \Omega$ subject to the constraints

$$\sum_i n_i = N \qquad \sum_i n_i \varepsilon_i = U$$

Because $S = k \ln \Omega$, the total differential $d \ln \Omega$ is exact, where $\Omega = \Omega(N, U, V)$; thus at constant V

$$d \ln \Omega = \left(\frac{\partial \ln \Omega}{\partial N} \right)_{U,V} dN + \left(\frac{\partial \ln \Omega}{\partial U} \right)_{N,V} dU \tag{14-21}$$

Defining

$$\alpha = \left(\frac{\partial \ln \Omega}{\partial N} \right)_{U,V} \qquad \beta = \left(\frac{\partial \ln \Omega}{\partial U} \right)_{N,V} \tag{14-22}$$

we can rewrite Eq. (14-21) as

$$d \ln \Omega = \alpha\, dN + \beta\, dU = k^{-1}\, dS \tag{14-23}$$

The functions α and β are intensive because N and U are extensive and $d \ln \Omega$ is extensive. If we now increase the size of the system to m times the original size, while keeping all the intensive variables fixed, then we obtain, on integration of Eq. 14-23,

$$\int_{\ln \Omega}^{m \ln \Omega} d \ln \Omega = \alpha \int_N^{mN} dN + \beta \int_U^{mU} dU$$

$$\ln \Omega = \alpha N + \beta U \tag{14-24}$$

We now seek the maximum in Eq. 14-24 when the n_i values are varied; that is,

$$\frac{\partial}{\partial n_i} (\ln \Omega - \alpha N - \beta U) = 0 \tag{14-25}$$

but

$$N = \sum_i n_i \quad \text{and} \quad U = \sum_i n_i\varepsilon_i$$

Thus

$$\frac{\partial \ln \Omega}{\partial n_i} - \alpha - \beta\varepsilon_i = 0 \tag{14-26}$$

The expression for $\ln \Omega$ in terms of n_i is obtained from Eq. 14-18 (note that the logarithm of a product is equal to the sum of the logarithms):

$$\ln \Omega_n = \sum_i \{\ln g_i + \ln(g_i - \delta) + \cdots + \ln[g_i - (n_i - 1)\delta]\}$$
$$- \sum_i [\ln n_i + \ln(n_i - 1) + \cdots + \ln 2 + \ln 1] \tag{14-27}$$

The change in $\ln \Omega_n$ produced by an increase in n_i by one unit is obtained from Eq. 14-27:

$$\frac{\Delta \ln \Omega_n}{\Delta n_i} = \ln(g_i - n_i\delta) - \ln(n_i + 1) \tag{14-28}$$

Substitution of Eq. 14-28 into Eq. 14-26 yields

$$\ln(g_i - n_i\delta) - \ln(n_i + 1) - \alpha - \beta\varepsilon_i = 0 \tag{14-29}$$

Neglecting 1 with respect to n_i in Eq. 14-29 yields

$$\frac{g_i}{n_i} - \delta = e^{\alpha + \beta\varepsilon_i} \tag{14-30}$$

or

$$n_i = g_i(e^{\alpha + \beta\varepsilon_i} + \delta)^{-1} \tag{14-31}$$

The values given by Eq. 14-31 are the *equilibrium* (most probable) *values* because they correspond to the distribution $(n_1, n_2, n_3, \ldots)$ for which the entropy S is a maximum.

We now summarize our results up to this juncture. There is a set of closely spaced states available to the particles in the system. Neighboring states have essentially the same energy (not degenerate—just very close). Therefore, the energy spectrum can be divided arbitrarily into intervals. There are g_i states and n_i particles in the ith interval. The interval energy is approximated by ε_i:

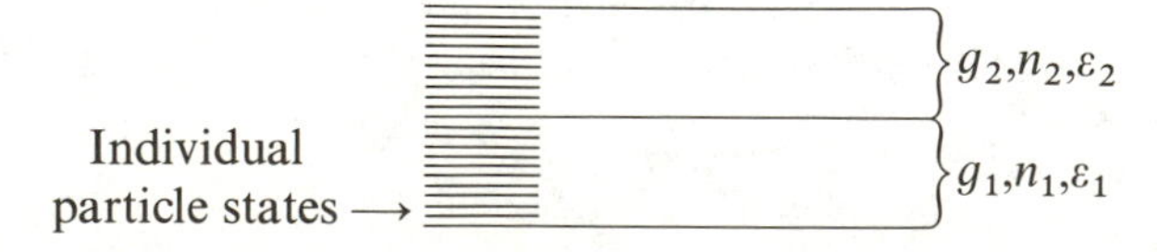

The most probable number of particles in the ith interval is given by

$$n_i = g_i(e^{\alpha + \beta\varepsilon_i} + \delta)^{-1}$$

Because the most probable number of particles in the ith interval depends on the energy of the interval and on the number of states in the ith interval (which is arbitrary), the result in Eq. 14-31 implies that the distribution (which is essentially dependent on only ε_i) for each particle state, that is, the number of particles in a *particle energy state* n_p, is given by

$$n_p = (e^{\alpha + \beta\varepsilon_p} + \delta)^{-1} \tag{14-32}$$

where

$$\sum_p n_p = n_i \tag{14-33}$$

For boltzons ($\delta = 0$) we obtain from Eqs. 14-33 and 14-31

$$n_p = e^{-\alpha}e^{-\beta\varepsilon_p}$$

If we now write the energies in terms of levels and not states, and if ω_j is the degeneracy of a particular level ($\omega_j \neq g_i$), we have for the set of degenerate energy levels

$$\left.\begin{aligned} n_p &= e^{-\alpha}e^{-\beta\varepsilon_p} \\ n_p &= e^{-\alpha}e^{-\beta\varepsilon_p} \\ n_p &= \cdots \\ n_p &= \cdots \end{aligned}\right\} \quad \omega_j \text{ levels}$$

Summing the n_p values for the above ω_j levels yields

$$\sum_p n_p = n_j = \omega_j e^{-\alpha}e^{-\beta\varepsilon_j} \tag{14-34}$$

EXAMPLE 14-6

Use Eq. 14-22 and $S = k \ln \Omega$ to evaluate α and β by comparison with general thermodynamic equations.

Solution: From the First and Second Laws of Thermodynamics, we have

$$dU = T\,dS - P\,dV + \mu\,dN \tag{14-35}$$

Solving Eq. 14-35 for dS yields

$$dS = \frac{dU}{T} + \frac{P}{T}dV - \frac{\mu}{T}dN$$

but

$$dS = d(k \ln \Omega)$$

Thus

$$k\beta = \left(\frac{\partial k \ln \Omega}{\partial U}\right)_{V,N} = \left(\frac{\partial S}{\partial U}\right)_{V,N} = \frac{1}{T}$$

and

$$\beta = \frac{1}{kT} \tag{14-36}$$

Further

$$k\alpha = \left(\frac{\partial k \ln \Omega}{\partial N}\right)_{U,V} = \left(\frac{\partial S}{\partial N}\right)_{U,V} = -\frac{\mu}{T}$$

and thus

$$\alpha = -\frac{\mu}{kT} = -\mu\beta \tag{14-37}$$

Substituting Eq. 14-36 and 14-37 into Eq. 14-31 yields for the distribution laws

$$n_i = \frac{g_i}{e^{\beta(\varepsilon_i - \mu)} + \delta} \tag{14-38}$$

14-3 *The Maxwell-Boltzmann Distribution Law Can Be Used When $n_i \ll g_i$*

Recall that corrected Boltzmann statistics is used as an approximation to Bose-Einstein or Fermi-Dirac statistics. The Boltzmann distribution law, which is often called the *Maxwell-Boltzmann* (MB) distribution law, is obtained from the general distribution law

$$n_i = g_i(e^{\alpha + \beta\varepsilon_i} + \delta)^{-1} \tag{14-31}$$

by setting $\delta = 0$. Because $\delta = +1$ (fermions) or $\delta = -1$ (bosons), the general condition for the applicability of the MB distribution law is

$$e^{\alpha + \beta\varepsilon_i} \gg 1 \tag{14-39}$$

The MB distribution law is

$$n_i = g_i e^{-\alpha} e^{-\beta\varepsilon_i} \tag{14-40}$$

From Eqs. 14-39 and 14-40 we have

$$\frac{n_i}{g_i} = \frac{1}{e^{\alpha + \beta\varepsilon_i}} \ll 1 \tag{14-41}$$

Thus, the condition for the applicability of the MB distribution law can be written as $n_i \ll g_i$. If $n_i \ll g_i$, then the number of particles n_i with energy ε_i is much less than the number of microcells g_i in the ith interval for all i. Such a system is said to be a *dilute system.*

Summation of Eq. 14-40 over all i intervals yields

$$\sum_i n_i = N = e^{-\alpha} \sum_i g_i e^{-\beta\varepsilon_i}$$

or

$$e^{-\alpha} = \frac{N}{\sum_i g_i e^{-\beta\varepsilon_i}} \tag{14-42}$$

We define the *partition function q* as

$$q = \sum_i g_i e^{-\beta\varepsilon_i} = \sum_j \omega_j e^{-\beta\varepsilon_j} \tag{14-43}$$

The partition function is thus a *weighted sum over states*; the weighting factor for each state is *the Boltzmann factor* $e^{-\beta\varepsilon_i}$ for that state.

EXAMPLE 14-7

(a) Show that N/q is an intensive quantity.
(b) Show that $N/q \ll 1$.

Solution:

(a) From Eqs. 14-43 and 14-42 we have

$$e^{-\alpha} = \frac{N}{q} \tag{14-44}$$

$e^{-\alpha}$ is intensive ($e^{-\alpha}$ is dimensionless), and thus N/q must be intensive.

(b) From Eq. 14-41 we have

$$e^{-\alpha}e^{-\beta\varepsilon_i} = \frac{n_i}{g_i} \ll 1$$

and thus

$$\frac{N}{q}e^{-\beta\varepsilon_i} = \frac{n_i}{g_i} \ll 1$$

Because $\beta = 1/kT > 0$, and $\varepsilon_i \geqslant 0$, the largest value of $e^{-\beta\varepsilon_i}$ is $e^{-0} = 1$; thus

$$\frac{N}{q} \ll 1 \tag{14-45}$$

Substitution of Eq. 14-44 into Eq. 14-40 yields for the fraction of particles in the ith interval

$$\frac{n_i}{N} = \frac{g_i e^{-\beta\varepsilon_i}}{q} \tag{14-46}$$

The fraction of particles in the jth particle energy level is

$$\frac{n_j}{N} = \frac{\omega_j e^{-\beta\varepsilon_j}}{q} \tag{14-47}$$

where q in Eq. 14-47 is given by

$$q = \sum_j \omega_j e^{-\beta\varepsilon_j} \tag{14-48}$$

The ratio of the numbers of particles in two different energy levels is obtained by applying Eq. 14-47 twice and then dividing:

$$\frac{n_j}{n_k} = \frac{\omega_j}{\omega_k} e^{-\beta(\varepsilon_j - \varepsilon_k)} \tag{14-49}$$

Note that at high temperatures

$$\lim_{T\to\infty} \exp\left[\frac{-(\varepsilon_j - \varepsilon_k)}{kT}\right] \to 1$$

and thus

$$\lim_{T\to\infty} \frac{n_j}{n_k} = \frac{\omega_j}{\omega_k} \tag{14-50}$$

Thus at high temperatures the ratio of the populations of any two energy levels is equal to the ratio of the degeneracies of the two levels. Note further that if the levels are nondegenerate, then in the high-temperature limit the populations of all levels become equal. In the low-temperature limit we obtain from Eq. 14-49

$$\lim_{T \to 0} \frac{n_j}{n_k} \to 0 \qquad \text{for } \varepsilon_j - \varepsilon_k > 0$$

In other words, as $T \to 0$, all the particles are in the lowest available energy level.

In general, at a finite nonzero temperature, statistical thermodynamic equilibrium represents a balance between the randomizing forces of thermal agitation that tend to populate all the levels equally (with $\omega_i = 1$) and the tendency of a mechanical system to go into the lowest available energy level.

We can obtain a statistical thermodynamic interpretation of heat and work as follows. The internal energy of the system is given by

$$U = \sum_i n_i \varepsilon_i \tag{14-51}$$

Differentiation of Eq. 14-51 yields

$$dU = \sum_i \varepsilon_i \, dn_i + \sum_i n_i \, d\varepsilon_i = \delta q + \delta \omega \tag{14-52}$$

The value of $\delta\omega$ is given by

$$\delta\omega = -\sum_i n_i F_{i\ell} \, d\ell = -\sum_i n_i \left[-\left(\frac{\partial \varepsilon_i}{\partial \ell} \right) \right] d\ell \tag{14-53}$$

where $F_{i\ell}$ is the force that acts over the displacement $d\ell$; thus

$$\delta\omega = \sum_i n_i \, d\varepsilon_i \tag{14-54}$$

and therefore

$$\delta q = \sum_i \varepsilon_i \, dn_i \tag{14-55}$$

From Eq. 14-54 we conclude that the statistical thermodynamic interpretation of work is that work done on the system corresponds to a change in the energy levels with fixed n_i values. The input of energy as heat corresponds to a change in the distribution of the particles over the energy levels at fixed values of the ε_i (Equation 14-55).

14-4 *The Thermodynamic State Functions Can Be Expressed in Terms of the Partition Function*

If we have the partition function for a system, then we can compute all the equilibrium thermodynamic properties of the system. In this section we derive relationships between the thermodynamic state functions and the partition functions for boltzons. Our starting point is the Boltzmann-Planck equation:

$$S = k \ln \Omega \tag{14-56}$$

There are two distinct cases of interest, namely, *indistinguishable particles* and *distinguishable particles*.

Substitution of Eq. 14-11 into Eq. 14-56 yields

$$S = k \ln\left(\prod_i \frac{g^{n_i}}{n_i!}\right) \tag{14-57}$$

From Eq. 14-57 we obtain

$$S = k \sum_i (n_i \ln g_i - n_i \ln n_i + n_i) \tag{14-58}$$

where we have used Stirling's approximation for the factorial ($\ln N! = N \ln N - N$; see Problem 2). Rearrangement of Eq. 14-58 yields

$$S = k \sum_i n_i \left(\ln \frac{g_i}{n_i} + 1\right) \tag{14-59}$$

Substitution of Eq. 14-46 into Eq. 14-59 yields

$$S = k \sum_i n_i \left(\ln \frac{q}{N} + \beta\varepsilon_i + 1\right) \tag{14-60}$$

Rearrangement of Eq. 14-60 yields

$$S = k \ln\left(\frac{q}{N}\right) \sum_i n_i + k\beta \sum n_i \varepsilon_i + k \sum_i n_i \tag{14-61}$$

Thus

$$S = kN \ln\left(\frac{q}{N}\right) + \frac{U}{T} + kN \tag{14-62}$$

or

$$S = k \ln q^N - k(N \ln N - N) + \frac{U}{T} \tag{14-63}$$

Using Stirling's approximation in Eq. 14-63 yields ($N \ln N - N = \ln N!$)

$$S = k \ln \frac{q^N}{N!} + \frac{U}{T} \qquad \text{(indistinguishable particles)} \tag{14-64}$$

We can obtain an expression for U by substituting Eq. 14-46 into the equation

$$U = \sum_i n_i \varepsilon_i \tag{14-65}$$

Thus

$$U = \frac{N}{q} \sum g_i \varepsilon_i e^{-\beta\varepsilon_i} \tag{14-66}$$

Comparison of Eq. 14-66 with the expression (recall that

$$q = \sum_i g_i e^{-\beta\varepsilon_i}$$

and that $\beta = 1/kT$)

$$\left(\frac{\partial \ln q}{\partial T}\right)_{N,V} = \frac{1}{q}\left(\frac{\partial q}{\partial T}\right)_{N,V} = \frac{1}{q} \sum_i \frac{g_i \varepsilon_i}{kT^2} e^{-\varepsilon_i/kT} \tag{14-67}$$

shows that

$$U = NkT^2\left(\frac{\partial \ln q}{\partial T}\right)_{N,V} = kT^2\left[\frac{\partial \ln (q^N/N!)}{\partial T}\right]_{N,V} \tag{14-68}$$

EXAMPLE 14-8

Show that for distinguishable particles

$$S = k \ln q^N + \frac{U}{T}$$

Solution: Substitution of Eq. 14-10 into the Planck-Boltzmann equation yields

$$S = k \ln \Omega = k \ln\left(N! \sum_i \frac{g_i^{n_i}}{n_i!}\right) \tag{14-69}$$

and thus (using Eq. 14-46)

$$S = k(N \ln N - N) + k \sum_i n_i\left(\ln \frac{q}{N} + \beta\varepsilon_i + 1\right)$$

Comparison with Eqs. 14-60 to Eq. 14-63 shows that

$$S = k(N \ln N - N) + k \ln q^N - k(N \ln N - N) + \frac{U}{T}$$

and thus

$$S = k \ln q^N + \frac{U}{T} \quad \text{(distinguishable particles)} \tag{14-70}$$

We define the *canonical partition function Q* ("big *Q*") as follows:

$$Q = \frac{q^N}{N!} \quad \text{(indistinguishable particles)} \tag{14-71}$$

$$Q = q^N \quad \text{(distinguishable particles)} \tag{14-72}$$

We summarize our key results for boltzons as follows:

$$q = \sum_i \omega_i e^{-\varepsilon_i/kT} \tag{14-73}$$

$$\frac{n_i}{N} = \frac{\omega_i e^{-\varepsilon_i/kT}}{q} \tag{14-74}$$

$$S = k \ln Q + \frac{U}{T} \tag{14-75}$$

$$U = kT^2\left(\frac{\partial \ln Q}{\partial T}\right)_{N,V} \tag{14-76}$$

The equations for the state functions obtained from Eq. 14-73 through Eq. 14-76 are

$$A = U - TS = -kT\ln Q \tag{14-77}$$

$$G = A + PV = -kT\ln Q + PV \tag{14-78}$$

$$P = -\left(\frac{\partial A}{\partial V}\right)_{T,N} = kT\left(\frac{\partial \ln Q}{\partial V}\right)_{T,N} \tag{14-79}$$

$$\mu = \left(\frac{\partial A}{\partial N}\right)_{T,V} = -kT\left(\frac{\partial \ln Q}{\partial N}\right)_{T,V} \tag{14-80}$$

$$C_V = \left(\frac{\partial U}{\partial T}\right)_{N,V} = 2kT\left(\frac{\partial \ln Q}{\partial T}\right)_{N,V} + kT^2\left(\frac{\partial^2 \ln Q}{\partial T^2}\right)_{N,V} \tag{14-81}$$

We used the relationship

$$dA = -S\,dT - P\,dV + \mu\,dN = -d(kT\ln Q)$$

to obtain Eqs. 14-79 and 14-80.

When no two identical particles are permitted in the same quantum state of the system (Fermi-Dirac statistics) or when many particles are placed in the same quantum state of the system (Bose-Einstein statistics), then additional complications arise whenever the number of available states is not very much larger than the number of particles, and Eqs. 14-71 and 14-72 cannot be used. All elementary particles, and hence all atoms and molecules, are either fermions or bosons. They are fermions when the number of protons plus neutrons plus electrons is odd (e.g., D and ^{3}He) and bosons when this number is even (e.g., H, H_2, and D_2). We shall confine our analysis to particles that obey Maxwell-Boltzmann statistics. However, it must be realized that if the criterion for the applicability of MB statistics ($N \ll q$) does not hold, then either BE or FD statistics, as appropriate, must be used.

14-5 *Energy Level Expressions Are Obtained from Quantum Theory*

The molecular partition function q is given by the equation

$$q = \sum_i \omega_i e^{-\varepsilon_i/kT} \tag{14-82}$$

where the ε_i values are the energy levels and the ω_i values are the degeneracies ($\omega_i \geqslant 1$) of the corresponding energy levels. If we know ε_i, ω_i for all i, then we can compute all the equilibrium thermodynamic properties of the system by substituting the resulting expression for q into Eq. 14-73 through Eq. 14-81. The energy level expressions used to obtain an expression for q are obtained from quantum theory together with spectroscopic measurements, which give the numerical values of the required molecular spectroscopic constants.

We shall restrict our treatment to ideal gases. (Recall that the standard state for a gaseous species is the 1-atm ideal gas.) To simplify the analysis further we shall assume that the various molecular degrees of freedom (*t*ranslation, *r*otation, *v*ibration, *e*lectronic, and *n*uclear) are separable. That is, we assume that various contributions to the total energy of the molecule ε_{tot} are simply additive (no cross terms); thus

$$\varepsilon_{tot} = \varepsilon_0 + \varepsilon_t + \varepsilon_r + \varepsilon_v + \varepsilon_e + \varepsilon_n \qquad (14\text{-}83)$$

where ε_0 fixes the zero of energy for the molecule. If the energies are additive, then the partition function can be written as a product of terms:

$$q = q_0 q_t q_r q_v q_e q_n \qquad (14\text{-}84)$$

Equation 14-84 follows from Eq. 14-68 because the logarithm of a product is equal to the sum of the logarithms.

We shall now proceed to evaluate the individual partition function factors in Eq. 14-84 using quantum theory results for the energy level expressions.

14-6 *The Translational Partition Function*

The three-dimensional translational partition q_t can be written as a product of three one-dimensional partition functions,

$$q_t = q_x q_y q_z \qquad (14\text{-}85)$$

because the total translational energy ε_t is the sum of the translational energies in the x, y, and z directions:

$$\varepsilon_t = \varepsilon_x + \varepsilon_y + \varepsilon_z$$

The energy levels for a one-dimensional particle in an infinite rectangular potential well of width x are given by

$$\varepsilon_n = \frac{n^2h^2}{8mx^2} \qquad n = 1, 2, 3, \ldots \qquad (14\text{-}86)$$

where m is the particle mass, h is Planck's constant, and n is the translational quantum number. The one-dimensional translational energy levels are nondegenerate ($\omega_n = 1$), and thus the one-dimensional partition function for translation is

$$q_x = \sum_{n=1}^{\infty} e^{-n^2h^2/8mx^2kT} \qquad (14\text{-}87)$$

The separation between translational energy levels is very small compared to kT, provided that the length x is macroscopic in magnitude and that T is not close to zero.

EXAMPLE 14-9

Compare the difference in energy between the $n = 1$ and $n = 2$ translational energy levels for an $H_2O(g)$ molecule in a 10-cm rectangular potential well at 300 K with the value of kT at 300 K.

Solution: From Eq. 14-86 we have

$$\Delta\varepsilon_t = (2^2 - 1^2)\frac{h^2}{8mx^2} = \frac{3h^2}{8mx^2}$$

and

$$\Delta\varepsilon_t = \frac{3 \times (6.63 \times 10^{-34}\ \mathrm{J\cdot s})^2}{8 \times (18.0\ \mathrm{g\cdot mol^{-1}}/6.02 \times 10^{23}\cdot \mathrm{mol^{-1}}) \times (0.100\ \mathrm{m})^2 \times (1\ \mathrm{kg}/1000\ \mathrm{g})}$$

$$= 5.49 \times 10^{-40}\ \mathrm{J}$$

The value of kT at 300 K is

$$kT = (1.38 \times 10^{-23}\ \mathrm{J\cdot K^{-1}})(300\ \mathrm{K}) = 4.14 \times 10^{-21}\ \mathrm{J}$$

and thus $\Delta\varepsilon_t \ll kT$.

If $\Delta\varepsilon_t \ll kT$, then we can treat n as a continuous variable and change the sum in Eq. 14-87 to an integral

$$q_x = \int_0^\infty e^{-n^2h^2/8mx^2kT}\, dn \tag{14-88}$$

The conversion of a quantum number function to a continuous variable is called the *classical* (nonquantized) *approximation;* the classical translational energy is not quantized but rather is continuous.

Figure 14-1 shows the comparison between the sum in Eq. 14-87 and the integral in Eq. 14-88. The integral in Eq. 14-88 is of the form

$$\int_0^\infty e^{-bn^2}\, dn = \frac{1}{2}\left(\frac{\pi}{b}\right)^{1/2} \qquad (b\ \text{constant})$$

where $b = h^2/8mx^2kT$, and therefore

$$q_x = \left(\frac{2\pi mkT}{h^2}\right)^{1/2} x \tag{14-89}$$

Using Eq. 14-89 together with analogous results for q_y and q_z we obtain for q_t

$$q_t = \left(\frac{2\pi mkT}{h^2}\right)^{3/2} V \tag{14-90}$$

where the volume $V = xyz$. The condition for the applicability of Eq. 14-90 is obtained by substituting Eq. 14-90 into Eq. 14-45:

$$\frac{N}{q} = \left(\frac{h^2}{2\pi mkT}\right)^{3/2} \frac{N}{V} \ll 1 \tag{14-91}$$

Setting $m = M/N_0$, where M is the molecular mass ($\mathrm{g\cdot mol^{-1}}$) and N_0 is Avogadro's number, in Eq. 14-90 we obtain

$$q_t = \frac{(2\pi MkT)^{3/2} V}{h^3 N_0^{3/2}} \tag{14-92}$$

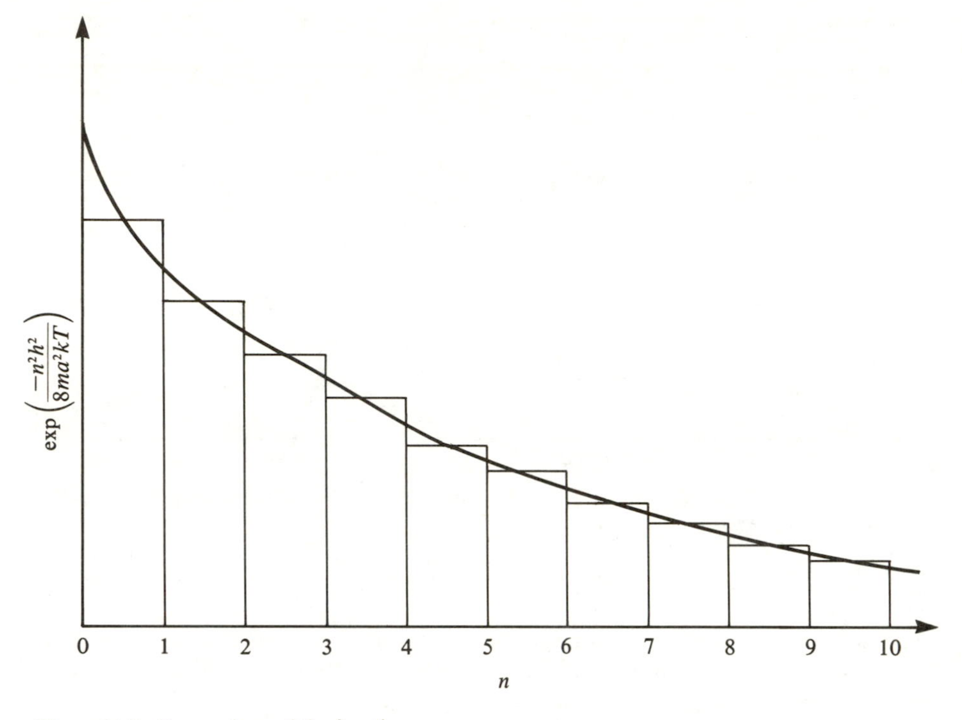

Figure 14-1. Comparison of the function

(a) $e^{-n^2h^2/8mx^2kT}$ $\quad n = 1, 2, 3, 4, \ldots$

and the function

(b) $e^{-n^2h^2/8mx^2kT}$ $\quad n$ continuous

plotted against n; $h^2/8mx^2kT$ is a constant. If function (a) changes only very slowly with n, then the sum of the areas under the rectangles is approximately equal to the area under the curve defined by function (b) with n a continuous variable.

Further, because $Q_t = q_t^N/N!$, we are now in a position to calculate the translational contributions to the thermodynamic properties of a gas. Noting first that

$$\ln Q_t = N \ln q_t - N \ln N + N = N \ln q_t + N \ln\left(\frac{e}{N}\right) = N \ln\left(\frac{q_t e}{N}\right)$$

we obtain using Eq. 14-92

$$\ln Q_t = N \ln\left[\frac{(2\pi MkT)^{3/2} Ve}{h^3 N_0^{3/2} N}\right] \tag{14-93}$$

Substitution of Eq. 14-93 into Eq. 14-76 yields for the translation energy of an ideal gas relative to its energy at absolute zero

$$U_t = (U - U_0)_t = kT^2\left(\frac{\partial \ln Q_t}{\partial T}\right)_{N,V} = \frac{3}{2} NkT \tag{14-94}$$

For 1 mol of gas we have from Eq. 14-94 with $N_0 k = R$

$$\bar{U}_t = (\bar{U} - \bar{U}_0)_t = \tfrac{3}{2}RT$$

Because U_t is independent of P and the gas is ideal, $\bar{U}_t = \bar{U}_t^\circ = (\bar{U}^\circ - \bar{U}_0^\circ)_t$; therefore,

$$(\bar{U}^\circ - \bar{U}_0^\circ)_t = \tfrac{3}{2}RT \tag{14-95}$$

Furthermore, because $\bar{H} = \bar{U} + P\bar{V} = \bar{U} + RT$ and $\bar{H}_0^\circ = \bar{U}_0^\circ$, we have from Eq. 14-95

$$(\bar{H}^\circ - \bar{H}_0^\circ)_t = \tfrac{5}{2}RT \tag{14-96}$$

From Eqs. 14-95 and 14-96 we obtain

$$\bar{C}_{V,t}^\circ = \left(\frac{\partial \bar{U}_t^\circ}{\partial T}\right)_V = \frac{3R}{2} \tag{14-97}$$

$$\bar{C}_{P,t}^\circ = \left(\frac{\partial \bar{H}_t^\circ}{\partial T}\right)_P = \frac{5R}{2} \tag{14-98}$$

respectively. For the translational entropy of 1 mol of the gas, we have from Eqs. 14-75, 14-92, and 14-95

$$\bar{S}_t = k \ln Q_t + \frac{\bar{U}_t}{T} = R \ln\left[\frac{(2\pi MkT)^{3/2} Ve}{h^3 N_0^{5/2}}\right] + \frac{3}{2}R \tag{14-99}$$

or

$$\bar{S}_t = \frac{3}{2} R \ln M + \frac{3}{2} R \ln T + R \ln V + \frac{3}{2} R + R \ln\left[\frac{(2\pi k)^{3/2} e}{h^3 N_0^{5/2}}\right] \tag{14-100}$$

Eliminating V from Eq. 14-100 using the relation $PV = RT$, we obtain (on collecting constants into one term)

$$\bar{S}_t = \frac{3}{2} R \ln M + \frac{5}{2} R \ln T - R \ln P + R \ln\left[\frac{(2\pi k)^{3/2} R e^{5/2}}{h^3 N_0^{5/2}}\right] \tag{14-101}$$

Equation 14-101 is known as the *Sackur-Tetrode equation.* Evaluation of the last term in Eq. 14-101 yields

$$\bar{S}_t = \tfrac{3}{2}R \ln M + \tfrac{5}{2}R \ln T - R \ln P - 9.686\ \mathrm{J \cdot K^{-1} \cdot mol^{-1}} \tag{14-102}$$

Because the gas is ideal, we have $\bar{S}_t = \bar{S}_t^\circ$ at $P = 1$ atm; and from Eq. 14-102 we obtain

$$\bar{S}_t^\circ = \tfrac{3}{2}R \ln M + \tfrac{5}{2}R \ln T - 9.686\ \mathrm{J \cdot K^{-1} \cdot mol^{-1}} \tag{14-103}$$

For a *monatomic gas* without electronic excitation that also has a nondegenerate electronic ground state ($q_e = 1$), Eq. 14-103 gives the total entropy of the gas in its standard state.

EXAMPLE 14-10

The experimental value for the molar entropy of Ne(g) in its 1-atm ideal gas standard state at 298.15 K is $\bar{S}^\circ = 146.5 \pm 0.4\ \mathrm{J \cdot K^{-1} \cdot mol^{-1}}$. Calculate $\bar{S}_{298.15}^\circ$ for Ne(g) using Eq. 14-103.

Solution: For Ne(g), $M = 20.183$; thus

$$\overline{S}^\circ_{298.15} = \tfrac{3}{2}R\ln M + \tfrac{5}{2}R\ln T - 9.686\ \mathrm{J\cdot K^{-1}\cdot mol^{-1}}$$

$$\overline{S}^\circ_{298.15} = \left(\frac{8.3144\ \mathrm{J\cdot K^{-1}\cdot mol^{-1}}}{2}\right)(3\ln 20.183 + 5\ln 298.15) - 9.686\ \mathrm{J\cdot K^{-1}\cdot mol^{-1}}$$

and

$$\overline{S}^\circ_{298.15} = 146.22\ \mathrm{J\cdot K^{-1}\cdot mol^{-1}}$$

In this case the calculated value of $\overline{S}^\circ_{298.15}$ is more accurate than the experimental value, and in fact the value of $\overline{S}^\circ_{298.15}$ for Ne(g) that is found in standard tabulations of thermodynamic data (e.g., NBS Technical Note 270-3) is a *calculated value.*

From the relation $A = U - TS$ together with Eqs. 14-95 and 14-103 we obtain

$$(\overline{A}^\circ - \overline{A}^\circ_0)_t = (\overline{A}^\circ - \overline{U}^\circ_0)_t = (\overline{U}^\circ - \overline{U}^\circ_0)_t - T\overline{S}^\circ_t$$

$$(\overline{A}^\circ - \overline{U}^\circ_0)_t = T(\tfrac{3}{2}R - \tfrac{3}{2}R\ln M - \tfrac{5}{2}R\ln T + 9.686)$$

or

$$(\overline{A}^\circ - \overline{U}^\circ_0)_t = T(22.158 - \tfrac{3}{2}R\ln M - \tfrac{5}{2}R\ln T) \qquad (\mathrm{J\cdot mol^{-1}}) \qquad (14\text{-}104)$$

From the relation $G = H - TS$ together with Eqs. 14-96 and 14-103 we obtain

$$(\overline{G}^\circ - \overline{G}^\circ_0)_t = (\overline{G}^\circ - \overline{H}^\circ_0)_t = (\overline{H}^\circ - \overline{H}^\circ_0)_t - T\overline{S}^\circ_t$$

$$(\overline{G}^\circ - \overline{H}^\circ_0)_t = T(30.472 - \tfrac{3}{2}R\ln M - \tfrac{5}{2}R\ln T) \qquad (\mathrm{J\cdot mol^{-1}}) \qquad (14\text{-}105)$$

From Eq. 14-105 we find for the free energy function

$$-\left[\frac{(\overline{G}^\circ - \overline{H}^\circ_0)}{T}\right]_t = R\left(\frac{3}{2}\ln M + \frac{5}{2}\ln T\right) - 30.472 \qquad (\mathrm{J\cdot K^{-1}\cdot mol^{-1}}) \qquad (14\text{-}106)$$

TABLE 14-1

Thermodynamic Functions for Translation

$$\overline{S}^\circ_t = R\left(\frac{3}{2}\ln M + \frac{5}{2}\ln T\right) - 9.686\ \mathrm{J\cdot K^{-1}\cdot mol^{-1}}$$

$$-\left(\frac{\overline{G}^\circ - \overline{H}^\circ_0}{T}\right)_t = R\left(\frac{3}{2}\ln M + \frac{5}{2}\ln T\right) - 30.472\ \mathrm{J\cdot K^{-1}\cdot mol^{-1}}$$

$$(\overline{H}^\circ - \overline{H}^\circ_0)_t = \frac{5}{2}RT$$

$$\overline{C}^\circ_{P,t} = \frac{5}{2}R$$

In Table 14-1 we have collected the more useful of the thermodynamic functions for translation. The expressions in Table 14-1 can be taken as generally valid even though they were obtained assuming that the translational energy levels are continuous (conversion of the sum to an integral), because the translational energy levels are so closely spaced relative to kT that quantum effects can be ignored for translation except as $T \rightarrow 0$ or as $V \rightarrow 0$.

14-7 *Nuclear and Electronic Partition Functions*

The translational partition function is rigorously separable from the electronic and nuclear partition functions (i.e., there is no mixing of translational states with electronic and nuclear states), and the electronic and nuclear partition functions also are separable to a high degree of accuracy (Born-Oppenheimer approximation); therefore, in the case of a monatomic gas,

$$Q = \frac{q^N}{N!} = \frac{(q_t q_n q_e)^N}{N!} \tag{14-107}$$

We have already discussed $(q_t)^N/N!$, and we now turn our attention to the nuclear and the electronic partition functions q_n and q_e, respectively. Owing to the very large energy separation between the two lowest nuclear energy levels (about 1 MeV) at all ordinary temperatures, the nucleus is essentially certain to be in its ground state (see Problem 7). For a nucleus of spin i the degeneracy of the ground nuclear energy level is $2i + 1$; therefore, the nuclear partition function of a single nucleus can be written as

$$q_n = \sum_i \omega_i e^{-\varepsilon_i/kT} = \omega_0 = 2i + 1 \tag{14-108}$$

where we have taken $\varepsilon_0 = 0$ for the lowest nuclear energy state. If we are dealing with a molecule of r atoms, then

$$q_n = \prod_{j=1}^{r} (2i_j + 1) \tag{14-109}$$

The nuclear partition function gives rise to a molar nuclear spin entropy of

$$\bar{S}_n^\circ = N_0 k \ln q_n + \frac{\bar{U}_n^\circ}{T} = R \ln q_n \tag{14-110}$$

where $\bar{U}_n = 0$, with all nuclei in their ground states. Substitution of Eq. 14-109 into Eq. 14-110 yields

$$\bar{S}_n^\circ = R \ln \prod_{j=1}^{r} (2i_j + 1) \tag{14-111}$$

The nuclear spin entropy given by Eq. 14-111 is not included in the statistical thermodynamic calculation of third-law entropies because the nuclear spin entropy is not ordinarily removed at the lowest temperatures of the calorimetric measurements (see Chapter 6).

EXAMPLE 14-11

The nuclear spin of ^{14}N is $i = 1$, and the nuclear spin of ^{1}H is $i = \frac{1}{2}$. Compute the nuclear spin entropy of $^{14}N^{1}H_3(g)$ at 25°C.

Solution: Using Eq. 14-111 we have

$$\bar{S}_n^\circ = R \ln \prod_{j=1}^{4} (2i_j + 1) = R \ln\{[2(1)+1][2(\tfrac{1}{2})+1]^3\} = 26.42\ \text{J}\cdot\text{K}^{-1}\cdot\text{mol}^{-1}$$

We now turn our attention to the electronic partition function q_e. The separation between electronic energy levels is usually of the order of 100 $\text{kJ}\cdot\text{mol}^{-1}$; therefore, for most cases at ordinary temperatures (<500 K), electronic excitation can be ignored. In any event, we have for q_e

$$q_e = \omega_0 + \omega_1 e^{-\varepsilon_1/kT} + \omega_2 e^{-\varepsilon_2/kT} + \cdots \tag{14-112}$$

Given the term symbol for the electronic state of an atom, one can readily compute the electronic degeneracy of the state as $2J + 1$, where J is the subscript on the term symbol. Thus, for a $^1S_0$ state, $\omega = 2(0) + 1 = 1$; whereas, for a $^2S_{1/2}$ state, $\omega = 2(\frac{1}{2}) + 1 = 2$. From Eq. 14-74 we have

$$\frac{n_i}{N} = \frac{\omega_i e^{-\varepsilon_i/kT}}{q_e} \tag{14-113}$$

where Eq. 14-113 is useful in assessing the relative importance of an excited state.

For the noble gases, $\varepsilon_1 \sim 100\ \text{kJ}\cdot\text{mol}^{-1}$, the ground electronic state is $^1S_0$, and $\omega_0 = 1$; therefore, $q_e = 1$. For the alkali metals, $\varepsilon_1 \sim 10\ \text{kJ}\cdot\text{mol}^{-1}$ and the ground state is $^2S_{1/2}$, therefore, at ordinary temperatures, $q_e = 2$. Thus for Na(g) the electronic partition function will make a contribution of $R \ln 2$ to the entropy and free-energy function but does not affect the heat capacity and the enthalpy because the excited states are negligibly populated below 500 K.

For the gaseous halogen atoms the two lowest-energy electronic states are $^2P_{3/2}$ (ground state) and $^2P_{1/2}$ (first excited state); thus (ignoring higher-energy states) we have for q_e

$$q_e = 4 + 2e^{-\varepsilon_1/kT} \tag{14-114}$$

The $^2P_{1/2}$ state lies 401 cm^{-1} above the ground state for F(g). From the Planck equation, $\varepsilon = h\nu$ and $\nu = c/\lambda$; thus

$$\frac{\varepsilon}{kT} = \frac{h\nu}{kT} = \frac{hc}{\lambda kT} = \left(\frac{1.4388}{T}\right)\lambda^{-1} \tag{14-115}$$

where c is the velocity of light ($3.00 \times 10^8\ \text{m}\cdot\text{s}^{-1}$) and λ^{-1} is the reciprocal wavelength (usually expressed in wave numbers, cm^{-1}). Thus, substituting

$\lambda^{-1} = 401\ \text{cm}^{-1}$ into Eq. 14-115 and combining Eq. 14-114 yields for the q_e of F(g)

$$q_e = 4 + 2e^{-577/T} \tag{14-116}$$

Note from Eq. 14-116 that the maximum possible value of q_e is (let $T \to \infty$) $q_e = 6$; whereas the minimum possible value of q_e is (let $T \to 0$) $q_e = 4$. We can use Eq. 14-116 to compute the electronic contributions to the thermodynamic properties of F(g) at various temperatures.

EXAMPLE 14-12

Compute q_e, $(\bar{U}^\circ - \bar{U}_0^\circ)_e$, and $\bar{S}_e^\circ$ for F(g) at 298.15 K and 1000 K.

Solution: An expression for the electronic energy per mole $(\bar{U} - \bar{U}^\circ)_e$ is computed using Eq. 14-76:

$$(\bar{U}^\circ - \bar{U}_0^\circ)_e = RT^2\left(\frac{\partial \ln q_e}{\partial T}\right)_{N,V}$$

Taking $q_e = 4 + 2e^{-577/T}$, we have

$$(\bar{U}^\circ - \bar{U}_0^\circ)_e = \frac{RT^2(2 \times 577/T^2)e^{-577/T}}{4 + 2e^{-577/T}} = \frac{9594\ \text{J}\cdot\text{mol}^{-1}}{4e^{577/T} + 2}$$

The molar electronic entropy is given by Eq. 14-75 (note that $Q_e = q_e^N$, because the $1/N!$ term in Q_{tot} is included with translation $Q_t = q_t^N/N!$):

$$\bar{S}_e^\circ = R \ln q_e + \frac{(\bar{U}^\circ - \bar{U}_0^\circ)_e}{T}$$

Using the above equations for q_e, $(\bar{U} - \bar{U}_0^\circ)_e$, $\bar{S}_e^\circ$, we obtain the following results:

T/K	q_e	$(\bar{U}^\circ - \bar{U}_0^\circ)_e/J\cdot mol^{-1}$	$\bar{S}_e^\circ/J\cdot K^{-1}\cdot mol^{-1}$
298.15	4.289	323.0	13.19
1000	5.123	1051.7	14.64

Note that the maximum value of the electronic entropy is $\bar{S}_e^\circ = R \ln 6 = 14.90\ \text{J}\cdot\text{K}^{-1}\cdot\text{mol}^{-1}$. We also could obtain an expression for $\bar{C}_{V,e}^\circ = \bar{C}_{P,e}^\circ$ using the available expression for $(\bar{U}^\circ - \bar{U}_0^\circ)_e$.

The degeneracies of electronic states for molecules can be calculated from the term symbols for the respective electronic states. Thus, the degeneracies of electronic states of *diatomic* molecules are as follows (where n is the number of unpaired electrons):

Type of state	*Degeneracy of the state*
Σ	$n + 1$
π, Δ, or Φ	$2(n + 1)$

The three lowest molecular electronic states for $O_2(g)$ are

State	*Energy/cm^{-1}*	*Degeneracy, ω*
$^3\Sigma_g^-$	0	3
$^1\Delta_g$	7882	2
$^1\Sigma_g^+$	13,121	1

Note that the left-hand superscript on the molecular term symbol is equal to $n + 1$. The electronic partition function for $O_2(g)$ is

$$q_e = 3 + 2e^{-11,340/T} + e^{-18,878/T}$$

and, at 298.15 K,

$$q_e = 3.000, (\bar{U}^\circ - \bar{U}_0^\circ)_e = 0, \quad \text{and} \quad \bar{S}_e^\circ = R \ln 3 = 9.134 \text{ J}\cdot\text{K}^{-1}\cdot\text{mol}^{-1}$$

Note that a degenerate ground state makes a significant contribution to the entropy, even though such degeneracy has no effect on the other thermodynamic functions. In the $O_2(g)$ case the $R \ln 3$ contribution to $\bar{S}^\circ$ arises because the $O_2(g)$ molecules are randomly distributed among the three electronic energy levels of the same energy.

If a molecule is diamagnetic (no unpaired electron spins) and has zero orbital angular momentum ($L = 0$), then $\omega = 1$ and $q_e = 1$. Thus for $N_2(g)$ at ordinary temperatures, $q_e = 1$, $(\bar{U}^\circ - \bar{U}_0^\circ)_e = 0$, and $\bar{S}_e^\circ = 0$. In general, for nonlinear molecules the degeneracy of a molecular electronic energy state is given by

$$\omega = (n + 1)(2L + 1) \tag{14-117}$$

where n is the number of unpaired electrons and L is the orbital angular momentum.

EXAMPLE 14-13

The ground electronic state of $NO_2(g)$ has one unpaired electron and $L = 0$ (zero orbital angular momentum). Determine the degeneracy of the ground electronic state.

Solution: For the ground state of $NO_2(g)$ we have $n = 1$ and $L = 0$; thus using Eq. 14-117 we compute

$$\omega = (n + 1)(2L + 1) = (2)(1) = 2$$

Note that the degeneracy of the ground state gives rise to an $R \ln 2$ contribution to the molar entropy of $NO_2(g)$.

14-8 *Rotational Partition Functions*

The rotational energy levels of a rigid linear molecule are given by

$$\varepsilon_J = \frac{J(J + 1)h^2}{8\pi^2 I} \qquad J = 0, 1, 2, 3, \ldots \tag{14-118}$$

where J is the rotational quantum number and I is the *moment of inertia*. The moment of inertia I of a diatomic molecule is given by the expression

$$I = \mu r^2 = \left(\frac{m_1 m_2}{m_1 + m_2}\right) r^2 \tag{14-119}$$

where m_i is the mass of atom i in the molecule and r is the bond distance (see Problems 38 and 39 for the calculation of moments of inertia of polyatomic molecules). The degeneracy, ω_J, of the Jth rotational energy level is given by

$$\omega_J = 2J + 1 \tag{14-120}$$

and thus the rotational partition function q_r for a rigid linear molecule is given by

$$q_r = \sum_{J=0}^{\infty} (2J + 1) \exp\left[\frac{-J(J + 1)h^2}{8\pi^2 IkT}\right] \tag{14-121}$$

We define the *characteristic rotational temperature* θ_r of a linear molecule as

$$\theta_r = \frac{h^2}{8\pi^2 Ik} = \frac{hc}{k}\left(\frac{h}{8\pi^2 Ic}\right) \tag{14-122}$$

The *rotational constant* B of a linear molecule is defined as

$$B = \frac{h}{8\pi^2 Ic} = \frac{27.993 \times 10^{-40}\ \text{g}\cdot\text{cm}}{I} \tag{14-123}$$

From Eqs. 14-122 and 14-123 we obtain

$$\theta_r = \frac{hcB}{k} = (1.4388\ \text{K}\cdot\text{cm})B \qquad (B \text{ in cm}^{-1}) \tag{14-124}$$

Substitution of Eq. 14-122 into Eq. 14-121 yields

$$q_r = \sum_{J=0}^{\infty} (2J + 1) \exp\left[\frac{-J(J + 1)\theta_r}{T}\right] \tag{14-125}$$

EXAMPLE 14-14

Show that the energy separation $\Delta\varepsilon_J$ between successive rotational energy levels is given by

$$\Delta\varepsilon_J = 2(J + 1)k\theta_r$$

where k is Boltzmann's constant; compute $\Delta\varepsilon_J/kT$ for $O_2(g)$ at 300 K given $\theta_r = 2.07$ K for oxygen.

Solution: From Eq. 14-125 we have

$$\frac{\varepsilon_J}{kT} = \frac{J(J + 1)\theta_r}{T}$$

The value of $\Delta\varepsilon_J$ for the transition $J \rightarrow J + 1$ is given by

$$\frac{\Delta\varepsilon_J}{kT} = [(J + 1)(J + 2) - J(J + 1)]\frac{\theta_r}{T} = 2(J + 1)\frac{\theta_r}{T}$$

where $J \geqslant 1$. For $\Delta\varepsilon_J$ we have

$$\Delta\varepsilon_J = 2(J + 1)k\theta_r$$

and for $O_2(g)$ at 300 K we compute

$$\frac{\Delta\varepsilon_J}{kT} = 2(J + 1)\frac{2.07\ \text{K}}{300\ \text{K}} = 0.014(J + 1)$$

Thus, $\Delta\varepsilon_J < kT$, except for large J.

If $\Delta\varepsilon_J \ll kT$, then we can convert the sum in Eq. 14-121 to an integral by taking J to be a continuous variable (classical approximation); thus

$$q_r = \int_0^\infty (2J + 1)\exp\left[\frac{-J(J + 1)\theta_r}{T}\right]dJ \qquad (14\text{-}126)$$

Note from the result in Example 14-14 that as $J \rightarrow \infty$, we necessarily have $\Delta\varepsilon_J \gg kT$; but this is not a real problem, because for very large J values the terms in the sum of Eq. 14-125 make a negligible contribution to the value of q_r. In Figure 14-2 we have plotted the function

$$(2J + 1)\exp\left[\frac{-J(J + 1)\theta_r}{T}\right]$$

versus J for the case in which J is continuous and the case in which $J =$ 0, 1, 2, 3, ... Provided that $\theta_r \ll T$, the sum in Eq. 14-125 can be replaced by the integral in Eq. 14-126.

There is a complication that must be considered before we evaluate the integral in Eq. 14-126; namely, for symmetric molecules, not all the rotational energy levels are allowed. For symmetric diatomic molecules (e.g., $^{16}O_2$ or

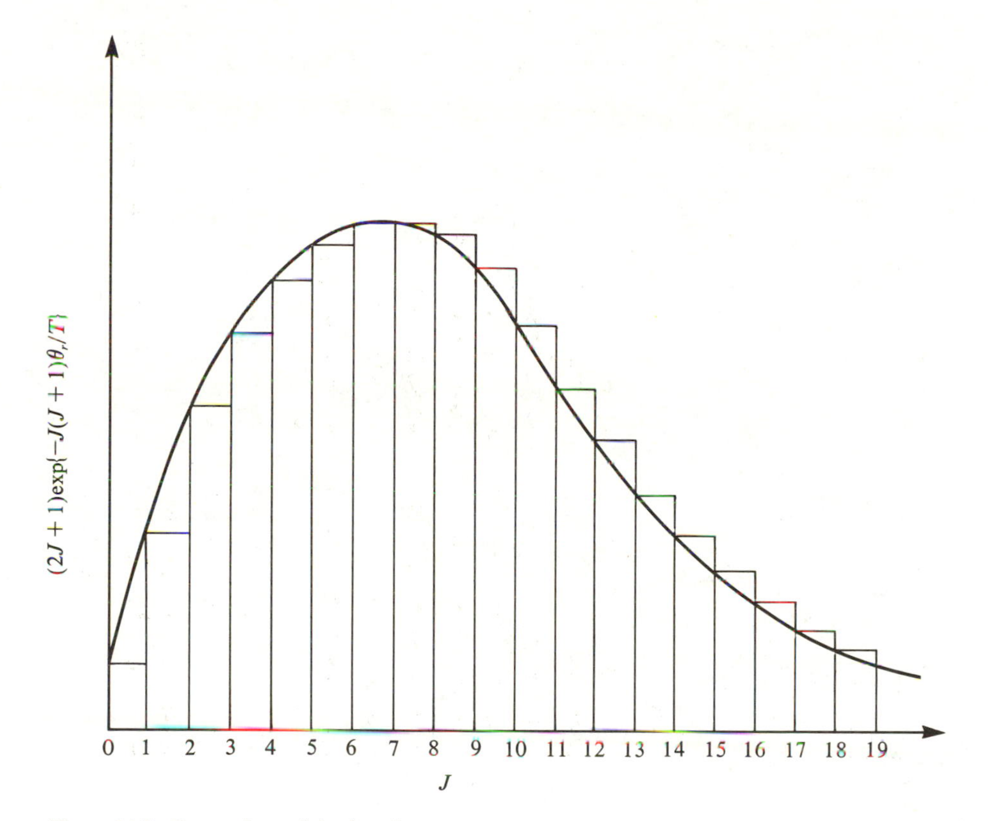

Figure 14-2. Comparison of the function

$$(2J + 1)\exp\left[\frac{-J(J + 1)\theta_r}{T}\right]$$

plotted against n; $h^2/8mx^2kT$ is a constant. If function (a) changes only very slowly with n, then the $J = 0, 1, 2, 3, \ldots.$ Note that the smooth curve passes through a maximum and thus the population of the rotational levels passes through a maximum.

The fractional population of the rotational levels is given by (see Eq. 14-74) the equation

$$\frac{n_J}{N} = \frac{(2J + 1)\exp[-J(J + 1)\theta_r/T]}{q_r}$$

$^1H_2$) either all the J even *or* all the J odd rotational levels are forbidden by symmetry. Thus for q_r we write

$$q_r = \frac{1}{\sigma}\int_0^\infty (2J + 1)\exp\left[\frac{-J(J + 1)\theta_r}{T}\right]dJ \qquad (14\text{-}127)$$

where σ is called *the symmetry number;* $\sigma = 2$ for a symmetric diatomic molecule and $\sigma = 1$ for an asymmetric diatomic molecule.

To evaluate the integral in Eq. 14-127, we first note that

$$d[J(J + 1)] = (2J + 1)\,dJ$$

and thus from Eq. 14-127 we obtain

$$q_r = \frac{1}{\sigma}\int_0^\infty \exp\left[\frac{-J(J+1)\theta_r}{T}\right] d[J(J+1)] \tag{14-128}$$

Setting $x = J(J+1)$, the integral in Eq. 14-128 takes the form

$$\int_0^\infty e^{-ax}\,dx$$

and thus

$$q_r = -\frac{T}{\sigma\theta_r}\int_0^\infty de^{-\theta_r x/T} = -\frac{T}{\sigma\theta_r} e^{-\theta_r x/T}\Big|_{x=0}^{x=\infty} = \frac{T}{\sigma\theta_r}$$

The diatomic-molecule rotational partition function expression

$$q_r = \frac{T}{\sigma\theta_r} \tag{14-129}$$

gives q_r to within 0.1% error provided that $\theta_r/T \leqslant 0.01$. Some examples of θ_r and θ_r/T for diatomic molecules are as follows:

	H_2	HCl	CO	O_2	I_2
θ_r/K	87.5	15.2	2.77	2.07	0.054
θ_r/298.15 K	0.29	0.051	0.0092	0.0069	0.00018

Note from the results given above that Eq. 14-129 is not reliable at the 0.1% level for $H_2(g)$ and $HCl(g)$ at 300 K. The value of the sum in Eq. 14-125 can be estimated to within 0.1% if $\theta_r/T \leqslant 0.3$, by using the approximation

$$q_r = \frac{T}{\sigma\theta_r}\left(1 + \frac{\theta_r}{3T} + \frac{\theta_r^2}{15T^2}\right) \tag{14-130}$$

If θ_r/T is not less than 0.3, then we can always directly evaluate the sum in Eq. 14-125.

EXAMPLE 14-15

For HF(g), $B = 20.939\ \text{cm}^{-1}$. Evaluate q_r for HF gas at 298.15 K using Eq. 14-129, and compare the result obtained with the result for q_r calculated from Eq. 14-130.

Solution: Using Eq. 14-124 we obtain for θ_r

$$\theta_r = (1.4388\ \text{K}\cdot\text{cm})B = 1.4388 \times 20.939\ \text{K} = 30.127\ \text{K}$$

From Eq. 14-129 we compute for the q_r of HF(g) ($\sigma = 1$) at 298.15 K

$$q_r = \frac{T}{\theta_r} = \frac{298.15\ \text{K}}{30.127\ \text{K}} = 9.8964$$

whereas from Eq. 14-130 we compute for q_r

$$q_r = 9.8964\left[1 + \frac{1}{3(9.8964)} + \frac{1}{15(9.8964)^2}\right]$$

$$q_r = 9.8964(1.03436) = 10.2364$$

or about a 3.5% difference in q_r.

The energy level scheme for polyatomic molecules is much more complicated than for linear molecules, but with J treated as a continuous variable only the number of rotational degrees of freedom (three for nonlinear molecules versus two for linear molecules) and the symmetry of the molecule are important. The result for the rotational partition function of a nonlinear molecule is

$$q_r = \frac{\pi^{1/2}T^{3/2}}{\sigma(\theta_{r1}\theta_{r2}\theta_{r3})^{1/2}} \tag{14-131}$$

where θ_{r1}, θ_{r2}, θ_{r3} are the three characteristic rotational temperatures, which are related to the three principal moments of inertia of the molecule

$$\theta_{ri} = \frac{h^2}{8\pi^2 kI_i} \tag{14-132}$$

and σ is the symmetry number of the molecule.

The symmetry number σ is equal to the number of physically indistinguishable spatial orientations of the molecule. Thus we have for the following examples

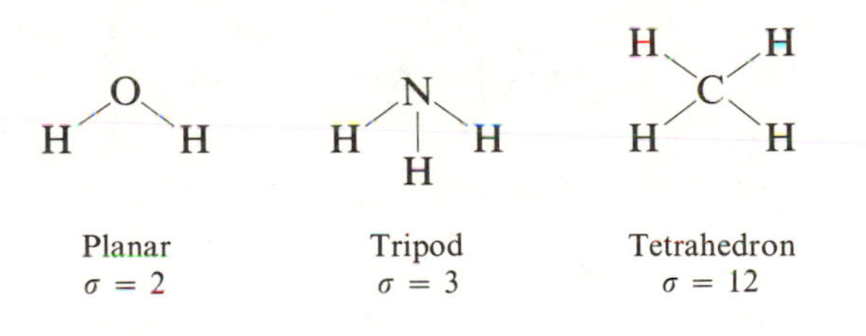

EXAMPLE 14-16

Determine the value of σ for the planar trigonal molecule $BF_3(g)$.

Solution: The geometry of the molecule is

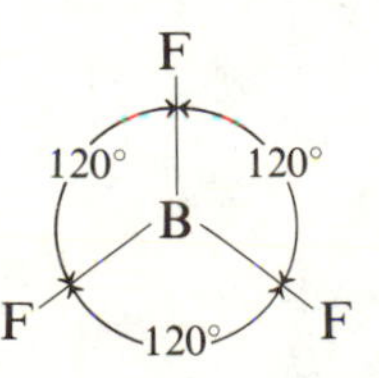

There are three indistinguishable orientations for the molecule in the plane, plus three more that result when the molecule is flipped over in the plane; thus $\sigma = 6$.

The thermodynamic functions for rotation are obtained using the appropriate expressions for the rotational partition functions, together with Eq. 14-75 through Eq. 14-81. The results are given in Table 14-2.

TABLE 14-2

Thermodynamic Functions for Rotation (Rigid Molecules, $T \gg \theta_{ri} = h^2/8\pi^2 I_i k$; $\theta_{ri} = 1.4388\ B_i$)

A. *Linear molecules*

$$(\bar{H}^\circ - \bar{H}_0^\circ)_r = (\bar{U}^\circ - \bar{U}_0^\circ)_r = RT$$

$$\bar{C}_{P,r}^\circ = \bar{C}_{V,r}^\circ = R$$

$$\bar{S}_r^\circ = R\ln\left(\frac{T}{\sigma\theta_r}\right) + R$$

$$-\left(\frac{\bar{G}^\circ - \bar{H}_0^\circ}{T}\right)_r = R\ln\left(\frac{T}{\sigma\theta_r}\right)$$

B. *Nonlinear molecules*

$$(\bar{H}^\circ - \bar{H}_0^\circ)_r = (\bar{U}^\circ - \bar{U}_0^\circ)_r = \frac{3RT}{2}$$

$$\bar{C}_{P,r}^\circ = \bar{C}_{V,r}^\circ = \frac{3R}{2}$$

$$\bar{S}_r^\circ = R\ln\left[\frac{\pi^{1/2}T^{3/2}}{\sigma(\theta_{r1}\theta_{r2}\theta_{r3})^{1/2}}\right] + \frac{3R}{2}$$

$$-\left(\frac{\bar{G}^\circ - \bar{H}_0^\circ}{T}\right)_r = R\ln\left[\frac{\pi^{1/2}T^{3/2}}{\sigma(\theta_{r1}\theta_{r2}\theta_{r3})^{1/2}}\right]$$

14-9 *Vibrational Partition Functions*

The energy levels of a one-dimensional harmonic oscillator are given by

$$\varepsilon_n = (n + \tfrac{1}{2})h\nu \qquad n = 0, 1, 2, 3, \ldots \tag{14-133}$$

where n is the vibrational quantum number and ν is the vibrational frequency. For a diatomic molecule, ν is given by

$$\nu = \frac{1}{2\pi}\left(\frac{k}{\mu}\right)^{1/2} \tag{14-134}$$

where μ is the reduced mass

$$\mu = \frac{m_1 m_2}{m_1 + m_2} \tag{14-135}$$

and k is the force constant for the bond-stretching motion; that is,

$$k = \frac{1}{2}\left(\frac{\partial^2 U}{\partial r^2}\right)_{r=r_e} \qquad (14\text{-}136)$$

where U is the energy, r is the interatomic separation, and r_e is the bond distance when U is a minimum.

The vibrational energy levels are nondegenerate ($\omega_n = 1$), and thus the vibrational partition function for a one-dimensional harmonic oscillator is given by

$$q_v = \sum_n \omega_n e^{-\varepsilon_n/kT} = \sum_n \exp\left[-\left(n + \frac{1}{2}\right)hv \bigg/ kT\right] \qquad (14\text{-}137)$$

We define the characteristic temperature for vibration θ_v as

$$\theta_v = \frac{hv}{k} = \frac{hc\tilde{v}}{k} = (1.4388\ \text{K}\cdot\text{cm})\tilde{v} \qquad (14\text{-}138)$$

where $\tilde{v}$ is the so-called vibrational frequency in reciprocal centimeters. Some examples of θ_v and θ_v/T for diatomic molecules are as follows:

	H_2	N_2	O_2	HCl	Cl_2	I_2
θ_v/K	6331	3340	2230	4140	810	310
θ_v/298.15 K	21.23	11.20	7.479	13.89	2.717	1.040

Note that $\theta_v/T > 1$, and thus the conversion of the sum in Eq. 14-137 to an integral (the classical approximation) will give unacceptable results for q_v unless $\theta_v/T \ll 1$, which is not true for any diatomic molecule at 25°C.

Using Eq. 14-138, we can rewrite Eq. 14-137 as

$$q_v = e^{-\theta_v/2T} \sum_{n=0}^{\infty} e^{-n\theta_v/T} \qquad (14\text{-}139)$$

The sum in Eq. 14-139 is equal to

$$\sum_{n=0}^{\infty} e^{-n\theta_v/T} = \frac{1}{1 - e^{-\theta_v/T}} = 1 + e^{-\theta_v/T} + e^{-2\theta_v/T} + \cdots \qquad (14\text{-}140)$$

as readily verified by long division. It is fortunate for calculational purposes that the sum in Eq. 14-139 can be expressed in closed form, because the sum cannot be replaced by an integral.

Combination of Eqs. 14-139 and 14-140 yields for the vibrational partition function

$$q_v = e^{-\theta_v/2T}(1 - e^{-\theta_v/T})^{-1} \qquad (14\text{-}141)$$

Inspection of Eq. 14-133 shows that, for $n = 0$, $\varepsilon_0 = hv/2$. The quantity $hv/2$ is called the *zero-point energy*. Note that in the lowest possible vibrational energy level ($n = 0$) the atoms vibrate, and they continue to do so down to

TABLE 14-3
Thermodynamic Functions for a One-Dimensional Harmonic Oscillator

$$[\theta_v = (1.4388\ \text{K}\cdot\text{cm})\tilde{\nu}]$$

$$(\bar{H}^\circ - \bar{H}_0^\circ)_v = (\bar{U}^\circ - \bar{U}_0^\circ)_v = \frac{R\theta_v}{e^{\theta_v/T} - 1}$$

$$\bar{C}_{P,v}^\circ = \bar{C}_{V,v}^\circ = \frac{R(\theta_v/T)^2 e^{\theta_v/T}}{(e^{\theta_v/T} - 1)^2}$$

$$\bar{S}_v^\circ = -R\ln(1 - e^{-\theta_v/T}) + \frac{R\theta_v/T}{e^{\theta_v/T} - 1}$$

$$-\left(\frac{\bar{G}^\circ - \bar{H}_0^\circ}{T}\right)_v = -R\ln(1 - e^{-\theta_v/T})$$

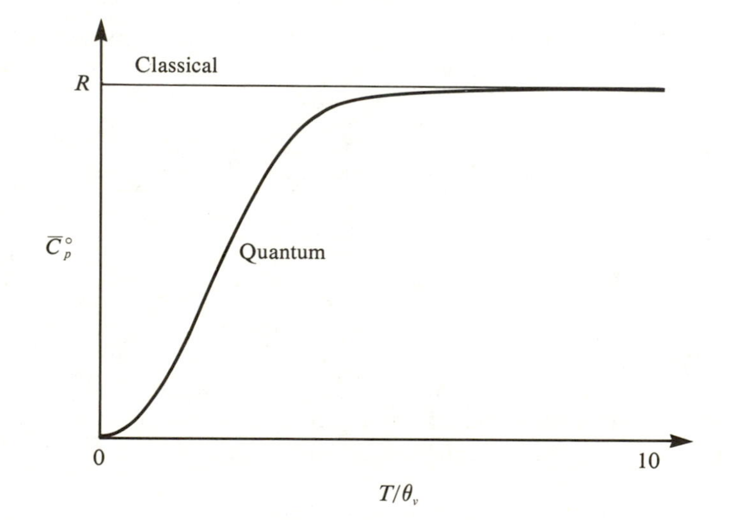

Figure 14-3. Comparison of $\bar{C}_P$ given by the equation

$$\bar{C}_P^\circ = \frac{R(\theta_v/T)^2 e^{\theta_v/T}}{(e^{\theta_v/T} - 1)^2}$$

with the classical result $\bar{C}_P^\circ = R$ for a one-dimensional harmonic oscillator. Note that $\bar{C}_P^\circ$ goes to a maximum value of R as T becomes large and that $\bar{C}_P^\circ$ goes to zero as T goes to zero.

absolute zero. Taking our zero of energy as the lowest vibrational energy level, we have

$$\varepsilon_n - \varepsilon_0 = nh\nu \tag{14-142}$$

With such a choice for the zero of energy we obtain for the vibrational partition function

$$q_v = (1 - e^{-\theta_v/T})^{-1} \tag{14-143}$$

All the thermodynamic functions for a one-dimensional harmonic oscillator are obtained from Eq. 14-143 using Eq. 14-75 through Eq. 14-81; the results are given in Table 14-3.

It is of interest to compare the classical result for $\bar{C}^\circ_{P,v}(=R)$ with the statistical thermodynamic expression for $\bar{C}^\circ_{P,v}$ given by the expression for $\bar{C}^\circ_{P,v}$ in Table 14-3, as done in Figure 14-3.

The number of vibrational degrees of freedom for a polyatomic molecule is

$$3N - 6 \qquad \text{nonlinear}$$

$$3N - 5 \qquad \text{linear}$$

where N is the number of atoms in the molecule. For each vibrational mode (degree of freedom) there is a corresponding $\theta_{v,i}$ value and a corresponding contribution to the thermodynamic properties of the molecule. Thus we apply the equations in Table 14-3 to each $\theta_{v,i}$ value for the molecule and sum the results to obtain the total contribution to the thermodynamic properties arising from the vibrational motions of the molecule.

EXAMPLE 14-17

Show that the fraction of a sample of diatomic molecules that are in excited vibrational states is given by $e^{-\theta_v/T}$.

Solution: From Eq. 14-74 we obtain for the fraction of the molecules in the vth vibrational state f_v

$$f_v = \frac{n_v}{N} = \frac{\exp[-(n + \frac{1}{2})\theta_v/T]}{e^{-\theta_v/2T}(1 - e^{-\theta_v/T})^{-1}} = e^{-n\theta_v/T}(1 - e^{-\theta_v/T})$$

The total fraction of the molecules in excited states is f_{ex}, where f_0 is the fraction in the ground state; thus

$$f_{ex} = 1 - f_0 = 1 - (1 - e^{-\theta_v/T}) = e^{-\theta_v/T}$$

The value of θ_v for $N_2(g)$ is 3340 K, and thus, for N_2 at 300 K, $f_{ex} = 1.46 \times 10^{-5}$. The corresponding value for $I_2(g)$ ($\theta_v = 310$ K) at 300 K is $f_{ex} = 0.356$. The lower the value of θ_v, the greater the value of f_{ex}.

14-10 *Comparison of Calculated and Experimental Molar Entropies*

Using the expressions for the translational, rotational, vibrational, and electronic contributions to the molar entropy developed in the preceding sections, we can calculate a value for $\bar{S}^\circ_{stat}$, the statistical thermodynamic entropy of a gas in its 1-atm standard state at the temperature of interest. The calculated value of $\bar{S}^\circ_{stat}$ is an approximation because we have neglected such factors as *vibrational anharmonicity, centrifugal stretching, vibration-rotation interactions,* and the *mixing of electronic states with rotational and vibrational states.* However, the equations for the thermodynamic properties that we have derived are a good first approximation for most substances, and, in any event, these equations are used as first-order terms in more accurate calculations involving higher order corrections.

As an example we have, for $O_2(g)$ at its normal boiling point, 90.13 K (W. F. Giauque and H. L. Johnston, *J. Amer. Chem. Soc.* 51, 2300 (1929)), the following:

Calorimetric entropy/cal·K^{-1}·mol^{-1}

0 K–11.75 K	Debye extrapolation	0.321
11.75–23.66	Graphical	1.697
23.66	Phase transition	0.948
23.66–43.76	Graphical	4.661
43.76	Phase transition	4.058
43.76–54.39	Graphical	2.397
54.39	Fusion	1.954
54.39–90.13	Graphical	6.462
90.13	Vaporization	18.08
90.13	(correction from real gas at 1 atm to standard state)	0.17
		40.74

For $O_2(g)$ at 90.13 K,

$\bar{S}^\circ_{cal} = 40.74 \pm 0.10$ cal·K^{-1}·mol^{-1}

Statistical entropy/cal·K^{-1}·mol^{-1}

Translation ($M = 32.00$)	30.36
Rotation ($I = 19.6 \times 10^{-40}$ g·cm^2)	8.11
Vibration ($\tilde{\nu} = 1556$ cm^{-1})	0.00
Electronic degeneracy ($R \ln 3$)	2.18
Electronic-rotational mixing	0.04
	40.69

$\bar{S}^\circ_{stat} = 40.69 \pm 0.02$ cal·K^{-1}·mol^{-1} (90.13 K)

As is clear from the foregoing results, the values of $\bar{S}^\circ_{cal}$ and $\bar{S}^\circ_{stat}$ obtained for $O_2(g)$ are the same within the experimental error. The calculations for $O_2(g)$ show that, at the lowest temperatures of the calorimetric measurements, the $R \ln 3$ electronic degeneracy is removed; that is, the electrons that are unpaired in the ground electronic state of the gas ($^3\Sigma_g$) are coupled in the solid. This

TABLE 14-4

Comparison of $\bar{S}^\circ_{cal}$ and $\bar{S}^\circ_{stat}$ for Gases (T = 298.15 K Unless Otherwise Noted and Entropies Are in Calories per Kelvin per mole)

(a) Molecules without frozen-in configurational entropy in the solid phase

Gas	$\bar{S}^\circ_{cal}$	$\bar{S}^\circ_{stat}$	Gas	$\bar{S}^\circ_{cal}$	$\bar{S}^\circ_{stat}$
Ne	35.01	34.95	CO_2	51.11	51.07
Ar	36.95	36.99	SO_2	59.24	59.27
Kr	39.17	39.20	$COCl_2$	67.81	67.81
Xe	40.7 ($\pm$0.3)	40.54	NH_3	45.94	45.91
Cd	40.0	40.1	PH_3	50.35	50.37
Zn	38.4	38.5	AsH_3	53.15	53.18
Hg	41.3 ($\pm$0.6)	41.8	CH_3Cl	55.94	55.98
N_2	45.9	45.79	CH_3Br (276.7 K)	57.9	58.0
O_2	49.09	49.03	CH_3NO_2	65.73	65.73
Cl_2	53.32	53.31	C_2H_4	52.48	52.47
HCl	44.5	44.64	Cyclopropane (240.3 K)	54.17	54.26
HBr	47.6	47.53	Benzene	64.46	64.36
HI	49.5	49.4	Toluene	76.77	76.68
H_2S	49.1	49.17			

(b) Molecules with presumed frozen-in configurational entropy (see Chapter 6) in the solid phase

Gas	T/K	$\bar{S}^\circ_{cal}$	$\bar{S}^\circ_{stat}$	$\bar{S}^\circ_{stat} - \bar{S}^\circ_{cal}$	Frozen-in entropy per mole (calculated)
CO	298.15	46.2	47.31	1.1	$R \ln 2 = 1.38$
N_2O	298.15	51.4	52.58	1.2	$R \ln 2 = 1.38$
NO	298.15	49.7	50.43	0.7	$\frac{1}{2}R \ln 2 = 0.69$
H_2O	298.15	44.28	45.104	0.82	$R \ln \frac{6}{4} = 0.81$
D_2O	298.15	45.89	46.66	0.77	$R \ln \frac{6}{4} = 0.81$
$H_2C{=}CD_2$	169.40	48.48	49.84	1.36	$R \ln 2 = 1.38$
CH_3D	99.7	36.72	39.49	2.77	$R \ln 4 = 2.75$
H_2	298.15	29.74	31.208	1.47	$\frac{3}{4}R \ln 3 = 1.63$
D_2	298.15	33.9	34.62	0.7	$\frac{1}{3}R \ln 3 = 0.72$

has been confirmed by independent magnetic measurements that show that the solid is diamagnetic.

In Table 14-4 we have collected a number of comparisons between $\bar{S}^\circ_{cal}$ and $\bar{S}^\circ_{stat}$, including several cases involving frozen-in configurational disorder in the solid phase. For small molecules $\bar{S}^\circ_{\text{stat}}$ is probably more accurate than $\bar{S}^\circ_{\text{cal}}$ in most cases, but for larger molecules the reverse obtains, owing to the difficulties associated with the extraction of reliable and complete molecular data from the complex spectra of large molecules.

14-11 *The Calculation of Equilibrium Constants for Gas-Phase Reactions from Partition Functions*

In analyzing the thermodynamics of chemical reactions from a statistical thermodynamic viewpoint (i.e., via the partition function), we must choose a common zero of energy for the various reactants and products. In applying statistical thermodynamics to a particular compound, the zero-point energy is just an additive term in the total energy of the species (it does not appear in $\bar{S}^\circ_{\text{stat}}$), and as such it is of little consequence because we are interested in $\bar{U}^\circ_T - \bar{U}^\circ_0$, $\bar{H}^\circ_T - \bar{H}^\circ_0$, and $(\bar{G}^\circ_T - \bar{H}^\circ_0)/T$ and not in $\bar{U}^\circ_T$, $\bar{H}^\circ_T$, or $\bar{G}^\circ_T$. On the other hand, for the calculation of equilibrium constants, we require $\Delta G^\circ_{\text{rxn}}$, because this is the thermodynamic quantity from which K is calculated:

$$\Delta G^\circ_{\text{rxn}} = -RT \ln K \qquad (14\text{-}144)$$

The evaluation of $\Delta G^\circ_{\text{rxn}}$ for a chemical reaction involves the *combination* of $\bar{G}^\circ_i$ terms for the various species involved, and for this reason differences in zero-point energies of the various species *must* be taken into account.

By convention, the zero of energy for a compound is taken as the energy of the constituent atoms at rest, infinitely separated from one another and at 0 K. Taking a diatomic molecule as an example we have (see Figure 14-4)

$$\varepsilon^\circ_0 = -D^\circ_0 = -[D^\circ_e - (\tfrac{1}{2})h\nu] \qquad (14\text{-}145)$$

where the quantities in Eq. 14.145 are defined in Figure 14-4.

For a chemical reaction we have

$$\Delta\varepsilon^\circ_0 = -\sum_i \phi_i D^\circ_{i0} = -\sum_i \phi_i [D^\circ_{ie} - h\nu_i/2] \qquad (14\text{-}146)$$

where the ϕ_i values are the reaction-balancing coefficients that are positive for products and negative for reactants. For any particular species in the reaction, we interpret D°_0 to be the energy necessary to dissociate the ground-state gaseous *molecule* at 0 K into its ground-state constituent gaseous atoms infinitely separated at 0 K ($N_0 D^\circ_0$ is the energy per mole). The energy zero terms are incorporated into the total partition function by the relation

$$q_0 = e^{-\varepsilon^\circ_0/kT} = e^{D^\circ_0/kT} \qquad (14\text{-}147)$$

where the total partition function of a gas is given by

$$Q = \frac{(q_t q_r q_v q_e q_n q_0)^N}{N!} \qquad (14\text{-}148)$$

The Helmholtz energy A of a gas is given by Eq. 14-77 as

$$A = -kT \ln Q \qquad (14\text{-}77)$$

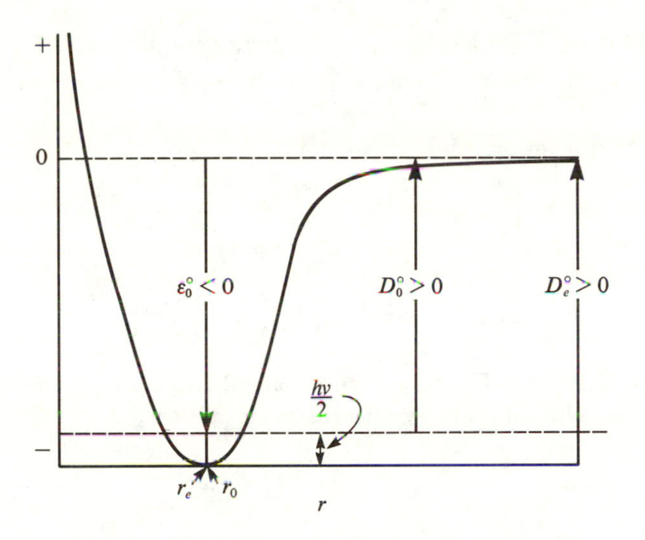

Figure 14-4. Energy diagram for a diatomic molecule. The value of r_e is taken as r at the minimum of the potential well, and r_0 is the value of r at the middle of the $n = 0$ vibrational level.

Factoring the zero-point energy term out of Q gives

$$A = -NkT \ln q_0 - kT \ln Q' = N\varepsilon_0^\circ - kT \ln Q' \tag{14-149}$$

Furthermore, because $Q' = q^N/N!$ for a gas, we obtain for gas i

$$A_i = N\varepsilon_{0i}^\circ - NkT \ln q_i + kT(N \ln N - N)$$

or

$$A_i = N\varepsilon_{0i}^\circ - NkT - NkT \ln\left(\frac{q_i}{N}\right) \tag{14-150}$$

For 1 mol of the gas in the 1-atm ideal gas standard state we have $N = N_0$ and $A = \bar{A}^\circ$; thus

$$\bar{A}_i^\circ = N_0\varepsilon_{0i}^\circ - RT - RT \ln\left(\frac{q_i^\circ}{N_0}\right) \tag{14-151}$$

The molar Gibbs energy function of i, $\bar{G}_i^\circ$, can be obtained from $\bar{A}_i^\circ$ using the relationship

$$\bar{G}_i^\circ = \bar{A}_i^\circ + P\bar{V}_i^\circ \tag{14-152}$$

Substituting Eq. 14-151 in Eq. 14-152 yields

$$\bar{G}_i^\circ = N_0\varepsilon_{0i}^\circ - RT \ln\left(\frac{q_i^\circ}{N_0}\right) \tag{14-153}$$

Application of Eq. 14-153 to ϕ_i mol of species i in a balanced gas-phase reaction yields

$$\phi_i\bar{G}_i^\circ = N_0\phi_i\varepsilon_{0i}^\circ - RT \ln\left(\frac{q_i^\circ}{N_0}\right)^{\phi_i} \tag{14-154}$$

For the complete balanced reaction we have

$$\Delta G_{\text{rxn}}^\circ = \sum_i \phi_i\bar{G}_i^\circ = N_0 \sum_i \phi_i\varepsilon_{0i}^\circ - RT \ln \prod_i \left(\frac{q_i^\circ}{N_0}\right)^{\phi_i} \tag{14-155}$$

where in obtaining Eq. 14-155 we used the following property of logarithms:

$$\sum_i \ln X_i = \ln \prod_i X_i$$

We define $\Delta\varepsilon_0^\circ$ for the reaction as

$$\Delta\varepsilon_0^\circ = \sum_i \phi_i \varepsilon_{0i}^\circ \tag{14-156}$$

and thus Eq. 14-155 can be rewritten as

$$\Delta G_{\text{rxn}}^\circ = -RT \ln \prod_i \left(\frac{q_i^\circ}{N_0}\right)^{\phi_i} e^{-\Delta\varepsilon_0^\circ/kT} \tag{14-157}$$

Comparison of Eqs. 14-157 and 14-144 yields the following expression for the equilibrium constant for a gas-phase reaction:

$$K = e^{-\Delta\varepsilon_0^\circ/kT} \prod_i \left(\frac{q_i^\circ}{N_0}\right)^{\phi_i} \tag{14-158}$$

where

$$q_i^\circ = (q_t^\circ q_r^\circ q_v^\circ q_e^\circ)_i \tag{14-159}$$

The translational partition function q_t° is proportional to the volume of the gas in the 1-atm ideal gas standard state. We can factor the volume out of q_i° as follows:

$$\frac{q_i^\circ}{N_0} = \frac{V}{N_0} \cdot \frac{q_i^\circ}{V} = \frac{kT}{P} \cdot \frac{q_i^\circ}{V}$$

where we have used $V = N_0 kT/P$. In the standard state, $P = 1$ atm (exactly), and thus

$$\left(\frac{q_i^\circ}{N_0}\right)^{\phi_i} = (kT)^{\phi_i} \left(\frac{q_i^\circ}{V}\right)^{\phi_i} \tag{14-160}$$

Substitution of Eq. 14-160 into Eq. 14-158 yields

$$K = (kT)^{\Delta\phi} e^{-\Delta\varepsilon_0^\circ/kT} \prod_i \left(\frac{q_i^\circ}{V}\right)^{\phi_i} \tag{14-161}$$

where

$$\Delta\phi = \sum_i \phi_i \tag{14-162}$$

Equation 14-161 is the equation that we shall use to compute K values for gas-phase reactions. In applying Eq. 14-161, the following points should be remembered.

1. We must take $k = 1.3626 \times 10^{-22}$ cm^3·atm·K^{-1} in the $(kT)^{\Delta\phi}$ factor to obtain a numerical value of K that corresponds to the conventional 1-atm ideal gas standard states.
2. In the vibrational partition function we have for each vibrational frequency a term of the type $(1 - e^{-\theta_v/T})^{-1}$, and *not* $e^{-\theta_v/2T}(1 - e^{-\theta_v/T})^{-1}$, because the zero-point vibrational energy factor $e^{-\theta_v/2T}$ is accounted for in $e^{-\Delta\varepsilon_0^\circ/kT}$.
3. As a result of our choice of the zero of energy, we have for the electronic

partition function

$$q_e = \omega_0 + \omega_1 e^{-\varepsilon_1/kT} + \omega_2 e^{-\varepsilon_2/kT} + \cdots$$

wherein all the ε_i values are positive.

4. We do not include the nuclear parition functions in q_i°, because for a balanced chemical reaction the q_n factors cancel out.
5. Except for the $(kT)^{\Delta\phi}$ factor we use $k = 1.3807 \times 10^{-23}\ \text{J}\cdot\text{K}^{-1}$ for Boltzmann's constant in the K expression.

Some examples of the use of Eq. 14-161 to calculate K for chemical reactions are given below.

EXAMPLE 14-18

Calculate K at 2000 K for the ionization of gaseous cesium:

$$\text{Cs}(g,\ ^2\text{S}_{1/2}) \rightleftharpoons \text{Cs}^+(g,\ ^1\text{S}_0) + e^-(g)$$

Solution: Application of Eq. 14-161 to the above reaction yields ($\Delta\phi = 1$)

$$K = kTe^{-\Delta\varepsilon_0^\circ/kT}\frac{(q^\circ/V)_{\text{Cs}^+}(q^\circ/V)_{e^-}}{(q^\circ/V)_{\text{Cs}}}$$

Only the translational and electronic partition functions have to be considered for $\text{Cs}(g)$, $\text{Cs}^+(g)$, and $e^-(g)$; thus

$$\left(\frac{q^\circ}{V}\right)_{e^-} = \left(\frac{q_e q_t}{V}\right)_{e^-} = 2\left(\frac{2\pi m_{e^-} kT}{h^2}\right)^{3/2}$$

$$\left(\frac{q^\circ}{V}\right)_{\text{Cs}^+} = \left(\frac{q_e q_t}{V}\right)_{\text{Cs}^+} = \left(\frac{2\pi m_{\text{Cs}^+} kT}{h^2}\right)^{3/2}$$

$$\left(\frac{q^\circ}{V}\right)_{\text{Cs}} = \left(\frac{q_e q_t}{V}\right)_{\text{Cs}} = 2\left(\frac{2\pi m_{\text{Cs}} kT}{h^2}\right)^{3/2}$$

Note that $q_e = 2$ for $e^-(g)$ ($\omega_0 = n + 1 = 2$). Taking $m_{\text{Cs}^+} = m_{\text{Cs}}$ and substituting the above $(q^\circ/V)_i$ expressions into the K expression we obtain

$$K = kTe^{-\Delta\varepsilon_0^\circ/kT}\left(\frac{2\pi m_{e^-} kT}{h^2}\right)^{3/2}$$

From spectroscopic tables (C. E. Moore, NBS Circular 467), we find for the ionization energy of $\text{Cs}(g)$, $\Delta U_0^\circ = 89.77\ \text{kcal}\cdot\text{mol}^{-1}$; thus at 2000 K we have for K ($m_{e^-} = 9.11 \times 10^{-28}$ g)

$$K = (1.36 \times 10^{-22})(2000)$$

$$\times \left[\frac{2 \times 3.14 \times 9.11 \times 10^{-28} \times 1.38 \times 10^{-16} \times 2000}{(6.62 \times 10^{-27})^2}\right]^{3/2}$$

$$\times \exp\left(\frac{-89.77 \times 10^3}{1.987 \times 2000}\right)$$

and

$$K = 58.83 \times 1.55 \times 10^{-10} = 9.12 \times 10^{-9}$$

The above expression for $K(T)$ can be combined with the classical thermodynamic expression for the ionization reaction, expressed in terms of the fraction of Cs(g) that is ionized, to yield an expression on which an *ionization thermometer* can be based. Thus we have for the reaction

$$\underset{n(1-\alpha)}{\mathrm{Cs}(g)} \rightleftharpoons \underset{\alpha n}{\mathrm{Cs}^{+}(g)} + \underset{\alpha n}{e^{-}(g)}$$

where α is the fraction of the n mol of Cs(g) that is ionized at T. The total number of moles is

$$(1 - \alpha)n + 2\alpha n = (1 + \alpha)n$$

and therefore

$$P_{\mathrm{Cs}^+} = \left[\frac{\alpha n}{(1 + \alpha)n}\right] P_{\mathrm{tot}} \qquad P_{e^-} = \left[\frac{\alpha n}{(1 + \alpha)n}\right] P_{\mathrm{tot}}$$

$$P_{\mathrm{Cs}} = \left[\frac{(1 - \alpha)n}{(1 + \alpha)n}\right] P_{\mathrm{tot}}$$

Thus

$$K = \frac{\alpha^2 P_{\mathrm{tot}}}{1 - \alpha^2} = kTe^{-\Delta\varepsilon_0^\circ/kT}\left(\frac{2\pi m_e kT}{h^2}\right)^{3/2}$$

and at 2000 K, with $P_{\mathrm{tot}} = 1.00 \times 10^{-3}$ atm, $\alpha = 3.02 \times 10^{-3}$. If we know P_{tot} and measure α, by measuring the current I through the partially ionized gas (α is proportional to I), then we can use the above expression relating α to T to compute the temperature of the partially ionized gas. The equation relating α to T for the ionization of a monatomic gas is known as the *Saha equation.*

EXAMPLE 14-19

Compute the equilibrium constant at 500 K for the reaction

$$\mathrm{I}_2(g, {}^1\Sigma_g^+) \rightleftharpoons 2\,\mathrm{I}(g, {}^2P_{3/2})$$

given the following data:

For $I_2(g)$: $D_0^\circ = 12{,}453\ \mathrm{cm}^{-1}$ $\quad B_e = 0.03736\ \mathrm{cm}^{-1}$

$\tilde{\nu}_e = 214.5\ \mathrm{cm}^{-1}$

The first excited electronic state for $I_2(g)$, a triply degenerate state, lies $8124\ \mathrm{cm}^{-1}$ above the ground state; the first excited electronic state for I(g), a ${}^2P_{1/2}$ state, lies $90.89\ \mathrm{kJ \cdot mol^{-1}}$ above the ground state.

Solution: It is easily verified, using the equation

$$\frac{n_i}{N} = \frac{\omega_i e^{-\varepsilon_i/kT}}{q_e}$$

that the excited electronic states for $I_2(g)$ and $I(g)$ make completely negligible contributions at 500 K to the electronic partition functions, and thus these excited states can be ignored in the calculation of K.

The statistical thermodynamic equilibrium constant expression for the $I_2(g)$ dissociation is obtained using Eq. 14-161:

$$K = kTe^{-\Delta\varepsilon_0^\circ/kT}\frac{(q^\circ/V)_I^2}{(q^\circ/V)_{I_2}}$$

The partition function expression for $I(g)$ is equal to $[(q_e = 2(\frac{3}{2}) + 1)]$

$$\left(\frac{q^\circ}{V}\right)_I = \left(\frac{q_e q_t}{V}\right) = 4\left(\frac{2\pi m_I kT}{h^2}\right)^{3/2}$$

whereas for $I_2(g)$ we have

$$\left(\frac{q^\circ}{V}\right)_{I_2} = \left(\frac{q_e q_t q_r q_v}{V}\right) = (1)\left(\frac{2\pi m_{I_2} kT}{h^2}\right)^{3/2}\left(\frac{T}{2\theta_r}\right)(1 - e^{-\theta_v/T})^{-1}$$

where $q_e = 1$ and $\sigma = 2$ for $I_2(g)$. Further for $I_2(g)$

$$\theta_r = 1.4388B_e \qquad \theta_v = 1.4388\tilde{\nu}_e$$

The value of $\Delta\varepsilon_0^\circ$ for the reaction with D_0° values given in cm^{-1} units is

$$\Delta\varepsilon_0^\circ = -hc\sum\phi_i D_{0i}^\circ = -hc[2(0) - D_{0,I_2}^\circ] = hcD_{0,I_2}^\circ$$

Thus K is given by $(m_{I_2} = 2m_I)$

$$K = \frac{(kTe^{-hcD_{0,I_2}^\circ/kT})(16)(2\pi m_I kT/h^2)^3(1 - e^{-\theta_v/T})}{2^{3/2}(2\pi m_I kT/h^2)^{3/2}(T/2\theta_r)}$$

or

$$K = (5.567kT)(e^{-hcD_{0,I_2}^\circ/kT})\left(\frac{2\pi m_I kT}{h^2}\right)^{3/2}\left(\frac{2\theta_r}{T}\right)(1 - e^{-\theta_v/T})$$

At 500 K we compute $(m_I = 126.93/N_0)$

$$K = (3.853 \times 10^{-19})(2.736 \times 10^{-16})(3.005 \times 10^{27})(2.150 \times 10^{-4})(0.4606)$$
$$= 3.14 \times 10^{-11}$$

The above calculation clearly shows the magnitudes of the various factors that give rise to the value of K for the $I_2(g)$ dissociation reaction.

EXAMPLE 14-20

Compute K at 298.15 K for the reaction

$$H_2(g) + I_2(g) \rightleftharpoons 2\,HI(g)$$

given the following data

	$H_2(g)$	$I_2(g)$	$HI(g)$
D_0°/eV	4.4763($^1\Sigma$)	1.5417($^1\Sigma$)	3.0514($^1\Sigma$)
$\tilde{\nu}_e/\text{cm}^{-1}$	4400.4	214.5	2309.5
B_e/cm^{-1}	60.809	0.03736	6.551

Solution: From Eq. 14-161 we have for K ($\Delta\phi = 0$)

$$K = e^{-\Delta\varepsilon_0^\circ/kT} \frac{(q^\circ/V)_{\text{HI}}^2}{(q^\circ/V)_{\text{H}_2}(q^\circ/V)_{\text{I}_2}}$$

For $\Delta\varepsilon_0^\circ$ we compute

$$\Delta\varepsilon_0^\circ = -[2D_0^\circ(\text{HI}) - D_0^\circ(\text{H}_2) - D_0^\circ(\text{I}_2)]$$

$$\Delta\varepsilon_0^\circ = -2(3.0514) + 4.4763 + 1.5417 = -0.0848 \text{ eV}$$

Using the conversion factor $1 \text{ eV} = 1.6022 \times 10^{-19}$ J we compute at 298.15 K

$$e^{-\Delta\varepsilon_0^\circ/kT} = \exp\left[+\frac{(0.0848 \text{ eV})(1.6022 \times 10^{-19} \text{ J}\cdot\text{eV}^{-1})}{(1.3807 \times 10^{-23} \text{ J}\cdot\text{K}^{-1})(298.15 \text{ K})}\right] = 27.13$$

We now compute separately the factors in the K expression arising from the translational, rotational, vibrational, and electronic partition functions.

Translation (Eq. 14-90):

$$\frac{(q_t^\circ/V)_{\text{HI}}^2}{(q_t^\circ/V)_{\text{H}_2}(q_t^\circ/V)_{\text{I}_2}} = \frac{M_{\text{HI}}^3}{M_{\text{H}_2}^{3/2}M_{\text{I}_2}^{3/2}} = \frac{(127.912)^3}{(2.0159 \times 253.809)^{3/2}} = 180.83$$

Rotation (Eq. 14-129 for I_2 and Eq. 14-130 for H_2 and HI):

$$\theta_{r,\text{HI}} = (1.4388 \text{ K}\cdot\text{cm})6.551 \text{ cm}^{-1} = 9.426 \text{ K}$$

$$\theta_{r,\text{I}_2} = (1.4388)(0.03736) = 0.05375 \text{ K}$$

$$\theta_{r,\text{H}_2} = (1.4388)(60.809) = 87.492 \text{ K}$$

$$\frac{(q_r^\circ)_{\text{HI}}^2}{(q_r^\circ)_{\text{H}_2}(q_r^\circ)_{\text{I}_2}} = \frac{\{T/\theta_r[1 + (\theta_r/3T) + (\theta_r^2/15T^2)]\}_{\text{HI}}^2}{\{T/2\theta_r[1 + (\theta_r/3T) + (\theta_r^2/15T^2)]\}_{\text{H}_2}(T/2\theta_r)_{\text{I}_2}}$$

$$= \frac{(31.966)^2}{(1.8803)(2773.5)} = 0.1959$$

Vibration (Eq. 14-143);

$$\theta_{v,\text{H}_2} = (1.4388 \text{ K}\cdot\text{cm})(4400.4 \text{ cm}^{-1}) = 6331.3 \text{ K}$$

$$\theta_{v,\text{I}_2} = (1.4388)(214.5) = 308.62 \text{ K}$$

$$\theta_{v,\mathrm{HI}} = (1.4388)(2309.5) = 3322.9 \text{ K}$$

$$\frac{(q_v^\circ)^2_{\mathrm{HI}}}{(q_v^\circ)_{\mathrm{H_2}}(q_v^\circ)_{\mathrm{I_2}}} = \frac{(1 - e^{-\theta_{v,\mathrm{H_2}}/T})(1 - e^{-\theta_{v,\mathrm{I_2}}/T})}{(1 - e^{-\theta_{v,\mathrm{HI}}/T})^2} = \frac{(1.0000)(0.6448)}{(0.99997)} = 0.6448$$

Electronic: The ground states of $H_2(g)$, $I_2(g)$, and $HI(g)$ are all $^1\Sigma$, and thus $q_e = 1$. There are no low-lying excited electronic states.

Our result for K at 298.15 K is therefore

$$K = 27.13 \times 180.83 \times 0.1959 \times 0.6448 \times 1.0000 = 620$$

compared with a measured value of $K = 616$. Uncertainties in D_0° values are usually the limiting factor on the accuracy of calculated K values.

The methods of calculation of K for gas-phase reactions outlined in the preceding examples is not difficult to extend to more complicated reactions such as the *ammonia synthesis reaction*

$$N_2(g) + 3\,H_2(g) \rightleftharpoons 2\,NH_3(g)$$

the *methanol synthesis reaction*

$$CO(g) + 2\,H_2(g) \rightleftharpoons CH_3OH(g)$$

the *water-gas shift reaction*

$$CO(g) + H_2O(g) \rightleftharpoons CO_2(g) + H_2(g)$$

and the *catalytic methanation reaction*

$$CO(g) + 3\,H_2(g) \rightleftharpoons H_2O(g) + CH_4(g)$$

The only essential difference from a statistical thermodynamic point of view between the above reactions and those in Examples 14-20 and 14-19 is the presence of polyatomic molecules, which have more complicated vibrational and rotational partition functions. For example, for $H_2O(g)$, which is a non-linear three-atom molecule and thus has $3(=3 \times 3 - 6)$ normal vibrational modes, we have

$$q_v = (1 - e^{-\theta_{v1}/T})^{-1}(1 - e^{-\theta_{v2}/T})^{-1}(1 - e^{-\theta_{v3}/T})^{-1}$$

where $\theta_{v1} = 1.4388\tilde{\nu}_1$ and so forth.

14-12 *Isotope-Exchange Reactions*

The Born-Oppenheimer approximation involves the assumption that nuclear motions can be neglected in the quantum-mechanical solution for the electronic energies of a molecule. That is, because the electron velocities are much larger than the nuclear velocities, the nuclei can be assumed to be fixed

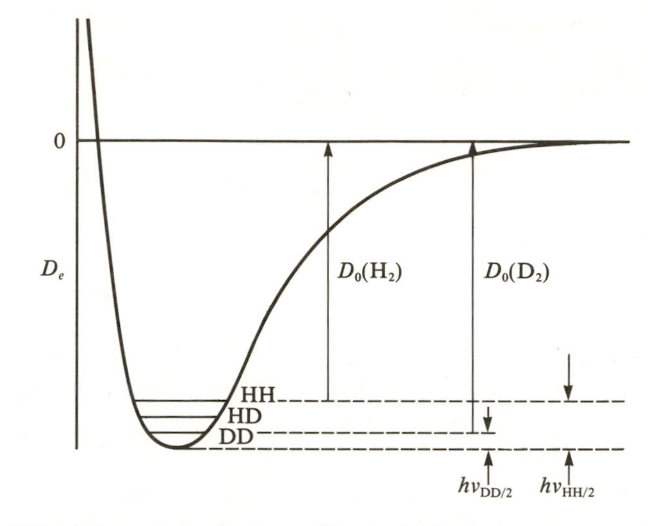

Figure 14-5. Different isotopic species lie at different depths in the same potential well (Born-Oppenheimer approximation) with the heavier isotope lying lower (lower zero-point energy). Note that the dissociation energy D_0 for the isotopic species with greater mass is greater than for the isotopic species with the lesser mass.

at the equilibrium internuclear separation(s) when the Schrödinger equation is solved. If the Born-Oppenheimer approximation is valid, then two molecules that differ only by isotopic substitution, for example, D_2 and H_2 or $D^{35}Cl$ and $H^{35}Cl$, have the same electronic potential energy surface; thus they have the same values of D_e, the dissociation energy measured from the bottom of the potential energy well (see Figure 14-5), and the same values of the bond force constants.

However, because the zero-point vibrational energy of a diatomic-molecule harmonic oscillator is given by

$$\varepsilon_0 = \frac{h\nu}{2} = \frac{h}{4\pi}\left(\frac{k}{\mu}\right)^{1/2} \tag{14-163}$$

the energy of the lowest vibrational energy level depends on the masses of the vibrating atoms:

$$\mu = \frac{m_1 m_2}{m_1 + m_2}$$

From Eq. 14-163 we conclude that the heavier isotope lies deeper in the potential well; the heavier isotope thus forms the stronger bond (larger D_0 value, see Figure 14-5). Note that the vibrational frequencies of two isotopic diatomic molecules are related as follows:

$$\frac{\nu_2}{\nu_1} = \frac{(k/\mu_2)^{1/2}}{(k/\mu_1)^{1/2}} = \left(\frac{\mu_1}{\mu_2}\right)^{1/2} = \frac{\tilde{\nu}_2}{\tilde{\nu}_1} \tag{14-164}$$

EXAMPLE 14-21

Given that $\tilde{\nu}_{H_2} = 4400.39\ \text{cm}^{-1}$, compute $\tilde{\nu}_{D_2}$. Take $m_H = 1.00783$ and $m_D = 2.01400$.

Solution: From Eq. 14-164 we have

$$\tilde{\nu}_{D_2} = \tilde{\nu}_{H_2}\left(\frac{\mu_{H_2}}{\mu_{D_2}}\right)^{1/2} = \tilde{\nu}_{H_2}\left[\frac{(m_H^2/2m_H)}{(m_D^2/2m_D)}\right]^{1/2}$$

and thus

$$\tilde{\nu}_{D_2} = \tilde{\nu}_{H_2}\left(\frac{m_H}{m_D}\right)^{1/2} = (4400.39\ \text{cm}^{-1})\left(\frac{1.00783}{2.01400}\right)^{1/2} = 3112.83\ \text{cm}^{-1}$$

Note that $\tilde{\nu}_{D_2} < \tilde{\nu}_{H_2}$.

Isotope-exchange reactions have received considerable attention in statistical thermodynamics owing to the enormous theoretical and practical importance of isotope chemistry and also because such reactions are particularly amenable to a statistical thermodynamic analysis. The principal reason that isotope-exchange reactions are so amenable to a statistical treatment is that the only data necessary for the calculation of K are the symmetry numbers and the fundamental vibrational frequencies of the isotopic molecules. (For higher approximations the vibrational anharmonicities and vibration-rotation interaction terms are also needed.)

Because isotope-exchange reactions always have $\Delta\phi = 0$ (i.e., no change in the number of molecules) and, in addition, because the electronic structure of an atom and the potential energy curves for the molecules are unchanged (Born-Oppenheimer approximation) when the atomic mass of the nucleus with a fixed atomic number is changed (i.e., the D_e values but not the D_0 values remain unchanged), we have, assuming harmonic oscillators, for $\Delta\varepsilon_0^\circ$

$$\Delta\varepsilon_0^\circ = -\sum \phi_i(D_{ie}^\circ - \tfrac{1}{2}h\nu_i) = \frac{h}{2}\sum_i \phi_i\nu_i \tag{14-165}$$

Substitution of Eq. 14-165 in Eq. 14-161 yields

$$K = \prod_i \left(\frac{q_i^\circ}{V}\right)^{\phi_i} \exp\left(-\frac{h}{2kT}\sum_i \phi_i\nu_i\right) \tag{14-166}$$

The repeat product in Eq. 14-166 can be expressed as a product of ratios of the type

$$\frac{q_2^\circ}{q_1^\circ} = \frac{(q_t q_r q_v)_2}{(q_t q_r q_v)_1} \tag{14-167}$$

where the subscripts 1 and 2 denote the two isotopic species and where the V's cancel because an isotope-exchange reaction is necessarily symmetric in the mole numbers.

As an example, consider the isotope-exchange reaction

$$D_2(g) + 2\,H^{35}Cl(g) \rightleftharpoons H_2(g) + 2\,D^{35}Cl(g) \qquad (14\text{-}168)$$

The equilibrium constant expression for the reaction in Eq. 14-168 is

$$K = \left(\frac{q^\circ_{H_2}}{q^\circ_{D_2}}\right)\left(\frac{q^\circ_{DCl}}{q^\circ_{HCl}}\right)^2 \exp\left\{-\frac{hc}{2kT}\left[(\tilde{\nu}_{H_2} - \tilde{\nu}_{D_2}) + 2(\tilde{\nu}_{DCl} - \tilde{\nu}_{HCl})\right]\right\} \qquad (14\text{-}169)$$

For an isotopic pair of diatomic molecules we have from Eq. 14-167 and the q_t, q_r, and q_v expressions (see Eq. 14-90, 14-129 and 14-143)

$$\frac{q^\circ_2}{q^\circ_1} = \left(\frac{M_2}{M_1}\right)^{3/2}\left(\frac{\mu_2}{\mu_1}\cdot\frac{\sigma_1}{\sigma_2}\right)\left(\frac{1 - e^{-1.4388\tilde{\nu}_1/T}}{1 - e^{-1.4388\tilde{\nu}_2/T}}\right) \qquad (14\text{-}170)$$

where the M_i values are the molecular masses, the μ_i values are the reduced masses, and the σ_i values are the symmetry numbers.

The equilibrium constant expression for the reaction in Eq. 14-168 is

$$K = \left(\frac{m_H}{m_D}\right)^{1/2}\left(\frac{m_D + m_{Cl}}{m_H + m_{Cl}}\right)\left(\frac{1 - e^{-u_{DD}}}{1 - e^{-u_{HH}}}\right)\left(\frac{1 - e^{-u_{HCl}}}{1 - e^{-u_{DCl}}}\right)^2 e^{-\Delta\varepsilon^\circ_0/kT} \qquad (14\text{-}171)$$

$$\frac{\Delta\varepsilon^\circ_0}{kT} = \frac{1}{2}\left[2(u_{DCl} - u_{HCl}) + (u_{HH} - u_{DD})\right] \qquad (14\text{-}172)$$

where

$$u_i = \frac{hc\tilde{\nu}_i}{kT} = \frac{1.4388\tilde{\nu}_i}{T}$$

Taking $m_H = 1.00783$, $m_D = 2.01400$, $\tilde{\nu}_{HH} = 4400.39\ \text{cm}^{-1}$, and $\tilde{\nu}_{H^{35}Cl} = 2990.99\ \text{cm}^{-1}$ and using Eq. 14-164, we compute, for the equilibrium constant for the reaction in Eq. 14-168 at 25.00°C, $K = 1.93$. This value of K is only a rough approximation, because Eq. 14-129 is not strictly applicable. Use of Eq. 14-130 for q_r gives $K = 2.04$.

Equilibrium effects are inherently quantum mechanical in origin. At high temperatures, the value of K for the reaction in Eq. 14-168 approaches unity (see Problem 29). At high temperatures, differences in zero-point energies of isotopic molecules are no longer important because most of the molecules are in highly excited vibrational states.

The calculation of K values for isotope-exchange reactions outlined above has several simplifying assumptions built into the calculations, namely, the Born-Oppenheimer approximation, the harmonic oscillator approximation, the rigid rotor approximation, and the use of $T/\sigma\theta_r$ for q_r. All of these approximations introduce significant errors (1% to 5%) into the calculated value of K, but there is some error cancellation, and the value of K calculated by the simple method used above is usually good to within about 5%.

All the above noted approximations, including the Born-Oppenheimer approximation (which introduces about a 3% error in the calculated value of K), can be removed, and when this is done, the calculated value of K agrees to within the experimental error ($<1\%$ in K) of the most accurate experimental data available. The excellent agreement between the experimental and calculated values of K provides strong support for the statistical thermodynamic theory of equilibrium isotope effects.

Problems

1. Calculate the $n = 1$ to $n = 2$ translation energy change for an argon atom trapped in a crystalline cavity of a size such that the argon atom can move a maximum of 100 pm. Compare your result for $\varepsilon_2 - \varepsilon_1$ to kT at 300 K.

2. Compare the value of $N!$ calculated exactly and calculated using Stirling's approximation,
$$\ln N! = N \ln N - N$$
for $N = 5$, 10, 100, 1000, and 6.0×10^{23}. Compute the percent error in each case.

3. Determine the rotational symmetry numbers for the molecules

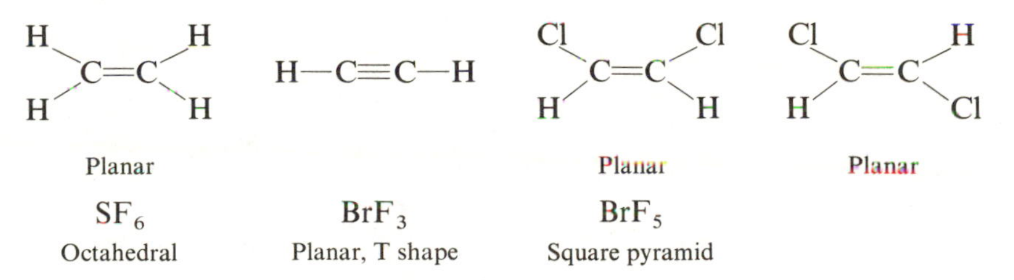

4. Using only symmetry numbers, estimate the equilibrium constants for the gas reactions
$$C^{18}O_2 + C^{16}O_2 \rightleftharpoons 2\,C^{16}O^{18}O$$
$$C^{18}O_2 + 2\,H_2^{16}O \rightleftharpoons C^{16}O_2 + 2\,H_2^{18}O$$
$$PH_2D + H_2O \rightleftharpoons PH_3 + HDO$$
$$2\,NH_3 + ND_3 \rightleftharpoons 3\,NH_2D$$

5. Show by expanding the exponentials ($e^x \approx 1 + x$ for small x) in Eq. 14-171 that $K \to 1$ as $T \to \infty$.

6. Given
$$Q_t = \frac{q_t^N}{N!} \quad \text{and} \quad q_t = \frac{(2\pi M k T)^{3/2} V}{h^3 N_0^{3/2}}$$
and using the relation
$$P = -\left(\frac{\partial A}{\partial V}\right)_{N,T}$$
show that $PV = NkT$.

7. Calculate the value of n_1/N, that is, the ratio of the number of nuclei in the first excited nuclear energy level to the total number of nuclei for a nucleus with $\varepsilon_1 - \varepsilon_0 = 1$ MeV at 25°C. At what T is kT of the order of $\varepsilon_1 - \varepsilon_0$?

8. Show that the ratio of the rotational partition function for $^{18}O_2$ and $^{16}O_2$ is

$$\frac{q_r(^{18}O_2)}{q_r(^{16}O_2)} = \frac{m_{18}}{m_{16}}$$

where m_{18} is the mass of ^{18}O and m_{16} is the mass of ^{16}O.

9. Show that the ratio of rotational partition functions for $^{12}CH_4$ and $^{13}CH_4$ is equal to unity.

10. Plot f_v versus v for $I_2(g)$ at 300 K. Take $\theta_v = 310$ K. Also plot f_{ex}, the fraction of molecules in excited states, versus T for $I_2(g)$ over the range 200 K to 1000 K.

11. Plot n_J/N and n_v/N versus T for $I_2(g)$. Take $\theta_r = 0.0538$ K and $\theta_v = 310$ K.

12. Derive the equations in Table 14-2.

13. Derive the equations in Table 14-3.

14. Compute the entropies at 25°C for Ar, Kr, and Xe. The experimental values are 36.95 ± 0.2 cal·K^{-1}·mol^{-1}, 39.17 ± 0.1 cal·K^{-1}·mol^{-1}, and 40.7 ± 0.3 cal·K^{-1}·mol^{-1}, respectively. Are your calculated values more accurate than the measured values? Also, calculate $\bar{U}^\circ - \bar{U}_0^\circ$, $\bar{A}^\circ - \bar{U}_0^\circ$, $\bar{H}^\circ - \bar{H}_0^\circ$, $\bar{C}_V^\circ$, $\bar{C}_P^\circ$, and $-(\bar{G}^\circ - \bar{H}_0^\circ)/T$ for these gases at 25°C.

15. Show that $\bar{C}_{P,v}^\circ \to R$ as $T \to \infty$ and that $\bar{C}_{P,v} \to 0$ as $T \to 0$. See Table 14-3 for the $\bar{C}_{P,v}^\circ$ expression.

16. Show that the classical vibrational partition function for a one-dimensional harmonic oscillator is given by

$$q_v = e^{-\theta_v/2T}\frac{T}{\theta_v}$$

Compare T/θ_v with $(1 - e^{-\theta_v/T})^{-1}$ for $H_2(\theta_v = 6331\text{ K})$ and $I_2(\theta_v = 310\text{ K})$ at $T = 300$ K and 1000 K.

17. Compute the thermodynamic functions for translation and rotation of HBr ($r_e = 1.414$ Å) and N_2 ($r_e = 1.095$ Å) at 25°C. The total entropies at 25°C are 47.5 cal·K^{-1}·mol^{-1} and 45.8 cal·K^{-1}·mol^{-1}. What do you estimate for $\bar{S}_v^\circ$ for these molecules? Compare with the values calculated from $\bar{S}_v^\circ$ in Table 14-3. (θ_v is 3812 K for HBr and 3395 K for N_2.)

18. Compute the thermodynamic functions of $CO_2(g)$ at 298 K, given that $I_{CO_2} = 71.9 \times 10^{-40}$ g·cm^2 and that the θ_{vi} values are 1890 K, 3360 K, 954 K, and 954 K. Compare your results with the measured value of $\bar{S}_{298}^\circ = 51.1$ cal·K^{-1}·mol^{-1}.

19. For an asymmetric diatomic molecule, show that the J value for the most highly populated rotational level J_{max} is given by

$$J_{max} = \left(\frac{T}{2\theta_r}\right)^{1/2} - \frac{1}{2}$$

20. The ground electronic state of NO(g) is a $^2\Pi$ state. The $^2\Pi$ ground state is *split* by spin-orbital coupling into $^2\Pi_{1/2}$ ($\varepsilon = 0$) and $^2\Pi_{3/2}$ ($\varepsilon = 121.1\ \text{cm}^{-1}$) components. Show that

$$q_e = 2 + 2e^{-174.2/T}$$

and compute q_e, $(\bar{U}^\circ - \bar{U}_0^\circ)_e$, $\bar{S}_e^\circ$, $(\bar{H}^\circ - \bar{H}_0^\circ)_e$, $(\bar{G}^\circ - \bar{G}_0^\circ)_e$, and $-(\bar{G}^\circ - \bar{H}_0^\circ)/T$ at 298.15 K, 500 K, and 1000 K.

21. Derive an expression for $\bar{C}_{P,e}^\circ$, the electronic heat capacity at constant pressure, for F(g), and plot $\bar{C}_{P,e}^\circ$ versus T over the range 200 K to 1500 K. At what value of T is $\bar{C}_{P,e}^\circ$ a maximum? (See Example 14-12.)

22. Show that K for the reaction

$$Na_2(g, {}^1\Sigma_g) \rightleftharpoons 2\ Na(g, {}^2S_{1/2})$$

is given by

$$K = kTe^{-\Delta\varepsilon_0^\circ/kT}\left(\frac{4}{2^{3/2}}\right)\left(\frac{2\theta_r}{T}\right)(1 - e^{-\theta_v/T})\left(\frac{2\pi m_{Na}kT}{h^2}\right)^{3/2}$$

Given that $\tilde{\nu} = 159.23\ \text{cm}^{-1}$, $r_0 = 3.078 \times 10^{-8}$ cm, and $\Delta U_0^\circ = 17.50$ kcal·mol^{-1} for Na_2,

(a) Compute the value of K at 100.0°C.

(b) Derive expressions for ΔH_{rxn}°, ΔS_{rxn}°, and ΔG_{rxn}°, and determine the value of these quantities at 25°C.

23. Compute K at 25°C for the reaction

$$2\ H^{35}Cl(g) + {}^{37}Cl_2(g) \rightleftharpoons 2\ H^{37}Cl(g) + {}^{35}Cl_2(g)$$

using only the following data and atomic masses:

Species	$^{35}Cl_2$	$^{37}Cl_2$	$H^{35}Cl$	$H^{37}Cl$
$\tilde{\nu}$/cm^{-1}	565	549	2989	2988

24. Hydrogen molecules H_2 exist in one of two forms: *ortho* H_2 or *para* H_2. The two forms differ in the relative orientations of the nuclear spins. The *ortho* hydrogen molecules are restricted to rotational energy levels with odd J values ($J = 1, 3, 5, 7, \ldots$), whereas *para* hydrogen molecules are restricted to rotational levels with even J values ($J = 0, 2, 4, 6, \ldots$).

(a) Given that at 14 K solid H_2 consists of $\frac{3}{4}$ *ortho* and $\frac{1}{4}$ *para* molecules, estimate the frozen-in rotational entropy of $H_2(s)$ at 14 K.

(b) Derive the expressions for $q_n q_r$ given in Table 14-5 for H_2 and D_2.

25. The energy levels of a particle in a parabolic potential well are given by

$$\varepsilon_n = \frac{h^2}{8\pi^2 m}(n + b)^2 \qquad \begin{array}{l} b = \text{constant} \\ n = 1, 2, 3, \ldots \end{array}$$

Derive the partition function for the particle.

26. Suppose a thermodynamic system is composed of particles that can exist in either one of two states: a lower state of energy 0 and degeneracy g_0 and an upper state of energy ε and degeneracy g_1.
(a) Write the partition function for this system.
(b) Derive an expression for the total energy of the system as a function of ε, T, g_0, g_1, and N (the number of particles).
(c) Derive expressions for the entropy, the energy, and C_V for this system.
(d) Now assume that $g_0 = g_1 = 1$ and $\varepsilon = 100\ \text{cm}^{-1}$, and compute (n_1/n_0) at $T = 0$ K, 100 K, and ∞ K.
(e) Sketch curves of (n_1/n_0), $\bar{S}$, $\bar{U}$, and $\bar{C}_V$ as a function of T on the same graph.

27. Repeat parts (a) through (e) of Problem 26 for a thermodynamic system composed of particles that are restricted to energy levels given by

$$\varepsilon_n = (n + \tfrac{1}{2})\varepsilon \qquad n = 0, 1, 2, 3, \ldots, \infty$$

Discuss the origin of the different thermodynamic behavior as T increases for the two systems. What can you conclude about the number of available energy levels of a system that exhibits a maximum in the $\bar{C}_V$ versus T curve?

28. I. Langmuir obtained the following experimental results for the percent dissociation of $H_2(g)$:

$$H_2(g) \rightleftharpoons 2\,H(g)$$

T/K	2000	2500	3000	3500
% dissociated (1 atm)	0.17	1.6	7.2	21

Calculate the percent dissociation of $H_2(g)$ at the above temperatures and compare with the measured values. For $H_2(g)$, take $D_0^\circ = 4.4763$ eV.

29. Consider the general isotope-exchange reaction

$$X_2(g) + Y_2(g) \rightleftharpoons 2\,XY(g)$$

where X and Y are isotopes with atomic masses of m_X and m_Y, respectively. Show that the equilibrium constant expression for this reaction is given by (rigid rotor, harmonic oscillator approximation)

$$K = 4\left[\frac{(m_X + m_Y)}{2m_X^{1/2}m_Y^{1/2}}\right] \cdot \frac{(1 - e^{-u_{XX}})(1 - e^{-u_{YY}})}{(1 - e^{-u_{XY}})^2} \cdot e^{-\Delta\varepsilon_0^\circ/kT}$$

where $u_i = h\nu_i/kT$ and

$$\frac{\Delta\varepsilon_0^\circ}{kT} = \frac{hc\tilde{\nu}_{XY}}{kT}\left[1 - \frac{(m_X^{1/2} + m_Y^{1/2})}{2^{1/2}(m_X + m_Y)^{1/2}}\right]$$

Also show that
(a) If $u \gg 1$, then

$$K = 4\left[\frac{(m_X + m_Y)}{2m_X^{1/2}m_Y^{1/2}}\right]e^{-\Delta\varepsilon_0^\circ/kT}$$

(b) If $u \ll 1$, then

$$K = 4e^{-\Delta\varepsilon_0^\circ/kT}$$

30. For molecules containing groups of atoms separated by single bonds (e.g., the C—C bond in C_2H_6 or the O—O bond in H_2O_2) the possibility of rotation about these bonds must be considered. If the barrier to rotation is large compared to kT, then this internal motion can be treated like a vibration. On the other hand, for very small barriers the motion can be treated as a free rotation (when the intermediate case prevails, the problem is more complex but has been solved). For free rotation the energy levels are given by

$$\varepsilon_{\text{f rot}} = \frac{h^2n^2}{8\pi^2 I_r}$$

where n is the quantum number that can have all values from $-\infty$ to $+\infty$, except that if one or both of the rotating groups have σ equivalent orientations, then only $1/\sigma$ of the n values are allowed. The quantity I_r is the reduced moment of inertia given by $I_r = I_A I_B/(I_A + I_B)$, where I_A and I_B are the moments of inertia of the two rotating groups about the rotation axis. Show that the partition function for free rotation is given by

$$q_{\text{f rot}} = \frac{1}{\sigma}\left(\frac{8\pi^3 I_r kT}{h^2}\right)^{1/2}$$

and obtain all the thermodynamic functions for free rotation.

31. Show that for bosons

$$\lim_{T\to 0}\frac{n_i}{n_0} \to 0$$

where the subscript zero denotes the lowest energy level. (*Hint:* First find the value of $-\mu/kT$ as $T \to 0$.) Note that at $T = 0$ all the bosons are in the lowest available energy level.

32. Show that as $T \to 0$ for fermions, we obtain the following results:

$\varepsilon_i < \mu$	$\varepsilon_i = \mu$	$\varepsilon_i > \mu$
$\frac{n_i}{g_i} \to 1$	$\frac{n_i}{g_i} \to \frac{1}{2}$	$\frac{n_i}{g_i} \to 0$

Sketch n_i/g_i versus μ at $T = 0$ and at $T > 0$. The level for which $\varepsilon_i = \mu$ is called the *Fermi level*. Note that at $T = 0$ all levels up to $\varepsilon_i = \mu$ are equally populated, whereas all levels with $\varepsilon_i > \mu$ are empty.

33. The Teller-Redlich *product rule* relates the ratio of products of frequencies of a pair of isotopic molecules to a ratio of products of atomic masses and moments of inertia. Defining $u_i = h\nu_i/kT$, the product rule takes the forms

$$\left(\frac{M'}{M}\right)^{3/2}\left(\frac{I'_a I'_b I'_c}{I_a I_b I_c}\right)^{1/2} = \prod_{i=1}^{N}\left(\frac{m'_i}{m_i}\right)^{3/2}\prod_{j=1}^{3N-6}\frac{u'_j}{u_j} \qquad \text{(nonlinear molecule)}$$

$$\left(\frac{M'}{M}\right)^{3/2}\left(\frac{I'}{I}\right) = \prod_{i=1}^{N}\left(\frac{m'_i}{m_i}\right)^{3/2}\prod_{j=1}^{3N-5}\frac{u'_j}{u_j} \qquad \text{(linear molecule)}$$

The product rule is especially useful in the calculation of isotope-exchange equilibrium constants, because it removes the need to know the moments of inertia of the isotopic molecules. In other words, the product rule enables us to calculate K solely from the masses and frequencies of the isotopic molecules. Use the product rule to compute K for the reaction

$$D_2(g) + H_2O(g) \rightleftharpoons D_2O(g) + H_2(g)$$

at 25°C given the following data:

Species	$\tilde{\nu}/cm^{-1}$
D_2	3112.8
H_2	4400.4
H_2O	3825.3, 1653.9, 3935.6
D_2O	2758.1, 1210.2, 2883.6
$m_H = 1.0078$	$m_D = 2.0140$

34. Suppose a crystal of monatomic species [e.g., Pb(s)] is viewed as a collection of N independent three-dimensional harmonic oscillators (Einstein crystal). Show that the partition function of an Einstein solid is given by

$$Q = q^{3N} = e^{-N\varepsilon_0/kT}\prod_{i}^{3N} e^{-h\nu_i/2kT}(1 - e^{-h\nu_i/kT})^{-1}$$

where $\varepsilon_0 = -D_e$ is the energy of an atom in the crystal relative to the isolated gaseous atoms at rest at 0 K.

35. Using the result from Problem 34, show that for an Einstein crystal

$$S = 3Nk\left[\frac{\theta_E/T}{e^{\theta_E/T} - 1} - \ln(1 - e^{-\theta_E/T})\right]$$

where $\theta_E = h\nu_E/k$ (all oscillators have the same frequency); and

$$C_V = \frac{3Nk(\theta_E/T)^2 e^{\theta_E/T}}{(e^{\theta_E/T} - 1)^2}$$

Plot $\bar{S}$ and $\bar{C}_V$ versus T, and show that $\bar{C}_V \to 3R$ as $T \to \infty$.

36. Can the population of an excited state ever exceed that of a ground state under equilibrium conditions? (Explain using examples.)

37. Calculate the equilibrium constant at 500 K for the reaction

$$N_2(g) + O_2(g) \rightleftharpoons 2\,NO(g)$$

given the following data:

	N_2	O_2	NO
B_e/cm^{-1}	1.998	1.445	1.704
$\tilde{\nu}_e/\text{cm}^{-1}$	2357	1580	1904 (in both $^2\Pi_{1/2}$ and $^2\Pi_{3/2}$)
D_0°/eV	9.756	5.080	6.480
Term symbol	$^1\Sigma_g^+$	$^3\Sigma_g^-$	$\left.\begin{matrix}^2\Pi_{1/2}\\ ^2\Pi_{3/2}\end{matrix}\right\}\Delta\varepsilon = 121\ \text{cm}^{-1}$

38. Let x_i', y_i', z_i' be the Cartesian coordinates of atom i of mass m_i in a molecule. The coordinates of the center of mass of the molecule are given by

$$x_{cm} = \frac{1}{M}\sum_i m_i x_i' \qquad y_{cm} = \frac{1}{M}\sum_i m_i y_i' \qquad z_{cm} = \frac{1}{M}\sum_i m_i z_i'$$

where M is the total molecular mass

$$M = \sum_i m_i$$

The coordinates of an atom relative to the center of mass are given by

$$x_i = x_i' - x_{cm} \qquad y_i = y_i' - y_{cm} \qquad z_i = z_i' - z_{cm}$$

The moments of inertia are defined by the equations

$$I_{xx} = \sum_i m_i(y_i^2 + z_i^2)$$

$$I_{yy} = \sum_i m_i(x_i^2 + z_i^2)$$

$$I_{zz} = \sum_i m_i(x_i^2 + y_i^2)$$

By a suitable choice of coordinate axes about the center of mass, we make $I_{xy} = I_{xz} = I_{yz} = 0$, where

$$I_{xy} = \sum_i m_i y_i z_i$$

with analogous expressions for I_{xz} and I_{yz}. The values of I_{xx}, I_{yy}, and I_{zz} when I_{xy}, I_{xz}, I_{yz} are all equal to zero are called the *principal moments of inertia.* In symmetrical molecules the direction of the principal axes of inertia usually can be determined by inspection. A symmetry axis in the molecule coincides with one of the principal axes. One of the planes of symmetry in a molecule must contain two of the principal axes and be perpendicular to the third principal axis. Calculate the principal moments of inertia and the rotational partition functions at 25.00°C for the following molecules:

H_2O	NH_3	CH_4
∡HOH = 104.45°	∡HNH = 106.67°	∡HCH = 109.50°
$r_{OH} = 0.971$ Å	$r_{NH} = 1.012$ Å	$r_{CH} = 1.093$ Å
	(*Note:* $I_{xx} = I_{yy} \neq I_{zz}$)	(*Note:* $I_{xx} = I_{yy} = I_{zz}$)

39. Derive Eq. 14-119 using the equations given in Problem 38.

40. Calculate $\bar{S}^\circ$ at 25.00°C for CH_3Cl given the following data:

$$\tilde{\nu}_1 = 2966.2 \text{ cm}^{-1} \qquad \tilde{\nu}_4 = 3041.8 \text{ cm}^{-1} \qquad \tilde{\nu}_7 = 1454.6 \text{ cm}^{-1}$$
$$\tilde{\nu}_2 = 1354.9 \text{ cm}^{-1} \qquad \tilde{\nu}_5 = 3041.8 \text{ cm}^{-1} \qquad \tilde{\nu}_8 = 1015.0 \text{ cm}^{-1}$$
$$\tilde{\nu}_3 = 732.0 \text{ cm}^{-1} \qquad \tilde{\nu}_6 = 1454.6 \text{ cm}^{-1} \qquad \tilde{\nu}_9 = 1015.0 \text{ cm}^{-1}$$
$$r_{\text{C-Cl}} = 1.77 \text{ Å} \qquad r_{\text{C-H}} = 1.10 \text{ Å}$$
$$∡\text{HCH} = 109^\circ \qquad ∡\text{ClCH} = 110^\circ$$

41. Obtain q_e for a hydrogen atom given that the energy levels relative to the ground state are (m and e are the electron mass and charge)

$$\varepsilon_n = \frac{2\pi^2 m e^4}{h^2}\left(1 - \frac{1}{n^2}\right) \qquad \begin{array}{l} \omega_n = 2n^2 \\ n = 1, 2, 3, \ldots \end{array}$$

42. Show on the basis of symmetry numbers alone that $K_{2a} < K_{1a}/4$ for the successive acid dissociation constants of the diprotic acid

$$\text{HOOC(CH}_2)_n\text{ COOH}(aq)$$

I

Thermodynamic Data for Various Substances

	In $kJ \cdot mol^{-1}$				In $J \cdot K^{-1} \cdot mol^{-1}$				
							$-(\bar{G}^\circ_T - \bar{H}^\circ_{298})/T$		
Substance	$\Delta\bar{H}^\circ_{f,0}$	$\Delta\bar{H}^\circ_{f,298}$	$\Delta\bar{G}^\circ_{f,298}$	$\bar{H}^\circ_{298} - \bar{H}^\circ_0$	$\bar{S}^\circ_{298}$	$\bar{C}^\circ_{P,298}$	*At* 500 K	*At* 700 K	*At* 1000 K
$H_2(g)$	0	0	0	8.468	130.57	28.82	133.867	138.712	145.427
$H^+(aq)$	—	0	0	—	0	0	—	—	—
$O_2(g)$	0	0	0	8.680	205.03	29.35	208.413	213.501	220.769
$OH^-(aq)$	—	−229.99	−157.29	—	−10.75	−148.5	—	—	—
$H_2O(\ell)$	—	−285.83	−237.18	—	69.91	75.29	—	—	—
$H_2O(g)$	−238.91	−241.82	−228.59	9.902	188.72	33.58	192.573	198.347	206.614
$Br_2(\ell)$	—	0	0	24.514	152.23	75.69	—	—	—
$Br_2(g)$	45.70	30.91	3.14	9.719	245.35	36.02	—	—	—
$HBr(g)$	28.56	−36.40	−53.43	8.648	198.59	29.14	201.903	206.782	213.635
$S(s, rh)$	0	0	0	4.410	31.80	22.64	—	—	—
$SO_2(g)$	−294.29	−296.83	−300.19	10.548	248.11	39.87	252.868	260.316	271.228
$SO_3(g)$	−389.99	−395.72	−371.08	11.698	256.65	50.67	262.889	272.843	287.663
$H_2S(g)$	−17.71	−20.63	−33.56	9.954	205.69	34.23	209.618	215.668	224.488
$H_2S(aq)$	—	−39.7	−27.87	—	121.	—	—	—	—
$HS^-(aq)$	—	−17.5	12.05	—	62.8	—	—	—	—
$S^{2-}(aq)$	—	33.0	85.8	—	−14.6	—	—	—	—
$S_2O_3^{2-}(aq)$	—	−652.3	—	—	—	—	—	—	—
$SO_4^{2-}(aq)$	—	−909.27	−744.63	—	20.1	−293.0	—	—	—
$N_2(g)$	0	0	0	8.669	191.50	29.12	194.811	199.707	206.598
$NH_3(g)$	−39.08	−46.11	−16.48	9.991	192.34	35.06	196.845	203.539	213.648

$NH_3(aq)$	—	−80.29	−26.57	—	111.3	—	—	—	—
$NH_4^+(aq)$	—	−132.51	−79.37	—	113.4	79.91	—	—	—
$NO(g)$	89.75	90.25	86.57	—	210.65	29.84	214.041	219.074	226.204
$NO_2(g)$	35.98	33.18	51.30	10.201	239.95	37.20	244.333	251.237	261.437
$N_2O(g)$	85.50	82.05	104.18	9.556	219.74	38.45	224.509	231.819	242.588
$NO_2^-(aq)$	—	−104.6	−37.2	—	140.2	−97.5	—	—	—
$NO_3^-(aq)$	—	−207.36	−111.34	—	146.4	−86.6	—	—	—
C(s, grp)	0	0	0	1.050	5.74	8.53	6.887	9.063	12.636
C(s, diam)	2.43	1.897	2.900	0.523	2.38	6.11	—	—	—
$CO(g)$	−113.80	−110.52	−137.15	8.668	197.56	29.12	200.857	205.777	212.735
$CO_2(g)$	−393.14	−393.51	−394.36	9.363	213.64	37.11	218.187	225.287	235.806
$CO_2(aq)$	—	−413.80	−386.02	—	117.6	—	—	—	—
$HCHO(g)$	−113.4	−117.0	−113.0	10.016	218.66	35.40	—	—	—
$CH_4(g)$	−66.82	−74.81	−50.75	9.991	186.15	35.31	190.510	197.736	209.267
$CH_3OH(g)$	−189.77	−200.66	−162.00	11.427	239.70	43.89	—	—	—
$C_2H_5OH(\ell)$	—	−277.69	−174.89	—	160.7	111.46	—	—	—
$C_2H_2(g)$	227.29	226.73	209.20	10.008	200.83	43.92	206.284	214.961	227.886
$C_2H_4(g)$	60.73	52.26	68.12	10.565	219.45	43.56	224.773	234.300	249.638
$C_2H_6(g)$	−69.13	−84.68	−32.89	11.950	229.49	52.63	—	—	—
$CH_3Br(g)$	−19.75	−35.1	−25.9	10.610	246.27	42.43	—	—	—
$H_2NCONH_2(aq)$	—	−317.79	—	—	—	—	—	—	—

The data given here were obtained primarily from NBS Technical Notes 270-3 through 270-8 and the JANAF tables.

	In kJ·mol^{-1}				In J·K^{-1}·mol^{-1}				
							$-(\bar{G}^\circ_T - \bar{H}^\circ_{298})/T$		
Substance	$\Delta\bar{H}^\circ_{f,0}$	$\Delta\bar{H}^\circ_{f,298}$	$\Delta\bar{G}^\circ_{f,298}$	$\bar{H}^\circ_{298} - \bar{H}^\circ_0$	$\bar{S}^\circ_{298}$	$\bar{C}^\circ_{P,298}$	*At* 500 K	*At* 700 K	*At* 1000 K
$CH_3COCH_3(\ell)$	—	−217.57	−153.06	—	294.93	—	—	—	—
$C_6H_6(\ell)$	—	49.03	124.50	—	172.79	—	—	—	—
Fe(*s*)	0	0	0	4.489	27.28	25.10	30.305	35.008	42.267
$Fe_2O_3(s)$	−817.80	−824.2	−742.2	15.560	87.40	103.85	100.282	121.022	152.360
$Fe_3O_4(s)$	−1112.10	−1118.4	−1015.5	24.56	146.4	143.43	163.594	193.673	240.484
$Fe_{0.947}O(s)$	−267.15	−266.27	−245.14	9.46	57.49	48.12	63.195	71.659	83.663
Zn(*s*)	0	0	0	5.648	41.63	25.40	—	—	—
ZnO(*s*)	—	−348.28	−318.32	—	43.64	40.25	—	—	—
ZnS(*s*) (sphalerite)	—	−205.98	−201.29	—	57.7	46.0	—	—	—
Ag(*s*)	0	0	0	5.745	42.55	25.35			
$Ag^+(aq)$	—	105.58	77.12	—	72.68	21.76	—	—	—
$Ag(NH_3)_2^+(aq)$	—	−111.29	−17.24	—	245.2	—	—	—	—
$Ag(S_2O_3)_2^{3-}(aq)$	—	−1285.7	—	—	—	—	—	—	—
AgCl(*s*)	—	−127.07	−109.80	—	96.2	50.79			
HCl(*g*)	−92.13	−92.31	−95.30	8.644	186.80	29.12	190.100	194.962	201.752
$Cl^-(aq)$	—	−167.16	−131.26	—	56.5	−136.4	—	—	—

The data given here were obtained primarily from NBS Technical Notes 270-3 through 270-8 and the JANAF tables.

II

SI Units

THE INTERNATIONAL System of Units (abbreviated SI from the French, *Le Système Internationale d' Unités*) was adopted by the General Conference of Weights and Measures (CGPM) in 1960 as *the* recommended units for use in science and technology.

The SI is constructed from seven base units for independent quantities II-1 to II-7, plus two supplementary units for plane and solid angles, the radian (rad), and steradian (sr), respectively.

The definitions of the seven SI base units follows.

II-1 *Unit of Length, the Meter (m)*

The meter is the length equal to 1,650,763.73 *wavelengths in vacuum of the radiation corresponding to the transition between the levels* $2p_{10}$ *and* $5d_5$ *of the krypton-86 atom.*

The SI unit of area is the square meter, m^2. The SI unit of volume is the cubic meter, m^3. Fluid volume is often measured by the liter, $1\ L = 10^{-3}\ m^3 = 1\ dm^3$.

II-2 *Unit of Mass, the Kilogram (kg)*

The kilogram is the unit of mass it is equal to the mass of the international prototype of the kilogram.

The "international prototype of the kilogram" is a cylinder of Pt-Ir alloy kept by the International Bureau of Weights and Measures at Paris. A duplicate in the custody of the U.S. National Bureau of Standards serves as the mass standard for the United States. Mass is the only base unit still defined by an artifact.

The SI unit of force is the newton (N), 1 N = 1 kg·m·s^{-2}. The SI unit of work and energy of any kind is the joule (J), 1 J = 1 N·m. The SI unit for power of any kind is the watt (W), 1 W = 1 J·s^{-1}.

II-3 *Unit of Time, the Second (s)*

The second is the duration of 9,192,631,770 *periods of the radiation corresponding to the transition between two hyperfine levels of the ground state of the cesium-133 atom.*

Originally, the second was defined as 1/86,400 of the mean solar day. Observations by astronomers have established that irregularities in the rotation of the earth make it impossible for that definition to guarantee the desired accuracy, hence the shift to the atomic clock.

The second is realized by tuning an oscillator to the resonant frequency of the ^{133}Cs atoms as they are passed through a system of magnets and a resonant cavity into a detector. The SI unit for frequency is the hertz (Hz), 1 Hz = 1 cycle per second = 1 s^{-1}. Standard frequencies and correct time are broadcast from NBS stations WWV, WWVB, WWVH, and WWVL, as well as stations of the U.S. Navy. Many shortwave receivers can pick up WWV on frequencies of 2.5, 5, 10, 15, 20, and 25 MHz.

II-4 *Unit of Electric Current, the Ampere (A)*

The ampere is that constant current that, if maintained in two straight parallel conductors of infinite length, of negligible circular cross section, and placed 1 meter apart in vacuum, would produce between these conductors a force equal to 2×10^{-7} *newton per meter of length.*

The force between the two wires results from the interaction of the magnetic fields around the current-carrying wires. The SI unit of voltage is the volt (V), 1 V = 1 W·A^{-1}. The SI unit of electrical resistance is the ohm (Ω), 1 Ω = 1 V·A^{-1}.

II-5 *Unit of Thermodynamic Temperature, the Kelvin (K)*

The kelvin, unit of thermodynamic temperature, is the fraction 1/273.16 *of the thermodynamic temperature of the triple point of water.*

The unit kelvin and its symbol K (not °K) should also be used to express an interval or a difference in temperature. The Celsius temperature (symbol t) is defined by the equation

$$t/°\mathrm{C} = T/\mathrm{K} - 273.15$$

where T is the thermodynamic temperature. Celsius temperatures are expressed as °C.

II-6 *Unit of Amount of Substance, the Mole (mol)*

The mole is the amount of substance of a system that contains as many elementary entities as there are atoms in 0.012 *kilogram of carbon-12.*

When the mole is used, the elementary entities must be specified; they may be atoms, molecules, ions, electrons, other particles, or specified groups of such particles.

II-7 *Unit of Luminous Intensity, the Candela (cd)*

The candela is the luminous intensity, in the perpendicular direction, of a surface of 1/600,000 *square meter of a blackbody at the temperature of freezing platinum under a pressure of* 101,325 $N \cdot m^{-2}$ (i.e., 1 atm).

Further Considerations

1. No restrictions are placed on the units employed for general descriptive information that does not enter into calculations or expression of results (e.g., "pressures in the range 1–50 torr").
2. All integral powers, positive or negative, of SI units are acceptable. Note that exponents operate also on prefixes, as in cm^2, mm^3, which are $10^{-4}\ \mathrm{m}^2$, $10^{-9}\ \mathrm{m}^3$, and not $10^{-2}\ \mathrm{m}^2$, $10^{-3}\ \mathrm{m}^3$.
3. Unit combinations should be designated by means of a dot or dots (as in m·K for meter-kelvin, which avoids confusion with millikelvin, mK).
4. Words and symbols should not be mixed; if mathematical operations are indicated, only symbols should be used. For example, one may write "joules per mole," "J/mol," "$\mathrm{J \cdot mol^{-1}}$," but not "joules/mole," "joules$\cdot \mathrm{mol}^{-1}$," etc.
5. Roman type, in general lowercase, is used for symbols of units; however, if the symbols are derived from proper names, capital Roman type is used for the first letter. These symbols are not followed by a period. Unit symbols do not change in the plural.
6. Certain units not part of the SI are approved for a limited time during the changeover to SI units. Some of these follow:

Ångstrom	$1\ \text{Å} = 0.1\ \mathrm{nm} = 10^{-10}\ \mathrm{m}$
Standard atmosphere	$1\ \mathrm{atm} = 101{,}325\ \mathrm{N \cdot m^{-2}}$
Bar	$1\ \mathrm{bar} = 10^5\ \mathrm{N \cdot m^{-2}}$
Curie	$1\ \mathrm{Ci} = 3.7 \times 10^{10}\ \mathrm{s^{-1}}$
Röentgen	$1\ \mathrm{R} = 2.58 \times 10^{-4}\ \mathrm{Ci \cdot kg^{-1}}$

References

1. NBS Special Publication 330, 1972 ed., U.S. Department of Commerce, NBS.
2. Policy for NBS Usage of SI Units, *J. Chem. Educ.*, 48, 569, 1971.

We have relied heavily on these two publications in preparing this appendix.

III

References

Angrist, S. W., and Hepler, L. G. 1967. *Order and chaos.* New York: Basic Books. An interesting, introductory level discussion of some of the historical aspects of the laws of thermodynamics.

Bates, R. G. 1974. *Determination of pH* (2nd ed.). New York: John Wiley. The authoritative source on pH measurements.

Bent, H. A. 1965. *The second law.* New York: Oxford University Press. An interesting account of the second law presented from both the classical and statistical thermodynamic viewpoint.

Blinder, S. M. 1966. "Mathematical methods in elementary thermodynamics." *J. Chem. Educ.* 43: 85–92. An excellent discussion of the topic; highly recommended.

Bradley, R. S., and Monroe, D. C. 1965. *High pressure chemistry.* New York: Pergamon Press.

Bridgman, P. W. 1941. *The nature of thermodynamics.* New York: Harper Torchbook Reprint. An informative discussion of the operational approach to thermodynamic concepts.

Davidson, N. 1962. *Statistical mechanics.* New York: McGraw-Hill. An excellent text; strongly recommended.

Domb, C. 1968. "Thermodynamics of critical points." *Physics Today* 21: 23.

Durst, R. A. 1974. *Ion-selective electrodes.* Washington, D.C.: U.S. Government Printing Office, Special Publication No. 314.

Dyson, Freeman J. 1971. "Energy and the universe." In *Energy and Power*, Scientific American Reprints. San Francisco, Calif.: W. H. Freeman & Company Publishers, 1971. A fascinating article on the flow of energy in the universe in which Dyson introduces $\Delta S/\Delta U$ as a criterion of energy quality; highly recommended.

Eisenberg, D., and Kauzmann, W. 1969. *The structure and properties of water.* New York: Oxford University Press. A remarkably thorough and authoritative discussion of the properties of water.

Fenn, John B. 1982. *Engines, energy and entropy.* San Francisco, Calif.: W. H. Freeman & Company Publishers. A soft approach to the laws of thermodynamics from an engineer's perspective.

Ferguson, F. D., and Jones, T. K. 1966. *The phase rule.* Washington, D.C.: Butterworth. A brief and modern discussion.

Fermi, Enrico. 1936. *Thermodynamics.* New York: Dover. An introductory treatise on classical thermodynamics by a great scientist and great teacher.

Feynman, R. P., Leighton, R. B. and Sands, M. 1963. *The Feynman lectures in physics.* Reading, Mass.: Addison-Wesley. An exciting discussion of the foundations of physics at the introductory level; highly recommended for chemists.

Findlay, A. 1951. *The phase rule* (9th ed.). Rev. by A. M. Campbell and N. O. Smith. New York: Dover. A classic exposition of the qualitative aspects of phase equilibria.

Giauque, W. F. 1939. *Nature* 143:623. Discussion of the single-fixed-point thermodynamic temperature scale.

Gibbs, J. W. *The Collected Works*, vol. I. *Thermodynamics*, New York: Dover.

Hall, J. A. 1966. *The measurement of temperature.* New York: Barnes & Noble.

Harned, H. S. and Owen, B. B. 1958. *The physical chemistry of electrolytic solutions.* New York: Van Nostrand Reinhold. A classic exposition.

Ives, D. J. C. and Ganz, G. J. 1961. *Reference electrodes.* New York: Academic Press. A classic—essential reference for the experimental electrochemist.

Klein, Martin J. "Carnot's contribution to thermodynamics." *Physics Today* (August 1974), 23.

Latimer, W. M. 1952. *Oxidation potentials.* Englewood Cliffs, N.J.: Prentice-Hall. A classic by a master of solution thermodynamics.

Lewis G. N., and Randall, M. 1961. *Thermodynamics* (2nd ed.). Revised by K. S. Pitzer and L. Brewer. New York: McGraw-Hill.

MacInnes, D. A. 1939. *The principles of electrochemistry.* New York: Dover. An old, but still useful, reference for electrochemical thermodynamics.

Margenau, H. and Murphy, G. M. 1956. *The Mathematics of Physical Chemistry* (2nd ed.). New York: Van Nostrand. Chapter 1.

McQuarrie, D. A. 1973. *Statistical thermodynamics.* New York: Harper & Row. Pub. An excellent, clearly written treatment of the subject with numerous applications. Highly recommended.

Mendelssohn, K. 1966. *The quest for absolute zero.* New York: McGraw-Hill. A fascinating account of experimental efforts to reach ever lower absolute temperatures.

Pippard, A. B. 1966. *Elements of classical thermodynamics.* London: Cambridge University Press. Contains several interesting insights on the laws of thermodynamics. Good discussion of nonpressure-nonvolume work terms.

Planck, Max. 1926. *Treatise on thermodynamics* (3rd ed.). Translated by A. Ogg. New York: Dover. A very readable discussion of the foundations of thermodynamics by one of the most profound thinkers of the twentieth century.

Robinson R. A., and Stokes, R. H. 1959. *Electrolyte solutions* (2nd ed., revised). London: Butterworth's. A superb treatment.

Rossini, F. D. (ed.). 1956. *Experimental Thermochemistry*, vol. 2. New York: Wiley/Interscience.

Rubin, L. G. Temperature-concepts scales and measurement techniques. A Leeds and Northrup Co. reprint of *Technical Memorandum T-538* issued by the Research Division of Raytheon Co. This article is an excellent starting point for those contemplating high-precision temperature measurements.

Schrödinger, E. 1967. *Statistical thermodynamics*. Cambridge: Cambridge University Press. A marvelous little book devoted to the foundations of the subject.

Sengers J. V., and Sengers, A. L. 1968. The critical region. *Chem. Eng. News* 46: 104.

Skinner, H. A. (ed.). 1962. *Experimental Thermochemistry*, vol. 2. New York: Wiley/Interscience.

Stewart, J. W. 1967. *The world of high pressure*. Princeton, N.J.: Van Nostrand.

Temperature: Its measurement and control in science and industry, vols. I, II, and III, 1941, 1955, 1962. New York: Van Nostrand Reinhold.

The international practical temperature scale of 1968. *Metrologia*. 5:35–44.

Zemansky, M. W. 1964. *Temperatures: very low and very high*. Princeton, N.J.: Van Nostrand. An interesting discussion of temperature and its measurement.

IV

Answers to Selected Problems

Chapter 1

19. See Problem 4-18.

24. As water freezes, most of the dissolved gases are expelled. These gases become trapped between the surface ice and the unfrozen liquid and act as an insulator that slows the rate of freezing.

Chapter 2

2. 37.0°C, 310.15 K **3.** −40 **4.** −17.78°C **5.** 491.67
10. −38.87°C (freezing point of Hg) **15.** 0.01 K **17.** 6000 K
22. 9.6666°C, −4.7744°C, at 0°C, $dR_t/dt = 0.10168\ \Omega/°\text{C}$ **24.** $1 \times 10^{-5}\ \Omega$

Chapter 3

1. 9.8×10^5 J **2.** No **4.** (a) −5480 J, (b) 0, (c) $\Delta U = 0$ for both (a) and (b)
6. $C_P = \infty$ **7.** 1.480×10^6 J **9.** 321.4 kJ, No **10.** $161.8\ \text{J}\cdot\text{g}^{-1}$
13. (a) $\Delta U = 7.48$ kJ, $\Delta H = 12.47$ kJ, (c) 0
15. (a) −2.33 kJ, (b) −0.91 kJ, (c) −0.51 kJ; −0.81 kJ **17.** $\Delta U = 0, q = -w$
18. (a) $(A - BT)L_0^2/2$, (b) KTL_0 **19.** rises 0.33 cm **21.** −11.29 kJ vs −11.49 kJ
22. 17.7°C **25.** −78.9 J **26.** $c\mathscr{H}^2/2T$

Chapter 4

1. $175.1\ \text{J}\cdot\text{K}^{-1}$ **2.** $2.94\ \text{kJ}\cdot\text{K}^{-1}$ **3.** $9198\ \text{J}\cdot\text{K}^{-1}$
4. (a) 0, (b) $3.954\ \text{J}\cdot\text{K}^{-1}$, (c) 3.238 g, (d) $-0.293\ \text{cm}^3$, (e) $\Delta S_{\text{res}} = 3.51\ \text{J}\cdot\text{K}^{-1}$, $\Delta S_{\text{sur}} = 0$

6. $\Delta S = 46\ \text{J}\cdot\text{K}^{-1}$, $\Delta H = 19.37$ kJ, $\Delta U = 14.76$ kJ **7.** (a) yes. (b) no **8.** 1.0 ton
18. 7.41% **19.** 41.1% vs 33.3%
20. The stretched spring will have the greater heat of solution by an amount equal to ΔU for the stretching process.
21. Yes, provided the process is not cyclic; for example, isothermal expansion of an ideal gas.
23. -2.62×10^{12} J **26.** $1.64\ \text{J}\cdot\text{K}^{-1}$ **32.** $T = 273.16$ K

Chapter 5

2. Yes **5.** 13.8% **7.** $q = 0$, $w = -1247$ J, $\Delta H = -1746$ J, $\Delta S_{gas} = \Delta S_{tot} = 0$
16. $w = -11.49$ kJ, $q = 11.49$ kJ, $\Delta H = 0$, $\Delta S = 38.29\ \text{J}\cdot\text{K}^{-1}$, $\Delta A = -11.49$ kJ, $\Delta S_{tot} = 0$
18. $w = -40.64$ kJ, $\Delta U = 417$ J, $\Delta H = 328$ J, $\Delta S = 82.1\ \text{J}\cdot\text{K}^{-1}$, $\Delta G = -40.72$, $\Delta A = -40.64$ kJ
19. $w = \Delta U = -17.16$ kJ, $\Delta H = -24.30$ kJ, $\Delta S = 0$ **21.** 933 atm
31. $\Delta U = 0$, $w = 202$ J, $\Delta H = -202$ J, $\Delta S = 38.28\ \text{J}\cdot\text{K}^{-1}$
35. $\Delta H = -27.12$ kJ, $\Delta S = -43.60\ \text{J}\cdot\text{K}^{-1}$, $\Delta U = -19.59$ kJ **47.** (a) 0.33

Chapter 6

4. $-198.53\ \text{J}\cdot\text{K}^{-1}$, $-202.19\ \text{J}\cdot\text{K}^{-1}$ **7.** For $^{12}C^{1}H_4$, $\overline{S}_{spin} = 23.05\ \text{J}\cdot\text{K}^{-1}\cdot\text{mol}^{-1}$
8. $R \ln \left(\frac{6}{4}\right)$ **9.** $9.134\ \text{J}\cdot\text{K}^{-1}\cdot\text{mol}^{-1}$; no **10.** $R \ln 4$ **12.** 0, $R \ln 4$, $R \ln 8$
13. $R \ln 3 = 9.13\ \text{J}\cdot\text{K}^{-1}\cdot\text{mol}^{-1}$ **16.** 0 **19.** $\Delta\overline{H}_{ads} = -279\ \text{J}\cdot\text{mol}^{-1}$ at 4.23 K
21. $\Delta\overline{S} = 0.33\ \text{J}\cdot\text{K}^{-1}\cdot\text{mol}^{-1}$

Chapter 7

1. (a) $\Delta H^\circ_{rxn} = -98.69$ kJ, $\Delta S^\circ_{rxn} = -93.98\ \text{J}\cdot\text{K}^{-1}$, $\Delta G^\circ_{rxn} = -70.89$ kJ
2. $-1260.2\ \text{kJ}\cdot\text{mol}^{-1}$
3. (a) $\Delta H^\circ_{rxn} = -439.13$ kJ, $\Delta S^\circ_{rxn} = -73.5\ \text{J}\cdot\text{K}^{-1}$, $\Delta G^\circ_{rxn} = -417.22$ kJ
5. -35.10 kJ **6.** -4.2 kJ **16.** $78\ \text{kJ}\cdot\text{mol}^{-1}$ **23.** 4422 K

Chapter 8

1. (a) $c = 2, f = 3$; (b) 2, 2; (c) 2, 2 **3.** (a) 2, (c) 1, (e) 1
4. Aragonite **5.** 2 **6.** (a) $P_{Zn} = P_{CO} + 2P_{CO_2}$, (b) $f = 1$
46. (a) T, (b) F, (c) F, (d) F, (e) T, (f) F, (g) T, (h) F, (i) F, (j) T
47. False (a), (b), (e), (i), (j), (k), (l), (o), (p), (q), (r), (s), (t)

Chapter 9

1. 2.06 atm **4.** $133\ \text{kJ}\cdot\text{mol}^{-1}$ **5.** 172.34 K, 11.02 torr
10. $\Delta\overline{H}_{sub} = 31.41\ \text{kJ}\cdot\text{mol}^{-1}$, $\Delta\overline{H}_{vap} = 22.19\ \text{kJ}\cdot\text{mol}^{-1}$, $T = 170.0$ K, $P = 8.98$ torr
20. 3.66 atm **22.** 146 K **25.** (a) 1.076, 2.056
45. (b) $a = 89.9$, $\mu - \mu^0 = 18.70\ \text{kJ}\cdot\text{mol}^{-1}$

Chapter 10

2. 2.00 M **3.** 0.34 M **4.** $P_{CO} = 1.65$ atm **8.** 0.014 torr
11. $\Delta G^\circ = 32.21$ kJ, $\Delta S^\circ = 380\ \text{J}\cdot\text{K}^{-1}$ **12.** 6.35×10^{-14} atm **13.** 1.15×10^{19} atm
14. At 298 K, $P = 1.42 \times 10^4$ atm **23.** 313 J **24.** (a) -2.84 kJ

Chapter 11

1. (b) 0.050 m, (d) 0.0191 m **2.** (c) 4.50×10^{-2} **3.** 9×10^{-6} m **4.** 57.47 kJ
9. (d) $P_{Si_3N_4} = 0.50$ atm, $P_{N_2} = 1.39 \times 10^{-11}$ atm **10.** $19.54\ \text{kJ}\cdot\text{mol}^{-1}$
14. $K = 0.186$ **19.** (a) 4.11×10^{-5}

Chapter 12

2. (b) $Zn(s)|ZnSO_4(aq)|Hg_2SO_4(s)|Hg(\ell)|Hg_2SO_4(s)|CuSO_4(aq)|Cu(s)$
(c) $Ag(s)|AgNO_3(aq),NaNO_3(aq)|Na^+(glass)|NaCl(aq)|AgCl(s)|Ag(s)$
4. 0.404 V **7.** Na(s) reacts directly with $H_2O(\ell)$; Pb(Hg) reacts directly with Cu^{2+} (aq)
11. $K_{25} = 1.49 \times 10^4$ **23.** $\Delta G^\circ_{500} = -1008.75$ kJ

Chapter 13

1. 18.3 torr **3.** $\pi_{15} = 25.75$ atm **7.** $\Delta S_{mix} = 18.27\ \text{J}\cdot\text{K}^{-1}$ **8.** $15.88\ \text{J}\cdot\text{K}^{-1}$
27. $-1.75\ \text{cm}^3$ and $-1.01\ \text{cm}^3$ **37.** -5.7 atm
40. $X_m = 0.6564$, $X_0 = 0.3436$, 65.32°C; $T = 390.1$ K for $X = \frac{1}{3}$ case

Chapter 14

1. 2.48×10^{-22} J **3.** $\sigma = 24$ for SF_6 **4.** $K = 4$ for first reaction
22. $K_{100} = 3.30 \times 10^{-7}$ **23.** 0.999 **24.** $(\frac{3}{4})\ R \ln 3$
38. For H_2O: $I_{xx} = 1.116 \times 10^{-47}\ \text{kg}\cdot\text{m}^2$, $I_{yy} = 2.092 \times 10^{-47}$, $I_{zz} = 3.207 \times 10^{-47}$, and $q_r = 47.43$

Index

The tables below and on the following two pages provide multiplicative factors for converting from one energy or power unit to another. For example, to convert a quantity from kilocalories to British thermal units, find the original unit at the left edge of the table and read across to the column headed by the desired final unit. This tells one to multiply by 3.968. Thus 4.5 kcal equals 4.5×3.968 Btu (17.86 Btu). Caution should be used in conversions involving the energy values of fuels (coal, petroleum, and natural gas). The energy content of such fuels varies by 10% or more according to the quality of the source and the method of production. When accuracy is required in calculations involving fuels, specific data should be obtained.

Power Conversion Factors*

	Watt	*GJ/yr*	*Quad/yr*	*Petroleum, tonne/yr*	*Petroleum, bbl/day*
watt	1	3.16×10^{-2}	2.99×10^{-11}	7.17×10^{-4}	1.42×10^{-5}
gigajoule (GJ)/yr	31.69	1	9.49×10^{-10}	2.27×10^{-2}	4.49×10^{-4}
quad/yr	3.34×10^{10}	1.05×10^{9}	1	2.40×10^{7}	4.73×10^{5}
petroleum, tonne/yr	1394	44.0	4.17×10^{-8}	1	1.98×10^{-2}
petroleum, bbl/day	7.06×10^{4}	2.23×10^{3}	2.11×10^{-6}	50.6	1

* From J. M. Hollander, M. K. Simmons, and D. O. Wood, *Annual Review of Energy*, Vol. 2, Palo Alto, Calif.: Annual Reviews, Inc., 1977.

Energy Conversion Factors*

	GJ[a]	*Btu*	*kW-hr*	*kcal*	*Coal, tonne (MT)*[b]	*Coal, ton*[c]
gigajoule (GJ)[a]	1	9.48×10^{5}	278	2.39×10^{5}	3.57×10^{-2}	3.94×10^{-2}
Btu	1.055×10^{-6}	1	2.93×10^{-4}	0.252	3.77×10^{-8}	4.15×10^{-8}
kW-hr	3.60×10^{-3}	3413	1	860	1.29×10^{-4}	1.42×10^{-4}
kcal	4.186×10^{-6}	3.968	1.163×10^{-3}	1	1.50×10^{-7}	1.65×10^{-7}
coal, tonne[b] (MT)	28	2.65×10^{7}	7.78×10^{3}	6.69×10^{6}	1	1.10
coal, ton[c]	25.4	2.41×10^{7}	7.06×10^{3}	6.07×10^{6}	0.907	1
petroleum, tonne (MT)	44	4.17×10^{7}	1.22×10^{4}	1.05×10^{7}	1.57	1.73
petroleum, bbl	6.1	5.8×10^{6}	1.70×10^{3}	1.46×10^{6}	0.218	0.240
gasoline, liter	0.034	3.22×10^{4}	9.45	8.13×10^{3}	1.21×10^{-3}	1.34×10^{-3}
gasoline, gal[d]	0.129	1.22×10^{5}	35.9	3.08×10^{4}	4.60×10^{-3}	5.07×10^{-3}
natural gas, tonne (MT)	56	5.31×10^{7}	1.56×10^{4}	1.34×10^{7}	2.00	2.20
natural gas, m^3	0.040	3.79×10^{4}	11.1	9.56×10^{3}	1.43×10^{-3}	1.57×10^{-3}
natural gas, ft^3	1.06×10^{-3}	1005	0.295	253	3.79×10^{-5}	4.17×10^{-5}

	Petroleum, tonne (MT)	*Petroleum, bbl*	*Gasoline, liter*	*Gasoline, gal*[d]	*Nat. gas, tonne (MT)*	*Nat. gas, m^3*[e]	*Nat. gas, ft^3*[f]
gigajoule (GJ)[a]	2.27×10^{-2}	0.164	29.4	7.75	0.0179	25.0	943
Btu	2.40×10^{-8}	1.73×10^{-7}	3.10×10^{-5}	8.18×10^{-6}	1.88×10^{-8}	2.64×10^{-5}	9.95×10^{-4}
kW-hr	8.18×10^{-5}	5.90×10^{-4}	0.106	2.79×10^{-2}	6.43×10^{-5}	9.00×10^{-2}	3.40
kcal	9.51×10^{-8}	6.86×10^{-7}	1.23×10^{-4}	3.24×10^{-5}	7.48×10^{-8}	1.05×10^{-4}	3.95×10^{-3}
coal, tonne[b] (MT)	0.636	4.59	824	217	0.50	700	2.64×10^{4}
coal, ton[c]	0.577	4.16	747	197	0.454	635	2.40×10^{4}
petroleum, tonne (MT)	1	7.21	1290	341	0.786	1100	4.15×10^{4}
petroleum, bbl	0.139	1	179	47.3	0.109	152	5.76×10^{3}
gasoline, liter	7.73×10^{-4}	5.57×10^{-3}	1	0.264	6.07×10^{-4}	0.850	32.1
gasoline, gal[d]	2.93×10^{-3}	2.11×10^{-2}	3.79	1	2.30×10^{-3}	3.22	122
natural gas, tonne (MT)	1.27	9.18	1.65×10^{3}	434	1	1.40×10^{3}	5.28×10^{4}
natural gas, m^3	9.09×10^{-4}	6.56×10^{-3}	1.18	0.310	7.14×10^{-4}	1	37.7
natural gas, ft^3	2.41×10^{-5}	1.74×10^{-4}	3.12×10^{-2}	8.22×10^{-3}	1.89×10^{-5}	2.65×10^{-2}	1

* From J. M. Hollander, M. K. Simmons, and D. O. Wood, *Annual Review of Energy*, Vol. 2, Palo Alto, Calif.: Annual Reviews, Inc., 1977.
[a] One GJ = 10^9 joules.
[b] One tonne = 1000 kilograms. The value of energy content given here is typical of a bituminous coal. MT denotes metric ton (tonne).
[c] One ton = 2000 pounds. The value of energy content given here is typical of a bituminous coal.
[d] U.S. gallon = 3.78 liters.
[e] At 0°C, and a pressure of 760 torr.
[f] At 60°F, and a pressure of 30 in. of mercury.

SI Base Units[a]

Physical quantity	*Name of unit*	*Symbol*
length	meter	m
mass	kilogram	kg
time	second	s
electric current	ampere	A
thermodynamic temperature	kelvin	K
luminous intensity	candela	cd
amount of substance	mole	mol

Special Names and Symbols for Certain SI Derived Units

Physical quantity	*Name of SI unit*	*Symbol for SI unit*	*Definition of SI unit*
force	newton	N	$kg \cdot m \cdot s^{-2}$
pressure	pascal	Pa	$kg \cdot m^{-1} \cdot s^{-2}$ $(= N \cdot m^{-2})$
energy	joule	J	$kg \cdot m^2 \cdot s^{-2}$
power	watt	W	$kg \cdot m^2 \cdot s^{-3}$ $(= J \cdot s^{-1})$
electric charge	coulomb	C	$A \cdot s$
electrical potential difference	volt	V	$kg \cdot m^2 \cdot s^{-3} \cdot A^{-1}$ $(= J \cdot A^{-1} \cdot s^{-1})$
electric resistance	ohm	Ω	$kg \cdot m^2 \cdot s^{-3} \cdot A^{-2}$ $(= V \cdot A^{-1})$
electric conductance	siemens	S	$kg^{-1} \cdot m^{-2} \cdot s^3 \cdot A^2$ $(= A \cdot V^{-1} = \Omega^{-1})$
electric capacitance	farad	F	$A^2 \cdot s^4 \cdot kg^{-1} \cdot m^{-2}$ $(= A \cdot s \cdot V^{-1})$
magnetic flux	weber	Wb	$kg \cdot m^2 \cdot s^{-2} \cdot A^{-1}$ $(= V \cdot s)$
inductance	henry	H	$kg \cdot m^2 \cdot s^{-2} \cdot A^{-2}$ $(= V \cdot A^{-1} \cdot s)$
magnetic flux density	tesla	T	$kg \cdot s^{-2} \cdot A^{-1}$ $(= V \cdot s \cdot m^{-2})$
luminous flux	lumen	lm	$cd \cdot sr$
illumination	lux	lx	$cd \cdot sr \cdot m^{-2}$
frequency	hertz	Hz	s^{-1} (cycle per second)

[a] See Appendix II for definitions of the SI base units.